# Lecture Notes in Artificial Intelligence     9969

Subseries of Lecture Notes in Computer Science

More information about this series at http://www.springer.com/series/1244

Ashok Goel · M. Belén Díaz-Agudo
Thomas Roth-Berghofer (Eds.)

# Case-Based Reasoning Research and Development

24th International Conference, ICCBR 2016
Atlanta, GA, USA, October 31 – November 2, 2016
Proceedings

 Springer

*Editors*
Ashok Goel
Georgia Institute of Technology
Atlanta, GA
USA

Thomas Roth-Berghofer
University of West London
London
UK

M. Belén Díaz-Agudo
Universidad Complutense de Madrid
Madrid
Spain

ISSN 0302-9743          ISSN 1611-3349   (electronic)
Lecture Notes in Artificial Intelligence
ISBN 978-3-319-47095-5        ISBN 978-3-319-47096-2   (eBook)
DOI 10.1007/978-3-319-47096-2

Library of Congress Control Number: 2016953331

LNCS Sublibrary: SL7 – Artificial Intelligence

Printed on acid-free paper

This Springer imprint is published by Springer Nature
The registered company is Springer International Publishing AG
The registered company address is: Gewerbestrasse 11, 6330 Cham, Switzerland

# Preface

This volume contains the papers presented at the 24th International Conference on Case-Based Reasoning Research and Development, ICCBR 2016, held October 31 to November 2, in Atlanta, Georgia, USA.

The International Conference on Case-Based Reasoning (ICCBR) is the premier, annual meeting of the CBR community and the leading international conference on this topic. The theme for ICCBR 2016 was "Creativity." ICCBR 2016 was co-located with the Fourth International Conference on Design and Creativity.

Previous ICCBR conferences have been held in Sesimbra, Portugal (1995), Providence, USA (1997), Seeon Monastery, Germany (1999), Vancouver, Canada (2001), Trondheim, Norway (2003), Chicago, USA (2005), Belfast, UK (2007), Seattle, USA (2009), Alessandria, Italy (2010), Greenwich, UK (2011), Lyon, France (2012), Saratoga Springs, USA (2013), Cork, Ireland (2014), and Frankfurt, Germany (2015).

For the 2016 conference, the published papers were carefully selected from 44 submissions from 18 countries; each was reviewed by at least three Program Committee members. The committee decided to accept 14 papers for oral presentation at the conference; an additional 15 papers were accepted for poster presentation. The 29 papers included in this book cover a wide range of CBR topics that are of interest both to researchers and practitioners, from foundations of case-based reasoning, novel retrieval and reuse approaches, advances in compositional adaptation, case generation and knowledge discovery, to CBR systems, applications, and lessons learned in specific areas of expertise.

The first day of ICCBR 2016 was given over to the selected workshops and the 8th Annual ICCBR Doctoral Consortium (DC), which is designed to nurture PhD candidates by providing them with opportunities to explore and obtain mutual feedback on their research, future work plans, and career objectives with senior CBR researchers, practitioners, and peers.

Three workshops were selected for this conference: Computational Analogy, Synergies Between CBR and Knowledge Discovery, and Reasoning About Time in CBR. We would like to thank all the co-chairs of these workshops for creating such a stimulating program.

Days two and three comprised presentations and posters on technical and applied CBR papers, as well as invited talks from two distinguished scholars: Pablo Gervás, of the Complutense University of Madrid, Spain, and Mehmet Goker, Vice President, Business Data Science Salesforce, San Francisco, USA.

Pablo Gervás's talk, "How Creative Can Reuse Be?," pointed to CBR as a favored technology for trying to model creative tasks such as story generation or music generation in artificial intelligence. Mehmet Goker gave an applied industry point of view of CBR in his talk entitled "The Business End of Data Science."

Many people participated in making ICCBR 2016 a great success: Ashok Goel, Georgia Institute of Technology, USA, as the general chair; program chairs, Belén Díaz-Agudo, Universidad Complutense de Madrid, Spain, and Thomas Roth-Berghofer, School of Computing and Engineering, University of West London, UK; publicity chairs, Santiago Ontañón, Drexel University, USA and Swaroop Vattam, MIT Lincoln Labs, USA.

We wish to thank Alexandra Coman, Ohio Northern University, USA, and Stelios Kapetanakis, University of Brighton, UK, for organizing the workshop program and Sarah Jane Delaney, Dublin Institute of Technology, Ireland, and Stefania Montani, Università del Piemonte Orientale, Italy, for the successful organization of the Doctoral Consortium.

We thank the Program Committee and all our additional reviewers for their thoughtful and timely participation in the paper selection process. We acknowledge the time and effort put in by the members of the local Organizing Committee at the Georgia Institute of Technology, USA: Stephen Lee-Urban, and Elizabeth Whitaker.

We are very grateful for the generous support of the ICCBR 2016 sponsors, including the US NSF, Knexus Corporation, Springer, Georgia Tech, and Georgia Tech GVU research center. Finally, we appreciate the help provided by EasyChair in the management of this conference and we thank Springer for its continuing support in publishing the proceedings of ICCBR.

August 2016                                                              Ashok Goel
                                                                   Belén Díaz-Agudo
                                                              Thomas Roth-Berghofer

# Organization

## Program Committee

| | |
|---|---|
| Agnar Aamodt | Norwegian University of Science and Technology, Norway |
| David Aha | Naval Research Laboratory |
| Klaus-Dieter Althoff | DFKI/University of Hildesheim, Germany |
| Kerstin Bach | Norwegian University of Science and Technology (NTNU), Norway |
| Ralph Bergmann | University of Trier, Germany |
| Isabelle Bichindaritz | State University of New York at Oswego, USA |
| Derek Bridge | University College Cork, Ireland |
| Alexandra Coman | Ohio Northern University, USA |
| Amélie Cordier | LIRIS |
| Sarah Jane Delany | Dublin Institute of Technology, Ireland |
| Belen Diaz-Agudo | Universidad Complutense de Madrid, Spain |
| Michael Floyd | Knexus Research |
| Ashok Goel | Georgia Institute of Technology, USA |
| Pedro González Calero | Complutense University of Madrid, Spain |
| Andrés Gómez de Silva Garza | Instituto Tecnológico Autónomo de México, Mexico |
| Stelios Kapetanakis | University of Brighton, UK |
| Joseph Kendall-Morwick | Capital University, USA |
| Deepak Khemani | Indian Institute of Technology, Madras, India |
| Luc Lamontagne | Laval University, USA |
| David Leake | Indiana University, USA |
| Stephen Lee-Urban | Georgia Tech Research Institute |
| Jean Lieber | LORIA |
| Ramon Lopez De Mantaras | IIIA - CSIC |
| Cindy Marling | Ohio University, USA |
| Stewart Massie | Robert Gordon University, UK |
| Lorraine Mcginty | University College Dublin, Ireland |
| Mirjam Minor | Goethe University Frankfurt, Germany |
| Stefania Montani | University of Piemonte Orientale, Italy |
| Nadia Najjar | University of North Carolina at Charlotte |
| Emmanuel Nauer | LORIA, France |
| Santiago Ontañón | Drexel University, USA |
| Miltos Petridis | Brighton University, UK |
| Enric Plaza | IIIA-CSIC |
| Luigi Portinale | Università Piemonte Orientale A. Avogadro, Italy |

# Contents

# Searching Museum Routes Using CBR

Jesús Aguirre-Pemán, Belén Díaz-Agudo, and Guillermo Jimenez-Diaz(✉)

Department of Software Engineering and Artificial Intelligence,
Universidad Complutense de Madrid, Madrid, Spain
{belend,gjimenez}@ucm.es

**Abstract.** In this paper, we describe a CBR solution to the route planning problem for groups of people. We have compared keyword coverage results for our CBR approach and heuristic search algorithms. User preferences are important for individual visits but when dealing with group visits there are other aspects to consider. In our case study a group of people plans a visit to MIGS (Museo de Informática Garcia Santesmases http://www.fdi.ucm.es/migs/), a museum about computer science history located at the Computer Science Faculty of Complutense University in Madrid. CBR results are promising and we discuss the benefits of the experience in the case base when planning a group visit. CBR has become specially appropriate given that it assists the knowledge discovery task when learning about subtle differences affecting the suitability of group plans over individual plans computed by traditional search algorithms.

## 1 Introduction

When planning a route visit, either in a museum, a city or a mall, user preferences are important and typically, the resulting route should aim to satisfy user preferences. Different users may weigh their preferences differently. The problem is different when considering an individual visit or a group visit. Group preferences are modelled as an aggregation (e.g. weighted average) of the individual preferences of the group members [9]. Besides, when dealing with groups there are other aspects to consider, like the number of people, physical features if there are space restrictions, average age, etc.

In this paper, we describe a CBR system to help in planning a route for a group of people. In our case study a group of people plans a visit to MIGS (Museo de Informática, Garcia Santesmases[1]), a museum about computer science history located at the Computer Science Faculty of Complutense University in Madrid. We characterize the museum map as an undirected graph where nodes symbolize *Points Of Interest* (POIs) tagged with keywords or labels that represent description features, and the edges symbolize transitions between the museum POIs. Besides, there are time labels, both in nodes and edges, representing the

---

Supported by UCM (Group 910494) and Spanish Committee of Economy and Competitiveness (TIN2014-55006-R).

[1] Professor Jose Garcia Santesmases built the first computer in Spain, between 1953 y 1954.

© Springer International Publishing AG 2016
A. Goel et al. (Eds.): ICCBR 2016, LNAI 9969, pp. 1–15, 2016.
DOI: 10.1007/978-3-319-47096-2_1

average time to visit the POI and the time to move from one node to another, respectively.

In the literature, we find different approaches to solve the route planning problem for individual users. In [12] the authors present the problem as a keyword coverage problem which finds an optimal route from a source location to a target location such that the keyword coverage is optimized and that a budget score satisfies a specified constraint. Although the authors prove that this problem is NP-hard, they propose an adapted version of the A* algorithm using an admissible heuristic to preserve the solution optimality.

While the topic of group route planning in tourism applications has been widely studied, there is comparatively less research on studying benefits of CBR in this specific area. In our research, we compare a variation of an optimal search algorithm with a CBR approach and discuss the benefits of the domain expert experience when planning a group visit. We adapt the A* algorithm and the heuristic described in [12] to our case study, using *time* as a *budget* score in nodes and edges. Additionally, the preferences are not for individual users but we employ an aggregated set of preferences for groups of people. Besides, our proposal includes other features that characterize the group like its size and average age, which are not easy to include in the original A* algorithm.

As A* has resulted inapplicable to our case study due to its high cost in memory and computation time, we compare some experimental results of our CBR approach with a greedy algorithm that uses the same heuristic function. We have compared our approach to heuristic search according to the keyword coverage. As heuristic search performs an exhaustive search, its results in keyword coverage outperform the CBR solution. However, CBR results are very promising. Besides, CBR has become specially appropriate given that it assists the knowledge discovery task when learning about subtle differences affecting the suitability of group plans over individual plans computed by heuristic search algorithms. Cases have resulted to be an excellent tool to elicit domain knowledge and to capture important knowledge from real visits, like the museum space limitations for big groups in several showcases or common sequences of showcases due to the dependencies between POIs.

The paper runs as follows. Section 2 describes some related work in recommendation in tourism, emphasizing on works for route planning. Section 3 describes the problem formalization of a map as a connected graph and common definitions used in the compared approaches. Section 4 describes a solution to the route planning problem using an heuristic search while Sect. 5 proposes the CBR approach. Section 6 describes our case study at the MIGS museum and explains the experimental results. Finally, Sect. 7 summarizes the main results achieved and describes forthcoming work.

## 2   Related Work

Recommender systems have been increasingly employed in the field of tourism, recommending attractions or *Points Of Interest* (POIs), travel services (restaurants, hotels, transportations...), routes and tours, or personalized multiple-days tour planning, among others [5].

The route search and planning problem in tourism domain has been tackled using different approaches. One of the most classic and well known is the shortest path problem, which does not take into account user preferences and constraints. Systems like MacauMap [3] mixes other algorithms that involve user interests to select the POIs contained in the route and generates the travel sequence using an A* search algorithm. Like in our work, this recommender considers time between POIs and stay times to generate an optimal schedule.

Other authors propose an adapted version of the A* algorithm using an admissible heuristic to preserve the solution optimality [12]. This work describes the problem as a keyword coverage problem that finds an optimal route from a source location to a target location such that the keyword coverage is optimized according to the user preferences and satisfying some budget constraints. Our contribution in this paper relates with the CBR approach. The route planning algorithm using heuristic search that we use for comparison (described in Sect. 4) is an adaptation of this work [12], considering time as a budget constraint, and enhancing it with group preferences and features that characterize the group, like group size and average age.

Other approaches address the problem of recommending tourism itineraries as a Constraint Satisfaction Problem: the recommender system generates a sequence of POIs to be visited, filtering data according to the user constraints. This is the approach employed by INTRIGUE [2], which recommends itineraries and destinations taking into account preferences of group members.

Several recommender systems tackle the problem as an Orienting problem with user constraints. PERSTOUR [8] recommends personalized tours using POI similarity and inferred user interest preferences. It recommends an itinerary, with a time limit and a start and end selected by the user, which maximizes the popularity and the user interest in the recommended POIs, adhering to a time restriction. One of the main contributions of this work is the personalization of the visit duration at each POI and the *Time-based user interest*, which measures the user's level of interest in a POI category based on the time the user spent at such POI, relative to the average user.

User interests and POI matching implies the existence of knowledge description about the POIs. MoreTourism [11] uses tags and weights to recommend POIs. The recommendation is performed comparing user's tag clouds with POI tag clouds. This work also proposes group recommendations creating a group tag cloud using the tag clouds that describe its members.

The use of CBR approaches to recommend tourist itineraries is not commonly employed although case-based planning is a field widely extended with successful results [6,7,10]. These works aim to find routes between an origin and a goal position reusing cases that represent previous planned routes and do not address tourist recommendation. TURAS [10] and PRODIGY [7] performs some personalization reusing the routes that the user who is using the system prefers. However, our approach is completely different because it fixes the origin and goal positions and selects the sequence of POIs that the users will visit according to the preferences requested to the users and other additional restrictions.

A specific case-based planning system for programming tourist routes is TOURIST GUIDE-USAL [4], which generates the route adapting previous cases to the user profile described by features, like the type of visit, its budget and the time duration. In this system, cases are previous successful routes (plans) that include the POIs to visit, the time to spend visiting each POI, the time required for going to one place to another and some route labels (like museum route, family route, Romanesque route, etc.). These works inspired our CBR route planning approach, described in Sect. 5.

# 3   Problem Formalization

The aim of the route planning algorithms described in our work is to find a route that traverses a sequence of POIs in a map for a group of visitors. The route has a fixed start and end and the POIs are labelled with some domain keywords. The route must maximize the group preference satisfaction, described as a set of weighted keywords, while complying with a given time restriction.

Let $V = \{v_1, \ldots, v_p\}$ be a non-empty set of POIs and let $K = \{k_1, \ldots, k_n\}$ be a set of keywords to describe both POI features and the group preferences. Keywords (also labels) are domain dependent. Every POI $v_i$ is characterized by a weighted vector $v_i.cov = \{\lambda_{k_1}, \ldots, \lambda_{k_n}\}$, with $\lambda_{k_j} \in [0,1]$, and a time value $t_{v_i}$ that represents the average time spent to visit $v_i$.

Every POI has a location. The map representing POI locations is characterized as an undirected graph $G = (V, E)$, where $E$ is a set of edges defined as pairs of POIs $< v_i, v_j >$ where $v_i, v_j \in V$. Each edge is annotated with a time value $t_{<v_i,v_j>}$. G is a strongly connected graph, where every POI is reachable from every other POI.

A visitor group (the query in the CBR approach) is characterized by a tuple $(size_q, age_q, pref_q, t_q)$ where: $size$ is the number of people in the group, $age$ is the average age of its members, $pref_k = \{\lambda_{k_1} \ldots \lambda_{k_q}\}$ with $\lambda_{k_i} \in [0,1]$ is a set of weights indicating the group preferences for each keyword, and $t_q$ is the time constraint, the maximum length of the visit. Note that, although we use this formalization for the sake of an easier comparison, one of the advantages of the CBR approach is that it simplifies the query elicitation process and the algorithm could work with incomplete input information.

The route planning solution will be a route or an ordered sequence of POIs R $= \{v_1, .., v_m\}$ with $m$ stops, where $v_i \in V$, and $t_R = \sum_{i=1}^{m} t_{v_i} + \sum_{i=1}^{m-1} t_{<v_i,v_{i+1}>}$, with $t_R \leq t_q$.

## 3.1   Keyword Coverage

The problem of measuring how well a route covers the user's preferences (defined as a set of weighted keywords) is non-trivial. According to [12] simply accumulating the keyword degree associated with the locations in a route cannot well reflect the satisfiability of the route. We define the **keyword coverage** function as the degree to which the user preferences are covered by a route planning solution. Given the set of keywords $K = \{k_1, \ldots, k_q\}$ and a set of user preferences $pref_q = \{\lambda_{k_1}, \ldots, \lambda_{k_q}\}$ with $\lambda_{k_i}$ the weight value associated to the keywords $k_i$, the **keyword coverage** function from a route $R = \{v_1, \ldots, v_m\}$ is defined as:

$$kc(R) = \sum\nolimits_{k_i \in K} \lambda_{k_i} \cdot cov(k_i, R)$$

where $cov(k_i, R)$ is the degree to which the label $k_i$ is covered in the path $R$:

$$cov(k_i, R) = 1 - \prod\nolimits_{v_j \in R}(1 - v_j.cov(k_i))$$

The value $kc(R)$ used in [12] is adequate for routes with a small number of items. However, we found that this measure is not interesting for large graphs because it produces indistinguishable values for routes with more than 30 nodes. For this reason, we redefine the keyword coverage function:

$$kc\_sum(R) = \sum\nolimits_{v_i \in R} v_i.cov(K) \tag{1}$$

where $v_i.cov(K)$:

$$v_i.cov(K) = \sum\nolimits_{k_j \in K} \lambda_j \cdot v_i.cov(k_j).$$

## 4   Route Planning Using Heuristic Search

Our first approach uses an algorithm to plan a route in the graph of POIs that is a modification of the A* algorithm with the heuristic described in [12], taking user preferences into account. The problem of optimizing the keyword coverage and time constraint is NP-hard. To solve this complex problem, like the authors in [12], we propose an admissible heuristic function that preserves the solution optimality. Although the experimental results are optimal, the main drawbacks of A* algorithm are its time and memory requirements for large graphs (over 15 nodes). In our case study, where the museum map will be a *complete graph* with every pair of nodes connected, the A* algorithm uses more time and memory resources than the expected ones.

As a solution we propose other approaches: a greedy algorithm (described later in Algorithm 1) and a CBR approach (related in Sect. 5). While basic A* algorithms only use costs in edges, our approaches use the input preferences and the node labels. We reuse some key concepts of the A* algorithm described in [12], such as using *time* as a *budget* score in nodes and edges. Instead of using the preferences of an individual user, we employ an aggregated set of preferences of a group of people. We use other additional features to characterize the group, like its size and the average age of its members.

A Greedy algorithm, unlike the aforementioned A*, just takes into account a heuristic function $h(n)$, the evaluation function that estimates the cost for the current node $n$ (meaning we are evaluating the best path traversing $n$). We have defined a greedy algorithm (Algorithm 1), whose main idea is to generate a list

---

**Algorithm 1.** Greedy Algorithm

---

**Function 1** *GREEDY(Graph G, double $t_q$)*
$R = \emptyset$;
$U = G.getAllItems()$;
**while** (*U.size() > 0*) **do**
   $max = -\infty$; $maxindex = 0$;
   **for** *(i = 0; i < U.size(); i + +)* **do**
      **if** $v_i.cov(K)/(t_{v_i} + v_i.cost()) > max$ **then**
         $maxindex = i$;
         $max = v_i.cov(K)/(t_{v_i} + v_i.cost())$;
      **end**
   **end**
   **if** $t_R + t_{v_{maxindex}} \leq t_q$ **then**
      $R.add(v_{maxindex})$;
   **end**
   $U.delete(maxindex)$;
**end**
$R = R.sort()$;
*return R;*

---

with the best nodes according to the user preferences. Finally, we sort the list to visit the museum showcases in order.

On each iteration, the greedy algorithm selects the node that best fits the user preferences in the least amount of time, considering the time spent to arrive at the node and the time visiting it:

$$h(v_i) = \frac{v_i.cov(K)}{t_{v_i} + v_i.cost()} \tag{2}$$

The time spent to arrive at node $v_i$ is computed using the following function:

$$v_i.cost() = \min_{v_j \in G} \{t_{<v_j, v_i>}\}$$

which denotes the minimum cost of reaching $v_i$ from any other node in $G$. $v_s.cost() = 0$ for the initial node $v_s$. Although in our case study this value is close to zero, this could be useful for a different graph where the POIs were more distant among them. Additionally, the partial solution computed on each iteration should be admissible, so the time spent in the partial computed route should not exceed the time constraint $t_q$.

### 4.1 A Motivating Example

The following example compares the A\* and the greedy algorithms with an equivalent heuristic function in action. Let us suppose that we define a map with 9 POIs (1), a time restriction $t_q = 15$ min and the user preferences and the keyword coverage for each node described in Table 1.

**Table 1.** User preferences (Q) and keyword coverage for each POI.

| | Q | 1 | 2 | 3 | 4 | 5 | 6 | 7 | 8 | 9 |
|---|---|---|---|---|---|---|---|---|---|---|
| Art | 0.40 | | | | | | | | | |
| Computer science | 0.90 | 0.95 | 0.95 | 0.70 | | 0.30 | 0.30 | | 0.70 | |
| Curiosity | 0.70 | | | | | | | | | |
| Hardware | 0.50 | | 0.80 | | | | | | | |
| History | 0.60 | 0.95 | 0.60 | 0.30 | 0.70 | | | | 0.30 | |
| Movies | 0.50 | | | | | | 0.30 | | | 0.20 |
| Networks | 0.60 | | | | | | | | | |
| Pcs | 0.80 | | 0.50 | | 0.80 | | | | | |
| Periferics | 0.50 | | | | | | | | | |
| Science | 0.70 | | | | | | | | | 0.50 |
| Servers | 0.60 | | | | | | | | | |
| Spain | 0.60 | 0.90 | 0.70 | 0.70 | 0.80 | | | 0.30 | 0.30 | |
| Storage | 0.40 | | | | | | | | | |
| Teaching | 0.50 | | | | | 0.20 | | 0.40 | 0.80 | |
| Videogames | 0.90 | | | | | | 0.30 | | | |

After running both algorithms, we get two different routes. Results are summarized in Table 2. We can see that both routes fulfil the time restriction $t_q$. We have computed the keyword coverage metrics described in previous sections and we see that A* algorithm achieves better results. However, the computational time shows that even in a small graph with 9 nodes, the Greedy algorithm is extremely faster than the A* algorithm (Algorithm 1).

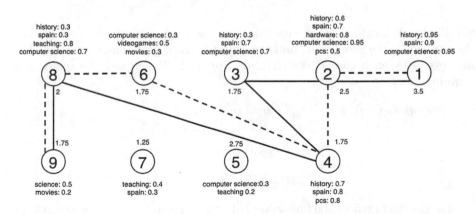

**Fig. 1.** Museum Graph. The dashed line represents the A* route and the solid line, the Greedy route.

**Table 2.** Routes calculated by A\* and Greedy algorithm. $t_R$ represents the time for completing the route (in both cases they do not exceed the $t_q = 15$ min). *Elapsed CPU time* confirms that Greedy runs faster than A\* even in a small graph.

|  | A\* | Greedy |
|---|---|---|
| $R$ | $\{1, 2, 4, 6, 8, 9\}$ | $\{1, 2, 3, 4, 8, 9\}$ |
| $t_R$ | 13.9 min | 14 min |
| *Ellapsed CPU time* | 272 ms | 4 ms |
| $kc(R)$ | 4.45 | 3.76 |
| $kc\_sum(R)$ | 9.01 | 8.47 |

# 5    Route Planning Using CBR

As we described in Sect. 4, the knowledge of the heuristic search algorithms is compiled in the evaluation function $h$ to deal with both time and preference restrictions given in the query. With a CBR (Case Based Reasoning) approach, we face the problem resolution with a different perspective. Given a query, CBR reuses previous solutions (i.e. routes) that are stored in a case base when a similar situation occurs [1]. With this approach, we need to deal with the definition of an appropriate similarity measure and a method to adapt a solution if the case solution does not fulfil the query restrictions.

We will use here the same problem formalization described in Sect. 3: given a query Q describing the group $(size_q, age_q, pref_q, t_q)$, with a set of preferences $(pref_q = \{\{\lambda_{k_1}, .., \lambda_{k_q}\}\})$ and the visit time restriction $(t_q)$, we aim to find a route that maximizes the group satisfaction on their preferences while complying with the given time restriction. CBR measures similarity between the query Q and the case base and retrieves the most similar case (1-Nearest Neighbour) to reuse its solution.

**Similarity Function.** Similarity takes into account every query feature, i.e., the time visit, the group size, and the group average preferences and age. The distance between a case $c = (size_c, age_c, pref_c, t_c, R_{sol_c})$ and the query $q$ is computed as:

$$distance(c, q) = |t_c - t_q| \cdot constTime$$
$$+ |size_c - size_q| \cdot constSize$$
$$+ |age_c - age_q| \cdot constAge$$
$$+ \left(\sum_{k_i \in K} |pref_c(k_i) - pref_q(k_i)|\right) \cdot constLabel$$

These weights reflect the importance of every feature. `constTime`, `constSize` and `constLabel` values may change depending on the domain, although the first two should be greater than the last one. For our specific case study, the chosen values are `constTime` $= 0.4$, `constSize` $= 0.2$, `constAge` $= 0.1$ and constLabel=0.5. constAge is lower because we observed a dependency between age and preference labels, i.e., some group preferences that may depend on the age are covered by the different labels in the query.

Distance is computed in the $[0, \infty]$ interval but we need a similarity measure in the $[0, 1]$ interval, with 1 for identical cases, and 0 when distance tends to infinity. For this reason, our similarity function is computed as:

$$Sim(c, q) = \frac{1}{e^{dist(c,q)}}. \tag{3}$$

**Adaptation.** Two different adaptation strategies are performed on the retrieved solution $R_{sol_c}$:

– Adaptation based on reducing the time spent in the nodes. If the time required to complete the recommended visit is slightly greater than the time available for the visit (the quotient lies between 0.8 and 1), the recommended visit is adapted reducing lightly the times in the items of the route.
– Modification of the route by deleting nodes of the solution. If the time required to complete the recommended visit is notably greater than the time available for the visit, the recommended visit is adapted by eliminating the nodes with worst coverage. To accomplish that, the nodes with lower $v_i.cov(K)$ values are deleted from the solution.

## 6   Case Study: MIGS

Our case study is MIGS museum, a museum about computer science history located at the Computer Science Faculty of Complutense University in Madrid. Located at the third and fourth floors of the Computer Science Faculty, this museum has more than 52 showcases related with history of computer science in Spain. It start with the Computer science origins in Spain with Professor García Santesmases and traverses the Spanish CS history until our current days, including all type of antiques, modern devices and gadgets.

During the knowledge acquisition phase, we revised the MIGS catalogue and the showcase organization of items, and we interviewed the domain expert to identify the set of keywords employed to describe both the POIs features and the query. In this case study we use a set of 15 labels or keywords: *storage, art, science, movies, curiosity, teaching, Spain, Computer Science, hardware, history, pcs, periferics, networks, servers* and *videogames*. Each label in a POI node is weighted with a [0,1] score indicating the keyword coverage value of the label in the node. For example, the MIGS showcase with the first computer made in Spain (dated 1952) will be annotated with labels *history* (0.8) and *computer science* (1) as we want to emphasize this features. Figure 2 shows the nodes in the museum. It is worth noting that we use a graph where every node is reachable from every other node. To simplify the representation, edges are not explicitly represented although there is an edge between every pair of nodes. Each edge is labelled with the time needed to walk between nodes. These time values are automatically computed using the geographical distance.

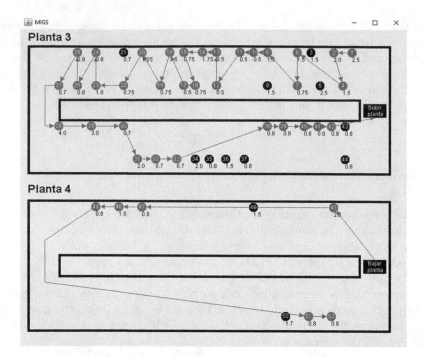

**Fig. 2.** MIGS museum Graph. Red arrows define a route proposed by CBR approach. (Color figure online)

**Case Base.** We acquire the cases using a process of observation of real visits to the MIGS during one month. As described in previous section, we also interviewed the museum guide (as the domain expert) to manually identify the keywords set and tag the museum POIs. We have acquired a prototypical set of 28 cases covering different situations for group of visitors: case with different size groups of different bachelors, masters or PhD students; quick visits of the most important items for different groups; children and elderly people visits; thematic visits with different interests (general, movies, games, history,..). Figure 2 shows the solution of a case of visit for a group of students, 70 min and preferences mainly on pcs, videogames, curiosities, and science (we obviate the specific values on the query features). Figure 3 visualizes the case base as a graph. Nodes represent the cases and edges represent the similarity between cases. Edge width correlates with the similarity value computed using Eq. 3.

## 6.1 Experimental Results

Our preliminary experiments with algorithm A* and the heuristic presented in [12] showed that A* is not applicable to the size of our map represented as an strongly connected graph. For this reason, we want to explore the pros and cons of using CBR against the use of a greedy algorithm with the heuristic described in Sect. 4.

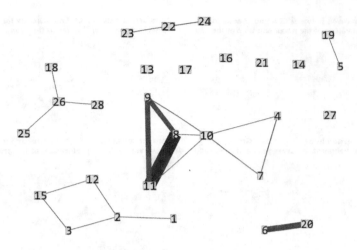

**Fig. 3.** Visualization of the case base as a graph. Line width correlates with the similarity value between linked cases.

The aim of the evaluation is to compare the coverage function value $kc\_sum(R)$ described in Sect. 3.1 according to the query attributes size, visit length, average age and group preferences. For each attribute, we fixed it to different values and generated random values for the rest of attributes. For the group preferences attribute, we fixed the preferences of three prototypical visiting groups, described later. We randomly generated 400 queries per each fixed value.

For each experiment we measured the keyword coverage value $kc\_sum(R)$ and we normalized the obtained result with the maximum value that can be obtained with an equivalent query that represents a visit that traverses all items and with $\lambda_i = i$ in all the preference labels employed in the experiment. Finally, we computed the average value for the 400 randomly generated queries, defining an average keyword coverage percentage.

As expected, the greedy algorithm offers better results in terms of keyword coverage, although we see how CBR also exhibits good results in coverage with a very efficient response time. Figure 4 (top-left) shows some experimental results obtained when fixing the visit duration ($t_q$). The results show, as expected, that the coverage increases with the time of the visit. 90 min give the visitor enough time to visit all the POIs in the museum, so preferences do not affect the route much. However, we observe that CBR keyword coverage is lower in large groups even in a 90 min visit where all nodes are chosen in the greedy solution. This fact occurs due to the space restrictions that have been captured as knowledge in the cases. The case routes collected show that big groups never stop at the small showcases. This explains why coverage is lower in CBR even when there is time enough to visit all the nodes. Greedy algorithm does not have this knowledge and it generates routes that will be uncomfortable for big groups.

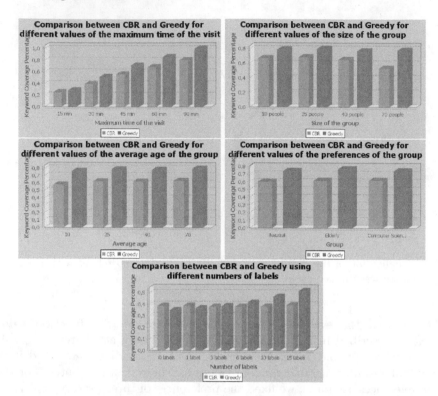

**Fig. 4.** Comparison of keyword coverage for different values of time, group size age and preferences.

In Fig. 4 (top-right) we observe coverage values for different group sizes ($size_q$). We see how keyword coverage almost does not vary with group size. As detailed before, coverage is lower in large groups due again to space restrictions in certain showcases that will never appear in cases with large groups.

Although the age ($age_q$) was considered as an attribute that characterizes well the visit group, its importance is lower than the importance of other attributes. Hence it has little influence in keyword coverage (Fig. 4 (middle-left)). This fact needs a proof revision to validate if this attribute is implicit in the group preferences or if it has impact only in relationship with other attributes.

Figure 4 (middle-right) shows a comparison among queries representing prototypical group preferences for thematic visits ($pref_q$). The first group (*neutral*) shows a middle interest in every label, the second group (*Elderly*) prefers items related with *history*, *spain*, and *curiosity*, showing low interest for categories like *hardware*, *servers*, *networks* or *videogames*. Finally, *Computer Science students* group is more interested in specific categories like *servers* and *networks*, showing a high overall interest in the contents exhibited in the museum. As *Elderly* has no interest for half of the labels, it is reasonable that the coverage of their recommended routes will be lower than the *Neutral* group. On the other hand, as the *Computer Science Students* group is so interested in almost every category, their coverage is notably better than the coverage of the *Neutral* group.

Coverage increases in both algorithms when more labels are included in the query (Fig. 4 (bottom)). For this experiment, we delimited the available time for visiting the museum between 15 and 30 min. For larger values the differences are not appreciated because both algorithms select almost every POI in the museum.

Coverage results in the CBR approach depend strongly on the quality of the case base. In these preliminary evaluation, the results of the CBR approach are promising with a small case base and an immediate response time. Optimal results in keyword coverage are not an issue in group visits where group preferences are obtained as a weighted average of its individual members. Heuristic search optimizes keyword coverage but relies on a detailed query definition process. CBR is applicable for queries where preferences in the queries are approximated with prototypical queries and even when no preferences are stated.

As we have pointed out before, some routes computed by the heuristic search algorithm have problems when used in real situations (like space in the showcases). We think that other evaluation measures, like user satisfaction, will benefit CBR as it captures real situations and other subtle aspects, like dependencies between nodes, which helps, for instance, to give coherence to the guide narrative when describing items in the POIs. We will study these aspects as future work. Next section summarizes our experience and advantages of using CBR in this case study and outlines some lines of future work.

## 7   Conclusions and Future Work

In this paper, we have described a CBR solution to the tourism route planning problem for groups of people. We have compared our approach to an heuristic search according to the keyword coverage that measures the goodness of the recommended route according to the group preferences. As long as the heuristic search perform an exhaustive search, its results in keyword coverage outperform the CBR solution. However, the differences are small, our results are very promising and the keyword coverage will be improved when including more visits in the case base. We expect the quality of the final solution measured as the keyword coverage will increase when we include more cases of real visits. We will also include a detailed cost evaluation study for bigger case bases.

Results from our experiments allow us to summarize the following conclusions on the advantages of using a CBR approach to solve this problem. First of all, cases have resulted to be an excellent tool to elicit domain knowledge. Cases in the CBR system represent real visits. First experiments showed that we can simplify the representation of the museum map where all nodes are connected to all other nodes: even if it is theoretically possible visiting a node after any other node, it is not happening in real visits, as in the route to the farther nodes, users always visit intermediate nodes. This knowledge has been captured observing the cases in the case base and could be used to simplify the map and apply A*.

We plan to do it as future work: use a two-step process where the routes stored in the CBR solutions are used as a map where the A* algorithm searches optimal routes in keyword coverage.

Another advantage of CBR regarding the query elicitation process is that CBR is applicable with less input knowledge. For the sake of an easier comparison between algorithms, we have used the same problem formalization. Meanwhile heuristic search algorithms rely on a detailed query description as it uses the individual preferences of all the group members (resulting in a very tedious query description process), CBR would allow a simpler alternative to query the system using an approximate description of the preference of the groups based on labelled prototypical visits (query example: children/teenager visit, very big group). This means that we could reuse case descriptions (complete group description) as well as reusing CBR solutions (routes). CBR could even work properly with an empty set of preferences in the query reusing the most typical visit for those groups that we do not have information about.

Regarding knowledge engineering we have noticed that cases capture some dependences between the nodes, i.e., certain nodes always appear together in all the real routes (cases). For example, every case solution includes the two most important items in the museum: the first computer built in Spain and the original *Enigma* machine. We will study the fact that cases captures important knowledge regarding narrative of the museum explanations given by the human guide during visits. Cases have captured important knowledge that is nearly impossible to elicit and include in a mathematical evaluation function. The museum expert labelled with the same label those museum items sharing a certain topic, like games or movies. However, when the museum guide describes the items, (s)he uses anecdotes and sometimes subtle dependencies between nodes. A narrative description on a piece can start in a showcase and ends in another (not necessary visited right after). These nodes are included together in all routes because there is a narrative dependency between them. It is only an observed fact from the CBR solutions, although we have not used it the evaluation of the algorithms. We study this as a future line of work. CBR uses knowledge from real and specific situations, like the fact that in big groups there are space limitation on the showcases. We would need a new annotated value for the heuristic search to include this knowledge in the evaluation function (like the node capacity). In the CBR approach, this node has been implicitly captured in the form of cases and has been elicited due to the experiments as this factor was not taken into account in the original version of the problem.

This research opens new lines of future work. We first plan to extend the case base using an automatic case acquisition procedure where we are tracking details for all the museum visit using location devices (beacons). Then we will study other similarity measures and more complex adaptation methods. We evaluate a 2 or 3-NN retrieval and an adaptation method based on mixing routes from different solutions. I addition, we will test an adaptation method where A* searches an optimal route in the map stored as the case solution, i.e., a two-step CBR+A* approach. We will evaluate the CBR approach using other measures that will benefit it, like coherence on the narrative of the explanations, or user satisfaction with the museum visit experience. We also plan to extend our research to other domains, like city tours.

# References

1. Aamodt, A., Plaza, E.: Case-based reasoning: foundational issues, methodological variations, and system approaches. AI Commun. **7**(1), 39–59 (1994)
2. Ardissono, L., Goy, A., Petrone, G., Segnan, M., Torasso, P.: Intrigue: personalized recommendation of tourist attractions for desktop and hand held devices. Appl. Artif. Intell. **17**(8–9), 687–714 (2003)
3. Biuk-Aghai, R.P., Fong, S., Si, Y.-W.: Design of a recommender system for mobile tourism multimedia selection. In: 2nd International Conference on Internet Multimedia Services Architecture and Applications, IMSAA 2008, pp. 1–6. IEEE (2008)
4. Corchado, J.M., Pavón, J., Corchado, E.S., Castillo, L.F.: Development of CBR-BDI agents: a tourist guide application. In: Funk, P., González Calero, P.A. (eds.) ECCBR 2004. LNCS (LNAI), vol. 3155, pp. 547–559. Springer, Heidelberg (2004). doi:10.1007/978-3-540-28631-8_40
5. Gavalas, D., Konstantopoulos, C., Mastakas, K., Pantziou, G.: Mobile recommender systems in tourism. J. Netw. Comput. Appl. **39**, 319–333 (2014)
6. Goel, A.K., Ail, K.S., Donnellan, M.W., de Silva Garza, G., Callantine, T.J.: Multistrategy adaptive path planning. IEEE Expert **9**(6), 57–65 (1994)
7. Haigh, K.Z., Veloso, M.: Route planning by analogy. In: Veloso, M., Aamodt, A. (eds.) ICCBR 1995. LNCS, vol. 1010, pp. 169–180. Springer, Heidelberg (1995). doi:10.1007/3-540-60598-3_16
8. Lim, K.H., Chan, J., Leckie, C., Karunasekera, S.: Personalized tour recommendation based on user interests and points of interest visit durations. In: Proceedings of the 24th International Conference on Artificial Intelligence, IJCAI 2015, pp. 1778–1784. AAAI Press (2015)
9. Masthoff, J.: Group recommender systems: combining individual models. In: Ricci, F., Rokach, L., Shapira, B., Kantor, P.B. (eds.) Recommender Systems Handbook, pp. 677–702. Springer, USA (2011)
10. McGinty, L., Smyth, B.: Personalised route planning: a case-based approach. In: Blanzieri, E., Portinale, L. (eds.) EWCBR 2000. LNCS, vol. 1898, pp. 431–443. Springer, Heidelberg (2000)
11. Rey-López, M., Barragáns-Martínez, A.B., Peleteiro, A., Mikic-Fonte, F.A., Burguillo, J.C.: Moretourism: mobile recommendations for tourism. In: IEEE International Conference on Consumer Electronics (ICCE), pp. 347–348. IEEE Computer Society (2011)
12. Zeng, Y., Chen, X., Cao, X., Qin, S., Cavazza, M., Xiang, Y.: Optimal route search with the coverage of users' preferences. In: Proceedings of the 24th International Conference on Artificial Intelligence, IJCAI 2015, pp. 2118–2124. AAAI Press (2015)

# Comparative Evaluation of Rule-Based and Case-Based Retrieval Coordination for Search of Architectural Building Designs

Viktor Ayzenshtadt[1,3]([✉]), Christoph Langenhan[4], Johannes Roith[4],
Saqib Bukhari[3], Klaus-Dieter Althoff[1,3], Frank Petzold[4],
and Andreas Dengel[2,3]

[1] Institute of Computer Science, University of Hildesheim,
Samelsonplatz 1, 31141 Hildesheim, Germany
{Viktor.Ayzenshtadt,Klaus-Dieter.Althoff}@dfki.de
[2] Kaiserslautern University, P.O. Box 3049, 67663 Kaiserslautern, Germany
Andreas.Dengel@dfki.de
[3] German Research Center for Artificial Intelligence,
Trippstadter Strasse 122, 67663 Kaiserslautern, Germany
Saqib.Bukhari@dfki.de
[4] Chair of Architectural Informatics, Faculty of Architecture,
Technical University of Munich, Arcisstrasse 21, 80333 Munich, Germany
{Langenhan,Roith,Petzold}@ai.ar.tum.de

**Abstract.** To support the early conceptualization phase in architecture with computer-aided solutions, in particular, with retrieval systems that can find similarly structured building designs in comprehensive collections of such designs, a number of approaches were presented to date. In the Metis project two retrieval coordination approaches (coordinators) were developed to govern the search of similar (sub-)structures of architectural designs. The main task of both coordinators is to select the retrieval method that is appropriate for the given user query. First approach is a standalone service that uses rules only to coordinate the retrieval and can use subgraph matching and database search methods, whereas the second one is rule- and case-based and is part of a distributed system for case-based retrieval of architectural designs. We compared both coordinators in a user study to find out which strengths and weaknesses both coordinators currently possess, and for which retrieval scenarios of the architectural conceptualization phase they could be appropriate. The results showed that the complexity of the particular scenario and the purpose of search are the main points that differentiate both coordinators. The rule-based coordinator performed better when a search for exact (sub-)structures was required, whereas the rule- and case-based coordinator is appropriate for queries aimed to be used for exploration and general search for inspiration. Visualization of the results of both coordinators is in need of improvement.

**Keywords:** CBR applications · CBR and creativity · Case-based agents · Case-based design · Case-based retrieval · Case-based coordination

© Springer International Publishing AG 2016
A. Goel et al. (Eds.): ICCBR 2016, LNAI 9969, pp. 16–31, 2016.
DOI: 10.1007/978-3-319-47096-2_2

# 1   Introduction

The early design phase in architecture is characterized by analysis of design ideas, that architects iteratively elaborate by themselves, and references found in corresponding specialist material. The traditionally established and nowadays still widely used conceptualization approach is a pen-and-paper-based design phase with iterative comparison of the progress with the references in the printed media. By comparing similar design ideas a design can be evaluated, used as inspiration or explicit design solution regarding different criteria. The computer-aided retrieval of similar design ideas in digital collections of such designs can be a significant improvement for the early conceptualization phase of architecture. It can help an architect to speed up the design process by immense reduction of time spent for search and so make it more efficient and productive. In cases, where the currently used retrieval system implements multiple search algorithms, a coordination approach is needed, that selects the proper algorithm and/or strategy based on the (user-generated) data contained in the search request.

In this work we compare two different retrieval coordination approaches, both developed, among other services, for the *Metis* project (see Sect. 2 for more information about the project). The first one is the rule-based only coordination, which selects a suitable retrieval method based on the implemented rulesets. This coordinator was developed by the KSD[1] research group of the Technical University of Munich *(TUM)* and will be referenced as the *KSD Coordinator* in further sections of this paper. The second is the case- and rule-based coordination agent of the distributed case-based retrieval system MetisCBR [3], developed by the German Research Center for Artificial Intelligence *(Deutsches Forschungszentrum für Künstliche Intelligenz, DFKI)*. This coordinator is a case-based agent *(CBA)* and is the main node of the retrieval process within the system. It will be referenced as the *MetisCBR Coordinator* in further sections of this paper.

This paper is structured as follows: first the related external and internal work of the Metis project will be presented. After that, in Sect. 3, both retrieval coordination approaches will be described in detail, their main features and abilities will be presented.

In the Sect. 4 we provide a comprehensive cross-evaluation of both coordination techniques. By means of applying a number of queries created during the study we will take into account the computed similarity values of the building designs retrieved by both coordinators, the subjective opinion on quality of the result set according to the architectural informatics experts, the overall number of results, and other details. The main purposes of this pilot evaluation is to find out which strengths and weaknesses both coordinators currently possess (in order to coordinate their development in the future) and to determine which coordination technique is currently the most suitable for which user scenario.

The conclusion and future work section closes this paper and give an overview of the presented study, following by a short description of work that is planned to be accomplished in the near future in the *Metis* project.

---

[1] Knowledge-Supported Design.

## 2  Related Work

To find essential solutions in order to provide the computer-aided support of the early conceptualization phase in architecture, an interdisciplinary basic research project *Metis – Knowledge-based search and query methods for the development of semantic information models for use in early design phases*[2] was initiated by the DFKI and the TUM[3]. The project unites following main research areas: case-based reasoning *(CBR)*, multi-agent systems *(MAS)*, computer-aided architectural design *(CAAD)*, and building information modelling *(BIM)*.

The work, which has been accomplished in the project to date, consists of implementation of different search techniques, query builder interfaces, and supporting services, such as databases. Besides of that, theoretical research has been conducted in the project as well. The two important theoretical approaches that were created during the project-related research are the *Semantic Fingerprint* [13] paradigm and *AGraphML* [12] (see Sect. 3). As databases, among others, the *Neo4j* graph database with building design graphs, the content management system (CMS) *mediaTUM* for graphical representations of the designs, and the *Open Source BIM Server* can be named. The query builder interfaces include the web-based floor plan editor *(Metis WebUI)* [4], together with a touch-table application, and iPad and Android apps. The currently implemented search techniques include the subgraph matching algorithms together with the case-based retrieval techniques [2] of MetisCBR and an index-based retrieval method with the *Cypher* language queries of Neo4j. The subgraph isomorphism techniques include the implementations of the VF2 approach [7] and of the enhancement [21] of the original Messmer-Bunke algorithm [16], implemented under the name *GML Matcher*. The study presented in this paper is the first direct comparison of retrieval techniques implemented by different working groups of the project.

Prior to Metis, much essential work has been done in the domain of support of the early phase of architectural design. These projects and research initiatives left a legacy of methodologies and applications that we could build on. Maher et al. provide a description of application of CBR to design problems in [15]. In [9,18] overviews of the applications for architectural domain are provided. An essential work of Richter [17] enhances this research by providing detailed in-depth descriptions and analysis of the approaches in particular and of CBR in architecture in general. Noticeable approaches are FABEL [20], CaseBook [10], DIM [11] or CBArch [6] inter alia. For case representation, VAT (Visual Architectural Topology) [14] provides a semantic way of representation, based on ontological expressions of floor plan topologies.

For the MAS area, the work of Anumba et al. [1] is one of the essential publications that provides insights into embedding of MAS in construction, architecture, and related domains. Application examples, as well as theoretical foundations, are presented and described in detail.

---

[2] *Metis – Wissensbasierte Such- und Abfragemethoden für die Erschließung von Informationen in semantischen Modellen für die Recherche in frühen Entwurfsphasen.*

[3] The project is funded by the German Research Foundation *(DFG)*.

# 3    Examined Coordinators and Their Features

In this section we present the features of the coordinators we are evaluating in this work, using the categories of *retrieval-related aspects* (such as *fingerprints, weighting,* and *query protocol*) and *additional aspects* (such as *extensibility*). First, we provide an overview over foundations that are common for both coordinators.

– *Building Information Modeling (BIM)* is an approach for machine-interpretable modeling of buildings for the purpose of storing building information across the lifecycle of a building. BIM's object-oriented concept is based on *parametric* objects that contain information about attributive geometric data of a building. A comprehensive information source about BIM is a handbook [8].
– *Semantic Fingerprint* [13] is a paradigm that describes a pattern structure for flexible, hierarchical and index-based definition of building design metadata. Semantic Fingerprint is related to BIM and is intended to extend BIM for use in modern computer-aided architectural applications. The fingerprint patterns can be used for description as well as for querying or comparison of building designs in such an application. The characteristics of a fingerprint *(FP)* can use semantic information of a building, topology, or relations (direct or adjacent) between rooms. Commonly, FPs can be represented as labeled floor plan graphs with rooms as nodes, and room connections as edges. For the Metis project a list of implementable FPs was defined. The Table 1 provides an overview over the FPs currently available for *both* coordinators.
– *AGraphML* is the extension of GraphML [5] that has an architectural specification [12] as its underlying structure. AGraphML was developed by TUM

Table 1. Fingerprint patterns currently available for both coordinators.

| Fingerprint | Description | Specifics |
|---|---|---|
| Room Count | Number of rooms | No connections between rooms and no labels specified |
| Edge Count | Number of edges | No room information specified |
| Room Graph | Anonymous representation of rooms and edges | No labels specified for rooms and edges |
| Room Functions | Labels of rooms | No connections between rooms, room labels are specified |
| Room Semantics | Emphasis on room semantics | Rooms information is complete, no edge labels specified |
| Passages Semantics | Emphasis on edge semantics | Edges information is complete, no room labels or names specified |
| Semantic Connections | Complete graph | All information about rooms and edges is available |

and DFKI for the projects of the KSD research group (like the Metis project). It is used for the XML-based representation of semantic fingerprint-based floor plan graphs (e.g., as part of query format, see also Sects. 3.1 and 3.2).

## 3.1  KSD Coordinator

The KSD Coordinator Web Service [19] provides a comprehensive *middleware* application for search requests from compatible front-end clients to different retrieval approaches in order to find similarly structured building designs. The key features of the KSD Coordinator are the specific *language for query construction* (with support of search constraints and similarity assessment definitions) and the underlying *rule engine* that is aimed to determine which retrieval method is the most suitable for the given query. The KSD Coordinator also provides an own user interface with configuration and query playground among other things. The next sections provide an overview over the KSD Coordinator's functionalities.

**Retrieval-Related Aspects.** The KSD Cooordinator is able to trigger a number of search techniques, where for each of these techniques a special particular thread/process, an *agent*, is responsible. The selection of agents depends on the fingerprints determined in the query. For each fingerprint an assigned agent type exists. Currently three of these types are available:

- *Mediatum Agent:* an agent that retrieves the database of the CMS mediaTUM directly and is suitable for FPs where a floor plan metadata attribute is the main search criterion (Room Count and Edge Count FPs).
- *GML Matcher:* an agent that uses the extended Messmer-Bunke algorithm [21] for retrieval of isomorphic subgraphs using the complete graph information. This agent is used for the Semantic Connections FP.
- *Neo4J Agent:* a flexible agent for retrieval of similar graphs, where information can be incomplete. Queries the Neo4j database directly with *Cypher* queries and is used for Room Graph, Room Functions, Room Semantics, and Passages Semantics FPs.

Retrieval results are weighted during the *Ranking* process of the KSD Coordinator. The final rank of a result is computed by the sum of all conditional ranks of this result. A conditional rank is a product between an indicator function and the sum of a fixed value and the product of the weight and similarity value.

The query protocol (or *query language*) of the KSD Coordinator is XML-based and defines a pipeline of components that will be decomposed and simplified during the retrieval process. Three types of such components (blocks) exist:

- let: Provides AGraphML-based graph definitions that can be declared by a variable. Optional block.

- select: Defines data contained in the result set. Optional block.
- where: A mandatory block. Defines which retrieval *conditions* should be applied to the query. Represented as a Boolean expression in XML format. A condition can contain a fingerprint reference (if a graph was provided in the let block), a defined retrieval method, or a specific metadata attribute.

The conditions from the where block will be later decomposed and parsed by the *rule engine* of the KSD Coordinator. The assessment of the conditions depends on the *rule scripts* contained in the rule engine. The rule scripts define which search agents should be triggered for a given condition.

**Additional Aspects.** The KSD Coordinator can be extended with new agents and new fingerprints, inter alia. A new agent definition can be added by providing a corresponding agent configuration, agent lifetime class, and the data source class. For addition of a new fingerprint, a definition of a corresponding graph equivalence concept and an associated agent are required.

A feature of *caching* is available for the KSD Coordinator that allows for saving intermediate results of previous queries. A set of caching rules includes the caching and reuse options for *full query caching* and *data source caching*. The *timeout* feature allows for setting a maximum query execution time for an agent.

## 3.2  MetisCBR Coordinator

In MetisCBR, the general task of retrieval coordination is distributed among the *Coordinator* agent and its helper agents *SubCoordinator* and *Timeout*. The two helpers were created to reduce the overload of work of the coordination agent. Given this context, we can also speak of the *coordination team* of MetisCBR.

The actual coordination in MetisCBR is divided into two steps: *rule-based* and *case-based*. In the first, rule-based step, the ruleset of the MetisCBR Coordinator determines, based on the *user-generated* data from the query, which FPs will be used for the current retrieval. Then, in the case-based step the myCBR-based mechanism of the MetisCBR Coordinator tries to find the most similar QUERY case instance in its case base. If the similarity value is sufficient, the results achieved by this previously saved request, will be presented to the user, without starting a new retrieval process. After the evaluation of the results, if the user prefers to conduct a new retrieval anyway, the retrieval will take place. The results (old/saved in the similar case, or achieved by a new search) will be added to the new QUERY case instance and saved in the coordinator's case base. If the user has not determined which FPs should be applied during the search, but the most similar QUERY case instance had FPs applied, then the user will be informed about this fact during the output of the results. The task of retrieval of the most similar QUERY and saving the results is optional and can be disabled in the configuration of the system. It can also be seen as the *caching* or *indexing* process. The complete procedure of query processing is shown in Fig. 1.

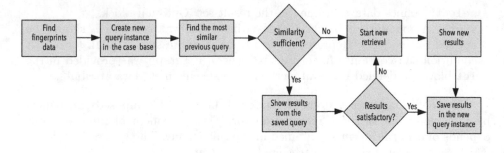

**Fig. 1.** Query processing steps of the MetisCBR Coordinator.

**Retrieval-Related Aspects.** The MetisCBR Coordinator has access to different retrieval methods implemented in the MetisCBR core. The methods can be classified into two main classes: *generic* and *fingerprint-related*. Both of these classes use the same set of retrieval strategies, the main difference between the classes is the purpose of use: the category of generic methods is used if no FPs were applied to the query, whereas the fingerprint-related methods are obviously applied if FP information is available in the query. Another, more practical, difference is the set of attributes used for the comparison during the search: for non-fingerprint methods all of the attributes are used, for each fingerprint method a set of suitable attributes is used to find the most similar cases. The underlying CBR domain model [2] defines the structure of the cases (building designs). The currently available retrieval strategies include two following types:

- The multi-step *Basic strategy* (described in [2] as well) for fingerprints considered complex (`Room Graph`, `Room Semantics`, `Passages Semantics`, or `Semantic Connections`) and deep search for queries without fingerprints.
- The single-step *metadata strategy* that uses the floor plan metadata information only and is applied for fingerprints considered simple (`Room Count`, `Edge Count`, `Room Functions`) and for the fast search without fingerprints.

The weighting of fingerprints is applied according to the weights set by the user during the query building process. The weights should sum to 1 and will be multiplied with each similarity value of the result set of the corresponding fingerprint query. The weights assignment influences positioning of results in the final result set, where result sets of all of the fingerprints are combined.

The query format of the MetisCBR Coordinator makes use of the protocol developed for query construction in the MetisWebUI. In this protocol the AGraphML representation of the query is wrapped by the request tags that include the information about FPs as their child elements as well (see Listing 1.1).

```
<?xml version="1.0"encoding="UTF-8"?>
<searchrequest>
    <agraphml>
        <graphml>...</graphml>
    </agraphml>
    <fingerprint name="Room_Count"weight="0.4"/>
    <fingerprint name="Passages"weight="0.6" />
</searchrequest>
```

**Listing 1.1.** Query protocol of MetisCBR (fingerprints are exemplary).

**Additional Aspects.** Currently, the MetisCBR coordination team possesses the *open-box* extensibility feature. New behaviours and functions can be added and later changed inside the source code of the agent. Also, the addition of new helper agents is possible: either by extension of the source code of MetisCBR or by communication with the coordinator from other compatible agent platforms.

Also, a *timeout* feature is available for the MetisCBR Coordinator. The helper agent Timeout checks periodically the activity of the retrieval processes (containers), i.e., if it has finished its task within the defined amount of time.

## 4  Comparative Evaluation

To analyse the current state of both coordinators we conducted a comparative user study that consisted of creating different floor plan queries with the Metis WebUI and evaluating the results returned by each of the coordinators. To accomplish this task, two CAAD experts of the Metis project were asked to take part in the study as representatives of the architectural research area. For the study, a special setting was prepared: the previously mentioned Metis WebUI, and the input and output interfaces of both coordinators. The search options of both coordinators were set to equalize the retrieval abilities best possible:

– No caching or indexing should be used. Expired queries count as 0 results.
– The FPs Room Count, Edge Count, Room Functions, and Semantic Room Connections can be used (due to some technical problems of the KSD Coordinator other FPs could not be used).
– The weighting can be set for queries with multiple FPs.

The datasets of both coordinators included the building design graphs imported from the Neo4j database. Due to the technical restrictions and differences in the nature of import, the exact number varied for both coordinators, but can be estimated to have 200 as the approximate value for all FPs of the MetisCBR Coordinator and the Semantic Connections FP of the KSD Coordinator. For the FPs of the KSD Coordinator that were mapped to the Mediatum Agent $\approx$ 400, and to the Neo4j Agent $\approx$ 500 building designs were used, as no import was needed, that is, the technical restrictions were also not applicable.

The actual process of the evaluation was divided into two phases. In the first phase a *storyline query* was created: during an iterative multistep process the initial/previous query was modified according to the view of the architect, the inspiration from the results, and requirements of this type of floor plan, and then used in the following step. Each of the iterations was related to the initial purpose (i.e., *scenario*) of the query. In the second phase separate *single queries*, which had no relation to each other, were used, each of these single queries had its own scenario. For each of the queries of both query phases the participants were asked to fill a questionnaire to review, estimate, or rank the overall results and the general behaviour of each of the examined coordinators, in order to provide their subjective opinion on the outcomes of the respective retrieval scenario. Following questions were included in the questionnaire:

- In general, are the results appropriate for this query? [scale 1(no)-5(yes)]
- Is an improvement of the results for this query required? [yes/no]
- Is it possible to mark one of the results as the *best* one? [yes/no]
- Can one of the results be used as a *template* for the next query? [yes/no, applicable only for the storyline query]
- Do the results contain a certain pattern or model? [yes/no].

For each of the answers it was also possible to leave an additional comment to provide an explanation of the opinion.

Besides of the subjective analysis, the *computed similarity values* (more precisely, the average similarity of all returned results and of the first 10 results), the *total number of results*, and the *inclusion of results of all selected FPs* in the returned result sets were also part of the evaluation.

The results of the retrieval were presented with the visualization method currently implemented in the corresponding coordinator interface (see Fig. 2): the *graph representation* with information about nodes and edges in the KSD Coordinator, and the *pre-rendered graph* (with information about nodes only) along with the separate *graphical representation* in the MetisCBR Coordinator.

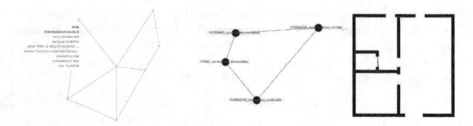

**Fig. 2.** The result visualization methods implemented in the coordinators at the time of the study. On the left side the graph representation of the KSD Coordinator with information about a node. On the right side the pre-rendered graph and the separate graphical representation of the MetisCBR Coordinator.

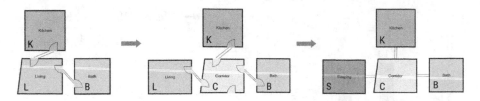

**Fig. 3.** The iterative scenario of an apartment for elderly married couple. K is used for the *KITCHEN* room type, L for the *LIVING* room, C for *CORRIDOR*, B for *BATH*, and S for the *SLEEPING* room.

## 4.1   Queries and Results

The results of the study will be presented divided into two parts, according to the number of query phases. We compare the outcomes of the questionnaire and the computed values by providing a summary for each of these result categories.

**Storyline Query.** For the storyline query (see Fig. 3) a scenario of an *apartment for elderly married couple* was selected by participants. For this query the participants decided to make two search requests in each iteration: first the request with the `Semantic Connections` FP and then the addition of the `Room Functions` FP in the next request (except the first iteration).

In general, the results of the storyline query (see Fig. 4 and the Table 2) confirmed the evident assumption that the exact matching approach of the GMLMatcher (Messmer-Bunke-Algorithm) would find the exact (sub-)structures among the graphs and can answer the question if such a structure is available in the database at all, so that the user can take a look how it is implemented in another floor plan. In contrast to this, the MetisCBR Coordinator returned quantitatively more results in each iteration, which provided much more space for exploration of similar designs. By adding the `Room Functions` FP in second and third iteration the KSD Coordinator increased the number of results and provided a sufficient number of explorable designs, where the room setting of the query and of the result are equal. Noticeable is also the fact that the average similarity of the first 10 results of the MetisCBR Coordinator is noticeably higher than this value of its all results, whereas for the KSD Coordinator the value remains (almost) equal in both measurements.

The questionnaire answers and corresponding comments and explanations confirmed that the number of results for the KSD Coordinator is currently an issue when using the `Semantic Room Connections` FP only. For example, this fact had the biggest influence on the question if the improvement of results is required. A similar problem also exists for the MetisCBR Coordinator: it has returned too much results and the most useful ones were not high positioned in the overall result set, so that a longer exploration was needed to find them. Regarding the question if a result can be considered a suitable template for the next iteration, both coordinators were equal, for both coordinators the separated/not graph-to-floor-plan mapped visualization was a problem in this particular

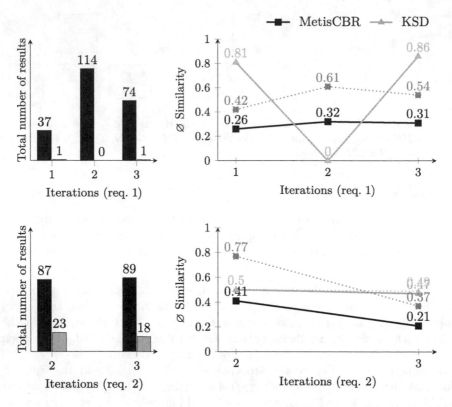

**Fig. 4.** The results of the computed values of *both* search requests of the *storyline query*. In the first row the results of the first request (`Semantic Connections` FP), and in the second row of the second request (`Semantic Connections` and `Room Functions` FPs), are presented. The dotted lines represent the average similarity of the *first 10* results. In the third iteration the participants applied weighting: 0.3 for the `Room Functions` FP and 0.7 for the `Semantic Connections` FP.

case. The patterns were recognized in all iterations, but not for each coordinator: in the first iteration a pattern of replacement of the `BATH` room type by `CORRIDOR` was noticed in the result set of the MetisCBR Coordinator, whereas no such patterns were seen in other iterations. In contrast to this, the result sets of the second and third iteration of the KSD Coordinator contained patterns, in particular, a pattern of greater number of rooms was recognized.

**Single Queries.** For the *single queries* part, three different scenarios were created by participants (see also Fig. 5). For this query type, the participants also decided to make two search requests for each of the queries: first the request with the `Semantic Connections` FP and then the addition of the `Room Functions` FP in the next request.

**Table 2.** Results of the questionnaire for the storyline query. Rank or estimation is accumulated or averaged if result sets for both requests were not empty.

| Query and Coordinator | Results appropriate? 1(no)-5(yes) | Results improvement required? | Best result available? | One of the results is suitable as template? | Pattern in results |
|---|---|---|---|---|---|
| *Iteration 1* | | | | | |
| MetisCBR | 2 | yes | yes | yes | yes |
| KSD | 5 | yes | yes | yes | no |
| *Iteration 2* | | | | | |
| MetisCBR | 2 | yes | no | no | no |
| KSD | 2 | yes | yes (req. 2) | no | yes |
| *Iteration 3* | | | | | |
| MetisCBR | 2 | yes | no | no | no |
| KSD | 2.5 | yes | yes | no | yes |

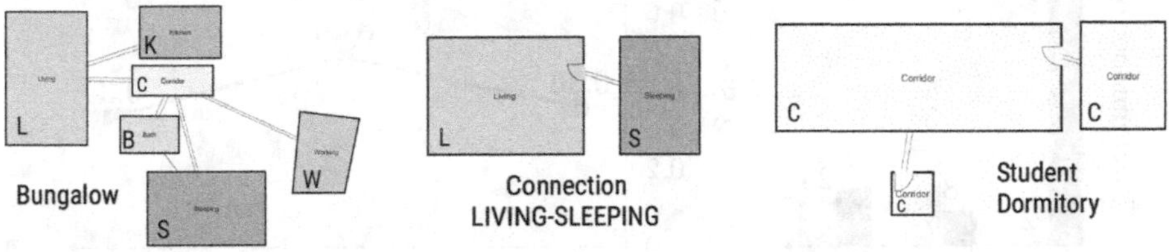

**Fig. 5.** Single queries. K is used for the *KITCHEN* room type, L for the *LIVING* room, C for *CORRIDOR*, B for *BATH*, S for *SLEEPING*, and W for the *WORKING* room.

- *Query 1:* **Bungalow** – A single-storey house with the working room placed at a slight distance to other rooms.
- *Query 2:* **Connection** LIVING–SLEEPING – The structure of living and sleeping room connected through a DOOR within other floor plan structures should be found to explore different implementations of this connection.
- *Query 3:* **Student Dormitory** – The building entry situation with three corridors connected through doors should be found. The architect is certain that a building design with such situation exists in the graph database.

The evaluation of results of the single queries (see Fig. 6) has shown that the advantages of the CBR-based retrieval for similar architectural designs are noticeable when it comes to queries with more complex structures. For example, for the *bungalow* query, the MetisCBR Coordinator returned much more results than the KSD Coordinator. Also, for this query, the MetisCBR Coordinator was able to return results for each FP, whereas the KSD Coordinator could only find results for the Room Functions FP (when it was added). In contrast to this, in the second query, which consisted of the very simple connection

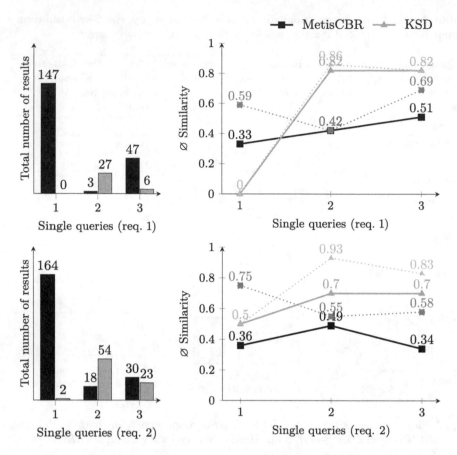

**Fig. 6.** The results of the computed values of *both* search requests for each of the single queries. In the first row the results of the first request (`Semantic Connections` FP), and in the second row of the second request (`Semantic Connections` and `Room Functions` FPs), are presented. The dotted lines represent the average similarity of the *first 10* results. Note the fact, that in the third query *(Student Dormitory)* the participants decided to apply weighting: 0.8 for the `Room Functions` FP and 0.2 for the `Semantic Connections` FP.

`LIVING-SLEEPING`, the KSD Coordinator could find more results for exactly this connection, using its advantages of the subgraph matching. The same applies to the *student dormitory* query, where the KSD Coordinator was able to find exactly the floor plan concept the architect was looking for. Though, the Metis-CBR Coordinator has found more similar cases, the exact match was not among them.

The subjective opinion of the architects, inferred from the questionnaires for single queries (see Table 3), showed that the MetisCBR Coordinator returned some satisfactory results for the bungalow query, only when the `Room Functions` FP was applied. Nevertheless, the question if there is a *working room* that is

**Table 3.** Results of the questionnaire for the single queries. Rank or estimation is accumulated or averaged if result sets for both requests were not empty.

| Query and Coordinator | Results appropriate? 1(no)-5(yes) | Results improvement required? | Best result available? | Pattern in results |
|---|---|---|---|---|
| *Query 1: Bungalow* | | | | |
| MetisCBR | 2 | yes | no | yes |
| KSD | 2.5 | yes | no | no |
| *Query 2: Connection LIVING-SLEEPING* | | | | |
| MetisCBR | 2 | yes | no | no |
| KSD | 4.5 | yes | no | yes |
| *Query 3: Student Dormitory* | | | | |
| MetisCBR | 2 | yes | no | yes |
| KSD | 4 | yes | yes | yes |

placed at a slight distance to other rooms could not be answered exactly. The same is applicable for the results from the KSD Coordinator, where one of the two results of the `Room Functions` FP was considered inspirational. For the *student dormitory* query, the usefulness of the results of the KSD Coordinator achieved by exact matching was noticeable. As mentioned above, the MetisCBR Coordinator was not able to find the exact concept, though some results could be considered very similar as they consisted mostly of the room setting given in the query (but not as part of a very complex floor plan). The visualization problem was noticed for the single queries as well as for the storyline queries. For example, the results of the *LIVING-SLEEPING* query could be evaluated more thoroughly, but the missing of concrete edge information in the visualized Metis-CBR results and missing representations of the actual floor plan for the KSD Coordinator results could not provide the needed information, so that no result could be considered best in the result sets for this query of both coordinators.

## 5    Conclusion and Future Work

In this paper we presented two retrieval coordination approaches: the rule-based KSD Coordinator and the rule- and case-based MetisCBR Coordinator. Both of them are integrated in the infrastructure of the KSD research group and able to access different retrieval methods for search for architectural designs with similar structures. Both coordinators use agents or related techniques for delegating of the actual retrieval tasks, the retrieval strategy depends on the user-defined architectural semantic fingerprint data in AGraphML-formatted queries.

We evaluated both coordinators in a comparative study that included two separate query types: an iterative storyline query where each iteration is related to a common scenario and is based on data from the previous iteration, and single queries where each of them has its own scenario. The results of the evaluation showed that the situations where an exact match of the structure is needed are more suitable for the KSD Coordinator as it can trigger the subgraph matching method based on the original Messmer-Bunke algorithm, whereas the Metis-CBR Coordinator can provide more space for exploration of the similar floor plans contained in the data-/case base. For both coordinators, a problem of visualization exists that has not allowed for more thorough evaluation of results in some cases. Our assumption is, that the combination of both coordinators, where each of them is responsible for special fingerprints, would currently provide the most comfortable way to support the early conceptualization phase in architecture using the techniques developed for the Metis project.

In future studies we are going to evaluate the implemented retrieval methods in a more comprehensive way taking more search techniques into account (including more exact and inexact graph matching methods). Also, a number of different result visualization methods will be further developed and evaluated.

# References

1. Anumba, C., Ren, Z., Ugwu, O.: Agents and Multi-agent Systems in Construction. Routledge, London (2007)
2. Ayzenshtadt, V., Langenhan, C., Bukhari, S.S., Althoff, K.D., Petzold, F., Dengel, A.: Distributed domain model for the case-based retrieval of architectural building designs. In: Petridis, M., Roth-Berghofer, T., Wiratunga, N. (eds.) Proceedings of the 20th UK Workshop on Case-Based Reasoning, UK Workshop on Case-Based Reasoning (UKCBR-2015), located at SGAI International Conference on Artificial Intelligence, 15–17 December, Cambridge, United Kingdom. School of Computing, Engineering and Mathematics, University of Brighton, UK (2015)
3. Ayzenshtadt, V., Langenhan, C., Bukhari, S.S., Althoff, K.D., Petzold, F., Dengel, A.: Thinking with containers: a multi-agent retrieval approach for the case-based semantic search of architectural designs. In: Filipe, J., van den Herik, J. (eds.) Proceedings of the 8th International Conference on Agents and Artificial Intelligence, International Conference on Agents and Artificial Intelligence (ICAART-2016), 24–26 February, Rome, Italy. SCITEPRESS (2016)
4. Bayer, J., Bukhari, S., Dengel, A., Langenhan, C., Althoff, K.D., Petzold, F., Eichenberger-Liwicki, M.: Migrating the classical pen-and-paper based conceptual sketching of architecture plans towards computer tools - prototype design and evaluation. In: 11th IAPR International Workshop on Graphics Recognition - GREC 2015, Nancy, France (2015)
5. Brandes, U., Eiglsperger, M., Herman, I., Himsolt, M., Marshall, M.S.: GraphML progress report structural layer proposal. In: Mutzel, P., Jünger, M., Leipert, S. (eds.) GD 2001. LNCS, vol. 2265, pp. 501–512. Springer, Heidelberg (2002). doi:10.1007/3-540-45848-4_59
6. Cavieres, A., Bhatia, U., Joshi, P., Zhao, F., Ram, A.: CBArch: a case-based reasoning framework for conceptual design of commercial buildings. In: Artificial Intelligence and Sustainable Design - Papers from the AAAI 2011 Spring Symposium (SS-11-02), pp. 19–25 (2011)
7. Cordella, L.P., Foggia, P., Sansone, C., Vento, M.: A (sub) graph isomorphism algorithm for matching large graphs. IEEE Trans. Pattern Anal. Mach. Intell. **26**(10), 1367–1372 (2004)

8. Eastman, C., Eastman, C.M., Teicholz, P., Sacks, R.: BIM Handbook: A Guide to Building Information Modeling for Owners, Managers, Designers, Engineers and Contractors. Wiley, Hoboken (2011)
9. Heylighen, A., Neuckermans, H.: A case base of case-based design tools for architecture. Comput. Aided Des. **33**(14), 1111–1122 (2001)
10. Inanc, B.S.: Casebook. An information retrieval system for housing floor plans. In: The Proceedings of 5th Conference on Computer Aided Architectural Design Research (CAADRIA), pp. 389–398 (2000)
11. Lai, I.C.: Dynamic idea maps: a framework for linking ideas with cases during brainstorming. Int. J. Architectural Comput. **3**(4), 429–447 (2005)
12. Langenhan, C.: A federated information system for the support of topological bim-based approaches. Forum Bauinformatik Aachen (2015)
13. Langenhan, C., Petzold, F.: The fingerprint of architecture-sketch-based design methods for researching building layouts through the semantic fingerprinting of floor plans. Int. Electron. Sci. Educ. J. Archit. Mod. Inf. Technol. **4**, 13 (2010)
14. Lin, C.J.: Visual architectural topology. In: Open Systems: Proceedings of the 18th International Conference on Computer-Aided Architectural Design Research in Asia, pp. 3–12 (2013)
15. Maher, M., Balachandran, M., Zhang, D.: Case-Based Reasoning in Design. Lawrence Erlbaum Associates, Hillsdale (1995)
16. Messmer, B.T., Bunke, H.: A decision tree approach to graph and subgraph isomorphism detection. Pattern Recogn. **32**(12), 1979–1998 (1999)
17. Richter, K.: Augmenting Designers' Memory: Case-Based Reasoning in Architecture. Logos-Verlag, Berlin (2011)
18. Richter, K., Heylighen, A., Donath, D.: Looking back to the future - an updated case base of case-based design tools for architecture. In: Knowledge Modelling-eCAADe (2007)
19. Roith, J., Langenhan, C., Petzold, F.: Supporting the building design process with graph-based methods using centrally coordinated federated databases. In: 16th International Conference on Computing in Civil and Building Engineering (2016)
20. Voss, A.: Case design specialists in FABEL. In: Issues and Applications of Case-based Reasoning in Design, pp. 301–335 (1997)
21. Weber, M., Langenhan, C., Roth-Berghofer, T., Liwicki, M., Dengel, A., Petzold, F.: Fast subgraph isomorphism detection for graph-based retrieval. In: Ram, A., Wiratunga, N. (eds.) ICCBR 2011. LNCS (LNAI), vol. 6880, pp. 319–333. Springer, Heidelberg (2011). doi:10.1007/978-3-642-23291-6_24

# Case Representation and Similarity Assessment in the SELFBACK Decision Support System

Kerstin Bach[1]([✉]), Tomasz Szczepanski[1], Agnar Aamodt[1],
Odd Erik Gundersen[1], and Paul Jarle Mork[2]

[1] Department of Computer and Information Science,
Norwegian University of Science and Technology, Trondheim, Norway
kerstin@idi.ntnu.no
[2] Department of Public Health and General Practice,
Norwegian University of Science and Technology, Trondheim, Norway
http://www.idi.ntnu.no, http://www.ntnu.no/ism

**Abstract.** In this paper we will introduce the SELFBACK decision support system that facilitates, improves and reinforces self-management of non-specific low back pain. The SELFBACK system is a predictive case-based reasoning system for personalizing recommendations in order to provide relief for patients with non-specific low back pain and increase their physical functionality over time. We present how case-based reasoning is used for capturing experiences from temporal patient data, and evaluate how to carry out a similarity-based retrieval in order to find the best advice for patients. Specifically, we will show how heterogeneous data received at various frequencies can be captured in cases and used for personalized advice.

**Keywords:** Case-based reasoning · Case representations · Data streams · Similarity assessment

## 1 Introduction

Low back pain is one of the most common reasons for activity limitation, sick leave, and disability. It is the fourth most common diagnosis (after upper respiratory infection, hypertension, and coughing) seen in primary care [23]. Cost of illness studies in different countries indicate that the total annual cost of low back pain in Europe is between 85 billion EUR and 291 billion EUR (equals approximately 0.4–1.2 % of the gross domestic product in the European Union) [3]. An expert group concluded that the most well-documented and effective approach to manage non-specific low back pain is to discourage bed rest, use over-the-counter pain killers in the acute stage if necessary, reassure the patient about the favorable prognosis, advise the patient to stay active both on and off work, and advise strength and/or stretching exercise to prevent a relapse [21].

Most patients (85 %) seen in primary care with low back pain have non-specific low back pain, i.e., pain that cannot reliably be attributed to a specific disease or pathology. Self-management in the form of physical activity and

© Springer International Publishing AG 2016
A. Goel et al. (Eds.): ICCBR 2016, LNAI 9969, pp. 32–46, 2016.
DOI: 10.1007/978-3-319-47096-2_3

strength/stretching exercises constitutes the core component in the management of non-specific low back pain; however, adherence to self-management programs is poor because it is difficult to make lifestyle modifications with little or no additional support. In the SELFBACK project we will develop and document an easy-to-use decision support system to be used by the patient him/herself in order to facilitate, improve and reinforce self-management of non-specific low back pain. The decision support system will be conveyed to the patient via a smart-phone app in the form of advice for self-management. A recent study [22] identified 283 pain-related apps available in the main app shops App Store, Blackberry App World, Google Play, Nokia Store and Windows Phone Store. However, none of these apps had effects documented through scientific publications, and none included a decision support system. In contrast, we will conduct a randomized control trial to evaluate the effectiveness of the SELFBACK decision support system.

## 1.1 Background

The SELFBACK system will constitute a data-driven, predictive decision support system that uses the Case-Based Reasoning (CBR) methodology to capture and reuse patient cases in order to suggest the most suitable activity goals and plans for an individual patient. This will be based on data from two sources. One is a questionnaire, presented to the patient at suitable intervals, in order to capture general information and progress of symptoms (e.g. disability and pain). The other is a stream of activity data collected using a wristband. The incoming data will be analyzed to classify the patients current state and recent activities, and matched against past cases in order to derive follow-up advices to the patient. Two main challenges are to detect the activity pattern represented at a suitable level of abstraction, and to match that structure against existing patient descriptions in the case base. Combined with the patient profile data from the questionnaire, and the current goal setting, this should enable the system to suggest the best next activity goal and plan for the patient.

Stratified care for patients with low back pain, based on initial pain intensity, disability related to low back pain, and fear-avoidance beliefs have been shown to improve patient outcomes as well as being cost-effective [9]. The SELFBACK system aims at further improving the stratified care approach by including data on the patients health and coping behaviour (i.e., the adherence to basic self-management principles) in order to support and prompt appropriate actions thereby empowering the patient to improve the self-management of their own low back pain. SELFBACK incorporates existing knowledge to recommend advice that is personalized based on the information input by the patient. Figure 1 shows the overall architecture of the SELFBACK system. The user initially uses a web page to sign up and provide answers to a set of screening questions (1), which are fed to the server (2) in order to initiate the smart phone app. All further interactions happen with the smart phone. It collects sensor data (3) from a wearable, subjective information from the user (4), pushes both the objective and

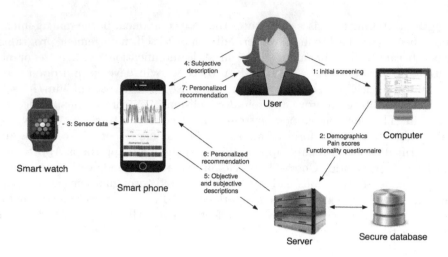

**Fig. 1.** SELFBACK has a distributed architecture in which data collection and user interaction mainly is done on the mobile devices. The case-based recommendation is performed remotely on the server.

subjective measurements (5) to the server and finally provides recommendations (6, 7) to the user.

In what follows we will discuss related work, before we describe the case representation and case content in Sect. 3. In Sect. 4 we introduce the applied similarity assessment. We have conducted experiments using already existing data sets from the domain, and discuss these in Sect. 5. The final section summarizes the paper and gives an outlook on future work.

## 2   Related Work

When reasoning with time in CBR, the temporal information can be dealt with at the feature level, the case level or a combination of these two. This idea was introduced by [15]. At the case level, history is described using temporally connected cases while at the feature level, the features of the cases contain temporal information. A combination of these two are temporally connected cases that contains temporal features. Temporal features could be of different types: (1) raw time series [18], (2) sequences of events [7,11], (3) graphs [10] and (4) piecewise interpretations of raw time-series [13].

The type of feature that represents temporal information directly influences the type of similarity metric that can be used to compare local similarity of the temporal features. In order to compare raw time series, the types of similarity metrics that are used include Euclidian Distance metrics, Fourier coefficient metrics, auto-regressive models, dynamic time warping, edit distance, time-warped edit distance and minimum jump cost dissimilarity metrics. See [20] for a review and empirical evaluation of these. Variants of these similarity metrics are used for both sequences and piecewise interpretations. [14] applies the Discrete Fourier Transform when comparing the similarity of time-series. In [13] time series are converted to temporal abstractions that describe the state or trend of a time-series. These temporal abstractions are organized in a tree based on the granularity of the abstractions, and similarity computations are conducted based on the distance in the tree structure. [4] compares the similarity of time-series using dynamic time-warping, and [7] uses edit distance on sequences of events. See [6] for a discussion on measuring the similarity of sequences of complex events.

Decision support systems that perform case-based reasoning on temporal data have been developed for a diverse set of domains that include weather prediction [8], buy and sell points prediction for stocks [2], oil well drilling [6] and fault diagnosis of industrial robots [17] among many others. One of the domains that has gotten the most attention is health care. Decision support systems that reason with temporal data in the medical domain include kidney function monitoring and prediction [19], long term follow up of stem cell transplantation patients [1], classification of respiratory sinus arrhythmia patterns [16], hemodialysis patient management [14], and Type 1 diabetes patient support [12]. The focus within SELFBACK is to monitor the different factors (pain, function, activity, etc.) in order to compare patients based on summaries and abstractions from collected raw data. At this stage, cases contain temporal information at the feature level only, the temporal data is piecewise interpretations of time-series which are compared using edit distance. The temporal features are piecewise interpretations of the patient activity stream over a day. The overall progress of the patient over larger time spans will be solved through temporally connected cases at the case history level, but this is future work.

## 3   SELFBACK Case Representation

Cases in SELFBACK consist of different types of information representing the patient description and the personalized advice. Data for the patient description is acquired with various frequency patterns in the SELFBACK life cycle, and the advice given through the app is updated accordingly. We can assume that the initial information, such as demographic data and information provided by the clinician, is somewhat static while other information is expected to change more frequently. At the extreme end of that scale is the continuous data stream from the wristband.

The case structure is shown in Table 1. As earlier described, we differentiate between subjective and objective measurements of a patient's situation. The subjective measurements are obtained by asking the patient about their level of pain, degree of functionality, etc. A particularly relevant piece of information is the patient's self-efficacy, i.e. the patient's degree of belief in that the bad condition will improve and eventually vanish. These questions are based on standardized questionnaires and screening tools such as the Pain Specific Function Scale (PSFS), Roland Morris Disability Questionnaire (RMDQ), Start-BACK, Pain Self Efficacy Questionnaire (PSEQ) or Quality of Life (EQ-5D).

**Table 1.** Overview of the case content in SELFBACK cases

|  | Case part | Content | Update frequency |
|---|---|---|---|
| Problem description | Subjective description | Demographics | Weekly/biweekly |
|  |  | Quality of life |  |
|  |  | Pain Level |  |
|  |  | Functionality |  |
|  | Objective description | Activity stream | Continiously |
| Solution | Advice | Activity plan | Weekly |
|  |  | Exercise plan |  |

Further on questions are asked regularly on a weekly basis, which provide a time series describing the course of pain and functionality. All those measures are captured using standardized screening tools applied in common practice. This information is enhanced by the objective measurement of a patient's activity. The objective measurement is obtained via a wristband worn by the patient providing continuous readings of activity parameters such as sleep, number of steps, the duration time of sedentary (e.g. lying, sitting, standing), moderate (e.g. walking), and vigorous (e.g. running) activities.

We are storing the raw data in a noSQL database and fetch it from there when cases are build or case matching is initiated. This approach allows us to keep a high level of detail, generate abstractions offline and extract from them on demand.

From research on non-specific low back pain, and the course of pain and functionality, we know that a severe episode of low back pain starts with an acute phase where a patient is in a lot of pain and basic movements are difficult. This phase can last from a few days up to four weeks. After that period one speaks about the sub-acute phase, followed by a chronic phase (pain lasts longer than 3 months), if the pain persists. In this paper we are focusing on the case content and similarity matching for patients in the acute phase.

### 3.1 Building Cases from Objective and Subjective Measurements

When looking into existing data collections, the pain level and functionality level changes in the first weeks of an acute phase. As part of assessing a patient's degree of pain, the patient is asked to mark the pain level on a scale from 0 to 10, where 0 is no pain at all and 10 is described as the worst pain possible to imagine. The reported pain (as shown in Fig. 2) usually decreases over time, but the timing differs. As pain goes down, usually functionality increases. When looking at the course of pain in more detail, one can see that the pain levels out at a certain point (see levelling in Fig. 2) and from this point onwards the patient can start rehabilitating with light exercises and activities.

However, there are different journeys until the patient reaches a point from which s/he can start the recovery phase, and the goal of SELFBACK in the acute phase is to provide suggestions to reach that point as fast as possible. In Fig. 2 we used the HUNT3 data set (described in Sect. 5.1), which captures the patients' pain levels for a few weeks. The patients were asked twice about their current pain levels (week 2 and 6 in Fig. 2) as well as a summary on how their pain was the last two weeks (week 0 and week 4 in Fig. 2). This information gives us an overall indication that there are different courses of pain development, and also that these are not bound to the pain level a patient has in the very beginning of the treatment (baseline).

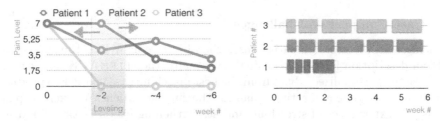

**Fig. 2.** Pain leveling (left) and data collection frequency for acute patients (right)

As *leveling* we describe the moment when the patient has reached a state, where the pain is bearable enough to start doing light strengthening and flexibility exercises. We aim at supporting the patient to reach that point as fast as possible by reusing successful advices from similar patients.

Driven by the reported pain in the SELFBACK application, we expect cases to cover flexible time spans. While a patient has a lot of pain the advice is given on a short term (usually a day or two), while afterwards we are targeting at "reporting" weekly phases. On the right in Fig. 2 we show possible update frequencies for the aforementioned patients. Patient 1 stays at a high pain level, hence s/he is asked very frequently whether there has been a change until the pain level goes down and the time between re-asking becomes longer as well. Patient 2 and especially patient 3 have a much faster decline of their pain and therefore the length of capturing new subjective information is longer. We base this approach on the assumption that low back pain patients with high pain levels usually experience a change in perceived pain level within a few days, but with medium and low pain levels the change takes longer.

## 4    Similarity of Cases

To recapture, the SELFBACK case contains objective and subjective data collected from one single patient and the advice given to that patient at specific timestamps (Table 1).

The subjective problem description is mostly static and collected at baseline and updated weekly to monthly throughout the course of the raw data. The objective problem description is based on a continuous data stream, which is interpreted and contains the activity pattern of a patient. From initial experiments we have conducted, we have seen that the collected data from a wristband contains around 700 entities per week when abstracted to the activity level. We distinguish between four main types of activity: sleep, sedentary, moderate and vigorous activities. An activity is described in terms of this activity type, the start time stamp, and the end time stamp. Depending on the confidence and level of detail, the SELFBACK application will include more detailed activities such as lying, sitting, standing, walking, running or biking, which will result in even more timestamps. This means that the activity stream data may constitute more than 99 % of the whole case content when stored.

### 4.1 Case Comparison Challenges

As mentioned, the case comparison uses the objective as well as the subjective problem description for case matching. Eventually, the SELFBACK application will reuse the personalized advice from the best matching case in order to produce a customized exercise and activity plan for the current patient. An exercise plan in this context is a set of stretching and strengthening exercises that build up gradually. Once an entry point is found, the patient is guided throughout specific workouts. An activity plan on the other hand sets goals for physical activities throughout a day or week such as reaching a specified number of steps/day or reducing continuous sitting time during the day to amount of minutes. The matching case may be a past case from the same patient or from another similar patient. The different characteristics of the problem descriptions require different approaches for similarity matching, hence we will compare the objective and subjective measurements.

Subjective measurements are matched by standard, simple similarity metrics for numerical and symbolic values. The activity stream, however, is converted into a string and several different string comparison methods are used. Each character in the string represents an activity the patient was doing at the $n^{th}$ (n is here the position of the character in the string) second from the time the case was stored. For instance, the string "SSSSRRR" represents a period of 4 seconds standing (S) followed by a period consisting of 3 seconds running (R). If we were to use all of the described data as attributes summed in a global similarity function, our approach would suffer from several logistical weaknesses. Firstly, 99 % of the data (the activity stream) would only be used for computing a few attributes in the global similarity function. Secondly, computing the similarity between those attributes is much more costly than computing the similarity between simpler and smaller data structures extracted from the questionnaire data. Therefore our next steps are to investigate which abstractions from the objective and subjective measurements are relevant for (a) predicting the course of a patient's convalescence and (b) which information is necessary for communicating the SELFBACK advice.

### 4.2 MAC/FAC Model and the Global Similarity Function Split

To solve the problems described in Sect. 4.1 we divide the retrieval of similar cases into two steps, as suggested by the MAC/FAC model [5]:

In the MAC phase, data from the subjective problem description is compared to the current situation of a SELFBACK patient. Simple and computationally cheap similarity metrics are used. In the FAC phase the activity streams from the most similar results are extracted from the database and compared with use of more expensive sequence similarity metrics. By doing this, we extract considerably less data from the database, and we save significant amounts of time when computing the similarity scores between cases. Another advantage with this MAC/FAC approach can be seen when a SELFBACK patient's MAC data remains unchanged (which often will be the case, because it mainly consists of reletively static data). If so, the whole retrieval from the MAC phase from last time can be reused.

One of the difficulties with the MAC/FAC model, is to ensure that the most similar cases retrieved after the FAC phase are indeed the most similar cases globally (during a retrieval when no cases are filtered out). Sometimes filters applied to cases in the MAC phase for selecting the globally most similar cases. Therefore, in our approach, we do not use metadata about the activity stream structure as a filter, but rather a part of the global similarity function itself. As we will see, this ensures that the case retrieved after the FAC phase is indeed the most globally similar one. Consider the properties (1) of a global similarity function which is a weighted sum of case attributes. Also assume that we are using it to compare cases containing $n$ attributes:

$$sim_i(x_i, c_i) \in [0, 1], \qquad \sum_{i=1}^{n} w_i = 1, \qquad \sum_{i=1}^{n} sim_i(x_i, c_i)w_i \in [0, 1] \qquad (1)$$

Here $sim_i(x_i, c_i)$ is an attribute similarity function, that compares the $i^{th}$ attribute from a case $x$ to the $i^{th}$ attribute of case $c$ (for example a SELFBACK patient case). The function returns a value between 0 and 1. Next, this value is multiplied by an attribute function weight $w_i$. Those weights reflect the importance of the attributes used in our case representation. In the current iteration of the SELFBACK system, the values of those weights are not known. However, we know that their sum is equal to 1.

Finally, the sum of all the $n$ attribute similarity functions multiplied with their corresponding weights constitutes the final global similarity function: the rightmost equation in (1). This function also returns a value between 0 and 1.

When these requirements are met, we can find a $k < n$ and rewrite the global similarity function, returning a similarity value $z$, into Eq. (2):

$$\sum_{i=1}^{k-1} sim_i(x_i, c_i)w_i + \sum_{i=k}^{n} sim_i(x_i, c_i)w_i = z \in [0, 1] \qquad (2)$$

In the SELFBACK system, all the inexpensive attribute comparisons are put into the leftmost sum of Eq. (2) and are computed during the MAC phase. The resulting sum, $M$, is then used as a filter for the FAC phase. Assume that the maximum value of the rightmost sum in Eq. (2) equals $F_{max}$. Suppose we compare a SELFBACK patient case to cases stored in the SELFBACK case base. However, instead of computing the whole global similarity function, we only compute the MAC phase sums (leftmost part of Eq. (2)). Let $M_{highest}$ be the highest such computed sum. To ensure that the retrieved most similar case after the FAC phase is also the globally most similar case, we have to consider all retrieved cases from the MAC phase for which $M$ is:

$$M \geqslant M_{highest} - F_{max} \tag{3}$$

For every case with $M$ satisfying Eq. (3), we compute the expensive FAC phase ($F$; rightmost sum in Eq. (2)). $M + F$ is then the global similarity of the compared case. Following this apporach the case with hte greatest similarity sum is also the most similar case overall. Splitting the global similarity function is only beneficial if $F_{max}$ is less than 0.5 (half of the maximum overall similarity). We assume this to be true, as it is sufficient for clinicians today to only monitor attributes that are part of the subjective case description, described in Table 2, for patient treatment.

## 5   Experiments and Results

In this section we will show that the suggested case representation can be used to describe patient cases that contain the development of pain, functionality and the activity of a patient. We will also show that the MAC/FAC approach can be used to carry out a similarity based retrieval on the given data sets.

Since this type of data collection is new, we do not have an existing data set to start experimenting. Therefore we use existing data sources which partially cover the SELFBACK target data in order to show how cases can be populated.

### 5.1   Case Base Population

In order to evaluate the applicability of the presented case structures within a CBR system, we created a data set from two already running projects, and we collected activity data from healthy people. The goal of the evaluation is (A) to test the representation of subjective measurements, and (B) to test the retrieval and similarity assessment of cases containing objective and subjective measurements.

**The FYSIOPRIM data set** has been created by a Norwegian medical research project focusing on capturing data about the treatments of musculoskeletal disorders in primary care[1]. A tablet app has been developed, which captures the status of the patient (while in the waiting area) along with the given treatments (inserted by the physiotherapist). The patients were also asked some of the questions again after a follow-up period. The project has finished its first phase including the development of the questionnaire and the app, as well as its integration in the electronic health record system. Since 2015 it is in the second

---

[1] http://www.med.uio.no/helsam/forskning/grupper/fysioprim/.

**Table 2.** Overview of the subjective problem description attributes currently captured in SELFBACK in comparison to the existing data sets from the FYSIOPRIM and HUNT3 study

| Attribute name | SELFBACK | FYSIOPRIM | HUNT3 | Example case |
|---|---|---|---|---|
| Gender | x | x | x | Male |
| Age | x | x | x | 45 |
| Height | x | x | - | 1.89 m |
| Weight | x | x | - | 82 |
| BMI | x | x | - | 23 |
| EQ.5D | x | x | - | 90 |
| RMDQ | x | x | - | 8 |
| NPRS | x | x | x | 8 |
| Work characteristics | x | x | x | Mostly sitting |
| Sleep breaks | x | - | - | Seldom-never |
| Sleep wake up | x | - | - | Seldom-never |
| Sleep difficulty | x | - | - | Seldom-never |
| Exercise frequency | x | x | x | 2–3 times per week |
| Exercise intensity | x | x | * | I push myself |
| Exercise length | x | x | - | 30–60 min |
| PSFS activity | x | - | - | Prolonged standing |
| PSFS difficulty | x | x | x | 8 |
| StartBACK screening | x | - | - | 2 |
| Pain medication | x | x | x | None |
| Pain history | x | - | x | None |

phase of collecting data in different places in Norway. We have used 45 patient cases from the FYSIOPRIM pilot study to build cases. As shown in Table 2, this data set covers the target representation of SELFBACK pretty well. The only major part that is missing is subjective information regarding the sleep quality of a patient.

**The HUNT3 Data Set:** The HUNT3[2] data set we have used is a larger data set, which has been collected between 2006–2008. While HUNT3 is a cohort study within the larger area of Nord-Trøndelag in Norway, our data set comes from a spin-off study, which originates in a questionnaire for participants asking for more detailed information on musculoskeletal disorders if they indicated the occurrence of some type of shoulder, neck or back pain. After an initial questionnaire, they got asked follow up questions up until 5 times over a period of six months. This data set contains data from 219 patients.

---

[2] https://www.ntnu.edu/hunt/hunt3.

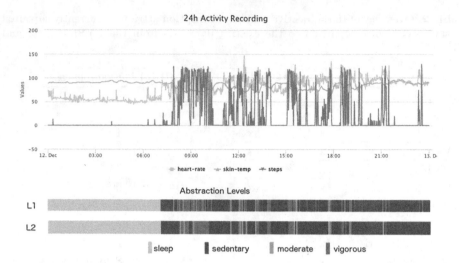

**Fig. 3.** Objective measurement and abstraction levels: on top you can see the first abstractions from accelerometer data: step counts together with heart rate and skin temperature measures. L1 and L2 show two abstraction levels for the activity recoding.

**Objective Measurements of Physical Activity:** In addition to the subjective case description, we collected objective measurements in order to achieve a complete set of data in terms of data types. We collected activity data of a healthy person over a period of a few weeks for 24 h per day. From that data we sampled out 27 days for this experiment focusing on the amount of data produced in one day. Figure 3 shows one of the recordings as well as the abstractions into the four main activity types included in the case representation. For the recording of the physical activity, we used the myBASIS Peak watch that provides the collected data as csv dumps. From these dumps we extracted 24 h periods and used simple rules to differentiate between sleeping, sedentary, moderate and vigorous activity.

## 5.2   Similarity-Based Case Retrieval: Subjective Measurements

We wanted to estimate how beneficial the use of the MAC/FAC model in our approach can be. Specifically, it is important to know how many of the cases can be filtered out before the costly FAC phase attribute comparisons. In order to do that, we investigated the similarity span when comparing the subjective problem description part of the SELFBACK cases populated with real data from patients with lower back pain. With similarity span we mean the difference in similarity between the most and least similar cases in a case base.

As shown in Fig. 2 in Sect. 3.1, the frequency of data updates within cases depends on how often a patient is reassessed. For the evaluation, we choose to compare each set of collected data to the current SELFBACK patient during the case retrieval. This gave us the opportunity to create 90 FYSIOPRIM cases and 1095 HUNT3 cases. From the set of created cases from the FYSIOPRIM data set, we took one case out (the query case) and compared it to the rest. This was then repeated for every case in the set. The same procedure was applied to the HUNT3 cases. For this comparison all attributes were weighted equally. The results are shown in Fig. 4.

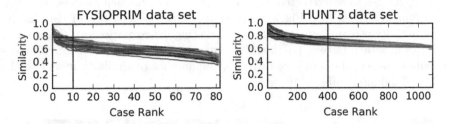

**Fig. 4.** MAC phase similarity decline within the FYSIOPRIM and HUNT3 data set. (Color figure online)

Each colored line in Fig. 4 represents a MAC phase case retrieval. The cases matched with the query case (one query case for each colored line) are sorted by the similarity on the x-axis. This means that for every retrieval/line, the x-value does not represent a specific case, but rather the $x^{th}$ case when sorted by similarity from greatest to lowest. The maximum value of the MAC phase part of the global similarity function, that will be used by the SELFBACK system, is not yet known. Thus a simplification is made and the value is set to 1.0.

As we can see, by increasing the number of attributes, from eight attributes in the HUNT3 data set to eleven attributes in the FYSIOPRIM data set (see Table 1), the similarity covered by the cases on the y-axis in Fig. 4 (the similarity span) also increases. Since both the HUNT3 and the FYSIOPRIM data sets only uses a portion of the attributes that are planned to be included by the SELFBACK system, it seems reasonable to assume that the similarity span would increase even further. The bigger the similarity span, the more cases will be filtered out by the MAC phase and the more beneficial will the usage of the MAC/FAC model be.

For example, take the maximum possible attribute similarity sum computed during the FAC phase in the final global similarity function, $F_{max}$, to be 0.2. We can see from the plots that we would only have to consider, in the worst case scenario (the colored lines at the top), 10 of the 95 FYSIOPRIM cases and only 400 of the 1095 HUNT3 cases during the expensive FAC phase. This is because only those cases satisfy Eq. (3) from Sect. 4.2 ($M \geqslant M_{highest} - F_{max} = 1.0 - 0.2$).

## 5.3  Similarity-Based Case Retrieval: Objective Measurements

It is important that the SELFBACK system will be able to generate a meaningful retrieval based on comparing activity streams that are abstracted into strings.

In order to get a picture on how such a retrieval might look like, we filled 27 cases with collected activity stream data. During retrieval, we only computed the similarity resulting after the FAC phase. The result is shown in Fig. 5 where the query is at the bottom, and the matched activity streams are ordered from the most similar at bottom to the least similar at the top.

After converting the activity streams into strings, as described in Sect. 4.1, twelve attributes were extracted when computing the similarity scores. For each of the four activities (see Fig. 3) we computed the percentage of each activity per activity stream and the number of distinguished periods (yielding eight attributes in total). Between each activity stream we computed the longest common sub-sequence, the sequence distance, the number of similar k-mers and the number of unique similar k-mers.

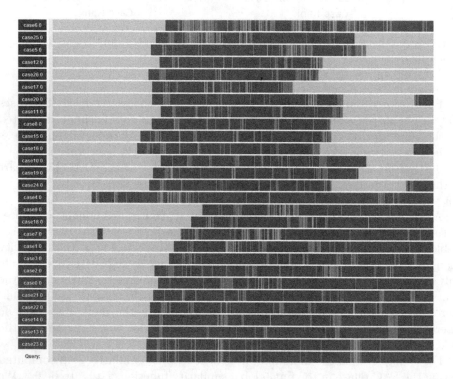

**Fig. 5.** Activity streams ordered by the FAC phase similarity to the query activity stream. (Color figure online)

The number of distinguished periods per activity type is in Fig. 5 shown as the number of blocks having the same color (number of consecutive substrings containing the same character). The sequence distance is the Levenshtein distance which is the least number of single character insertions, deletions or substitutions that is required to transform one string into the other. A k-mer is a consecutive sub-string of length k. For the similarity calculations in Fig. 5, every

30-mers were extracted from the compared abstracted activity stream sequences. The resulting 30-mer sets were then compared based on their greatest common subset and uniqueness.

# 6   Conclusion and Future Work

In this paper we introduced the overall SELFBACK system and how a case-representation that provide personalized advice to patients with non-specific low back pain can be designed. We focus on the case representation and similarity assessment within a temporal domain, because we base the advice on the course of pain, function, efficacy and activity. In order to evaluate the case representation and the accompanying similarity assessment we used existing data sets that match the target data in SELFBACK and show that we can populate and match cases effectively. Next steps are: completing the case representation; performing a qualitative evaluation of the retrieval; and optimizing the interplay between the objective and subjective measurements.

**Acknowledgement.** The work has been conducted as part of the SELFBACK project, which has received funding from the European Unions Horizon 2020 research and innovation programme under grant agreement No. 689043.

# References

1. Bichindaritz, I., Kansu, E., Sullivan, K.M.: Case-based reasoning in CARE-PARTNER: gathering evidence for evidence-based medical practice. In: Smyth, B., Cunningham, P. (eds.) EWCBR 1998. LNCS, vol. 1488, pp. 334–345. Springer, Heidelberg (1998). doi:10.1007/BFb0056345
2. Chang, P., Liu, C., Lin, J., Fan, C., Ng, C.S.P.: A neural network with a case based dynamic window for stock trading prediction. Expert Syst. Appl. **36**(3), 6889–6898 (2009)
3. Crow, W.T., Willis, D.R.: Estimating cost of care for patients with acute low back pain: a retrospective review of patient records. J. Am. Osteopath. Assoc. **109**(4), 229–233 (2009)
4. Fritsche, L., Schlaefer, A., Budde, K., Schroeter, K., Neumayer, H.: Recognition of critical situations from time series of laboratory results by case-based reasoning. J. Am. Med. Inform. Assoc. **9**(5), 520–528 (2002)
5. Gentner, D., Forbus, K.D.: Mac/fac: a model of similarity-based retrieval. Cogn. Sci. **19**, 141–205 (1991)
6. Gundersen, O.E.: Toward measuring the similarity of complex event sequences in real-time. In: Agudo, B.D., Watson, I. (eds.) ICCBR 2012. LNCS (LNAI), vol. 7466, pp. 107–121. Springer, Heidelberg (2012). doi:10.1007/978-3-642-32986-9_10
7. Gundersen, O.E., Sørmo, F., Aamodt, A., Skalle, P.: A real-time decision support system for high cost oil-well drilling operations. AI Mag. **34**(1), 21–32 (2013)
8. Hansen, B.K.: A fuzzy logic-based analog forecasting system for ceiling and visibility. Weather Forecast. **22**, 1319–1330 (2007)

9. Hill, J.C., Whitehurst, D.G.T., Lewis, M., Bryan, S., Dunn, K.M., Foster, N.E., Konstantinou, K., Main, C.J., Mason, E., Somerville, S., Sowden, G., Vohora, K., Hay, E.M.: Comparison of stratified primary care management for low back pain with current best practice (STarT Back): a randomised controlled trial. Lancet 378(9802), 1560–1571 (2011)

10. Jære, M.D., Aamodt, A., Skalle, P.: Representing temporal knowledge for case-based prediction. In: Craw, S., Preece, A. (eds.) ECCBR 2002. LNCS (LNAI), vol. 2416, pp. 174–188. Springer, Heidelberg (2002). doi:10.1007/3-540-46119-1_14

11. Juarez, J.M., Campos, M., Palma, J., Marin, R.: T-care: temporal case retrieval system. Expert Syst. 28(4), 324–338 (2011)

12. Marling, C., Shubrook, J., Schwartz, F.: Toward case-based reasoning for diabetes management: a prelimenary clinical study and decision support system prototype. Comput. Intell. 25(3), 165–179 (2009)

13. Montani, S., Leonardi, G., Bottrighi, A., Portinale, L., Terenziani, P.: Flexible and efficient retrieval of haemodialysis time series. In: Lenz, R., Miksch, S., Peleg, M., Reichert, M., Riaño, D., Teije, A. (eds.) KR4HC/ProHealth-2012. LNCS (LNAI), vol. 7738, pp. 154–167. Springer, Heidelberg (2013). doi:10.1007/978-3-642-36438-9_11

14. Montani, S.: Case-based decision support in time dependent medical domains. In: Bramer, M. (ed.) IFIP AI 2010. IAICT, vol. 331, pp. 238–242. Springer, Heidelberg (2010). doi:10.1007/978-3-642-15286-3_24

15. Montani, S., Portinale, L.: Case based representation and retrieval with time dependent features. In: Muñoz-Ávila, H., Ricci, F. (eds.) ICCBR 2005. LNCS (LNAI), vol. 3620, pp. 353–367. Springer, Heidelberg (2005). doi:10.1007/11536406_28

16. Nilsson, M., Funk, P., Olsson, E.M.G., von Schéele, B., Xiong, N.: Clinical decision-support for diagnosing stress-related disorders by applying psychophysiological medical knowledge to an instance-based learning system. Artif. Intell. Med. 36(2), 159–176 (2006)

17. Olsson, E., Funk, P., Xiong, N.: Fault diagnosis in industry using sensor readings and case-based reasoning. J. Intell. Fuzzy Syst. 15(1), 41–46 (2004)

18. Ram, A.: Continuous case-based reasoning. Artif. Intell. 90(1–2), 25–77 (1997)

19. Schmidt, R., Gierl, L.: A prognostic model for temporal courses that combines temporal abstraction and case-based reasoning. Int. J. Med. Inform. 74(2), 307–315 (2005)

20. Serr, J., Arcos, J.L.: An empirical evaluation of similarity measures for time series classification. Knowl. Based Syst. 67, 305–314 (2014)

21. van Tulder, M., Becker, A., Bekkering, T., Breen, A., del Real, M.T.G., Hutchinson, A., Koes, B., Laerum, E., Malmivaara, A.: Chapter 3 european guidelines for the management ofacute nonspecific low back painin primary care. Eur. Spine J. 15(2), s169–s191 (2006)

22. de la Vega, R., Miró, J.: mhealth: a strategic field without a solid scientific soul. A systematic review of pain-related apps. PLoS One 9(7), e:101312 (2014)

23. Wändell, P., Carlsson, A.C., Wettermark, B., Lord, G., Cars, T., Ljunggren, G.: Most common diseases diagnosed in primary care in stockholm, sweden, in 2011. Fam. Pract. 30(5), 506–513 (2013)

# Accessibility-Driven Cooking System

Susana Bautista[✉] and Belén Díaz-Agudo

Department of Software Engineering and Artificial Intelligence,
Universidad Complutense de Madrid, Madrid, Spain
{subautis,belend}@ucm.es

**Abstract.** Research area of designing recipes is an attractive problem for the CBR community. In this paper we deal with the problem of presenting the recipe information in an understandable format for a certain user. As different users have different presentation needs, we discuss the suitability of taking the user profile into account to personalize the presentation of a suggested recipe in a cooking system. Our system relies on text simplification processes that were born from the need of people who have difficulties reading and understanding textual contents. Our system collects a case base with the best choice of *presentation* for a certain collective. Given a recipe plus details on the user profile (age, genre, educational level, languages, disability and special needs) the system retrieves from a case base the best presentation and modify the recipe presentation according to the specific user needs. The system includes learning capacity as if the final presentation is difficult for the specific user the system can easily provide her with alternate presentations. Results on the preliminary experiments are very promising and show the applicability of a CBR approach to personalize and simplify textual recipe presentations for different collectives.

## 1 Introduction

The Computer Cooking Contest (CCC) is an open competition where participants submit software that creates recipes. The system relies on a database of basic recipes from which appropriate recipes must be selected, modified, or even combined. Given a query describing different aspects, like the must have or desired ingredients, the type of dish (dessert, main course,..), allergies, diet, among others, a CBR approach retrieves and reuses similar recipes. For most of the queries there is no single correct or best answer. That is, many different solutions are possible, depending on the creativity of the software[1]. Our research group has participated in this contest with different systems JaDaCook [12,13], JADAWeb [1] solving different challenges: single dish, negation and menu composition.

---

Supported by Complutense University Madrid, Spain (Group 910494) and Spanish Ministery of Economy and Competitivity (TIN2015-66655-R and TIN2014-55006-R projects).

[1] http://ccc2015.loria.fr/.

© Springer International Publishing AG 2016
A. Goel et al. (Eds.): ICCBR 2016, LNAI 9969, pp. 47–61, 2016.
DOI: 10.1007/978-3-319-47096-2_4

Although we have previously obviated this aspect, one problem with knowledge formalization in the cooking domain is the use of different measure units, different math representations of the quantities and different presentations of the information, for example pictograms or videos. Case acquisition, comparison and reuse needs to deal with the different ways of measuring different ingredients: liquid ingredients and small quantities are measured by volume units, dry bulk ingredients, such as sugar and flour, are measured typically by weight ("125 g sugar"), although it is measured by volume in North America and Australia ("1/2 cup flour"). Many ingredients are measured by count: "two pieces of bread", "one chicken", "three eggs", "two carrots", although it is imprecise due to the variability of size and weight on the individual units.

Recipes as cases require a preprocessing phase to homogenize the information of the recipe (units and quantities for ingredients). Comparison between query and cases requires normalizing measure units, as there are different possibilities.

If we take different recipes from different countries, we can observe that both cuisine and bakery there is a tradition to use the system of measure units "The Imperial System of units" from English-speaking countries where they use "cups', "tablespoon" (tbsp), "teaspoon" (tsp), "ounce" (oz) such as volume units. However in Europe, they use "The International System of units" and they have different measures such as "litre"(l), "milillitre" (ml), "grams" (gr) or "kilograms" (kg). In Table 1 we can see some examples of conversion of measure units from the different systems. In addition, we can observe that not only the measure units changes in the recipes, we can see that the math representations of the quantities can be different, sometimes fractions are used (3/4 cup), other times real numbers express the measure (5.5 gr) and even percentages (50 % of water) are used to represent different ingredients in the recipe.

**Table 1.** Some examples of conversion of measure units

| Cups | ml | Teaspoon | ml | Pounds | Ounce | gr |
|---|---|---|---|---|---|---|
| 1 cup | 240 ml | 1 tsp | 5 ml | 1 lb | 16 oz | 454 gr |
| 1/2 cup | 120 ml | 1/2 tsp | 2.5 ml | 2 lb | 32 oz | 908 gr |
| 1/3 cup | 80 ml | 1/3 tsp | 1.66 ml | 3 lb | 48 oz | 1362 gr |
| 1/4 cup | 60 ml | 1/4 tsp | 1.25 ml | 4 lb | 64 oz | 1816 gr |

In addition, we can observe other kinds of presentation of the information in the recipes. There is a specific collective, people with special needs, who have some problems to understand the information because they have any kind of disability or cognitive problems such as people with autism, people with aphasia, people with dyslexia or other kind of people such as children with ADHD (Attention-Deficit/ Hyperactivity Disorder) or elderly people. For many of these kinds of people, pictograms are used to represent the information that we want to offer. Pictograms are pictures that represents an object or concept, e.g. a picture of an envelope used to represent an e-mail message, it is a familiar pictograms nowadays. If we can represent different information using pictograms, we can provide access to the information for people with special needs. All these changes of presentations both of units of measure and math representations of quantities are simplifications of the original recipe for different kind of people with special needs. Text simplification process is important to adapt the information for persons who have problems to understand the information that they access. There are different type of text simplification systems to carry out this task automatically (different approaches are presented in Sect. 3). Nowadays, a lot of information is generated without considering the needs and factors of such target users.

We personalize different aspects to generate a human understandable presentation modifying for instance measure units, math representations of the quantities and different presentations of the information, for example pictograms or text. Variations are used to generate different presentations of the same representation of the recipe (how the recipe is manipulated by a computer) for different kind of people. In this paper we only deal with variations on the textual (and images in the case of pictograms) presentations for the recipes. Each recipe is internally represented using an structured sequence of ingredients, quantities and units of measure. We do not consider presenting the information in video format.

Therefore if we want to simplify the recipes for different end users, we have to consider different factors like age, cultural level, English level and some kind of disability to personalize the original recipe to the user presentation needs. We propose a CBR approach to personalize cooking recipes to the user presentation needs. Our system relies on a case base that collects what is the best choice of *presentation* for a certain collective. Our system proposes enriching the query with details on the user profile: age, genre, educational level, languages, disability, special needs and others, and we personalize the recipe representation to the specific needs using simplification strategies. The system includes learning capacity as if the final representation is difficult for the specific user the system can easily provide her with alternate presentations.

The rest of the paper is organized as follows. Section 2 describes existing related work on cooking systems and representation. Section 3 presents related work on text simplification systems for improving accessibility. Section 4 introduces the different representations we are using in our approach and motivates the decisions. Section 5 describes a CBR approach that combines cases of representation with the results of the cooking system JaDaCook. Section 6 presents some initial experimental results and Sect. 7 concludes the paper and describes the lines of future work.

## 2   Cooking Systems

In CBR research area choosing and designing recipes is an attractive problem. Two of the more well-known CBR cookery systems, CHEF [10] and Julia [14], have proven popular over the years. CHEF is a case-based planner which represents recipes as cases using goals and problems, while Julia is a case-based

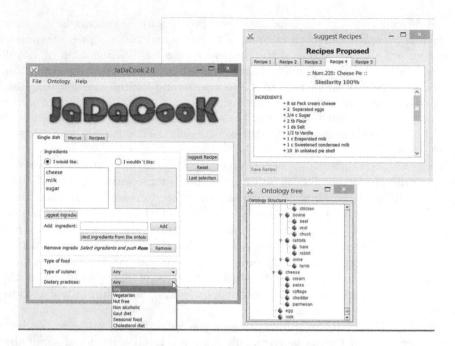

**Fig. 1.** JaDaCook 2 interface

design system. Both systems include an extensive semantic memory to hold the definitions of the terms needed, using frames organized into a semantic network in CHEF, and a large taxonomy of concepts in Julia.

Several systems have been presented at the last years in the Computer Cooking Contest (CCC). CookIIS [11] recommends and adapts recipes or creates a complete menu regarding to the user's preferences like explicitly excluded ingredients or previously defined diets. Taaable system [9] participated to the 2015 CCC. A formal concept analysis approach was used to improve the ingredient substitution, which must take into account a limited set of available foods. In addition, the adaptation of the ingredient quantities had improved in order to be more realistic with a real cooking setting. The adaptation of the ingredient quantities is based on a mixed linear optimization. CookingCAKE [18] is a framework for the adaptation of cooking recipes represented as workflows. CookingCAKE integrates and combines several workflow adaptation approaches applied in process-oriented case based reasoning (POCBR) in a single adaptation framework, thus providing a capable tool for the adaptation of cooking recipes. To our knowledge these previous systems do not consider different kinds of presentations of the recipe information. Most of the systems save all ingredients with the same units and they do not modify the math representations of the quantities. The simplification process in order to adapt the presentation of the recipe information for improving accessibility has not been previously considered in the systems developed in the CCC.

Our research group has developed previous approaches, namely JaDaCook [12,13], JADAWeb [1]. Both systems rely on an ontology with reusable knowledge about ingredients, types of ingredients, types of cuisine and dietary practices. The ontology is used as background knowledge to measure similarity between ingredients and single dishes, and to substitute ingredients during adaptation. Ontology has more than 300 ingredients, organized in types and classes that cover the 1484 recipes in the case base (originally provided by the CCC-09 organizers). JaDaCook 2 reasons using different knowledge sources: (1) a case base of recipes (available from textual sources), (2) a cooking ontology, (3) a set of association rules, obtained using data mining techniques, capturing co-occurrences of ingredients in the recipes. JaDaCook 2 has been implemented using the jCOLIBRI [7] framework. Regarding presentation (see Fig. 2), the JaDaCook uses "The Imperial System of units" where "cups', "tablespoon" (Tbsp), "teaspoon" (tsp), "ounce" (oz) are used such as units (see Fig. 2). Each recipe describes ingredients, and quantities, plus a natural language description of the detailed preparation steps. In this paper, we propose connecting JaDaCook solution with a CBR system that simplify the recipe presentation to the user features. However, our CBR system is not dependent on JaDaCook (see Fig. 5) and it could be used with other recipe recommendation system that generates on structured recipe representation with ingredients, quantities and units (see Sect. 4). Before describing our system in Sect. 4 we review some state of art in text simplification systems for improving accessibility.

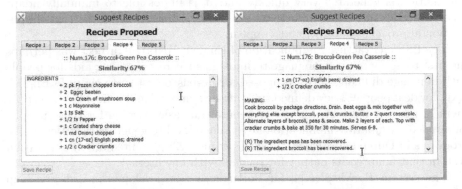

**Fig. 2.** Retrieved recipes, ingredients + preparation

# 3 Text Simplification Systems for Improving Accessibility

The text simplification process was born of the need to adapt content texts for people who have difficulties reading and understanding a text in order to be part of society because access to information is a right for all persons. Text simplification consists of the transformation of a text into a similar text, but easier to read and understand. The objective is to achieve more accessible, attractive and communicative texts so that they are interesting and motive people with difficulties to read them. Access to reading is a social need and a recognized right and reading is a pleasure that lets people share ideas, thoughts and experiences.

30 % of the population [19] has reading difficulties that can be caused by different factors and this group of people needs a simplified version of texts in order to have access to information. These factors may be intercultural difficulties, complex daily texts and cognitive aspects of the reader. People who may need a text adapted from the original version in order to understand its content are older people, people learning other languages, people with cognitive problems and a range of people with special educational needs (autism, aphasia, dyslexia, etc.).

In order to communicate using written texts, it is important to use simple, clear and direct expressions in order to ensure better comprehension of the texts, to achieve good communication with the target user, to work towards an inclusive social model. By carrying out certain operations at the lexical and syntactic level, linguistic complexity is taught, thus obtaining a simplified text for the final user.

There are several initiatives designed to develop the manual processing of text simplification following the European guidelines established by the IFLA [8], published by Inclusion Europe Association [15]. All of these initiatives work in the area of Easy-to-read, a movement to create special material (books, documents, website, etc.), while tending the content and layout (format, margins, fonts, spacing, etc.) so that people with reading difficulties can read and understand the material.

Simplifying a text manually is hard work in time and resources. Nowadays, information is generated very quickly and it is impossible to manually adapt accessible real-time texts. In order to solve this problem, automatic text simplification approaches have begun to appear.

The PSET (Practical Simplification of English Texts) [5] project was perhaps the first to apply natural language technologies to create reading aids for people with language difficulties. [16] proposed a rule-based system to simplify English texts for deaf people. They defined rules at the syntactic and lexical level to apply them in the original text in order to generate a easier version for these people. [6] applied automatic simplification at sentence level to generate subtitles in TV programs in Dutch and English for deaf people. [22] presented a text generation system that adapted its output for readers with low literacy.

The system *PorSimples* [4] for Portuguese was developed in order to help low-literacy readers process documents on the web. With the development of the *Guidelines for materials in readable* IFLA [8], in the work of [3] a subset of these guidelines was used to design and implement automatic rules at the syntactic and lexical level. The main objective of the project *Simplext*[2] [20] was to develop the product support for text simplification in Spanish for groups of people with special reading and comprehension needs.

The *FIRST (Flexible Interactive Reading Support Tool)* project [2] is developing a tool to assist people with autism spectrum disorders to adapt written documents into a format that is easier for them to read and understand. [21] developed a project focused on the treatment of educational texts in Spanish

---

[2] http://www.simplext.es/.

in order to reduce language barriers to reading comprehension for the hearing-impaired, or even people who are learning a language other than their mother tongue.

We can see that in all systems of automatic simplification developed so far both the language with which they work, the target user, the kind of text and the level of difficulty to adapt the texts to play a key role in one way or another. Each system considers a set of operations to simplify at various levels, syntactic or lexical, to carry out the adaptation of the original text. In our system, we consider a special kind of texts, recipes. We need to determine the level of difficulty to simplify the text and what kind of choices for the end user, measure units, math representation of the quantity and so on, which we have to make in the original recipe to adapt for a new presentation.

## 4    Variability on the Presentations

We need to chose different presentations to evaluate them. Inside the variability of the presentations we select textual representation for the ingredients, from the original language, in this case in English, in addition, in current language where we are doing the experiment, in our case, in Spanish. We select a recipe with high frequency of use, "The Cheesecake", with typical measures used in the "The Imperial System of units", such as "cups" or "tsp". This kind of recipe is easy to find in all recipes books and online webpages related to cooking. We introduce the use of pictograms to represent the ingredients of the recipe because we would like to evaluate different presentations considering different kind of people, included people with special needs who use pictograms to communicate between them. Figures 3 and 4 show some of the different presentations of the recipe used in the experiment (presented in Sect. 6).

We can identify different conversion rules to adapt the original recipe to final presentation. We consider rules to convert measure units, math representation of quantities and different kind of ways to present the information, using only text, using pictograms or adding video. In Table 2 we can see some examples of these conversion rules considered in our system.

---

Ingredients
1 cup and 3/4 cups graham cracker crumbs
1/3 cup butter - softened
1 cup and 1/3 cups granulated sugar - keep the 1/3 cup separate from the 1 cup
3 packages cream cheese - (8 ounces per pack - 24 ounces total)
2 teaspoons vanilla extract
3 eggs
1 cup sour cream

**Fig. 3.** Original recipe of the Cheesecake using the "The Imperial System" of units representation. Corresponding to representation 8 (R8).

Ingredientes

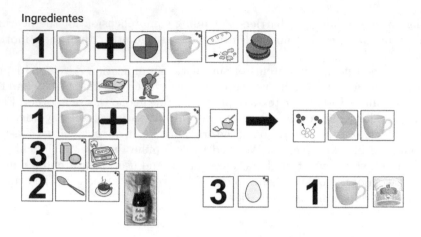

**Fig. 4.** Cheesecake recipe using pictograms representation for the quantities and for the ingredients. Corresponding to representation 7 (R7).

**Table 2.** Conversion rules from units, quantities and text.

| Original units | Adapted units |
|---|---|
| 1 cup | 240 ml |
| 1 tsp | 5 ml |
| 1 lb | 454 gr |
| Original quantities | Adapted quantities |
| 1/2 | 0.5 |
| 25 % | 1/4 |
| 0.75 | 3/4 |
| Original text | Adapted text |
| Single String | Pictograms |
| Single String | Video |
| Pictograms | Video |

We choose the following types of presentations for the case of study of "The cheesecake" recipe:

– R1: The Imperial System of units (cups, teaspoon), using fractions, Spanish language.
– R2: The Imperial System of units (cups, teaspoon), using decimals numbers, Spanish language.
– R3: The International System of units (litre, milillitre), using fractions, Spanish language.
– R4: The International System of units (litre, milillitre), using decimals numbers, Spanish language.

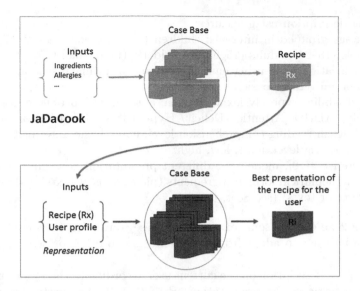

**Fig. 5.** Diagram of the CBR system to retrieve the best presentation of the recipe depending on the input user profile.

- R5: The Imperial System of units (cups, teaspoon), using pictograms for quantities, fractions, Spanish language.
- R6: The Imperial System of units (cups, teaspoon), using pictograms for ingredients, fractions, Spanish language.
- R7: The Imperial System of units (cups, teaspoon), using pictograms for quantities and ingredients, fractions.
- R8: The Imperial System of units (cups, teaspoon), using fractions, English language.
- R9: The Imperial System of units (cups, teaspoon), using fractions, Spanish language.

## 5    Accesibility Driven Personalization with CBR

Our proposal uses the JaDaCook CBR system to discover a recipe that better fits the input query. The retrieved recipe is used itself as the input of our CBR system in order to simplify the recipe to the presentation needs depending on the user profile. Figure 5 shows our ideas to combine two CBR systems to personalize cooking recipes to the user presentation needs. The second CBR system relies on a case base that collects what is the best choice of presentation for a certain collective. This case base has been compiled from the answers of people who participated in the experiment (see Sect. 6). We have created a case base CB of around 200 cases with the following $c = <P_c, Rc_i>$ case structure:

- $P_c$ - User description using features:
    - $age_c$: age grouped in intervals: children (6–12), teenagers (13–17), young people (18–24), adults (25–67) and elderly (more than 67).
    - $ed_c$: education level using enumerated values: primary school, secondary school, university, basic.
    - $dis_c$: disability type: dyslexia, aphasia, autism, deaf people, cognitive disability, ADHD (Attention-Deficit/ Hyperactivity Disorder) or None.
    - $en_c$: English reading comprehension level: none, basic, intermediate, high.
    - $cook_c$: cooking level: none, low, high.
- $Rc_i$, where i $\in [1, 9]$, models the different presentation according to the unit system, mathematical format, ingredients (pictograms or texts) and quantities (pictograms or texts) (see Sect. 4).

In Table 2 we can see some examples of these conversion rules considered in our system in order to adapt from a presentation to other presentation of the recipes.

Query Q is a description on the current user according to the $P_i$ features. The CBR system retrieves the presentation $(R_i), i \in [1..9]$ that better fits the user needs and uses the representation features to present the recipe information.

Similarity function is a weighted average between the case description $P_c = < age_c, ed_c, dis_c, en_c, cook_c >$ and the query $Q = < age_q, ed_q, dis_q, en_q, cook_q >$.

*Equality* has been used as the local similarity function between attribute values and weights have been adjusted for our experiment using a Genetic Algorithm (GA). We have calculated the set of weights that maximizes the performance of the system right answer rate. This is a satisfactory solution to learn the optimal weights [17]. This algorithm uses a population of individuals representing different weights. This population evolves until the algorithm obtains the individual (i.e. the weights) that returns the best performance. We obtain higher weights in order for attributes $ed_c$, $dis_c$, $en_c$, $age_c$ and $cook_c$ for the case base population used in the experiments described in Sect. 6. Weights reflects importance on the different features but the obtained values reflects dependencies between attributes. $ed_c$ is higher as there is a dependency with $dis_c$ and the

**Fig. 6.** Original recipe presentation (R8) and simplified recipe (R7)

**Fig. 7.** Recovered recipes (R5 and R9)

type of presentation needs. $cook_c$ is lower because the cheese pie recipe is easy and it doesn't affect much the presentation needs.

We use some examples to illustrate the operation of our system. We take the recipe retrieved by JaDaCook system in the query of the example shown in Fig. 6 like original recipe corresponding to "Cheese Pie" recipe (R8). Then, depending on the input profile user, our system recover different personalized recipes, based on our experiment and using our collected data. For example:

- If the input profile user is a child between 6–12 years old, at primary school, with some cognitive disability, low level of English and no cooking skills, then the best case of representation of the recipe is using pictograms (corresponding to R7, see Fig. 6).
- If the input profile user is the same but without some cognitive disability, then the best case of representation of the recipe is using pictograms only for quantities (pictograms to represent 1/4) and the kind of ingredient expressed in words (milk)(corresponding to R5, see Fig. 7).
- If the input profile user is an adult between 25–67 years old, with university level, without disability, basic English level and few cooking skills, then the best case of representation is the original recipe but translated to Spanish (corresponding to R9, see Fig. 7).

User profile is represented as $P_u$ using the features described above. Input recipe is represented as a set of structured ingredient lines with quantity and units. Besides, a textual field is used to provide recipe preparation details. An example is given in Fig. 1. Output of JaDaCook is used as the input of the CBR system that personalize the recipe presentation.

## 6   Experimental Results

We carried out an experiment with different kind of users. The total of participants was 202 people, 17 persons with some special needs $(8, 41\%)$ and 185 without special needs $(91, 58\%)$. In total, we collected 21 answers of children from 6 to 12 years old. 98 answers of young people from 12 to 17 years old. 2

answers of people between 18 and 24 years old. 75 answers of people from 25 to 67 years old. Finally, we collected 6 answers of people with more than 67 years old. We defined a questionnaire using Google Form in order to use it online and collect the vast of the answers. In addition, we carried out the experiment inside of different schools in Madrid[3] where we collected the answers from children and young people who participated in the experiment. The experiment is available online in the following link[4].

We defined a questionnaire with a first part of demographic data: age, education level, disability, English-level and cooking-level. In the second part we present 9 different representations of a selected recipe, in our recipe "Cheese-cake" (presented in Sect. 4). Each user ranked the representations and chose which ones are more easy understood and would be preferred to cook the recipe. From the collected data, we have compiled a case base $CB_p$ with 202 cases, where each $case_i$ has a solution set with two possible presentations $S_i = R_{i1}, R_{i2}$. First presentation $R_{r1}$ in the solution $S_r$ is used to present the retrieved recipe in the cooking system JaDaCook (see Fig. 5). Only when the user chooses an alternate presentation $R_{r2}$ is used to present the recipe. As we have described case description uses 5 features: age, education level, disability, English-level and cooking-level. We have run an experiment using leave one out cross-validation. Each case $c_i \in CB_p$ is used to query the system. CBR retrieves the most similar case $c_r$ and propose its solutions in order, $S_r = \{R_{r1}, R_{r2}\}$. We compare solution $S_r$ with the already known solutions of the query case: $S_i = \{R_{i1}, R_{i2}\}$. $R_{r1}$ success is evaluated as $success@n$ $(R_{r1})$ for $n = 1, 2$, where $success@1$ (R) = 1 if $R = R_{i1}$ $success@2$ (R) = 1 if $R \in S_i$ and is 0 otherwise. Besides $R_{r2}$ success is also evaluated. We obtained the following experimental results:

- $success@1$ $(R_{r1}) = 77.7228\%$
- $success@2$ $(R_{r1}) = 81.2178\%$
- $success@1$ $(R_{r2}) = 62\%$
- $success@2$ $(R_{r2}) = 73\%$

We have obtained good preliminary results (around 80% in the percentage of hits) on the first presentation $R_{r1}$. However, results with presentation $(R_{r2})$ are lower and it has not resulted on useful alternative presentations. In future experiments we will consider diversity measures to propose alternative presentations that are different from the first option and let the user acquire diverse visions from the same recipe.

Solutions distribute evenly according to the different age groups, educational level, disability type, English comprehension level and cooking level. We have observed that R9 is the preferred solution for most of the age groups, although children chose R7 (pictograms for quantities and ingredients) and disable people chose R5 (pictograms for quantities) as the best presentation.

---

[3] Thanks for the participation of students of schools "La Asunción Santa Isabel" and "La Asunción Cuestablanca".

[4] http://bit.ly/1SV0fJ3.

# 7   Conclusions and Future Work

Different people have different needs of information presentation. That is specially important in collectives of people with any kind of disability or cognitive problems, and also for children or elderly people where information needs to be simplified to be understood. In this paper we have reviewed alternative textual representations in the cooking domain. Using solutions from the cooking system JaDaCook [13] we experimented with different measure units, different formats for numeric quantities, and pictograms are used to represent recipe ingredients. In this paper we propose a CBR approach to personalize cooking recipes to the user presentation needs. Our system relies on a case base that collects what is the best choice of *presentation* for a certain collective. This case base has been collected from a questionnaire where each participant indicated her/his profile features and ranked the different choices of recipe presentations.

The CBR system uses as query a description of the user profile (age, cultural level, English level, disability,..), compares the query with the stored cases, retrieves the most similar and reuse as solution the most suitable recipe presentation. Then the cooking system use the case solution, i.e. user preferred presentation, to personalize the original recipe to the user presentation needs.

This paper discussed the suitability of taking the user needs into account when trying to simplify and personalize the presentations of the ingredients of the recipe. Results on the preliminary experiments are very promising and show the applicability of a CBR approach to personalize and simplify presentations for different collectives. According to the obtained results people with similar features share similar presentation preferences. Of course, there are personal variations that have not been captured by the cases in the case base. We will experiment further with more people. We plan to extend the experiment to more people with special needs, as we hypothesize that the system would improve results in those collectives. We would like to test our system with a variety of collectives. The work presented in this paper opens several lines of future work. Our future work involves not only the complete development and improvement of our system to include the part of preparation of the recipe but also the development of evaluation techniques that validate the suitability and portability of our proposal to personalize the presentation of the ingredients of the recipe for different target users. We would like to compare our approach with others, for example, we could compare our result with a basic classification approach by building group (set of users) profiles for each Rx presentations, and use these 9 profiles to classify a new user according to her features and to determine the best visual presentation for her, and discuss if our CBR approach gives better results. In addition, we plan to design a GUI where the user can select the easiest presentation, included the language and possible visual presentation for preparation part of the recipe.

# References

1. Ballesteros, M., Martín, R., Díaz-Agudo, B.: Jadaweb: A CBR system for cooking recipes. In: 18th International Conference on Case-based Reasoning, Computer Cooking Contest Workshop, pp. 179–188 (2010)
2. Barbu, E., Martín-Valdivia, M.T., Ureña-López, L.A.: Open Book: a tool for helping ASD users' semantic comprehension. In: Proceedings of the Workshop on Natural Language Processing for Improving Textual Accessibility (NLP4ITA) (2013)
3. Bautista, S., Gervás, P., Madrid, R.: Feasibility analysis for semiautomatic conversion of text to improve readability. In: Proceedings of the Second International Conference on Information and Communication Technology and Accessibility (ICTA) (2009)
4. Candido Jr., A., Maziero, E., Gasperin, C., Pardo, T.A.S., Specia, L., Aluisio, S.M.: Supporting the adaptation of texts for poor literacy readers: a text simplification editor for Brazilian Portuguese. In: Proceedings of the Fourth Workshop on Innovative Use of NLP for Building Educational Applications, Stroudsburg, PA, USA, pp. 34–42. Association for Computational Linguistics (2009)
5. Carroll, J., Minnen, G., Canning, Y., Devlin, S., Tait, J.: Practical simplification of english newspaper text to assist aphasic readers. In: Proceedings of the Workshop on Integrating Artificial Intelligence and Assistive Technology (AAAI), Madison, Wisconsin, pp. 7–10 (1998)
6. Daelemans, W., Hthker, A., Sang, E.T.K.: Automatic sentence simplification for subtitling in Dutch and English. In: Proceedings of the 4th International Conference on Languaje Resources and Evaluation, pp. 1045–1048 (2004)
7. Díaz-Agudo, B., González-Calero, P.A., Recio-García, J.A., Sánchez-Ruiz-Granados, A.A.: Building CBR systems with jcolibri. Sci. Comput. Program. **69**(1–3), 68–75 (2007)
8. Freyhoff, G., Hess, G., Kerr, L., Menzel, E., Tronbacke, B., Veken, K.V.: European Guidelines for the Production of Easy-to-Read Information for People with Learning Disability. Technical Report ISLMH (1998)
9. Gaillard, E., Lieber, J., Nauer, E.: Improving ingredient substitution using formal concept analysis and adaptation of ingredient quantities with mixed linear optimization. In: Proceedings of the ICCBR 2015 (2015)
10. Hammond, K.J.: CHEF: A model of case-based planning. In: Proceedings of the 5th National Conference on Artificial Intelligence, AAAI (1986)
11. Hanft, A., Newo, R., Bach, K., Ihle, N., Althoff, K.-D.: CookIIS–A successful recipe advisor and menu creator. In: Montani, S., Jain, L.C. (eds.) Successful Case-based Reasoning Applications - I. Studies in Computational Intelligence, vol. 305, pp. 187–222. Springer, Heidelberg (2010)
12. Herrera, P.J., Iglesias, P., Romero, D., Rubio, I., Díaz-Agudo, B., JaDaCook: Java application developed and cooked over ontological knowledge. In: Schaaf, M. (ed.) ECCBR Workshops, pp. 209–218 (2008)
13. Herrera, P.J., Iglesias, P., Sánchez, A.M.G., Díaz-Agudo, B.: Jadacook 2: Cooking over ontological knowledge. In: ICCBR Workshops, Computer Cooking Contest (2009)
14. Hinrichs, T.R.: Strategies for adaptation and recovery in a design problem solver. In: Proceedings of the Second Workshop on Case-Based Reasoning (1989)
15. Inclusion Europe Association: Inclusion Europe (1998). http://www.inclusion-europe.org

16. Inui, K., Fujita, A., Takahashi, T., Iida, R., Iwakura, T.: Text simplification for reading assistance: a project note. In: Proceedings of the 2nd International Workshop on Paraphrasing: Paraphrase Acquisition and Applications, pp. 9–16 (2003)
17. Jarmulak, J., Craw, S., Rowe, R.: Genetic algorithms to optimise CBR retrieval. In: Blanzieri, E., Portinale, L. (eds.) EWCBR 2000. LNCS, vol. 1898, pp. 136–147. Springer, Heidelberg (2000). doi:10.1007/3-540-44527-7_13
18. Muller, G., Bergmann, R.: CookingCAKE: A Framework for the adaptation of cooking recipesrepresented as workflows. In: Proceedings of the ICCBR 2015 (2015)
19. OECD. PISA 2012 Results: What Students Know and Can Do, vol. I, Revised edn., February 2014
20. Saggion, H., Gómez-Martínez, E., Etayo, E., Anula, A., Bourg, L.: Text simplification in simplext: making text more accessible. Procesamiento del Lenguaje Natural **47**, 341–342 (2011)
21. Saquete, E., Vázquez, S., Lloret, E., Llopis, F., Gómez, J., Mosquera, A.: Tratamiento de textos para mejorar la comprensión lectora enalumnos con deficiencias auditivas. Procesamiento del Lenguaje Natural **50**, 231–234 (2013)
22. Williams, S., Reiter, E.: Generating readable texts for readers with low basic skills. In: Proceeding of the 10th European Workshop on Natural Language Generation, Aberdeen, Scotland, pp. 140–147 (2005)

# Inferring Users' Critiquing Feedback on Recommendations from Eye Movements

Li Chen[✉], Feng Wang, and Wen Wu

Department of Computer Science, Hong Kong Baptist University,
224 Waterloo Road, Kowloon, Hong Kong
{lichen,fwang,cswenwu}@comp.hkbu.edu.hk

**Abstract.** In recommender systems, *critiquing* has been popularly applied as an effective approach to obtaining users' feedback on recommended products. In order to reduce users' efforts of creating critiquing criteria on their own, some systems have aimed at suggesting critiques for users to choose. How to accurately match system-suggested critiques to users' intended feedback hence becomes a challenging issue. In this paper, we particularly take into account users' eye movements on recommendations to infer their critiquing feedback. Based on a collection of real users' eye-gaze data, we have demonstrated the approach's feasibility of implicitly deriving users' critiquing criteria. It hence indicates a promising direction of using eye-tracking technique to improve existing critique suggestion methods.

**Keywords:** Recommender systems · Critiquing feedback · Eye movements · Fixation metrics · Feedback inference

## 1 Introduction

In current online environments, recommender systems have been widely applied in various scenarios to support users to make product choices (e.g., e-commerce, social media, tourism, finance). Especially, in the situations where it is difficult to obtain users' historical records like ratings for collaborative filtering, case-based or preference-based methods have mainly been used to generate recommendations by retrieving items that are similar to users' queries/preferences [2,8]. In these systems, one popular approach to obtaining users' feedback on recommendations is *critiquing*, which has become the core feedback mechanism in so called conversational recommenders [17,22] and critiquing-based recommender systems [10]. Specifically, the *critiquing* allows users to critique a recommended product in terms of its attribute values (e.g., *"I would like to see some laptops with different manufacture and higher processor speed"*), based on which the system will return new recommendations that satisfy their critiquing criteria. Previous studies show that this critiquing process is effective to assist users in exploring the product space, refining their requirements, and making more confident decisions [6,18,19].

© Springer International Publishing AG 2016
A. Goel et al. (Eds.): ICCBR 2016, LNAI 9969, pp. 62–76, 2016.
DOI: 10.1007/978-3-319-47096-2_5

So far, there are two major methods of acquiring users' critiquing feedback. One is *user-initiated critiquing* that requires users to specify critiquing criteria on their own [10]. For example, Fig. 1.a shows the screenshot of Example Critiquing interface [5], where users need to indicate which attributes to "keep" (keeping its existing value), "improve" (improving its existing value, e.g., cheaper), or "take any suggestion" (i.e., "compromise", accepting a compromised value). The advantage of this elicitation approach is that it can give users maximal freedom of creating any critiques they wish and stimulate them to make tradeoffs among attributes (i.e., sacrificing values on less important attributes for guaranteeing the intended improvements on more important ones), but it unavoidably demands extra user efforts and might hence be limited in real applications. Another method is *system-suggested critiquing* that proposes a set of critiques for users to choose [3,19]. For instance, the Dynamic Critiquing system generates several compound critiques (each operating over multiple attributes, e.g., *"Less Optical Zoom & More Digital Zoom & A Different Storage Type"*) according to remaining product cases' availability (see Fig. 1.b) [22]. Intuitively, the system-suggested critiquing method could reduce users' critiquing efforts, but when the suggested critiques cannot precisely match users' intended feedback, it is likely that users will be involved in longer interaction session by pursuing other ways to locate their target choice [6].

In this paper, we focus on investigating users' eye-movements on recommendations to infer their critiquing feedback, so as to potentially improve system-suggested critiquing. Based on a collection of real users' eye-gaze data and their true critiques, we have empirically verified two hypotheses: one is the feasibility of inferring what product a user tends to critique (i.e., *critiqued product*) from her/his fixations laid on different products; another is about deriving the user's concrete *critiquing criteria* for the product's attributes. That is, what attributes s/he may be inclined to keep, improve, or compromise. Furthermore, we have compared different fixation metrics, including *fixation count*, *total fixation duration*, and *average fixation duration*, in terms of their inference accuracy.

In the following, we first introduce related work (Sect. 2), and then give our research statement and hypotheses (Sect. 3). The experiment for data collection is described in Sect. 4, and in Sect. 5 we show the results. At the end, we conclude our findings and indicate future directions (Sect. 6).

## 2   Related Work

### 2.1   Critiquing-Based Recommender Systems

As mentioned before, existing critiquing-based recommender systems can be classified into two categories [10]: *user-initiated critiquing*, with ATA [16] and Example Critiquing [5] as representative systems; *system-suggested critiquing* that has been adopted in FindMe [3], Dynamic Critiquing [22], MAUT-based compound critiquing [28], and preference-based Organization [8].

Take Example Critiquing system as an example to illustrate user-initiated critiquing process [5]: it first presents some products to a user that best match

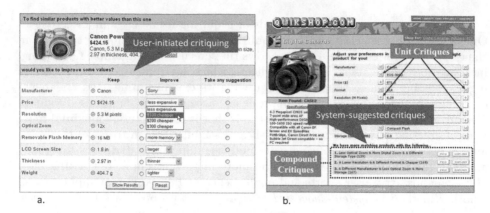

**Fig. 1.** a. User-initiated critiquing in Example Critiquing system [5]; b. System-suggested critiquing in Dynamic Critiquing system [19].

her/his initially specified preferences; then it stimulates the user to select a near-satisfactory product and critique it in terms of its attribute values; in the next recommendation cycle, the system will return a new set of recommendations according to the user's critiquing feedback. Experiments show that for a user to reach his/her target choice, a number of critiquing cycles are usually required [7].

As for system-suggested critiquing, some systems like FindMe pre-design some static critiques for users to pick, but since those fixed critiques cannot reflect available products' realistic status, other systems attempt to dynamically generate critique suggestions being adaptive to remaining product cases' characteristics [19,22] or users' attribute preferences [8,28]. However, an empirical user evaluation on a hybrid critiquing interface, which combines both user-initiated critiquing support and adaptive critique suggestions, shows that users more frequently created critiques on their own than choosing the suggested critiques, implying the latter approach's limited accuracy [6].

In order to save users' critiquing efforts, some researchers have also endeavored to adopt speech-based critique input interface [14], or harness other users' critiquing histories to guide the current user [20,26]. It shows though these methods are capable of enhancing system efficiency, the limitation of system-suggested critiques is still not well resolved.

## 2.2 Eye Tracking Studies in Recommender Systems

The development of eye tracking technology has enabled academic and commercial sectors to apply it in various interaction designs [21]. In recommender systems, it has mainly been adopted for two purposes. One was to evaluate the recommendation interface's usability. For instance, one user experiment measured the effect of interface layout design on users' visual searching pattern [9]. It shows users tend to fixate more on the top area if recommended products are displayed in a list layout, but will be directed to view more products if the recommendations are arranged in a category structure. Another experiment investigated whether users would gaze at recommendations during their entire product searching process [4]. Its results verify the important role of recommendations in users' purchase decision.

As the second purpose, some researchers have exploited eye-movement metrics to elicit users' implicit *relevance feedback* on recommendations, i.e., "positive" or "negative" (or called "like" or "dislike"). For instance, in [13], the documents that users consumed higher amount of fixations and longer average fixation time were regarded with "positive" feedback. They then used clustering and content based techniques to retrieve similar documents and recommended them to the user. Some studies emphasized developing algorithms to incorporate eye-based relevance feedback, such as interactive genetic algorithm [11], evolutionary programming [15], and attention prediction method [27].

However, little work attempts to infer users' critiquing criteria for product attributes (i.e., *critiquing feedback*) through eye tracking, which is in nature more challenging than that for relevance feedback.

## 3   Research Statement and Hypotheses

What a person is looking at is assumed to indicate her/his thought "on top of the stack" of cognitive processes. This "eye-mind" hypothesis means that eye-movement recordings can provide a dynamic trace of where a person's attention is directed in relation to a visual display. In practice, the process of inferring useful information from eye-movement recordings involves defining "Areas Of Interest" (AOI) over certain part of a display or interface under evaluation, and analyzing the eye movements that fall within those areas [21]. In our work, we define AOI at two levels (see Fig. 2): *product level* and *attribute level*. At product level, all descriptions about a recommended product, including its title, image, and major attributes' values (e.g., a laptop's price, operating system, processor speed, etc.), are comprised in one area. At attribute level, each attribute of the recommended product is treated as a specific area.

The metrics used to analyze eye-movement data are commonly related to *fixation*. Specifically, each fixation is a spatially stable gaze point, during which most information acquisition and processing occur. Its minimum duration is usually set as 200 ms [23]. We concretely adopt three popular fixation-derived metrics in our work, because they can represent users' relative engagement with the interface object [12,21,25]:

– **Fixation Count (FC)** - the number of times a user fixates on an AOI;
– **Total Fixation Duration (TFD)** - the sum of the duration of all fixations a user has laid in an AOI;
– **Average Fixation Duration (AFD)** - the average duration of a fixation in an AOI.

Given a user's fixation values at both product and attribute levels, the question we are interested in answering is whether they could be utilized to infer the user's critiquing feedback. At the first step, it is to infer what product within a set of $N$ recommendations the user would take as near-satisfactory and critique. Intuitively, we may hypothesize that the product with higher fixation values would be more likely to be selected, since more fixations on an object suggest that it is more important and engaging in some way [21,25].

**Hypothesis 1**: *Within a set of $N$ recommendations, users would tend to critique the product for which they have consumed higher fixation values.*

The second step is to infer the user's critiquing criterion for each attribute of the selected product. According to [5], there are three critique options: *keep*, *improve*, and *compromise* (as mentioned in Sect. 1). If a user *keeps* an attribute's existing value, it indicates s/he is satisfied with it, so we may assume the user would have spent certain fixations on this attribute when s/he evaluated the whole set of recommendations. If the user chooses to *improve* its value, it also implies the user has fixated on this attribute. The duration may be even longer than that on attributes for keeping, since the user might have compared different values of the attribute across different products and finally chosen one that is the best among all but still not fully satisfying her need. On the contrary, the attribute the user *compromises* may be of the fewest fixations, as it is what the user tends to sacrifice and hence less important than attributes for keeping or improving. Therefore, we can have the following hypothesis:

**Hypothesis 2**: *At attribute level, for the attributes of which users have consumed higher fixation values, they would be likely to improve, followed by some they may keep, and then the others with fewer fixations to compromise.*

## 4    Data Collection

With the purpose of verifying our hypotheses and comparing different fixation metrics (i.e., FC, TFD, and AFD), we conducted an experiment to collect users' eye movements on recommendations and their true critiquing criteria. In this section, we first introduce the experimental system, and then experimental procedure, participants, and data analysis results.

### 4.1    Experimental System

We chose Example Critiquing [5] as the experimental system, because its user-initiated critiquing support allows us to obtain users' true critiquing criteria for product attributes (see Fig. 1.a). To be specific, we adopted one of its prototypes, Laptop Finder, for conducting our experiment. Its laptop catalog was extracted from a commercial e-commerce website, each product described by 10 primary attributes (i.e., manufacturer, price, operating system, battery life, display size, hard drive capacity, installed memory, processor class, processor speed, and weight). There are four major steps during users' interaction with this system:

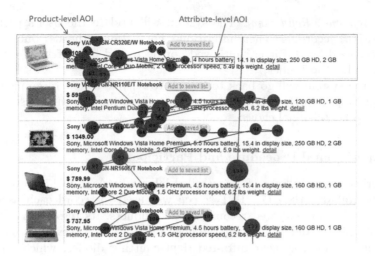

**Fig. 2.** Product-level and attribute-level Areas of Interest (AOI) for fixation analysis, and an example of a user's eye-gaze plot on recommended products (each fixation is represented by a blue circle). (Color figure online)

*Step 1: Initial Preference Elicitation.* The system first obtains a user's initial preferences by asking her/him to enter a product as query, or to state some specific preferences for product attributes. The system then builds a preference model for the user, which is formally represented as $Pref(u) = \{< V_i, W_i > |1 \leq i \leq A\}$, where $V_i$ denotes the user $u$'s value preference for attribute $a_i$ and $W_i$ is $a_i$'s relative weight.

*Step 2: Recommendation Generation.* Then, the system returns a set of $N$ products (e.g., $N = 25$ in Laptop Finder) that are most relevant to $Pref(u)$. Formally, a utility score is computed for each product to indicate its relevance: $U(P_j) = \sum_{i=1}^{A} W_i \times V_i(x_i)$, where a product $P_j$ contains attribute values $\mathbf{x} = \{x_i\}_{i=1}^{A}$. The products with higher utility scores are recommended.

In the recommendation interface, each product is described by three blocks of information (see Fig. 2): title, image, and ten primary attributes' values.

*Step 3: Critiquing Feedback Elicitation.* Within the set of $N$ recommended products, if the user cannot locate her/his target choice, s/he could select one product that is near-satisfactory and provide critiquing feedback on it.

Actually, users can initiate different types of critique. In terms of critiquing modality [5], there are two types: *similarity-based critique* (e.g., *"Find some products similar to this one"*) by "keeping" all of the critiqued product's attribute values, and *improvement-based critique* (e.g., *"Find some products that are cheaper"*) by "improving" some attribute values. For the latter, the user may even "compromise" the values of less important attributes. Regarding critiquing complexity [5], there are also two types: *unit critique* if the user "improves" or "compromises" only one attribute at a time, and *compound critique* if multiple attributes are involved in one critique (e.g., *"Find some products that are cheaper and bigger"*).

*Step 4: Preference Refinement.* The system will update the user's preference model $Pref(u)$ according to her/his critique. For instance, the attribute's weight will be increased by $\beta$ if it is "improved" or decreased by $\beta$ if "compromised" ($\beta = 0.25$ in Laptop Finder).

Then, the system will go back to Step 2 to resume a new recommendation cycle (from Steps 2 to 4). This interaction process continues until the user accepts a product as her/his final choice.

### 4.2   Experimental Procedure and Participants

The experiment was in form of a controlled lab study. A Tobii 1750 eye-tracker that is integrated with a 17″ TFT screen was used to record each user's eye movements when s/he viewed recommended products. Its resolution setting is $1290 \times 1024$ pixels, and can sample the position of a user's eyes by every 20 ms. The monitor frame has near infra-red light-emitting diodes, which allow for natural tracking without placing many restrictions on the user.

The user task was to "*find a product you would purchase if given the opportunity by using the Laptop Finder system.*" An administrator was present in each experiment. She debriefed the experiment's objective to the participant and asked her/him to fill in a demographic questionnaire at the beginning. Then, the participant was prompted to get familiar with the system's interface during a warm-up period. Subsequently, after the eye-tracker calibration was performed, the participant started to accomplish the given task.

During each recommendation cycle, in addition to recording each participant's eye movements, we retrieved the product s/he selected to critique (i.e., *critiqued product*) and actual *critiquing criteria* (i.e., "keep", "improve", or "compromise") for the product's attributes from her/his clicking actions.

We recruited 18 participants (2 females) to join the study, who were interested in buying a laptop at the time of experiment. According to [12], this scale is acceptable to conduct eye tracking experiment. They are from nine different countries (e.g., China, Switzerland, Italy, Spain, India, USA, etc.), and most of them were students pursuing Master or PhD degree at the university.

### 4.3   Data Analysis

It shows that every participant posted at least one critique before s/he made the final choice. The total number of critiques by all users is 38 (mean = 2.11 per user, st.d. = 1.45, min = 1, max = 6), among which the number of *improvement-based critiques* is largely higher than that of *similarity-based critiques* (36 vs. 2). Within those improvement-based critiques, 88.9 % (32 out of 36) are *compound critiques*, with average 2.69 attributes selected for "improving" and 1.94 for "compromising" in one critique. Through computing conditional probability[1], we find $P(\text{"improve"}|\text{"compromise"}) = 1$, whereas

---

[1] $P(h|e) = N(h \wedge e)/N(e)$, where $N()$ denotes the number of observations within all compound critiques.

$P(\text{"compromise"}|\text{"improve"}) = 0.72$, which implies that the appearance of "compromise" is always contingent on that of "improve", but not vice versa. All of the results hence indicate that users are inclined to *improve* certain attribute values of a product, which will (but not always) be at the cost of *compromising* some of other attributes' values.

As for users' eye movements on recommended products, Fig. 2 shows an example of a user's eye-gaze plot, where each fixation is represented by a blue circle with radius indicating its duration. Wish such eye-gaze plot, we are able to correspond each fixation point to the actual information shown on the interface. Specifically, two researchers first did the mapping independently. If it fell into a product-level AOI, they associated it with that product's ID; if it was placed on an attribute's value (attribute-level AOI), they associated it with both product ID and that attribute's name (e.g., price). They then met together to resolve any divergences. In this way, we identified 2,493 fixation points at product level, and 1,227 fixations associated with ten primary attributes.

Additional analysis shows that on average 9.87 products (st.d. = 5.73) were viewed per user within each set of 25 recommended products. The mean values of fixation count (FC), total fixation duration (TFD), and average fixation duration (AFD) on viewed products are respectively 6.57 (st.d. = 5.59), 2,308.87 ms (st.d. = 2,011.55), and 345.43 ms (st.d. = 50.95). As for attributes, there are 7.13 distinct attributes (st.d.= 2.64) viewed per user within each recommendation set, with mean FC per viewed attribute 3.83 (st.d. = 3.15), mean TFD 1,360.6 ms (st.d. = 1,199.9), and mean AFD 338.4 ms (st.d. = 54.1). Note that in this analysis the fixations on all values of an attribute in each recommendation set are counted together.

## 5    Inferring Critiquing Feedback

### 5.1    Inferring Critiqued Products

We are interested in first verifying Hypothesis 1 about the relationship between fixation values and critiqued products, for which *Hit-Ratio@K* (shortened as *H@K*) and *Mean Reciprocal Rank (MRR)*[2] are computed: (1) *Hit-Ratio@K* measures whether a user's critiqued product appears in the top-$K$ products that s/he has viewed, as ranked in the descending order of FC-p, TFD-p, or AFD-p values[3], and (2) *MRR* denotes the critiqued product's ranking position in this order.

From Table 1, we can see that Rank-by-FC-p and Rank-by-TFD-p are of higher accuracy than Rank-by-AFD-p and RAM (RAM refers to random ranking of viewed products), in terms of inferring critiqued products. For instance, when $K = 1$, the *Hit-Ratios* by Rank-by-FC-p and Rank-by-TFD-p are around 0.5,

---

[2] $H@K = \sum_{c \in C} \frac{1_{rank(p_c) \leq K}}{|C|}$ and $MRR = \sum_{c \in C} \frac{1}{rank(p_c)}$, where $rank(p_c)$ denotes the rank of critiqued product $p_c$ (in cycle $c$) within the top-$K$ viewed products as sorted by a certain fixation metric.

[3] We use FC-p, TFD-p, and AFD-p to respectively denote the measures of fixation count, total fixation duration, and average fixation duration at product level.

**Table 1.** Accuracy of inferring critiqued products based on different fixation metrics

|               | H@1   | H@2   | H@3   | H@4   | H@5   | MRR   |
| ------------- | ----- | ----- | ----- | ----- | ----- | ----- |
| Rank by FC-p  | 0.474 | **0.605** | **0.789** | 0.842 | **0.868** | 0.628 |
| Rank by TFD-p | **0.5** | **0.605** | 0.711 | **0.868** | **0.868** | **0.635** |
| Rank by AFD-p | 0.184 | 0.368 | 0.447 | 0.526 | 0.605 | 0.378 |
| RAM           | 0.316 | 0.342 | 0.342 | 0.553 | 0.5   | 0.36  |

**Table 2.** Relation between attribute-level fixations and critiquing criteria (*note: C* for "Compromise" and *K* for "Keep". The superscript indicates significant difference.)

|                    | FC-a       | TFD-a (ms)    | AFD-a (ms)      |
| ------------------ | ---------- | ------------- | --------------- |
| "Keep" attr.       | $3.165^C$  | $1,088.92^C$  | $289.23^{C,K}$  |
| "Improve" attr.    | $2.64^C$   | $1,038.19^C$  | $340.35^C$      |
| "Compromise" attr. | 1.42       | 448.42        | 143.96          |
| *ANOVA test*       | $F = 3.42, \mathbf{p = 0.036}$ | $F = 4.045, \mathbf{p = 0.02}$ | $F = 21.34, \mathbf{p < 0.001}$ |

showing that within about half of critiquing cycles, the product with the highest fixation count or total fixation duration was the one that the user selected to critique. When $K$ is increased to 5, the *Hit-Ratios* of Rank-by-FC-p and Rank-by-TFD-p both reach at 0.868. As for Rank-by-AFD-p, its hit ratio is relatively low (maximum 0.605 at $K = 5$). MRR results again indicate that Rank-by-FC-p and Rank-by-TFD-p are more predictive than Rank-by-AFD-p and RAM (0.635 and 0.628, vs. 0.378 and 0.36). Moreover, as the differences between Rank-by-FC-p and Rank-by-TFD-p are not obvious across all measures, they may be equivalent in terms of inferring users' critiquing propensity towards products.

Therefore, the above analysis shows that users' fixation values are helpful for inferring what products they tend to critique. Moreover, fixation count and total fixation duration are more effective than average fixation duration in achieving this goal. Concretely, it suggests that if a user more frequently views a product (with corresponding higher FC-p) or spends totally higher duration on a product (with higher TFD-p), the chance s/he selects it for critiquing will be higher than that of selecting others.

## 5.2   Inferring Critiquing Criteria for Attributes

Our second hypothesis is about inferring users' critiquing criteria (i.e., keep, improve, or compromise) for product attributes. Formally, each critiquing feedback can be represented as $(p_i, \{\langle a_j, c_j \rangle\})$, where $p_i$ is the critiqued product, $a_j \in A = \{a_1, ..., a_{10}\}$ ($A$ is the set of attributes), and $c_j \in \{$"keep", "improve", "compromise"$\}$. By comparing the fixation values among the three categories of attributes that were respectively selected to "keep", "improve", and "compromise", we find their differences are significant in terms

**Table 3.** Accuracy of inferring attributes' critiquing criteria through two alternative inference rules (1. *high* => "improve", *medium* => "keep", *low* => "compromise"; or 2. *high* => "keep", *medium* => "improve", *low* => "compromise")

| | Precision | | Recall | | F1 | | Hit-Ratio | |
|---|---|---|---|---|---|---|---|---|
| Classification by FC-a | $0.282_1$ | $0.38_2$ | $0.301_1$ | $0.366_2$ | $0.291_1$ | $0.373_2$ | $0.253_1$ | $0.297_2$ |
| Classification by TFD-a | $0.291_1$ | $0.392_2$ | $0.324_1$ | $0.393_2$ | $0.306_1$ | $0.392_2$ | $0.255_1$ | $0.303_2$ |
| Classification by AFD-a | $\mathbf{0.392_1}$ | $0.344_2$ | $\mathbf{0.417_1}$ | $0.416_2$ | $\mathbf{0.404_1}$ | $0.376_2$ | $\mathbf{0.355_1}$ | $0.289_2$ |

of FC-a, TFD-a, and AFD-a[4] by means of *ANOVA* test (see Table 2). Pairwise comparisons via *paired samples T-test* reveal that the fixation values of "keep" and "improve" attributes are significantly higher than those of "compromise" attributes. Specifically, the mean fixation count (FC-a) of "keep" attributes is 3.165 and that of "improve" attributes is 2.64, against 1.42 of "compromise" attributes ("keep" vs. "compromise": $t = 2.36$, $p = 0.02$; "improve" vs. "compromise": $t = 3.01$, $p < 0.01$). Similar trends are observed regarding total fixation duration (TFD-a) and average fixation duration (AFD-a). As for the difference between "keep" and "improve" attributes, it is moderately significant w.r.t. AFD-a (289.23 ms vs. 340.35 ms, $t = 1.75$, $p = 0.088$).

From the above results, we can derive two alternative inference rules: (1) *high* => "improve", *medium* => "keep", *low* => "compromise"; or (2) *high* => "keep", *medium* => "improve", *low* => "compromise". That is, suppose the fixation values on all attributes in one recommendation set are classified into three levels: *high, medium*, and *low*, so we may map each level to a specific critique criterion. For example, if the fixation count of an attribute is at relatively *high* level, we may infer the user would tend to "improve" it in her/his critique (*high* => "improve"). For this purpose, we applied 3-means clustering algorithm to automatically group 10 attributes into three clusters according to their fixation values in each recommendation cycle, and then classified the three clusters into *high, medium*, and *low* levels based on their centroids.

Next, we use *Precision, Recall, F1-measure*, and *Hit-Ratio*[5] to measure each inference rule's accuracy (see the results in Table 3). It shows as for FC-a and TFD-a, the second rule is more accurate in terms of all measures, and the classification by TFD-a achieves slightly higher accuracy than that by FC-a. In comparison, the accuracy of classification by AFD-a via the first rule is even higher, with the highest *Precision* 0.392, *Recall* 0.417, *F1* 0.404, and *Hit-Ratio*

---

[4] FC-a, TFD-a, and AFD-a respectively denote the measures of fixation count, total fixation duration, and average fixation duration at attribute level.

[5] $Precision = \sum_{k \in AC} \frac{|Pred(k) \cap R(k)|}{|Pred(k)|}/|AC|$, $Recall = \sum_{k \in AC} \frac{|Pred(k) \cap R(k)|}{|R(k)|}/|AC|$, $F1 = \sum_{k \in AC} \frac{2 \times Precision(k) \times Recall(k)}{Precision(k) + Recall(k)}/|AC|$, and $HitRaito = \frac{\sum_{k \in AC} |Pred(k) \cap R(k)|}{q}$, where $AC$ denotes the set of three critique options {"keep", "improve", "compromise"}, $Pred(k)$ denotes the set of attributes that are inferred with critique $k$, $R(k)$ contains attributes that are actually critiqued with $k$, and $q$ is the total number of attribute critiques (that is 380 in our data).

0.355 among all results. It hence suggests that, for inferring attributes' critiquing criteria, average fixation duration (AFD-a) behaves more effectively than the other two metrics fixation count (FC-a) and total fixation duration (TFD-a); and the attributes with relatively *high* level of AFD-a will be more likely to be "improved", followed by those at *medium* level to be "kept", and then the remainder at *low* level to be "compromised" (as per the first inference rule). Our Hypothesis 2 is thus verified.

## 5.3    Other Results: Refining Inference Rules

The derivation of inference rules in the previous section motivates us to consider more information to refine them. By matching each attribute fixation to its actual value (e.g., price $759.99), we can actually identify all values of an attribute the user had fixated (compared) across different recommended products before s/he made a critique. Therefore, in order to generate more precise inference rules, in this section, we investigate fixations on particular attribute values.

Specifically, for each attribute of the critiqued product, we can associate it with a *value comparison label* by comparing its value with the other values the user fixated. If it is a numerical attribute (e.g., price, processor speed, battery life), there are three possible labels: *Better than All* (in the case that the critiqued product's attribute value is better than or equal to the other viewed values in the same recommendation set), *Better than Some* (it is better than some of the other viewed values), and *Worse than All* (it is worse than the other viewed values). If it is a categorical attribute (e.g., manufacturer, processor class, operating system), there are two optional labels: *Equal to Others* (the critiqued product's attribute value is the same as the other viewed ones) and *Different from Others* (it is different from some of the other viewed values). If the user did not leave fixation on any of the attribute's values, it is labeled with *None*.

Figure 3 shows the distribution of all value comparison labels with respect to the three critiques "keep", "improve", and "compromise" that users posted to the corresponding attributes (the distribution is significant, i.e., $p < 0.01$, via *Pearson's Chi-square test*, relative to equal probabilities). Several phenomena can be observed from this figure: (1) *Better than All* and *Better than Some* attribute values appear more often in "keep" critiques. (2) Some *Better than All* attribute values are "compromised", implying that they may be less important to some users. (3) *Worse than All* values are more subject to be "improved" or "kept". These three observations imply that for a numerical attribute, if the user has viewed different values and then selected a product (to critique) that has better value than others, s/he may tend to "keep" it; otherwise, if its value is the worst, s/he may "improve" it. (4) *Different from Others* attribute values are more often "kept" or "improved", whereas *Equal to Others* attribute values are mostly "kept". This observation implies that if all fixations by a user regarding a categorical attribute (e.g., manufacturer) are laid on only one value (e.g., "Apple"), this value might be the user's target, so s/he will be likely to "keep" it. Otherwise, if s/he has fixated over different values of the attribute, s/he may either "keep" the value of her/his selected product (if it meets with

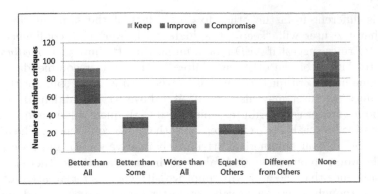

**Fig. 3.** Distribution of value comparison occurrences with respect to attribute critiques.

her/his requirement after comparison) or "improve" it (if none of viewed values are satisfactory). (5) The attributes without any fixations (labelled as *None*) are mostly "kept" or "compromised".

Then, we combine attribute fixations and value comparison labels to generate association rules. Concretely, we ran Apriori algorithm [1], which is a popular association rule mining tool, on the whole set of 380 attribute critiques, in order to derive high-confident rules in form of {*attribute fixation, value comparison*} => {*attribute critique*}. As for *attribute fixation*, there are three levels, *high*, *medium*, and *low*, as obtained via AFD-a based classification (see the previous section). As for *value comparison*, there are in total six different categories (as described above). The *attribute critique* takes any of the three options, "keep", "improve", and "compromise".

Among the returned rules, we first select those with *Lift*[6] greater than 1, because *Lift* > 1 (also called *Interest*) suggests that the occurrences of antecedent and consequence are dependent on each other, making the rule useful for predicting consequence in other data sets [24]. The selected rules are then sorted in descending order by *Confidence* value. As a result, there are six rules with *Confidence* bigger than or equal to 0.5[7]:

1. {*high AFD-a, Different from Others*} => "keep" ($Conf. = 0.857$);
2. {*medium AFD-a, Better than Some*} => "keep" ($Conf. = 0.826$);
3. {*medium AFD-a, Better than All*} => "keep" ($Conf. = 0.647$);
4. {*Equal to Others*} => "keep" ($Conf. = 0.633$);
5. {*high AFD-a, Worse than All*} => "improve" ($Conf. = 0.625$);
6. {*low AFD-a, Better than Some*} => "compromise" ($Conf. = 0.5$).

The 1st rule implies if a user's average fixation duration (AFD-a) on one categorical attribute is relatively *high* and this attribute's value in the critiqued

---

[6] $Lift(X \Rightarrow Y) = \frac{supp(X \cup Y)}{supp(X) \times supp(Y)}$, $Confidence(X \Rightarrow Y) = \frac{supp(X \cup Y)}{supp(X)}$, where $supp(X)$ gives the proportion of transactions that contain $X$.

[7] We set 0.5 as *Confidence* threshold, as it indicates a high probability that at least half of transactions contain the antecedent leading to the consequence.

product is different from the other fixated values of the same attribute, the chance that the user will "keep" it is high (with over 85 % confidence). The 2nd and 3rd rules suggest if AFD-a on a numerical attribute is at *medium* level (relative to AFD-a of the other attributes in the same recommendation set) and its value in the critiqued product is better than at least some of the other viewed values, the user may also "keep" it (above 64 % confidence). The 4th rule is about categorical attribute, which, if its critiqued value is equal to the other viewed values, is likely to be "kept" (with 63.3 % confidence). The 5th rule indicates if AFD-a on a numerical attribute is relatively *high* and its critiqued value is the worst among all viewed values of the same attribute, there is around 62.5 % confidence that the user will "improve" it; whereas for an attribute with *low* AFD-a, though its critiqued value is better than some compared ones, the probability that the user will "compromise" it is higher than that of "keeping" or "improving" it (the 6th rule).

## 6   Conclusions and Future Work

In conclusion, this work verifies our hypotheses about inferring users' critiquing feedback from their eye movements on recommended products. There are three major findings: (1) Based on products' fixation values, we can infer what product the user is inclined to critique within a set of recommendations. In particular, fixation count (FC-p) and total fixation duration (TFD-p) are more accurate than average fixation duration (AFD-p) for achieving this goal. (2) At attribute level, we find the fixation values of attributes that users choose to "keep" or "improve" are significantly higher than those of attributes they "compromise". On the other hand, average fixation duration (AFD-a) performs more effectively than FC-a and TFD-a in terms of inferring users' critiquing criteria for attributes. (3) We further attempted to derive some precise inference rules by incorporating users' value comparison behavior based on their fixations on attribute values. As a result, several high-confident association rules are generated. The findings are thus constructive for improving existing system-suggested critiquing methods in recommender systems. In addition to making the critique suggestions representative of remaining products [19, 22], we can make them more reflective of users' critiquing intentions so that the users will be more likely to accept them.

In the future, we will conduct more experiments to validate the association rules' inference accuracy. We will also investigate more fixation metrics, such as fixation spatial density, saccade/fixation ratio, and scanpath, in order to make the inference process more accurate. It is expected that we will eventually build a prediction model that can well unify all of the valuable eye-based metrics to infer users' critiquing feedback, which will enable the system to automatically adjust recommendations even without requiring users to explicitly make critiques.

**Acknowledgments.** We thank participants who took part in our experiment. We thank Dr. Weike Pan and Ms. Wai Yee Wong for assisting in data processing and analysis. We also thank Hong Kong RGC and China NSFC for sponsoring the described research work (under projects RGC/HKBU12200415 and NSFC/61272365).

# References

1. Agrawal, R., Imieliński, T., Swami, A.: Mining association rules between sets of items in large databases. In: Proceedings of the 1993 ACM SIGMOD International Conference on Management of Data, SIGMOD 1993, pp. 207–216. ACM, New York (1993)
2. Bridge, D., Göker, M.H., McGinty, L., Smyth, B.: Case-based recommender systems. Knowl. Eng. Rev. **20**(3), 315–320 (2005)
3. Burke, R.D., Hammond, K.J., Young, B.: The findme approach to assisted browsing. IEEE Expert Intell. Syst. Appl. **12**(4), 32–40 (1997)
4. Castagnos, S., Jones, N., Pu, P.: Eye-tracking product recommenders' usage. In: Proceedings of the Fourth ACM Conference on Recommender Systems, RecSys 2010, pp. 29–36. ACM, New York (2010)
5. Chen, L., Pu, P.: Evaluating critiquing-based recommender agents. In: Proceedings of the 21st National Conference on Artificial Intelligence, AAAI 2006, vol. 1, pp. 157–162. AAAI Press (2006)
6. Chen, L., Pu, P.: Hybrid critiquing-based recommender systems. In: Proceedings of the 12th International Conference on Intelligent User Interfaces, IUI 2007, pp. 22–31. ACM (2007)
7. Chen, L., Pu, P.: Interaction design guidelines on critiquing-based recommender systems. User Model. User-Adap. Inter. **19**(3), 167–206 (2009)
8. Chen, L., Pu, P.: Experiments on the preference-based organization interface in recommender systems. ACM Trans. Comput. Hum. Inter. **17**(1), 1–33 (2010)
9. Chen, L., Pu, P.: Eye-tracking study of user behavior in recommender interfaces. In: Bra, P., Kobsa, A., Chin, D. (eds.) UMAP 2010. LNCS, vol. 6075, pp. 375–380. Springer, Heidelberg (2010). doi:10.1007/978-3-642-13470-8_35
10. Chen, L., Pu, P.: Critiquing-based recommenders: survey and emerging trends. User Model. User-Adap. Inter. **22**(1–2), 125–150 (2012)
11. Cheng, S., Liu, X., Yan, P., Zhou, J., Sun, S.: Adaptive user interface of product recommendation based on eye-tracking. In: Proceedings of the 2010 Workshop on Eye Gaze in Intelligent Human Machine Interaction, EGIHMI 2010, pp. 94–101. ACM, New York (2010)
12. Ehmke, C., Wilson, S.: Identifying web usability problems from eye-tracking data. In: Proceedings of the 21st British HCI Group Annual Conference on People and Computers: HCI..But Not As We Know It, BCS-HCI 2007, vol. 1, pp. 119–128. British Computer Society, Swinton (2007)
13. Giordano, D., Kavasidis, I., Pino, C., Spampinato, C.: Content based recommender system by using eye gaze data. In: Proceedings of the Symposium on Eye Tracking Research and Applications, ETRA 2012, pp. 369–372. ACM, New York (2012)
14. Grasch, P., Felfernig, A., Reinfrank, F.: Recomment: towards critiquing-based recommendation with speech interaction. In: Proceedings of the 7th ACM Conference on Recommender Systems, RecSys 2013, pp. 157–164. ACM, New York (2013)
15. Jung, J., Matsuba, Y., Mallipeddi, R., Funaya, H., Ikeda, K., Lee, M.: Evolutionary programming based recommendation system for online shopping. In: Signal and Information Processing Association Annual Summit and Conference (APSIPA), 2013 Asia-Pacific, pp. 1–4, October 2013
16. Linden, G., Hanks, S., Lesh, N.: Interactive assessment of user preference models: the automated travel assistant. In: Jameson, A., Paris, C., Tasso, C. (eds.) UM 1997. ICMS, vol. 383, pp. 67–78. Springer, Heidelberg (1997). doi:10.1007/978-3-7091-2670-7_9

17. Mahmood, T., Mujtaba, G., Venturini, A.: Dynamic personalization in conversational recommender systems. IseB **12**(2), 213–238 (2014)
18. McCarthy, K., McGinty, L., Smyth, B., Reilly, J.: A live-user evaluation of incremental dynamic critiquing. In: Muñoz-Ávila, H., Ricci, F. (eds.) ICCBR 2005. LNCS (LNAI), vol. 3620, pp. 339–352. Springer, Heidelberg (2005). doi:10.1007/11536406_27
19. McCarthy, K., Reilly, J., McGinty, L., Smyth, B.: Experiments in dynamic critiquing. In: Proceedings of the 10th International Conference on Intelligent User Interfaces, IUI 2005, pp. 175–182. ACM (2005)
20. McCarthy, K., Salem, Y., Smyth, B.: Experience-based critiquing: reusing critiquing experiences to improve conversational recommendation. In: Bichindaritz, I., Montani, S. (eds.) ICCBR 2010. LNCS (LNAI), vol. 6176, pp. 480–494. Springer, Heidelberg (2010). doi:10.1007/978-3-642-14274-1_35
21. Poole, A., Ball, L.J.: Eye tracking in human-computer interaction and usability research: current status and future prospects. In: Ghaoui, C. (ed.) Encyclopedia of Human-Computer Interaction. Idea Group Inc., Pennsylvania (2005)
22. Reilly, J., McCarthy, K., McGinty, L., Smyth, B.: Dynamic critiquing. In: Funk, P., González Calero, P.A. (eds.) ECCBR 2004. LNCS (LNAI), vol. 3155, pp. 763–777. Springer, Heidelberg (2004). doi:10.1007/978-3-540-28631-8_55
23. Salvucci, D.D., Goldberg, J.H.: Identifying fixations and saccades in eye-tracking protocols. In: Proceedings of the 2000 Symposium on Eye Tracking Research & Applications, ETRA 2000, pp. 71–78. ACM, New York (2000)
24. Tan, P.-N., Kumar, V., Srivastava, J.: Selecting the right interestingness measure for association patterns. In: Proceedings of the 8th ACM SIGKDD International Conference on Knowledge Discovery and Data Mining, KDD 2002, pp. 32–41. ACM (2002)
25. Tullis, T., Albert, W.: Measuring the User Experience: Collecting, Analyzing, and Presenting Usability Metrics. Morgan Kaufmann Publishers Inc., San Francisco (2008)
26. Xie, H., Chen, L., Wang, F.: Collaborative compound critiquing. In: Dimitrova, V., Kuflik, T., Chin, D., Ricci, F., Dolog, P., Houben, G.-J. (eds.) UMAP 2014. LNCS, vol. 8538, pp. 254–265. Springer, Heidelberg (2014). doi:10.1007/978-3-319-08786-3_22
27. Xu, S., Jiang, H., Lau, F.C.: Personalized online document, image and video recommendation via commodity eye-tracking. In: Proceedings of the 2008 ACM Conference on Recommender Systems, RecSys 2008, pp. 83–90. ACM, New York (2008)
28. Zhang, J., Pu, P.: A comparative study of compound critique generation in conversational recommender systems. In: Wade, V.P., Ashman, H., Smyth, B. (eds.) AH 2006. LNCS, vol. 4018, pp. 234–243. Springer, Heidelberg (2006). doi:10.1007/11768012_25

# Eager to be Lazy: Towards a Complexity-guided Textual Case-Based Reasoning System

K.V.S. Dileep[✉] and Sutanu Chakraborti

Department of Computer Science and Engineering,
Indian Institute of Technology Madras, Chennai 600036, India
{kvsdilip,sutanuc}@cse.iitm.ac.in

**Abstract.** Finding an ideal representation for a case-base is important for a CBR system. This choice of an ideal representation is guided by the complexity of the cases. Based on the needs of each individual case, richer features are used for representation if required. While the framework is fairly general, this paper demonstrates its effectiveness on text classification due to the ease of evaluation. Each test case is treated differently by the classifier, in that if a shallow representation is deemed adequate for assigning a class label, the algorithm does away with a richer representation which is computationally expensive to generate. We also provided a time-budgeted evaluation of our framework which suggests that it holds promise in minimizing redundant or misleading comparisons and minimize time without compromising on effectiveness.

## 1 Introduction

Case-Based Reasoning (CBR) systems attempt to reuse past knowledge of problem solving (modelled as cases) to find solutions to new problems. One critical issue, that has perhaps been one of the most central themes in designing practical applications, is arriving at the right choice of representation. More often than not, CBR systems choose uniform representations for cases and queries. This paper proposes an alternative to such a uniform representation of cases, by arguing that, at training time each case can be stored using multiple representations, each capturing an aspect in which the case can potentially be used. Some of these representations can be shallow and facilitate faster comparisons with the query; others could be deep and computationally more expensive when used for similarity estimation. An example is the case of Textual CBR, where cases are crafted from unstructured representations. In such a context, bag of words provides a shallow representation, whereas a syntactically rich representation is deeper. At query time, the nature of the query case may be used to decide between different choices of representations. If a shallow representation is sufficient to solve a query, our proposed approach does away with creating deeper representations for it. The central thesis is that such an on-demand choice of representation can leverage on the nuances of the query, make best use of multiple representations and result in gains in both time efficiency and effectiveness.

© Springer International Publishing AG 2016
A. Goel et al. (Eds.): ICCBR 2016, LNAI 9969, pp. 77–92, 2016.
DOI: 10.1007/978-3-319-47096-2_6

The central ideas presented in the paper are cognitively inspired. Let us imagine the case of a person assigned the tedious job of assigning category labels (like sports, politics, religion) to news reports. It seems intuitive that she would not handle each new document uniformly. Rather, she would exploit the fact that while some documents are simple and can be unambiguously categorized using the titles alone, others are harder and may require a careful read of the body of the article. Also, for certain documents, surfing through a set of indicative words may suffice, for others she may need recourse to background knowledge (say Wikipedia) or detailed linguistic knowledge (such as that of syntax or semantic relatedness of words). The nature of the classification task also critically influences the classification process. Discriminating between Apple and IBM hardware documents needs recourse to richer knowledge sources than classifying between sports and politics, where few keywords are often adequate discriminators. Thus, dynamic integration of knowledge on an on-demand basis is an appealing idea. In this paper we investigate if a Textual CBR (TCBR) system can benefit from this idea as well.

While the idea of dynamic integration is generic in nature and can be applied across diverse CBR tasks, we discuss our approach with respect to text classification (Textual CBR) due to the ease of evaluation. Text classification algorithms often extract a set of discriminating features from the training data and use them to arrive at revised representations of test documents, which are then exploited to predict class labels. The central observation is that the process of arriving at revised representations of test documents is largely agnostic to the specific nuances of the test document in question. This is different from how a human would, for example, treat an easy document very differently from a hard one. Our central hypothesis is that a differential treatment of documents, and selective on-demand integration of knowledge may help in cutting down on redundant computations and also improve performance in select settings where indiscriminate revision of test documents adversely affects classification effectiveness. The gains, however, are expected to be sensitive to the complexity of the classification task.

## 2    Overview of Our Approach

In order to empirically test the hypothesis stated above, we need three building blocks. The first one involves identification of knowledge sources and features, which together define the revised representations. Secondly, we need a complexity measure that estimates the hardness of a given text document, in the context of a chosen representation. For the purpose of this paper, the terms 'document' and 'case' are used interchangeably. It is important to note that each instance in the training data is preprocessed to carry the knowledge of representations that are found to be most effective in classifying it. More precisely, a representation $R_1$ is said to be more effective than another representation $R_2$ with respect to a training instance T, if the neighbourhood of T obtained with $R_1$ is more predictive of its true class, compared to that obtained with $R_2$.

Thirdly, we need algorithms that can take as input a given test case, estimate its complexity with respect to a shallow representation, and decide what to do next. This component is built around the philosophy of two-step reminding. The first step is one where the nearest neighbours in a shallow representation predict the class label of the unseen test instance. If there is sufficient agreement between the neighbours, we decide to refrain from processing the test instance any further. A disagreement can be interpreted as the representation not being rich enough to disambiguate the test instance. In the event of a disagreement, we start the next level of reminding, which we call a process-oriented reminding. The name is motivated by the fact, that in this step, each neighbour in the shallow representation is used to predict the representation choice that is expected to yield highest gains in resolving the ambiguity. The outcome of the second step is a process of identifying a representation, where neighbours are likely to be more predictive of true class labels.

In the following three sections we elaborate on each of the building blocks mentioned above.

## 3   Knowledge Sources for Text Classification

Shallow representations of textual cases based on "Bag of Words" (BoW) are often supplemented with additional knowledge sources to facilitate more meaningful comparisons between cases. For the purpose of our paper, a knowledge source is any source of information from within and outside the document that helps in solving a given task. In practice, the following knowledge sources can be used to enrich the textual representations:

- Background knowledge - This knowledge corresponds to features generated using knowledge catalogued by humans. One of the often used resources is Wikipedia. To integrate this knowledge we use Explicit Semantic Analysis (ESA) proposed by Gabrilovich et al. [1]. ESA represents each document (as well as word) as a vector in a space spanned by Wikipedia article names, which are treated as concepts. Words that occur in the article are said to belong to the concept. For example, the words 'Barack Obama' and 'POTUS' have a non-zero projection onto the Wikipedia concept 'US President'.
- Linguistic knowledge - WordNet [2] is a lexical resource that expresses a concept as a synonym set or synset. WordNet is a semantic network of synsets. The semantic relatedness of any two words can be estimated using the shortest path between the senses of the first word and those of the second.
- Introspective knowledge - We can also obtain features by looking within the corpus itself, and exploiting co-occurrence patterns. We assume that the words originate from a set of latent concepts. The documents, as well as words, are represented in a space defined by latent concepts using Latent Semantic Analysis (LSA) [3], which are extracted using dual mode factor analysis realized by Singular Vector Decomposition (SVD).

We now define the notions of 'breadth' and 'depth' in knowledge sources. 'Breadth' refers to features across the document, while 'depth' is used to denote

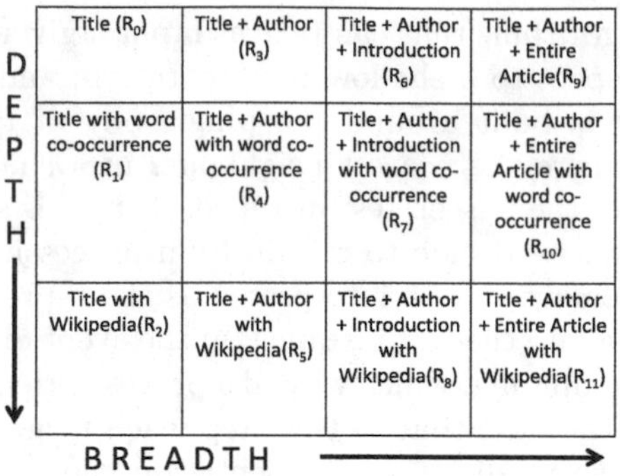

| Title ($R_0$) | Title + Author ($R_3$) | Title + Author + Introduction ($R_6$) | Title + Author + Entire Article($R_9$) |
|---|---|---|---|
| Title with word co-occurrence ($R_1$) | Title + Author with word co-occurrence ($R_4$) | Title + Author + Introduction with word co-occurrence ($R_7$) | Title + Author + Entire Article with word co-occurrence ($R_{10}$) |
| Title with Wikipedia($R_2$) | Title + Author with Wikipedia($R_5$) | Title + Author + Introduction with Wikipedia($R_8$) | Title + Author + Entire Article with Wikipedia($R_{11}$) |

D E P T H (vertical, left axis)

BREADTH ⟶

**Fig. 1.** An example representation matrix for documents. A traversal could be elementwise in a column-major fashion i.e. $R_0, R_1, ..., R_{11}$.

knowledge sources used for enriching the features. An article can be broken down into parts like title, first paragraph, half the article and the complete article, and this decomposition constitutes the breadth. To make a correct classification, for some articles reading the title and the first few lines are sufficient. Others might require reading through the entire article. At each level of breadth, we can augment representations using additional knowledge - background, introspective or linguistic. Knowledge sources define the depth of our representation. So, taking the combination of depth and breadth, we get a matrix of possible representations a document can take for classification.

An example representation matrix is shown in Fig. 1. We can progressively add these features in the order of traversal till we incorporate the complete article. The rows of the figure correspond to three knowledge sources that define depth: Bag of Words(BoW), word co-occurrence knowledge and Wikipedia. 'Title+word co-occurrence'($R_1$) means that introspective knowledge extracted from titles has been added. The notation $R_i$ means $i^{th}$ representation in matrix as shown.

## 4   Complexity Measures for Text Classification

To decide on the modality of knowledge integration, we estimate the complexity of the case we are trying to process. Alignment of a case-base tells us how well the CBR assumption - 'similar problems have similar solutions' holds true for the case-base. Hence, high alignment implies low complexity and vice versa. For areas in the case-base with low alignment, similar problems do not have similar solutions and hence, in these areas it will be difficult to answer a query. The problem part of the case corresponds to the document representation and the solution part corresponds to the class label. We use a neighbourhood approach to define complexity. Documents with same labels that are close to each other define 'easy' or well-aligned regions in the document space, while similar documents with differing labels define 'hard' or poorly-aligned regions in the space.

In terms of text classification, there is a higher chance of a document getting misclassified in poorly-aligned regions. Identifying such areas and progressively adding additional knowledge to simplify those regions would help to reduce the chances of a document getting misclassified.

Alignment can be global, in which case, it characterizes the case base as a whole. Local alignment, on the other hand, characterizes the immediate neighbourhood of a given case. A document collection with class labels can also be interpreted as a case-base. We use the alignment measure proposed by Massie et al. [4]. For a target case, we take the nearest cases based on the similarity on the problem side. Then we take the sum of similarity of target case to its neighbours on the problem side weighted by the similarity of target solution to the corresponding solution of the neighbours. This is normalized by the sum of just the problem side similarities of target to its neighbours. If "similar problems have similar solutions", then if the problem side similarities are high along with solution side similarities the alignment would be high. Also neighbours closer to the target case have more say in the alignment. We now define the alignment measure for a case-base with textual problems and class labels as solutions.

The local alignment of a case $d_i$ with class label $label_i$ is defined as:

$$local\,Align(\mathbf{d_i}) = \frac{\sum_{k \in nb(\mathbf{d_i})} ps(\mathbf{d_i}, \mathbf{d_k}) ss(label_i, label_k)}{\sum_{k \in nb(\mathbf{d_i})} ps(\mathbf{d_i}, \mathbf{d_k})} \tag{1}$$

$$ps(\mathbf{d_i}, \mathbf{d_k}) = \cos(\mathbf{d_i}, \mathbf{d_k}) \tag{2}$$

$$ss(label_i, label_k) = \begin{cases} 0 : label_i \neq label_k \\ 1 : label_i = label_k \end{cases} \tag{3}$$

where $N$ is the number of documents, $\mathbf{d_k}$ a document in the neighbourhood of $\mathbf{d_i}$ and $label_k$ its class label, $nb(x)$ is the set of nearest neighbours of a document $x$, $ps()$ is problem side similarity function which is the cosine similarity ($\cos(\mathbf{d_i}, \mathbf{d_k}) = \frac{\mathbf{d_i} \cdot \mathbf{d_k}}{|\mathbf{d_i}||\mathbf{d_k}|}$) when $\mathbf{d_i}$ and $\mathbf{d_k}$ are vectors in a space spanned by words). $ss()$ is solution side similarity as defined in Eq. 3. A value close to 1 indicates strong alignment, and a value close to 0 indicates weak alignment of the document. A weak alignment score in a representation suggests that it might not be suitable for classifying the document or any document in its neighbourhood. We cannot directly estimate the alignment of a test document, as its class label is unavailable. Thus, we indirectly estimate its alignment using the neighbouring training documents. More details regarding this procedure will be discussed in the next section.

## 5   Algorithms

Our procedure is based on the principles of reminding. Reminding happens when we access a structure in memory that is most likely to solve the new problem [5]. Our search for a solution to a new problem is related to solutions of similar problems that have been previously solved. The ideal representation of a new problem too might be related to representations of similar problems. Instead of looking at solutions directly, we try to find a representation of the problem that is best suited to arrive at a usable solution. We refer to this method as process-oriented reminding.

The factors that are important while choosing a representation are:

- Cost - This represents the effort (time or resources) needed to transform data from one representation to another. Transforming a document into a bag of words representation takes lesser effort than transforming the document into space spanned by Wikipedia concepts. We prefer representations that take less effort to transform into.
- Alignment - A representation in which a document is more aligned is preferred. There is a trade-off between alignment and cost. A representation with more alignment might take more time to transform into incurring more cost.

We transform the training data into each of the representations discussed earlier. We estimate the local alignment of each instance using Eq. 1 across all representations. We also estimate the cost of transforming a document into a given representation through cross-validation. The cost might not be sensitive to the properties of individual documents for representations like LSA. Yet, we consider it as a document related property (and not average of a collection) to keep our formulation general. Algorithm 1 outlines the steps in the training process.

We explain the notation we use in the algorithms - $R_i$ corresponds to $i^{th}$ representation, $\mathcal{N}$ refers to set of indices of representations, $\mathcal{M}$ refers to set of indices of documents in training collection $D$, and $R_{mi}$ means document $d$ of index $m \in \mathcal{M}$ is transformed into $R_i$. $\mathcal{M}'$ refers to set of indices of documents in test collection $D'$.

---

**Algorithm 1.** Training phase of AlignSelect algorithm

Given a collection of training documents, $D$ and possible representations $R_0, ..., R_N$

1. For every document $d \in D$, convert it into all representations $R_0, ..., R_N$.
2. For every converted document $R_{mn}$, calculate its alignment $a_{mn}$, where $m$ is index of document $d \in D$.
3. For every document $d \in D$, use a leave-one-out approach to determine the time taken to transform into all representations $R_0, ..., R_N$. This is the cost $c_{mn}$ of transforming $m^{th}$ document, $d \in D$ to representation $R_n$.

---

For an incoming test document, $d' \in D'$ with index $t \in \mathcal{M}'$, we transform it into $R_{t0}$, the basic representation. If we find the neighbourhood to be poorly-aligned, then we project it into a space with richer representation where we expect it to perform better. So for easy documents, we arrive at a representation with fewer features and knowledge than a difficult document. The inductive bias is that documents in the neighbourhood of highly aligned documents will also have high alignment. For easy documents which are located in well-aligned neighbourhoods, we prefer relatively shallow representations compared to difficult ones. This process is analogous to a human who treats every document on merit, thus ignoring the body of an article if the title or author names suffice for classification.

An important step in processing a test document is to estimate its difficulty in a given representation. An example is just the title in the form of BoW. For that purpose, we take $k$ neighbours and estimate the alignment of the test query by taking the average of the alignment of neighbours weighted by similarity to the query. The detailed procedure is described in Algorithm 2.

---

**Algorithm 2.** Procedure for calculating expected alignment(EA) of test document

---

Given a collection of training documents $D$, the number of neighbours $k$, and a test document $d' \in D'$.

1. Find the $k$ nearest training documents to $d'$, say $P$. For our experiments, we used $k = 3$.
2. Expected Alignment is calculated as $EA = \frac{\sum_{i=1}^{k} w_i a_i}{\sum_{i=1}^{k} w_i}$, where $a_i$ is the alignment of document $p_i$ in $P$ and $w_i$ its corresponding similarity to $d'$.

---

The next step is to decide whether to use the current representation or go to a richer representation. We set a threshold, against which the expected alignment of the test document is compared. To fix the threshold, we use a leave-one-out cross validation approach during training. If the expected alignment is above the threshold, we predict the class label using the neighbours of the query in the current representation. This is the primary problem of finding the label.

In case the expected alignment of the query falls below this threshold, we initiate the process-oriented reminding phase. This is the secondary problem, where we are interested in finding the right representation for the test document based on its neighbourhood. Instead of converting a test document $t$ into all representations $R_0, ..., R_N$ sequentially and do a column-wise traversal like in Fig. 1, we define a heuristic to choose the next representation. This leads to reduction of time during processing.

The choice of the representation depends on the cost, the alignment of the neighbours in the representation and distance to the query. But the neighbours of a document may change when the representation changes. Hence, we need an estimate of how much the neighbourhood of a document changes when the representation changes, which we call stability. More stability implies that the neighbourhood does not change much when transformed into another representation. Stability tells us how much we can trust the neighbours in the current representation to decide the choice of next representation.

A straight-forward notion of stability is to compare the set of top neighbours of each representation and check their intersection [6]. The more the intersection,

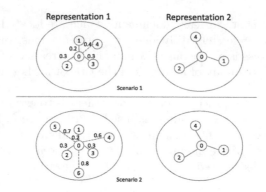

**Fig. 2.** An example to illustrate the notion of neighbourhood stability.

the greater the stability. We define a softer notion of stability, where we look at the distances of neighbours of both representations instead of a direct match. The stability of neighbourhood of document $d$ (index $m$ in $D$) in representation $R_i$ w.r.t representation $R_j$ is defined as

$$stab_{ij}(m) = 1 - \frac{\sum_{m'' \in K_j} dist(R_{mi}, R_{m''i}) - \sum_{m' \in K_i} dist(R_{mi}, R_{m'i})}{\sum_{m'' \in K_j} dist(R_{mi}, R_{m''i}) + \sum_{m' \in K_i} dist(R_{mi}, R_{m'i})} \qquad (4)$$

where $K_i$ is set of indices in $D$ of $k$ nearest neighbours of $d$ in $R_i$ and $K_j$ is set of indices of $k$ nearest neighbours of $d$ in $R_j$. $R_{m'i}$ corresponds to document of index $m'$ in $D$ in representation $R_i$.

We will explain our measure by means of an example. Let us assume we are measuring the stability of neighbourhood of document 0 in $R_1$ w.r.t $R_2$ as shown in Fig. 2. In scenario 1, the neighbour set for document 0 is $\{1, 2, 3\}$ in $R_1$ and $\{4, 2, 1\}$ in $R_2$. We take the sum of distances to document 0 for each of these sets separately in $R_1$ itself and find the difference. We find that the stability of doc 0 in scenario 1 is 0.91. In scenario 2, the neighbour sets are disjoint and we calculate the stability according to Eq. 4. We find the stability here to be 0.55. Clearly the stability is lower compared to scenario 1.

This notion of neighbourhood stability does not overly penalize the scenario where none of the neighbours in both sets of representations match. Like the cost factor, the stability of the documents also is estimated during training. With the cost, alignment and stability factors as weights, we estimate the next representation to transform into based on the neighbours of test document in the current representation. The detailed procedure is given in Algorithm 3.

## 6    Experiments and Results

For the purpose of our experiments, we consider different classification datasets. Two of these were created from the 20 Newsgroups(20NG) [7] corpus, RELPOL and HARDWARE. RELPOL contains 2 groups of discussion related to religion

---

**Algorithm 3.** Procedure to integrate knowledge sources during test phase for AlignSelect

---

1. Convert $d' \in D'$ with index $t \in \mathcal{M}'$ into the starting representation $R_0$ to get $R_{t0}$.
2. Calculate the expected alignment(EA) of $R_{t0}$, $ea_{t0}$ using Algorithm 2.
3. If $ea_{t0} \geq T$ threshold, predict class label using $k$-nn. Return.
4. Else,
   - For set of $k$ nearest neighbours of $R_{t0}$, say $L$. Let $K$ be the set of corresponding indices in $D$. For $m \in K$ retrieve $R_{mn}$, $\forall n | 0 < n \leq N$.
   - For $m \in K$, retrieve $c_{mn}$, the cost factor which corresponds to time involved in transforming to representation $n$.
   - For $m \in K$, retrieve $stab_{0n}(m)$, the stability of neighbourhoods of $m^{th}$ document in $D$
   - We calculate the next representation $i$ as

$$i = \underset{\substack{n \in \mathcal{N} \\ n \neq 0}}{\arg\max} \frac{\sum_{m \in K} w_{m0} \times c_{mn}^{-1} \times stab_{0n}(m) \times a_{mn}}{\sum_{m \in K} w_{m0} \times c_{mn}^{-1} \times stab_{0n}(m)}$$

   where $\mathcal{N}$ is set of indices of representations, $w_{m0}$ is distance of $m$ to $R_{t0}$, and $a_{mn}$ is alignment of $R_{mn}$, $m \in K$.
5. Convert $d'$ into $R_{ti}$ and calculate $ea_{ti}$, expected alignment of $R_{ti}$. Check with threshold $T$ and continue till the right representation is found or end of traversal.

---

and politics while HARDWARE contains 2 groups of discussion related to Apple hardware and IBM-PC hardware. The experimental setup is as follows. We create train and test splits, where each of the sets contains 20 % of the original corpus randomly chosen. Feature selection has been done using the Information Gain method. For repeated trials, 15 such train-test splits were created containing approximately 400 documents for each of the datasets. We also take data from Reuters-21578 distribution 1.0 dataset[1]. We take 3 datasets acq, corn and earn from this set and perform a one-vs-all binary classification over these datasets. Here too, we create train-test splits like earlier with 10 splits and feature selection through the chi-squared measure.

We also took raw data from imdb[2] (Internet Movie Database) and processed it into a MySQL database using the jmdb[3] (Java Movie Database) software. The database contains a series of tables such as - movies, actors, directors, writers, (plot) keywords, plot, distributors, editors, ratings etc. The complete list of tables provided by imdb is available online[4]. The features that we chose for the purpose of our work are actors, writers, directors, plot keywords, and finally plot description. From thousands of entries, we take a subset containing 1000 entries per class. We discard samples with missing features. Since a movie might belong to multiple genres, we ensure that the movie is not labelled with both

---

[1] http://kdd.ics.uci.edu/databases/reuters21578/reuters21578.html.
[2] www.imdb.com/interfaces.
[3] www.jmdb.de.
[4] ftp://ftp.fu-berlin.de/pub/misc/movies/database/.

**Table 1.** Hardness of the datasets as given by the average of local alignment values obtained using Eq. 1 for training data in BoW representation.

| Dataset | Alignment |
|---|---|
| RELPOL | 0.94 |
| HARDWARE | 0.71 |
| acq | 0.82 |
| earn | 0.70 |
| corn | 0.66 |
| imdb(a vs c) | 0.58 |
| imdb(a vs r) | 0.69 |

the genres we are classifying into. Currently, we perform 2 binary classification tasks - action vs romance (a vs r) and action vs comedy (a vs c). We make smaller datasets containing around 600 movies for training and 300 for test by stratified sampling and use 10 such datasets for our experiments.

The relative hardness of the datasets is given in Table 1. The higher the value of alignment, the easier it is to classify the corresponding dataset. We test the following hypotheses:-

- The choice of representation depends on the hardness of dataset.
- The algorithm compares favourably with uniform integration of knowledge.
- The algorithm requires lesser time on easy documents to get accurate results compared to harder documents.

We uniformly use 3 types of representations - Bag of Words(BoW), introspective (LSA) and background knowledge (ESA). For the RELPOL and HARDWARE datasets, we do the AlignSelect procedure with 3 representations - BoW, LSA and ESA with the entire document. For Reuters data, we choose 6 representations - titles of the article with BoW, LSA and ESA, and entire article with BoW, LSA and ESA. For imdb data, we choose 6 representations - actor, writer, director, plot keywords with BoW, LSA and ESA and actor,writer,director,plot keywords, and plot with BoW, LSA and ESA. For each of the datasets, we calculate the cost factor, stability and alignment during the training phase. For the AlignSelect procedure, we chose the threshold as described in the previous section.

First, we study how the relative difficulty of datasets influences the way the procedure selects which representation to choose for a given test document. Our hypothesis is that the richer the representation, the more the resources used for classifying the document which is analogous to a human experiencing a large cognitive load to classify the document. We construct the histograms of the percentage of documents chosen by AlignSelect procedure for various knowledge sources. The order of representations is the same order used in processing. The histograms are given in Fig. 3. We find that the procedure chooses richer representations for more percentage of documents in case of harder datasets

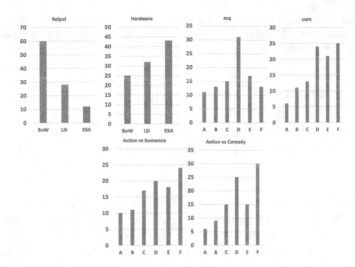

**Fig. 3.** Comparison of histograms of percentage of documents chosen by AlignSelect procedure for Relpol and Hardware(left), acq vs all and earn vs all(center), and action vs romance and action vs comedy datasets(right). Y axis denotes percentage of documents chosen by AlignSelect. X axis for center and right graphs is as shown in Tables 3 and 4 respectively.

**Table 2.** Classification results on a subset of 20 NG dataset with the AlignSel procedure. The threshold of alignment is given in brackets. Results of AlignSel highlighted in bold are statistically significant with $p < 0.05$ w.r.t rest of columns.

| Dataset | BoW | LSA | ESA | AlignSel(=0.7) |
|---------|-----|-----|-----|----------------|
| Relpol | 92 | 91.5 | 92 | **93.5** |
| Hardware | 73 | 73.4 | 74 | **76** |

especially when RELPOL and HARDWARE are compared as shown in Fig. 3. Similar trends can be observed in other histograms too. While choosing the next representation, it was seen that the cost factor ensured that a less rich but equally effective representation was chosen for some queries. For some of the harder queries, multiple hops had to be taken for arriving at the right representation. This led to some delay in processing. We can see that this method has promise in performing text classification the way humans do with a goal of minimizing cognitive load.

Now that we have seen how the hardness of the datasets influences the percentage of documents picked for richer representations, we will study the effectiveness of classification using the procedure as compared to individual representations. The classification results for AlignSelect procedure have been listed in Tables 2, 3 and 4 for the 20NG, Reuters data subgroups and imdb data respectively. The baseline we are comparing against is the approach where we integrate all the test documents with the BoW approach, LSA or ESA uniformly.

**Table 3.** Classification results on a subset of Reuters dataset with the different representations. Results of AlignSel highlighted in bold are statistically significant with $p < 0.05$ w.r.t rest of columns.

| Dataset | Title+ BoW(A) | Title+ LSA(B) | Title+ ESA(C) | Full BoW(D) | Full LSA(E) | Full ESA(F) | AlignSel (=0.7) |
|---------|------|------|------|------|------|------|------|
| acq | 25 | 20 | 35 | 77 | 78 | 79 | **81** |
| earn | 20 | 11 | 23 | 65 | 67 | 69 | **72** |
| corn | 17 | 10 | 19 | 60 | 59 | 65 | 64 |

The results show that our procedure is better than the baseline averaged over multiple trials and also statistically significant with two-tailed paired t-test ($p < 0.05$) for six of the seven datasets. This shows that indiscriminately integrating knowledge for documents that are easy to classify is not only inefficient and cognitively unintuitive, but also adversely affects classification effectiveness (accuracy).

**Table 4.** Classification results on imdb datasets with the different representations. AWDK means actors, writers, director and plot keywords features are used. AWDKP means with the aforementioned features, plot description is also used. Results of AlignSel highlighted in bold are statistically significant with $p < 0.05$ w.r.t rest of columns.

| Dataset | AWDK+ BoW(A) | AWDK+ LSA(B) | AWDK+ ESA(C) | AWDKP+ BoW(D) | AWDKP+ LSA(E) | AWDKP+ ESA(F) | AlignSel (=0.6) |
|---------|------|------|------|------|------|------|------|
| Action vs Romance | 69 | 56.5 | 66 | 69.6 | 60 | 72.4 | **73.5** |
| Action vs Comedy | 60 | 56 | 63 | 64 | 56 | 68.4 | **70** |

In the next set of experiments, we study how the procedure performs on harder documents as compared to easy documents within the same dataset. The dataset is partitioned on the basis of expected alignment on the test data. The partition value chosen is a median value so that we can get a fairly balanced set of easy and hard documents. Intuitively, documents harder to classify might need richer representation and hence require more time from the AlignSelect procedure. We run the procedure with time budgeting i.e. after the time elapses, the result using the current representation is returned. The longer the procedure is allowed to run, the more representations it can explore. The results for the time-budgeted run of the algorithm can be seen in Table 5.

An interesting point to note here is that for documents in poorly-aligned regions, a steady improvement can be seen as more time is given per document. For documents in well-aligned regions, a huge increase in accuracy initially is followed by minimal gains as more time is given per document. These trends are highlighted for the HARDWARE and Imdb (Action vs Comedy) datasets in Fig. 4. The other datasets too follow similar trends. This experiment shows that difficult documents require richer knowledge and hence longer time to get fairly accurate results.

**Table 5.** Classification results on datasets described earlier partitioned on the basis of alignment using the AlignSelect procedure with time budgeting. In brackets with each dataset is the expected alignment value used for creating the partitions. The time per query is in milliseconds.

| Dataset | Time budget per query (in millisecs) | | | |
|---|---|---|---|---|
| | 5 | 7 | 9 | 11 |
| Relpol($> 0.8$) | 92 | 94.5 | 95 | 95 |
| Relpol($< 0.8$) | 92.3 | 93.5 | 94 | 94 |
| Hardware($> 0.5$) | 71 | 74.5 | 76 | 77 |
| Hardware($< 0.5$) | 69 | 70 | 73.5 | 75 |
| acq($> 0.6$) | 63 | 73 | 75 | 82 |
| acq($< 0.6$) | 33 | 49 | 67 | 76 |
| earn($> 0.6$) | 47 | 60 | 68 | 73 |
| earn($< 0.6$) | 39 | 46 | 57 | 69 |
| corn($> 0.5$) | 31 | 47 | 57 | 65 |
| corn($< 0.5$) | 23 | 37 | 48 | 59.5 |
| Act vs Rom($> 0.6$) | 71 | 72.5 | 74 | 76 |
| Act vs Rom($< 0.6$) | 69 | 70 | 71 | 72.5 |
| Act vs Com($> 0.5$) | 63.5 | 66 | 67 | 69 |
| Act vs Com($< 0.5$) | 52 | 54 | 56 | 62 |

# 7   Related Work

The idea of using a subset of features (words) instead of the entire feature set for document classification has been explored by Koller et al. [8]. The idea was to decompose a huge hierarchical classification task into smaller problems at each node in the hierarchy. As the test document traverses through the hierarchy the features used to represent it change based on its location in the hierarchy. Though the feature set might be large as a whole, the number of features at each individual node would be few. The feature-set for each test query is sensitive to the node in the hierarchy, while for our method it is sensitive to the expected complexity of the query in its neighbourhood and not restricted to the hierarchy. Leveraging extra knowledge through heterogeneous data sources like Wikipedia and Yahoo! Answers have been shown to be beneficial for noisy text categorization by Gupta et al. [9]. Using various algorithms for the knowledge sources, different features were generated and combined into a large feature set. Our work tries to be more selective by trying to find documents that might be hard to classify and integrate knowledge only for those documents.

Recent trends in Information Retrieval have shown a selective treatment of queries with difficult queries getting processed differently than their easier counterparts. Liu et al. [10] showed that the fraction of query suggestions better than original query was lesser for easier queries than for difficult queries.

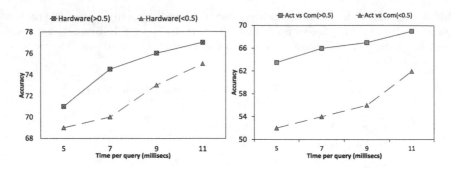

**Fig. 4.** Comparing accuracy and average time per query for HARDWARE and IMDB datasets.

Hassan et al. [11] showed that learning under-performing query groups from web search logs, and building specialized ranker for each of these groups performed better than a general ranker trained on a diverse set of queries. These contributions give assurance that our method of selectively integrating knowledge based on difficulty estimates might be a right direction to pursue.

SpeedBoost algorithm [12] automatically trades test prediction time with accuracy giving fast results initially using simple predictors and refine them with complex predictors as more time is available. The novelty of our approach is that the richness of features used is made sensitive to query difficulty. In our approach, a larger time budget allows the algorithm to pick richer features only if the query is found to be difficult.

Multistrategy learning is a paradigm that tries to use multiple learners and combine the predictions to make a single prediction. This method has been used in Information Extraction (IE) [13]. Each of the weak learners has a confidence associated with their predictions, and these confidences are mapped to probabilities of success using regression models. The probabilities are then combined to make a prediction. Closely related to this is multi-view learning [14], which takes multiple representations of the same data and learns from them. While our method too takes multiple views of data, we differ from these approaches from the fact that our approach is query sensitive and guided by its complexity. We do not use all the representations in predicting the class label. Instead, we let the approach pick a representation for a query that strikes the best trade-off between processing cost and retrieval effectiveness.

Our work can also be interpreted as a meta-classification problem where we search for an ideal representation for a test case that can, in turn, be the input to a classifier. Meta-classification has been explored in CBR by Cummins et al. [15] where a meta case base has been constructed from various case-bases for finding the ideal case-base maintenance algorithm for a given case-base.

# 8    Discussion

It can be noted that though the work is presented in the context of text classification, it can be extended to other CBR domains which can benefit from query sensitive on-?demand integration of knowledge. As with any case-base classification system, good coverage of cases improves the accuracy. This criteria is even more critical in our framework, since good coverage ensures good estimates of alignment and cost. During the retrieval phase of the CBR cycle, the test case can be enriched with knowledge guided by its expected complexity. This would also especially benefit time-sensitive applications. To summarize, on-demand integration makes sense when (a) the most effective representation is not known beforehand, and can be predicted by a process oriented reminding based on the specific query (b) integration of knowledge is computationally expensive at classification time (c) indiscriminate knowledge integration leads to deterioration in performance.

The price we pay for the benefits reaped at query time is largely in terms of extra effort during training in estimating alignments, cost and cross-representation stabilities. Thus the system eagerly estimates certain parameters that characterize the utility of a case in various representations, so that these can be harnessed at query time to flexibly choose representations based on query specific characteristics. Given that the system prefers a simpler representation over a more complex one unless it reckons that the latter is expected to generate significant improvements in effectiveness, we can say that it strives to be as lazy as possible at query time, by making use of its eagerness during training (recording of instances). This observation, incidentally inspires the title of the paper.

# 9    Conclusion

The paper presents a novel framework for knowledge integration which, unlike traditional approaches, estimates test case complexity to decide on the appropriate choice of knowledge representation. The basic premise is that shallow representations should be preferred over deep ones unless the latter promises substantial gains based on precomputed estimates of neighbourhood complexity. We have evaluated our framework w.r.t text classification. We have experimented with several textual datasets of varying complexity. It was encouraging to observe that not only were redundant query revisions avoided, there were also effectiveness gains in certain cases. This can be attributed to the process based reminding scheme which predicted the best representation, given a neighbourhood. A second reason is that some queries might get adversely affected when indiscriminately transformed to a revised feature space. As expected, our empirical studies revealed that, for simpler domains, shallow representations were preferred more often, whereas the trend was reversed with complex datasets. The time-budgeted evaluation revealed that it makes sense to quickly process easy documents with shallow features and process harder documents slowly with richer knowledge. The method presented has strong parallels to cognitive models of reminding, and has the potential to be applied across a cross section of CBR applications that can benefit from lazy selective knowledge integration.

# References

1. Gabrilovich, E., Markovitch, S.: Harnessing the expertise of 70,000 human editors: Knowledge-based feature generation for text categorization. J. Mach. Learn. Res. **8**, 2297–2345 (2007)
2. Fellbaum, C.: WordNet: An Electronic Lexical Database. Bradford Books, Cambridge (1998)
3. Deerwester, S., Dumais, S.T., Furnas, G.W., Landauer, T.K., Harshman, R.: Indexing by latent semantic analysis. J. Am. Soc. Inf. Sci. **41**(6), 391–407 (1990)
4. Massie, S., Wiratunga, N., Craw, S., Donati, A., Vicari, E.: From anomaly reports to cases. In: Weber, R.O., Richter, M.M. (eds.) ICCBR 2007. LNCS (LNAI), vol. 4626, pp. 359–373. Springer, Heidelberg (2007). doi:10.1007/978-3-540-74141-1_25
5. Riesbeck, C.K., Schank, R.C.: Inside Case-Based Reasoning. Lawrence Erlbaum Associates Inc., Hillsdale (1989)
6. Lamontagne, L.: Textual cbr authoring using case cohesion. In: Proceedings of 3rd Textual Case-Based Reasoning Workshop at the 8th European Conference on CBR (2006)
7. Lang, K.: Newsweeder: Learning to filter netnews. In: Proceedings of the Twelfth International Conference on Machine Learning, pp. 331–339 (1995)
8. Koller, D., Sahami, M.: Hierarchically classifying documents using very few words. In: Proceedings of the Fourteenth International Conference on Machine Learning, ICML 1997, pp. 170–178. Morgan Kaufmann Publishers Inc., San Francisco (1997)
9. Gupta, R., Ratinov, L.: Text categorization with knowledge transfer from heterogeneous data sources. In: Proceedings of the 23rd National Conference on Artificial Intelligence AAAI 2008, vol. 2, pp. 842–847. AAAI Press (2008)
10. Liu, Y., Song, R., Chen, Y., Nie, J.-Y., Wen, J.-R.: Adaptive query suggestion for difficult queries. In: Proceedings of the 35th International ACM SIGIR Conference on Research and Development in Information Retrieval, SIGIR 2012, pp. 15–24. ACM, New York (2012)
11. Hassan, A., White, R.W., Wang, Y.-M.: Toward self-correcting search engines: using underperforming queries to improve search. In: Proceedings of the 36th International ACM SIGIR Conference on Research and Development in Information Retrieval, SIGIR 2013, pp. 263–272. ACM, New York (2013)
12. Grubb, A., Bagnell, D.: Speedboost: Anytime prediction with uniform near-optimality. In: Proceedings of the Fifteenth International Conference on Artificial Intelligence and Statistics, AISTATS 2012, La Palma, Canary Islands, 21–23 April 2012, pp. 458–466 (2012)
13. Freitag, D.: Multistrategy learning for information extraction. In: Proceedings of the Fifteenth International Conference on Machine Learning, pp. 161–169. Morgan Kaufmann (1998)
14. Xu, C., Tao, D., Xu, C.: A survey on multi-view learning. CoRR, abs/1304.5634 (2013)
15. Cummins, L., Bridge, D.: Choosing a case base maintenance algorithm using a meta-case base. In: Bramer, M., Petridis, M., Nolle, L. (eds.) Research and Development in Intelligent Systems XXVIII, pp. 167–180. Springer, Heidelberg (2011)

# Personalized Opinion-Based Recommendation

Ruihai Dong$^{(\boxtimes)}$ and Barry Smyth

School of Computer Science and Informatics, Insight Centre for Data Analytics,
University College Dublin, Dublin, Ireland
ruihai.dong@ucd.ie

**Abstract.** E-commerce recommender systems seek out matches between customers and items in order to help customers discover more relevant and satisfying products and to increase the conversion rate of browsers to buyers. To do this, a recommender system must learn about the likes and dislikes of customers/users as well as the advantages and disadvantages (pros and cons) of products. Recently, the explosion of user-generated content, especially customer reviews, and other forms of opinionated expression, has provided a new source of user and product insights. The interests of a user can be mined from the reviews that they write and the pros and cons of products can be mined from the reviews written about them. In this paper, we build on recent work in this area to generate user and product profiles from user-generated reviews. We further describe how this information can be used in various recommendation tasks to suggest high-quality and relevant items to users based on either an explicit query or their profile. We evaluate these ideas using a large dataset of TripAdvisor reviews. The results show the benefits of combining sentiment and similarity in both query-based and user-based recommendation scenarios, and also disclose the effect of the number of reviews written by a user on recommendation performance.

**Keywords:** Recommender systems · Opinion mining · Sentiment analysis · Personalization

## 1 Introduction

Recommender systems help to provide users with the right information at the right time. They do this by profiling a user's interests and preferences over time and use these profiles to select and/or rank items for presentation by preferring those that are similar to the ones the user has liked in the past. In an e-commerce setting, recommender systems endeavour to learn from our past purchasing habits in order to identify new products that we may wish to purchase in the future. Getting this right can mean improved conversion rates for the retailer and improved satisfaction levels for the shopper. More satisfied customers are more loyal customers which improves the likelihood of repeat business, a win-win for customer and retailer alike.

Different types of recommendation techniques rely on different types of data. For example, collaborative filtering [1], either neighbourhood methods [2,3] or

© Springer International Publishing AG 2016
A. Goel et al. (Eds.): ICCBR 2016, LNAI 9969, pp. 93–107, 2016.
DOI: 10.1007/978-3-319-47096-2_7

latent factor models [4], relies on product ratings. These ratings may be explicitly provided by users, or they may be inferred from their behaviour. And patterns in these ratings are used to identify similar users and relevant products for recommendations.

Content-Based recommendation [5] is another important approach for building recommender systems, based on the availability of specific item knowledge. For example, product meta-data or catalog-data may be used to characterise products in terms of specific features (type, price, size, weight etc.). When this type of product data is available the preferences of users can be expressed in terms of the products that they have purchased and the features of these products. Recommendations can then be based on various forms of feature similarity, by recommending products that are *similar* to the user's profile. This approach obviously shares many commonalities with case-based reasoning methods and in fact many case-based recommendation techniques have been proposed based on representation, similarity, and retrieval ideas from the CBR community [6,7].

In the past, a big challenge when building recommender systems has been ensuring the availability of and access to these different sources of data. Sometimes product data is hard to come by, limiting the efficacy of content/case-based methods. And the type of rating data used by collaborative filtering approaches can be notoriously sparse. This has led researchers to develop hybrid models that make the best of both worlds [8] and also led to the exploit of auxiliary information; see [9,10].

In recent years, a new alternative data source has emerged with the ubiquity of user-generated reviews. The intuition is that these reviews contain important product knowledge and valuable customer preferences and insights for use in recommendation; see [11,12]. But the information contained within user-generated reviews can be noisy and unstructured. Nevertheless, opinion mining and natural language processing techniques (for example, see [13]) are now robust enough for researchers to use user-generated review content as an alternative (or complementary) source of recommendation knowledge [14].

User-generated reviews are plentiful and they contain rich product feature information including sentient information. For example, *The Thai red curry was delicious, ..., but price at $22, quite expensive*, tells us that the restaurant in question serves a very tasty Thai red curry (positive sentiment) that costs $22 but it is expensive (negative sentiment). This combination of traditional feature-value information and sentiment means that we can generate recommendations that are not only *similar* to those a user has liked in the past, but that are also *better* based on features that matter to the user; see [15,16].

The features mentioned by a user in her reviews may reflect her preferences. Intuitively, if a feature is mentioned many times by a user, it may indicate that it is an important one. Musat et al. [17] built a user interest profile for each user based on the topics mentioned in their reviews and used these profiles to produce personalized product rankings. Liu et al. [18] built user preferences based on the assumption that a user may have a higher requirement for a feature if she frequently gives a lower score for the feature compared to other users; see also the work of [19] on estimating reviewer weights at the aspect (feature) level.

In this paper, we focus on a pair of use-cases in the context of a hotel recommendation site such as TripAdvisor. Specifically, we imagine a traveller attempting to book a suitable hotel for an upcoming trip to a new city. In the first use-case (*more-like-this*) we consider the common scenario where the traveler has a previous hotel that they have liked but in a different city and they wish to look for something similar in their target city. In this use-case, the previous hotel becomes a target query and is used to select and rank hotels from the target city. We describe a *query-based* technique to show how the combination of similarity and sentiment can be used to recommend better hotels than similarity alone. In our second use-case (*personalization*) we use the user's profile as an implicit query in order to recommend hotels from the target city, again looking at the benefits of using similarity and sentiment; we refer to this as a *user-based* approach.

## 2   Mining User Profiles and Item Cases

This paper extends recent work on mining opinions from user reviews to generate user profiles and item descriptions. The work of [15, 20, 21] is especially relevant and describes how shallow NLP, opinion mining, and sentiment analysis can be used to extract rich feature-based product descriptions (product *cases*) based on the features that users mention in their reviews, and also the polarity of their opinions. In order to generate user profiles and hotel descriptions, we follow four basic steps:

1. Mine hotel features from user-generated reviews (extending the techniques described in [15]) to generate a set of *basic* hotel features and sentiment.
2. Apply clustering techniques to group related basic features together and use these clusters as *high-level* features. We refer to these as *clustered* features.
3. Aggregate clustered features and sentiment mined from the reviews of a hotel to generate *hotel cases*.
4. Aggregate clustered features for users to generate user profiles; that is, the features mined from the reviews of a given user form the basis of this user's profile.

This is summarized in Fig. 1 and produces a set of hotel cases and user profiles/cases which are used as the basic recommendation data for our recommender system as shown. In the next section we will describe the details of the recommendation process but first, in what follows, we will provide additional details about this feature mining and case generation process.

### 2.1   Features and Sentiment

The feature extraction and sentiment analysis approach adopted in this work are based closely on the approach taken by [15]. As such we mine *bi-gram* (adjective-noun) and *single-noun* features using a combination of shallow NLP and statistical methods.

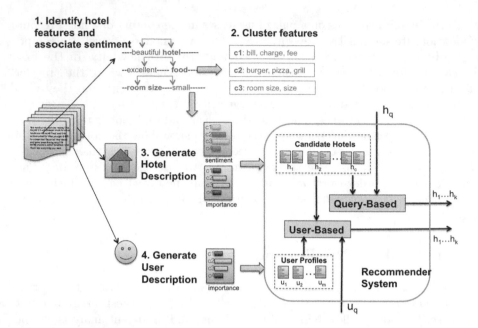

**Fig. 1.** High-level opinion mining architecture: basic features are extracted from user generated reviews; these features are clusterd into related sets; these clustered features are used as the basis for hotel/item cases and user profiles/cases.

In the current work we extend this approach to also consider additional *tri-gram* features by using the patterns described in the work of [22]. This allows us to find features like *member of staff, bottle of water*.

These single-noun, bi-gram, and tri-gram features are filtered using a number of statistical techniques as described previously in [15]. For example, we eliminate certain bi-gram features if the adjective is a common sentiment word; this way we can eliminate false bi-gram features such as *great location, excellent service*, which are really single-noun features.

Next, for each extracted (basic) feature, $f_i$, mentioned in some sentence $s_j$ of review $r_k$, we must evaluate its sentiment. We do this by finding the closest *sentiment word* $w_{min}$ to $f_i$ in $s_j$; sentiment words are available from a sentiment lexicon and for the purpose of this work we rely on the common Bing Lui sentiment lexicon [23]. Note, if no sentiment words are present in $s_j$ then the feature $f_i$ is labeled as *neutral*.

In order to further filter the basic features, we use the opinion pattern mining technique described by [15]. To do this we identify the part-of-speech (POS) tags for $w_{min}$, $f_i$, and any words that occur between them. These POS tags constitute an abstract opinion pattern. After a complete pass over all features, the frequency of occurrence of all of these opinion patterns is computed. Patterns that occur more than $k$ times are considered valid, on the grounds that less frequent patterns are likely to correspond to unusual or convoluted expressions of opinion that cannot be reliably understood. In this case, we set $k = 1$ and for each valid feature, $f_i$ (that is, a feature which corresponds to a valid opinion pattern) we assign it a sentiment label based on the sentiment of its $w_{min}$. Finally, we reverse the sentiment of $f_i$ if $s_j$ contains a negation term within a 4-word distance of $w_{min}$.

The end result of this process is, for each review, a set of valid features (mentioned in the review) and their sentiment labels. These reviews are associated with specific hotels and specific users and so their features and sentiment contribute to the descriptions of these hotels and the profiles of the users; we will return to this later.

## 2.2   Clustering Features

One of the problems with the above approach is that it tends to produce a proliferation of features. In reviews, people may refer to the same types of things in a variety of ways. For example, in hotel reviews, some reviewers will comment on the *elevators* while others will talk about *lifts*. One reviewer might comment on the quality of the *pillows* while another may comment on the comfort of the *bed* or the softness of the *matress*. Rather than treating these as separate and independent features it is more natural and useful to recognize that they are referring to the same aspect of a hotel.

```
Cluster 1  - location, walking distance, distance, minutes work, ...
Cluster 2  - bill, charge, fee
Cluster 3  - burger,pizza,grill
Cluster 4  - chip, chicken, salad, fish, steak, meat
Cluster 5  - egg, bacon, waffle, pancake, sausage
Cluster 6  - cake, afternoon tea, sandwich
Cluster 7  - bar, beer, drink, cocktail
Cluster 8  - bed, pillow, mattress
Cluster 9  - bathroom, tube, toilet, shower, bath
Cluster 10 - stair, lift, elevator
Cluster 11 - sofa, king-bed, couch, living room
Cluster 12 - desk staff, hotel staff
```

**Fig. 2.** Examples for clustered features

To do this we apply standard clustering techniques to group related features together based on their similarities. Firstly, we associate each basic feature with a term-vector made up of the set of words extracted from the sentences that refer to this feature; these words are converted to lowercase, stemmed, and stop words are removed. Thus, each feature $f_i$ is associated with a set of terms and the value of each term is a normalized term frequency weight. Next, we apply a standard clustering algorithm empirically setting the target number of clusters to be 35 % of the total number of features; we used CLUTO[1]. The result is a set

---

[1] http://glaros.dtc.umn.edu/gkhome/views/cluto.

of feature clusters such as the examples in Fig. 2. These clusters act as high-level features and they are used as the basis for generating hotel and user cases as we shall describe in the next sections.

## 2.3    Generating Hotel Cases

Each item/hotel ($h_i$) is associated with a set of customer reviews $reviews(h_i) = \{r_1, \ldots, r_n\}$ and the above processes extract a set of features and, ultimately, clusters, $c_1, \ldots, c_m$, from these reviews. Each cluster is comprised of a set of basic features and in effect, acts as a type of abstract feature, a *clustered feature*, such that the basic features it contains are related in some way. For example, in Fig. 2 we can see that the features of *Cluster 1* are all related to the location of the hotel. The features in clusters 4, 5, and 6 are all related to food but each is separately related to *dinner*, *lunch*, or *breakfast* meals.

Effectively each clustered feature is labelled with the cluster id, $c_j$, and it is assigned an *importance* score and a *sentiment* score as per Eqs. 2 and 3. Then a hotel/item case is made up of the clustered features associated with its reviews and their corresponding importance and sentiment scores; see Eq. 1. Note that $c_j \in reviews(h_i)$ means that any of the basic features in $c_j$ are present in $reviews(h_i)$.

$$item(h_i) = \{(c_j, s(c_j, h_i), imp(c_j, h_i)) : c_j \in reviews(h_i)\} \tag{1}$$

The importance score of $c_j$, $imp(c_j, h_i)$, is the relative number of times that $c_j$ (or rather its basic features) is mentioned in the reviews of hotel $h_i$.

$$imp(c_j, h_i) = \frac{\sum_{f_k \in c_j} count(f_k, h_i)}{|reviews(h_i)|} \tag{2}$$

The sentiment score of $c_j$, $s(c_j, h_i)$, is the degree to which $c_j$ (that is, its basic features) is mentioned positively or negatively in $reviews(h_i)$. Note, $pos(f_j, h_i)$ and $neg(f_j, h_i)$ denote the number of mentions of the basic feature $f_j$ labeled as positive or negative during the sentiment analysis phase.

$$s(c_j, h_i) = \frac{\sum_{f_k \in c_j} pos(f_k, h_i) - \sum_{f_k \in c_j} neg(f_k, h_i)}{\sum_{f_k \in c_j} pos(f_k, h_i) + \sum_{f_k \in c_j} neg(f_k, h_i)} \tag{3}$$

## 2.4    Generating User Profile Cases

Just as we generate hotel cases from the reviews written about a specific hotel we can generate user profile cases using the reviews written by a specific user, $u_q$. Equation 4 defines each user as a set of clustered features and each feature is associated with an importance score based on how frequently that user refers to that feature in her reviews; this is exactly analogous to the hotel cases. However, unlike the hotel cases, we do not associate any sentiment with the user profile features. The reason for this is that the sentiment that a user expresses for a given feature is a property of the feature in the context of a specific hotel, rather than a general property of the feature; we will revisit this in future work.

$$user(u_q) = \{(c_j, imp(c_j, u_q)) : c_j \in reviews(u_q)\} \qquad (4)$$

## 3   Ranking and Recommendation

Unlike traditional content-based recommenders — which tend to rely exclusively on similarity in order to rank products with respect to some user profile or query — our approach leverages feature sentiment *and* similarity, during recommendation. The starting point is the work of [15] which uses a linear combination of similarity and sentiment during recommendation ranking. Briefly, when it comes to recommending some candidate item $i$ we can compute a recommendation score based on how similar the item is to the query $q$, and based on the sentiment of the item. And as per Eq. 5 we can adjust the relative influence of similarity and sentiment using the parameter $w$.

$$Score(q, i) = (1 - w) \times Sim(q, i) + w \times Sent(q, i) \qquad (5)$$

In the present work we use clustered features as the basis of our item descriptions, rather than the basic features used by [15]; we will usually refer to these clustered features as just *features* in what follows. We will also describe a personalized version of ranking, which harnesses user profiles that are *better* with respect to features that matter to the query user.

### 3.1   Query-Based Recommendation

To begin with we will implement a standard non-personalized ranking approach, similar to [15], albeit based on clustered features. We imagine an user $u_q$ has a hotel in mind, $h_q$. Perhaps it is a hotel he has stayed in before and he is looking for something similar in a new city. We use $h_q$ as a query and we compare it to candidate items $h_c$, computing the similarity and sentiment values to score each $h_c$ for ranking and recommendation to $u_q$. For the purpose of similarity assessment we use a standard cosine metric; see [5]. Equation 6 demonstrates this for $h_q$ and $h_c$. Note that we use the importance scores of shared features as the feature values.

$$Sim_h(h_q, h_c) = \frac{\displaystyle\sum_{c_i \epsilon C(h_q) \cap C(h_c)} imp(c_i, h_q) \times imp(c_i, h_c)}{\sqrt{\displaystyle\sum_{c_i \epsilon C(h_q)} imp(c_i, h_q)^2} \times \sqrt{\displaystyle\sum_{c_i \epsilon C(h_c)} imp(c_i, h_c)^2}} \qquad (6)$$

Next, we need to calculate the sentiment score for $h_c$. As mentioned earlier, sentiment information is unusual in a recommendation context but it makes it possible to consider not only how similar $h_c$ is to $h_q$ but also whether it enjoys better sentiment; it seems reasonable to recommend items that are not only similar to $h_q$ but have also been more positively reviewed. We do this based on a feature-by-feature sentiment comparison as per Eq. 7. We can say that $c_i$ is *better* in $h_c$ than $h_q$ ($better(c_i, h_q, h_c) > 0$) if $c_i$ in $h_c$ has a higher sentiment score than it does in $h_q$.

$$better(c_i, h_q, h_c) = s(c_i, h_c) - s(c_i, h_q) \tag{7}$$

We calculate the sentiment score, $Sent(h_q, h_c)$ from the sum of these better scores for the features that are common to $h_q$ and $h_c$ as per Eq. 8; we use $C(i)$ to denote the clustered features of item $i$.

$$Sent(h_q, h_c) = \frac{\sum_{c_i \in C(h_q) \cap C(h_c)} better(c_i, h_q, h_c) \times imp(c_i, h_c)}{|C(h_q) \cap C(h_c)|} \tag{8}$$

Accordingly we can implement a non-personalized scoring function based on the above by combining $Sim_h$ and $Sent$ as per Eq. 9.

$$Score_{QB}(h_q, h_c) = (1 - w) \times Sim_h(h_q, h_c) + w \times Sent(h_q, h_c) \tag{9}$$

## 3.2    User-Based Recommendation

Rather than using a specific item ($h_q$) as a query to trigger recommendations, another common recommendation use-case, is to use the user profile $u_q$ as a query. To do this we need to implement a personalized version of the approach above, by introducing user preference information into both the similarity and the sentiment calculations.

Regarding similarity, we implement a version in which we use the importance weights from the query user $u_q$ instead of the weights from $h_q$ during similarity assessment as per Eq. 10. In this way, features that are more important to $u_q$ and $h_c$ play a greater role in the similarity computation.

$$Sim_u(u_q, h_c) = \frac{\sum_{c_i \in C(u_q) \cap C(h_c)} imp(c_i, u_q) \times imp(c_i, h_c)}{\sqrt{\sum_{c_i \in C(u_q)} imp(c_i, u_q)^2} \times \sqrt{\sum_{c_i \in C(h_c)} imp(c_i, h_c)^2}} \tag{10}$$

Regarding sentiment, we propose Eq. 11, which calculates a sentiment score based on the average sentiment of all hotels visited by $u_q$ with $h_c$. We use $H(u_q)$ to denote the hotels visited by $u_q$. Thus, the sentiment score of $h_c$ is influenced by $u_q$'s history. Obviously, this is just one way that we might make the sentiment calculation more personalized for $u_q$. In this work it serves as a useful test-case and future work will consider alternative approaches.

$$Sent_u(u_q, h_c) = \frac{\sum_{h_q \in H(u_q)} Sent(h_q, h_c)}{|H(u_q)|} \tag{11}$$

For now, this means we can implement Eq. 12 as the recommendation scoring function for generating personalized recommendations.

$$Score_{UB}(u_q, h_c) = (1 - w) \times Sim_u(u_q, h_c) + w \times Sent_u(u_q, h_c) \tag{12}$$

# 4    Evaluation

In this section, we will describe an initial evaluation of these query-based and user-based recommendation techniques using real-user data.

## 4.1    Datset and Setup

The dataset used in this work is based on the TripAdvisor dataset shared by the authors of [15]. This dataset covers 148,575 users for 1,701 hotels. We collected the member pages of these users from TripAdvisor website, we found that these users totally have written 1,008,585 hotel reviews until July, 2014; approximately 7 reviews per user, although some users have authored significantly more. For the purpose of this work, we focus on a subset of 1,000 users who have written at least 5 hotel reviews. This provides a test dataset of 11,993 reviews for 10,162 hotels. Finally, for each of these hotels, we collected up to its top 100 reviews for a total of 867,644 reviews; some hotels do not have 100 reviews.

For each of these users and hotels, we apply opinion mining to generate our feature-based descriptions. On average our test users have written 12 reviews resulting in profiles containing an average of 91 different clustered features. Likewise, the hotels are associated with an average of 89 reviews resulting in 189 clustered features per hotel on average. Clearly the opinion mining is capable of generating rich item descriptions. In what follows we will evaluate the query-based and user-based recommendation techniques using a standard leave-one-out style approach in which we attempt to recommend a specific target hotel given a query hotel or a user profile.

## 4.2    Evaluating Query-Based Recommendation

To evaluate the non-personalized, query-based recommendation strategy we produce a set of *test triples* of the form $(u_q, h_q, h_t)$ corresponding to a query user $u_q$, a query hotel from $u_q$'s profile and a target hotel, $h_t$, that is also in the user's profile. $h_q$ and $h_t$ are in different cities to simulate the use-case of the user looking for a hotel in some new city but based on a familiar hotel in a different city. For the purpose of this test we are careful to choose $h_t$ from a set of 8 cities in our dataset which have sufficient candidate hotels that are also in our dataset($>80$). In each triple $h_t$ is chosen only if it has been rated as 5-star by $u_q$; we assume the user is looking for a hotel they will like.

Furthermore, we note the user's rating for each of the query hotels in our test triples and distinguish between those query hotels that have a low (2-star or 3-star) rating by the user and those that have a much higher (4-star or 5-star) rating; we will refer to these as *23-queries* and *45-queries*, respectively. This allows us to compare the performance of the ranking based on how well the query user liked the query hotel; we might expect that it is easier to identify $h_t$ if we are starting from a 45-query than a 23-query. In this sense, the 23-queries correspond to more a challenging query-based recommendation task than the 45-queries.

This provides us with 888 test triples; 705 have 45-queries and the remaining 183 have 23-queries. Each triple is a recommendation test, the objective of which is to locate $h_t$ based on a ranking of hotels (from the same city as $h_t$) using $h_q$ as a query. In other words, for each triple we use $h_q$ as a query hotel and rank-order the other hotels in the same city as $h_q$ using $Score_{QB}$, varying $w$ to adjust the mix of similarity and sentiment.

The results are presented in Fig. 3(a) as a graph of the top-20 hit-rate – the percentage of times the target hotel is within the top-20 recommendations – versus $w$. As we increase $w$ (that is, increase the influence of sentiment) the hit-rate of the query-based algorithm improves for both 45- and 23-query groups. For example, at $w = 0$ (pure similarity-based scoring) we can see that the hit-rate is 0.20 for the 23-queries and 0.27 for the 45-queries. In other words $h_t$ is in among the top-20 recommendations between 20 % and 27 % of the time. As expected, we can more successfully recommend the target hotel when using a 45-query than a 23-query.

As we increase $w$ up to about 0.6–0.8 then this hit-rate increases to 0.35. That is, as we introduce more sentiment we improve hit-rates, from 27 % (45-queries) to 35 %, a relative increase of 30 % compared to the similarity-only setting at $w = 0$. Indeed, the improvement is even more striking for the more challenging 23-queries: we see a relative improvement of 75 % as hit-rate climbs from 20 % to as high as 35 % (at $w = 0.8$). Even when the query is a relatively poor starting point, as a 23-query is, the introduction of sentiment helps to produce a hit-rate that is comparable to the top hit-rate achieved by the 45-queries (with sentiment). Figure 3(a) also shows the average results across all queries (combining 23/45 query groups) where hit-rate climbs from 26 % ($w = 0$) to 35 % ($w = 0.6$) for a relative increase of about 35 %.

There is a point after which more sentiment causes a disimprovement in hit-rate as the influence of similarity is no longer felt and sentiment tends to dominate. This point is $w = 0.6$ for 45-queries (and, on average, for all queries). It occurs later ($w = 0.8$) for the 23-queries. This suggests that sentiment should be relied on more when the query hotel is not a strong reflection of a user's true interests, as is the case with 23-queries. This makes sense: if these query hotels are not a good reflection of a user's true interests then it may not be appropriate to rely on similarity, it would be much better to focus on recommendations that are better than the query hotels by as much as possible.

For comparison, in Fig. 3(b) we show the relative ranking of the target hotels in the recommendation lists. In this case, a *lower* relative ranking score means that the target hotel appears *higher* in the recommendation rankings. For example, across all queries we see that the relative ranking of the target hotel falls from being in the top 37 % of recommendations to the top 30 % of recommendations as $w$ is increased up to 0.6. These relative ranking results are entirely consistent with the previous hit-rate results. They show an improvement in recommendation performance with increasing $w$ up to $w = 0.6 - 0.8$ depending on the query set. Once again we see improved recommendation performance for the 45-queries compared to the tougher 23-queries, but the introduction of sentiment dramatically improves recommendation even for these 23-queries.

So far we have been looking for a single target hotel. That is not unusual in these types of offline, leave-one-out tests, but the results remain silent when it comes to the quality of the other recommendations. For example, what is the average rating of the top-20 recommended hotels? And how does this change with $w$? This is important because it tells us about the overall quality of the recommendation lists produced. This information is shown in Fig. 3(c) as the average TripAdvisor rating for the top-20 recommendations as we increase $w$.

Once again we can see that there is a benefit when it comes to introducing sentiment: as we increase $w$ the average rating of the recommended hotels increases from about 4-stars to over 4.5-stars; at $w = 0.6$, the turning point for hit-rate and relative ranking, the average rating is almost exactly 4.5-stars. As expected, we can see that the average rating for the 'easier' 45-queries is slightly higher than the average rating for the 23-queries. Note that there is no turning point on this average rating graph, as there was previously. The reason for this is that as we increase $w$ we are guiding recommendation towards hotels with more and more positive features and so we can expect their average ratings to increase accordingly. And as they do, the average ratings is up to a maximum of about 4.6-stars, which presumably is at, or close to, the maximum possible rating for a list of 20 hotels given the TripAdvisor rating distribution. Obviously, this does not mean that we should turn-up $w$ to its maximum level because if we do then there is a cost when it comes to retrieving a particular target case. And these results suggest a balancing point of $w = 0.6$ is close to optimal for this type of query-based recommendation, at least in this domain.

### 4.3   Evaluating User-Based Recommendation

We follow a similar approach to evaluate the personalized, user-based recommendation strategy. This time we use a set of user-item *test pairs*, each containing a user profile $u_q$, as a query, and a target item $h_t$. In each case $h_t$ is one of the hotels that $u_q$ has previously rated as 5-stars. This time our recommendation test will be to use $u_q$ as a query to recommend $h_t$.

There are 665 of these test pairs. Earlier we distinguished between hard and easy queries, we do similar here by dividing user profiles into *small*, *medium*, and *large* based on their number of reviews; see Table 1. For example, small profiles have up to 10 reviews and there are 274 pairs from these profiles involving 249 unique hotels and 211 unique users. The intuition is that small profiles

**Table 1.** Testing pairs for user-based recommendation

|                  | Pairs | Hotels | Users |
|------------------|-------|--------|-------|
| All              | 665   | 526    | 461   |
| Small (<=10)     | 274   | 249    | 211   |
| Medium (11–20)   | 238   | 215    | 166   |
| Large (>20)      | 153   | 142    | 84    |

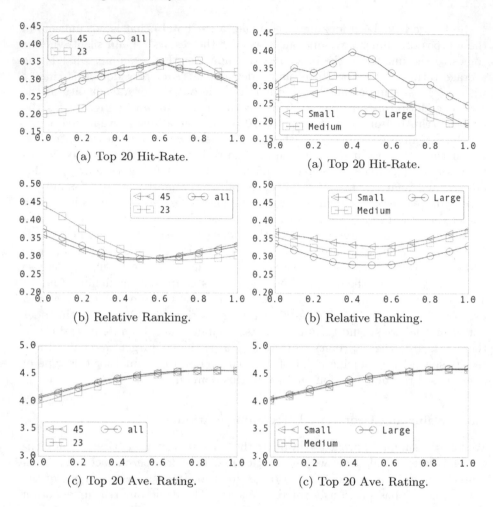

(a) Top 20 Hit-Rate.

(a) Top 20 Hit-Rate.

(b) Relative Ranking.

(b) Relative Ranking.

(c) Top 20 Ave. Rating.

(c) Top 20 Ave. Rating.

**Fig. 3.** Query-based recommendation.    **Fig. 4.** User-based recommendations.

will represent a tougher user-based recommendation test than medium or large profiles.

Each test pair defines a recommendation test in which $u_q$ is used to rank and recommend hotels from the same city as $h_t$ using $Score_{UB}$. And in this test, for each test pair, we re-generate $u_q$ without reviews from target hotel.

We calculate the top-20 hit-rate for different values of $w$, and the results are presented in Fig. 4(a). As before we can see how increasing $w$ tends to improve hit-rate. For example, at $w = 0$ the hit-rate for large profiles is about 30 % and this grows to 40 % (at $w = 0.4$), a relative improvement of about 33 %. A similar effect is noted for medium and small profiles but, as expected, the size of the effect is reduced.

Once again there is an optimal w in range of $w = 0.4 - 0.5$, that seems to deliver an optimum hit-rate. Beyond this, as sentiment dominates, the hit-rate falls sharply, eventually dropping below the hit-rate achieved at $w = 0$ (similarity only).

Relative ranking results are also presented in Fig. 4(b). They are consistent with the hit-rate results, showing a benefit to increasing $w$ up to about 0.4–0.5 and with the best relative ranking accruing to the larger profiles.

For completeness, we also show the average TripAdvisor rating for the full set of 20 recommendations as $w$ varies. Once again we see a gradual increase in recommendation quality for increasing $w$; the average rating of recommendation lists increases from 4-stars to about 4.6-stars.

### 4.4   Discussion

We have evaluated the performance of our query-based and user-based recommendation approaches on real-user data from TripAdvisor. In each case, we have explored the benefit of using sentiment during recommendation for queries of different difficulty levels. The results have been very consistent across a number of evaluation metrics. In each case, increasing sentiment can significantly improve recommendation performance to a point. Although it is likely that the best level of sentiment to include will likely be domain and task dependent, it is equally likely, we believe, that selecting a reasonable value for $w$, such as $w = 0.5$, should deliver significant recommendation improvements over more conventional similarity-based techniques. Fine-tuning might improve this further but it may not be necessary or particularly worthwhile in many settings.

## 5   Conclusions

This paper builds on recent work using product reviews as a novel source of recommendation knowledge; see [15]. Our main contribution has been to propose and evaluate a new, personalized, user-based recommendation approach that is capable of generating proactive recommendations for a user based on their mined preferences. In the process, we have modified the work of [15] by introducing tri-gram features and a feature-clustering step during opinion mining in order to better capture the relationship between basic features.

We have evaluated these approaches in an offline study using real-user evaluation data from TripAdvisor. The results show the benefits of mixing sentiment and similarity during recommendation in both query-based and user-based recommendation scenarios. Profile size is important and the size of the effect on recommendation performance is closely connected to the number of reviews that have contributed to a profile.

Extending the evaluation to larger datasets and/or new domains is one priority for future work. We also plan to explore new approaches to opinion mining by using topic-modeling techniques to better understand hotel preferences and trip purposes. Our intuition is that the probability distribution of topics in a review is a mixture of the distributions associated with the user interests, travel purpose, and the hotel features. We are currently exploring ways to learn these topic models from review texts and incorporate this into recommendations.

**Acknowledgments.** This work is supported by Science Foundation Ireland through the Insight Centre for Data Analytics under grant number SFI/12/RC/2289.

# References

1. Schafer, J.B., Frankowski, D., Herlocker, J., Sen, S.: Collaborative filtering recommender systems. In: Brusilovsky, P., Kobsa, A., Nejdl, W. (eds.) The Adaptive Web. LNCS, vol. 4321, pp. 291–324. Springer, Heidelberg (2007). doi:10.1007/978-3-540-72079-9_9
2. Resnick, P., Iacovou, N., Suchak, M., Bergstrom, P., Riedl, J.: Grouplens: an open architecture for collaborative filtering of netnews. In: Proceedings of the 1994 ACM Conference on Computer Supported Cooperative Work, CSCW 19694, pp. 175–186. ACM, New York (1994)
3. Sarwar, B., Karypis, G., Konstan, J., Riedl, J.: Item-based collaborative filtering recommendation algorithms. In: Proceedings of the 10th International Conference on World Wide Web, WWW 2001, pp. 285–295. ACM, New York (2001)
4. Koren, Y., Bell, R., Volinsky, C.: Matrix factorization techniques for recommender systems. Computer **42**(8), 30–37 (2009)
5. Pazzani, M.J., Billsus, D.: Content-based recommendation systems. In: Brusilovsky, P., Kobsa, A., Nejdl, W. (eds.) The Adaptive Web. LNCS, vol. 4321, pp. 325–341. Springer, Heidelberg (2007). doi:10.1007/978-3-540-72079-9_10
6. Bridge, D., Göker, M.H., McGinty, L., Smyth, B.: Case-based recommender systems. Knowl. Eng. Rev. **20**, 315–320 (2005)
7. Smyth, B.: Case-based recommendation. In: Brusilovsky, P., Kobsa, A., Nejdl, W. (eds.) The Adaptive Web. LNCS, vol. 4321, pp. 342–376. Springer, Heidelberg (2007). doi:10.1007/978-3-540-72079-9_11
8. Burke, R.: Hybrid recommender systems: survey and experiments. User Model. User-Adap. Interact. **12**(4), 331–370 (2002)
9. Wang, H., Chen, B., Li, W.-J.: Collaborative topic regression with social regularization for tag recommendation. In: IJCAI (2013)
10. Wang, H., Li, W.-J.: Relational collaborative topic regression for recommender systems. IEEE Trans. Knowl. Data Eng. **27**(5), 1343–1355 (2015)
11. Pero, Š., Horváth, T.: Opinion-driven matrix factorization for rating prediction. In: Carberry, S., Weibelzahl, S., Micarelli, A., Semeraro, G. (eds.) UMAP 2013. LNCS, vol. 7899, pp. 1–13. Springer, Heidelberg (2013). doi:10.1007/978-3-642-38844-6_1
12. Ganu, G., Elhadad, N., Marian, A.: Beyond the stars: improving rating predictions using review text content. In: WebDB, vol. 9, pp. 1–6. Citeseer (2009)
13. Liu, B.: Sentiment analysis and opinion mining. Synth. Lect. Hum. Lang. Technol. **5**(1), 1–167 (2012)
14. Chen, L., Chen, G., Wang, F.: Recommender systems based on user reviews: the state of the art. User Model. User-Adap. Inter. **25**(2), 99–154 (2015)
15. Dong, R., O'Mahony, M.P., Smyth, B.: Further experiments in opinionated product recommendation. In: Lamontagne, L., Plaza, E. (eds.) ICCBR 2014. LNCS (LNAI), vol. 8765, pp. 110–124. Springer, Heidelberg (2014). doi:10.1007/978-3-319-11209-1_9

16. Aciar, S., Zhang, D., Simoff, S., Debenham, J.: Informed recommender: basing recommendations on consumer product reviews. IEEE Intell. Syst. **22**(3), 39–47 (2007)
17. Musat, C.-C., Liang, Y., Faltings, B.: Recommendation using textual opinions. In: IJCAI International Joint Conference on Artificial Intelligence, IJCAI 2013, pp. 2684–2690. AAAI Press (2013)
18. Liu, H., He, J., Wang, T., Song, W., Du, X.: Combining user preferences and user opinions for accurate recommendation. Electron. Commer. Res. Appl. **12**(1), 14–23 (2013)
19. Wang, F., Chen, L.: Review mining for estimating users ratings and weights for product aspects
20. Dong, R., Schaal, M., O'Mahony, M.P., Smyth, B.: Topic extraction from online reviews for classification and recommendation. In: Proceedings of the 23rd International Joint Conference on Artificial Intelligence, IJCAI 2013, pp. 1310–1316. AAAI Press (2013)
21. Dong, R., O'Mahony, M.P., Schaal, M., McCarthy, K., Smyth, B.: Combining similarity and sentiment in opinion mining for product recommendation. J. Intell. Inf. Syst. 1–28 (2015)
22. Justeson, J.S., Katz, S.M.: Technical terminology: some linguistic properties and an algorithm for identification in text. Nat. Lang. Eng. **1**(01), 9–27 (1995)
23. Hu, M., Liu, B.: Mining opinion features in customer reviews. In: Proceedings of the 19th National Conference on Artifical Intelligence, AAAI 2004, pp. 755–760. AAAI Press (2004)

# Concept Discovery and Argument Bundles in the Experience Web

Xavier Ferrer[1,2](✉) and Enric Plaza[1]

[1] Artificial Intelligence Research Institute (IIIA), Spanish National Research Council (CSIC), Campus U.A.B., Bellaterra, Catalonia, Spain
{xferrer,enric}@iiia.csic.es
[2] Universitat Autònoma de Barcelona, Bellaterra, Catalonia, Spain
http://www.iiia.csic.es

**Abstract.** In this paper we focus on a particular interesting web user-generated content: people's experiences. We extend our previous work on aspect extraction and sentiment analysis and propose a novel approach to create a vocabulary of basic level concepts with the appropriate granularity to characterize a set of products. This concept vocabulary is created by analyzing the usage of the aspects over a set of reviews, and allows us to find those features with a clear positive and negative polarity to create the bundles of arguments. The argument bundles allow us to define a concept-wise satisfaction degree of a user query over a set of bundles using the notion of fuzzy implication, allowing the reuse experiences of other people to the needs a specific user.

**Keywords:** Experience web · Sentiment analysis · Arguments · Aspect extraction · Basic level concepts

## 1 Introduction

Our work is developed in the framework of the Experience Web [9]. This framework proposed to enlarge the paradigm of Case-based Reasoning (CBR), based on solving new problems by learning from past experiences, and include all forms of experiences about the real world expressed in the web as user-contributed content. The final goal is to *reuse* this collective experience in helping new individuals (the "users") in taking a more informed decision according to their preferences, which can be different from the preferences of the individuals who have expressed their experiences on the web. Relating these two extreme points, from numerous but varied individual experiences to a specific user request, is the overall goal of Experience Web approach, and this paper presents a complete instance of the approach.

In this approach, we focus on praxis and usage, and we want to analyze how users express their experiences about daily life; in this paper we will focus on the usage of digital cameras. A main goal is to discover the vocabulary they use, which need not be the same as the classical feature list describing the different

© Springer International Publishing AG 2016
A. Goel et al. (Eds.): ICCBR 2016, LNAI 9969, pp. 108–123, 2016.
DOI: 10.1007/978-3-319-47096-2_8

aspects of a camera (e.g. 4 GB RAM). Our goal is to use this vocabulary to elucidate the main pros and cons of each camera, according to the user reviews. To this end, we analyze textual reviews of user experiences with digital cameras and identify the set of aspects the users use and the polarity of the sentiment words associated with them [13,14]. Aspects are grouped in basic level concepts, creating a new concept vocabulary, to overcome the disparate granularity of the extracted aspects. Those concepts with a strong positive polarity over the set of reviews of a product are considered pros, while those with a strong negative polarity are considered cons.

We call a bundle of arguments the set of main pros and cons of a camera. We take this approach, already envisioned in [9], because the pros and cons allows us to reuse the knowledge for other users with other individual preferences. To support this reuse, we introduce the notion of query satisfaction by a bundle of arguments. The query expresses a new individual knowledge about her preferences (e.g. she's a travel photographer and needs long battery life).

The paper is organized as follows: Sect. 2 describes the discovery of basic level concepts from user reviews. Next in Sects. 3 and 4 we present the three different types of argument bundles and define a user query. Evaluation results are presented in Sect. 5, followed by related research in Sect. 6, and conclusions in Sect. 7.

## 2 Aspects and Basic Level Concepts

In our previous work on social recommender systems we harnessed knowledge from product reviews, and characterized every product by a set of aspect-sentiment pairs extracted from its reviews [13]. Based on these characterizations, we ranked and selected the most useful aspects for recommendation [14]. However, even after identifying the most useful aspects for recommendation, we still processed synonymous aspects and aspects referencing the same concept (such as *sensor* and *cmos*) as different aspects, adding noise to the recommendation process.

In this work, we use a similar approach to [13] in order to extract the set of salient aspects used to define important characteristics of photographic digital cameras. We call *aspect vocabulary* $\mathcal{A}$ the set of extracted aspects. However, instead of characterizing the products directly by the aspect vocabulary, we group them in *basic level concepts*. According to Rosch et al. [10], basic level concepts (BLC) are those that strike a tradeoff between two conflicting principles of conceptualization: inclusiveness and discrimination. They found that there is a level of inclusiveness that is optimal for human beings in terms of providing optimum cognitive economy. This level of inclusiveness is called the basic level, and concept or categories at this level are called basic-level concepts.

Research in the field of identifying basic level concepts is mostly oriented to improve the *word sense disambiguation* task. For instance, the class-based word sense disambiguation [6] approach requires to mark words by hand in a corpus as pertaining to one semantic class, that is interpreted as one BLC.

Once the corpus is marked, several supervised classifiers are trained to assign the proper semantic class to each ambiguous word. In our approach, we create a collection of basic level concepts in an unsupervised way from the review corpus, where each BLC assembles a set of aspects that, according to our analysis, are used in a similar way by the reviewers. As we show in Sect. 2.1, we estimate this similarity by taking into account semantic similarity and evaluating the coherence/incoherence of the sentiment values of the aspects assembled in a given BLC. Synonymy is a special case of aspects being semantically equivalent.

Consider, for instance, these three aspects in $\mathcal{A}$: *picture*, *pic* and *jpeg*. One may surmise people using those words in reviews are in fact referring to the same basic level concept, i.e. *the picture obtained by my digital camera*. Thus, we could consider that different reviews in the corpus using those words are referring to the same BLC, because they have the same intended meaning.

In this section we present a method to create a concept vocabulary $\mathcal{C}$ formed by a collection of BLCs. This concept vocabulary is useful to practically reuse other people's experiences with digital cameras because it abstracts the concrete terms used in the corpus as given by the aspect extraction approach. The creation of a collection of basic level concepts consist of three steps: (1) identifying synonymous aspects, (2) building a hierarchical clustering using the semantic, syntactic and sentiment similarities among aspects, and (3) creating a concept vocabulary $\mathcal{C}$ of basic level concepts from the hierarchical clusters.

## 2.1   Hierarchical Clustering of Aspects

The first step is to identify the synonyms of the aspects in the aspect vocabulary $\mathcal{A}$ using WordNet, a lexical database of English. Every aspect $a$ in $\mathcal{A}$ is mapped to the corresponding WordNet *synset* with the same noun word form, if it exists, and is disambiguated by identifying the synset with the shortest aggregated WordNet *Path Distance* [7] to a set of manually selected WordNet synsets formed by the top 5 most frequent aspects of the aspect vocabulary. The aspects that have a synonymy relation among them are grouped together into *aspect groups* $G_j$. Aspects without synonyms form a group of cardinality 1.

Next, we iteratively cluster the most similar groups of aspects and create a dendrogram. The set of basic level concepts will be selected from that dendogram. To cluster the aspect groups we use an unsupervised bottom-up hierarchical clustering algorithm that takes the most similar pair of groups at each stage and puts them together in a higher level group. We will define now similarity measures over aspects and over groups. The similarity measure between two aspects is:

$$Sim_A(a_i, a_j) = \alpha \cdot \Gamma(a_i, a_j) + \beta \cdot \Phi(a_i, a_j) + \gamma \cdot \Lambda(a_i, a_j)$$

where $\alpha$, $\beta$ and $\gamma$ are weighting parameters in $[0, 1]$ such that $\alpha + \beta + \gamma = 1$. The values of $Sim_A$ are in $[0, 1]$. Functions $\Gamma(a_i, a_j)$, $\Phi(a_i, a_j)$ and $\Lambda(a_i, a_j)$ estimate aspect similarity in three different dimensions:

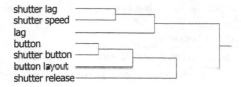

**Fig. 1.** Part of the dendrogram showing the clustering of concept *button*.

- Semantic Similarity ($\Gamma$): Compares two aspect co-occurrence vectors to estimate the similarity between aspects [11].
- String Similarity ($\Phi$): Uses the Jaro-Winkler distance to estimate the string similarity between two aspects.
- PhotoDict ($\Lambda$): *PhotoDict* is a small taxonomy of camera-related terms, where similarity is measured as the shortest path between two terms. The taxonomy is automatically generated from a camera related vocabulary existing in the Web, but its creation is out of the scope of this paper.

The similarity $Sim_G$ between two groups of aspects $G_i$ and $G_j$ is defined as:

$$Sim_G(G_i, G_j) = \frac{1}{|G_i||G_j|} \sum_{n=1}^{|G_i|} \sum_{m=1}^{|G_j|} Sim_A(a_n, a_m)$$

There is a special treatment of compound nouns in clustering. Since compound nouns are formed by two or more words (e.g. *image quality*), we group them with the most frequent aspect among the words forming the compound. The result of the hierarchical clustering is a dendrogram (or clustering tree) of aspects; Fig. 1 shows a small part of the resulting dendrogram for concept *button*. Since hierarchical clustering gives multiple partitions (clusterings) at different levels, next we have to select one partition to create our concept vocabulary.

## 2.2   Concept and Vocabulary Creation

We are interested in selecting a partition from the hierarchical clustering dendrogram that is able to describe the basic level concepts of digital cameras based on the user experiences of our corpus. The groups of aspects forming the selected partition will become our concept vocabulary $\mathcal{C}$.

To select the best partition, we cut the dendrogram at different levels. Then, for each partition, we analyze the coherence degree of the sentiment values in each aspect group. If the sentiments of the aspects of a group $G$ cohere into a clear positive, negative, or neutral value, we consider $G$ a potential basic level concept. For instance, let *picture*, *photo* and *image* be three aspects in a group. If those three aspects are used by people to refer to the same concept ('picture obtained by my digital camera'), then the sentiment values of those aspects with respect to the reviews of each product should have a high coherence degree.

**Table 1.** Three of the basic level concepts in $\mathcal{C}$ and their aspects.

| Concept name | Aspects in concept |
| --- | --- |
| Storage | storage, capacity, sd card, sdhc card, cf card |
| Button | lag, shutter release, shutter speed, shutter lag, shutter button, button, button layout |
| Battery | battery, battery life, battery pack |

The *Partition Ranking* score $R(K)$ of a partition $K$ is estimated as follows:

$$R(K) = \frac{1}{|K|} \sum_{i=1}^{|K|} IS(G_i)$$

where $|K|$ is the number aspect groups that form the partition K. The coherence degree is estimated by $IS(G_i)$, the average sentiment similarity among the aspects in a group $G_i$. The higher $R(K)$, the better the partition $K$.

The average sentiment similarity $IS$ of a group of aspects $G$ is the average cosine similarity among all pairs of aspects in $G$:

$$IS(G) = \frac{1}{|G| \cdot (|G| - 1)} \sum_{i=1}^{|G|} \sum_{j=1, j \neq i}^{|G|} cos(D(a_i), D(a_j))$$

where $cos(D(a_i), D(a_j))$ is the cosine of the angle between aspect vectors $D(a_i)$ and $D(a_j)$. An aspect vector is $D(a) = (S_{av}(p_i, a))_{i \in 1,...,|\mathcal{P}|}$, where $S_{av}(p_i, a) \in [0, 1]$ is defined as the normalized sentiment average over the set of sentences from the reviews of product $p_i$ in which aspect $a$ occurs.

In our experiments, we only considered partitions with 30 to 40 groups, a reasonable concept vocabulary size for our purposes. The partition $K$ with 36 groups, that had the highest $R(K)$, was selected. Each group of aspects is considered a basic level concept (BLC) and these 36 BLCs form the concept vocabulary $\mathcal{C}$. We will use $\mathcal{C}$ in Sect. 3 to create the bundles of arguments. Table 1 presents a small example of 3 concepts in $\mathcal{C}$ and their aspects. The *concept name* column corresponds to the most frequent aspect of each concept.

## 3    Bundle of Arguments

In this Section we characterize the set of products $p \in \mathcal{P}$ based on the concept vocabulary $\mathcal{C}$ created in previous section. Let $p \in \mathcal{P}$ be a product, $C \in \mathcal{C}$ a concept, and $Occ(p, C)$ the set of sentences from the reviews of product $p$ in which any of the aspects that form the concept $C$ appears. By analyzing the sentiment values of $Occ(p, C)$, we infer whether the people's experiences about a concept $C$ of a product $p$ have a positive or negative overall sentiment. If the overall polarity of the occurrences of a concept over the reviews of a product is positive, we consider that concept to be a *pro* argument for the product. If the overall polarity is negative, we consider that concept a *con* argument for the product. Finally, if the overall polarity of the occurrences of a concept over the reviews of a product is not clearly positive or negative, we consider the concept a *moot* argument of the product. By considering the pros, cons and moots of a product over the set of concepts in the concept vocabulary, we obtain a characterization about what people like or dislike of that product. The union of the pro, con, and moot arguments, considering all concepts in the concept vocabulary $\mathcal{C}$, form the bundle of arguments $B$ of a product $p$: $B(p) = Pros(p) \cup Cons(p) \cup Moots(p)$.

Let $Args(p) = \{Arg_i\}_{i=1,\ldots,|\mathcal{C}|}$ be the arguments of a product $p$, and let $Arg = \langle p, C, \mathbf{s} \rangle$ be an argument formed by a tuple of a product $p \in \mathcal{P}$, a concept $C \in \mathcal{C}$ and an aggregated sentiment $\mathbf{s}$ (calculated by aggregating the sentiment values of $Occ(p, C)$, to be defined later). The *Pros*, *Cons* and *Moots* are defined:

$$Pros(p) = \{Arg \in Args(p) | Arg.\mathbf{s} > \delta \}$$
$$Cons(p) = \{Arg \in Args(p) | Arg.\mathbf{s} < -\delta \}$$
$$Moots(p) = \{Arg \in Args(p) | -\delta \leq Arg.\mathbf{s} \leq \delta\}$$

where $\delta$ is a threshold that determines when an argument is considered *Pro*, *Con* or *Moot*; we will show later how $\delta$ depends on the bundle type $(\delta_G, \delta_\sigma, \delta_F)$.

In this work we consider three different methods to create a bundle of arguments: Gini $(B_G)$, Agreement $(B_\sigma)$, and Cardinality $(B_F)$ bundles. Each bundle type is built by a different sentiment aggregation measure; moreover, they share a parameter $\Delta$ that considers moot those arguments with a very small $Occ(p, C)$. We will now define the three types of argument bundles: $B_G$, $B_\sigma$ and $B_F$.

**Gini Bundle $(B_G)$:** An argument in $B_G$ has the form $\langle p, C, S_G(p, C) \rangle$, where the polarity value $S_G$ is calculated using the average sentiment $S_{av}(p, C)$ and then using the *Gini Coefficient* [15] to penalize the average sentiment according to the degree of dispersion of sentiment values: $S(p, C) = S_{av}(p, C)(1 - Gini(p, C))$.

$$S_G(p, C) = \begin{cases} 0 & \text{if } |Occ(p, C)| < \Delta \text{ or } -\delta_G > S(p, C) < \delta_G \\ S(p, C) & \text{otherwise} \end{cases}$$

Notice that, when $|Occ(p, C)| < \Delta$, we consider that we don't have enough reviews of product $p$ with concept $C$ and we assign a neutral sentiment value. Similarly, when $-\delta_G < S_{av}(p, C) \cdot (1 - Gini(p, C)) < \delta_G$, we consider that the polarity is not strong enough to define an argument as a pro or a con, and we assign a neutral sentiment value. Finally, the parameter $\delta_G$ (set to 0.1 in the experiments) determines when the argument is considered pro, con or moot.

**Agreement Bundle $(B_\sigma)$:** Let $Dev(p, C)$ be the standard deviation of the sentiment values of $Occ(p, C)$. The agreement sentiment measure $S_\sigma(p, C)$ is the sentiment average of the sentiment values of the sentences in $Occ(p, C)$, for those concepts whose $Dev(p, C) < \delta_{max}$. This measure uses two threshold parameters $\delta_{max}$ and $\delta_\sigma$. First, $\delta_{max}$ specifies the maximum acceptable standard deviation over the distribution of sentiment values in $Occ(p, C)$: when $Dev(p, C) > \delta_{max}$

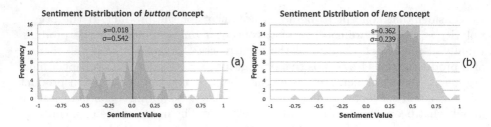

**Fig. 2.** Sentiment value distribution (a) *button* concept and (b) *lens* concept for Pentax K-5. Notice that values have a higher degree of dispersion in (a) than in (b).

we consider that we have no grounds for an informed decision on the overall polarity of $C$ with respect to product $p$. Second, $\delta_\sigma$ specifies the threshold for an argument sentiment value to be considered a pro, a con, or a moot argument. An argument in $B_\sigma$ has the form $\langle p, C, S_\sigma(p, C)\rangle$ where $S_\sigma$ is defined as follows:

$$S_\sigma(p, C) = \begin{cases} 0, & \text{if } Dev(p, C) > \delta_{max} \text{ or } |Occ(p, C)| < \Delta \\ S_{av}(p, C), & \text{otherwise} \end{cases}$$

Parameter $\delta_\sigma$ value is set to 0.1 in the experiments.

Figure 2 presents the sentiment value distribution of two arguments of Pentax K-5, *button (a)* and *lens (b)*. The *button* argument of the Pentax K-5 has a sentiment value deviation $\sigma = 0.542$, showing a high dispersion of sentiment values for concept *button* among the reviews of Pentax K-5. Since the deviation of the sentiment values of *button* is higher than $\delta_{max}$, we have no clear overall polarity. On the other hand, the deviation of the sentiment values of *lens* is lower than the threshold and has a positive average sentiment $(0.235 > \delta_\sigma)$. Therefore, argument *lens* is considered a pro argument with respect to Pentax K-5.

**Cardinality Bundle $(B_F)$:** The cardinality bundle is created by comparing the number of positive versus negative occurrences of a concept $C$ in $Occ(p, C)$. The number of positive $(O^+)$ and negative $(O^-)$ occurrences of a concept $C$ in the reviews of a product $p$ are defined as $O^+(p, C) = |\{x \in Occ(p, C) \,|\, s(C, x) > 0\}|$ and $O^-(p, C) = |\{x \in Occ(p, C) \,|\, s(C, x) < 0\}|$, where $s(C, x)$ is the sentiment value in $[-1, 1]$ of concept $C$ in sentence $x$.

An argument in $B_F$ has the form $\langle p, C, S_F(p, C)\rangle$ where $S_F$ is:

$$S_F(p, C) = \begin{cases} 0, & \text{if } \left(2 \cdot \frac{O^+}{O^+ + O^-}\right) - 1 = 0 \text{ or } |Occ(p, C)| < \Delta \\ \left(2 \cdot \frac{O^+}{O^+ + O^-}\right) - 1, & \text{otherwise} \end{cases}$$

where $O^+ = O^+(p, C)$ and $O^- = O^-(p, C)$. Notice that $S_F(p, C)$ takes values on $(0, 1]$ if $O^+ > O^-$, and in $[-1, 0)$ if $O^+ < O^-$. In the experiments we set $\delta_F = 0$ as the threshold that determines if an argument is pro, con or moot.

As a final step, we create three collections of bundles (one for each bundle type) considering the whole set of products and rescale the sentiment values of the arguments that form the bundles of the collection in a way that the most positive argument sentiment about a concept has a sentiment 1, and the most negative a sentiment $-1$. We rescale the rest of the sentiment values accordingly. This way, considering a collection of product bundles, the product with the best sentiment over a concept has a sentiment value of 1. When all arguments of a bundle $B$ are rescaled we call it a normalized bundle $\overline{B}$.

## 4   User Query over Product Bundles

A user query defines the requirements of a user expressed using the concept vocabulary $\mathcal{C}$. Since not all requirements are equally important for the user, every requirement over a concept has a utility value. Given a set of products characterized with the normalized bundles of arguments $\overline{B}(p)$, we can decide which is the product that has a higher level of query satisfaction.

We define a *user query* $Q = \{(C_j, U(C_j))\}_{j=1,\dots,k}$ and $k \leq |\mathcal{C}|$ as a set of concept utility pairs. Each concept utility pair $(C_j, U(C_j))$ expresses a requirement from the user over concept $C_j$ with a utility degree $U(C_j) \in [0.5, 1]$. For instance in a query $Q = \{(lens, 0.9), (video, 0.6)\}$, the user requires high quality lens and video, although the quality of the lens is more important than the quality of the video. Furthermore, a good lens or video are more important for the user than any other feature the camera could possess.

We will now define the degree of *Query Satisfaction*, $DS(Q, \overline{B}))$, that determines the degree in which a normalized bundle $\overline{B}$ satisfies a user query $Q$. Since t-norms and implications in fuzzy logic are defined in the interval $[0, 1]$, we need to rescale the sentiment values of all arguments that form all product bundles from $[-1, 1]$ to $[0, 1]$ by applying the linear mapping $f(\mathbf{s}) = \frac{\mathbf{s}+1}{2}$. For example, consider an argument $\langle p, lens, 0.83 \rangle \in \overline{B}(p)$, the sentiment of the argument will be $f(0.83) = 0.915$. Notice that the neutral value 0 in $[-1, 1]$ is mapped to the neutral value 0, 5 in $[0, 1]$.

We will first define a concept-wise satisfaction degree using the notion of fuzzy implication, specifically we will use fuzzy implication associated to the t-norm product ($\Rightarrow_\otimes$).

$$U(C_j) \Rightarrow_\otimes \mathbf{s}_j = \begin{cases} 1, & \text{if } U(C_j) \leq \mathbf{s}_j \\ \frac{\mathbf{s}_j}{U(\mathbf{s}_j)} & \text{otherwise} \end{cases}$$

where $\mathbf{s}_j$ is the rescaled sentiment value of argument $\langle p, C_j, \mathbf{s}_j \rangle$. We need now to aggregate these $k$ concept-wise satisfaction degrees into an overall degree of bundle satisfaction of a query $Q$. For this purpose, we will use the t-norm product as follows:

$$DS(Q, \overline{B}(p)) = \prod_{j=1}^{k} (U(C_j) \Rightarrow_\otimes \mathbf{s}_j)$$

where $\mathbf{s}_j$ is the rescaled sentiment value of argument $\langle p, C_j, \mathbf{s}_j \rangle$ of the argument bundle $\overline{B}(p)$ and $\overline{B}$ is a normalized argument bundle (either $\overline{B}_G$, $\overline{B}_\sigma$ or $\overline{B}_F$).

**Table 2.** Degree of satisfaction of two cameras for each requirement and the overall $DS$ for the query $Q_1$ and the $Q_2$.

| $Q_1$ Requirements | (*picture*, 0.7) | (*resolution*, 0.6) | $DS(Q_1, \overline{B}_F)$ | |
|---|---|---|---|---|
| $\overline{B}_F$(D7100) | 0.75 | 1.00 | | |
| $\overline{B}_F$(EOS70D) | 0.97 | 0.50 | | |
| $U(C_j) \Rightarrow_{\otimes} s_j$ for $\overline{B}_F$(D7100) | 1.00 | 1.00 | **1.00** | |
| $U(C_j) \Rightarrow_{\otimes} s_j$ for $\overline{B}_F$(EOS70D) | 1.00 | 0.83 | 0.83 | |
| $Q_2$ Requirements | (*picture*, 0.7) | (*resolution*, 0.6) | (*video*, 0.9) | $DS(Q_2, \overline{B}_F)$ |
| $\overline{B}_F$(D7100) | 0.75 | 1.00 | 0.64 | |
| $\overline{B}_F$(EOS70D) | 0.97 | 0.50 | 1.00 | |
| $U(C_j) \Rightarrow_{\otimes} s_j$ for $\overline{B}_F$(D7100) | 1.00 | 1.00 | 0.72 | 0.72 |
| $U(C_j) \Rightarrow_{\otimes} s_j$ for $\overline{B}_F$(EOS70D) | 1.00 | 0.83 | 1.00 | **0.83** |

Table 2 shows the degree of satisfaction of two user queries $Q_1$ and $Q_2$ against the cardinality bundles of two cameras: Nikon D7100 and Canon EOS70D (sentiment values are rescaled). The first query is created by a user who likes to go hiking and that is looking for a camera to capture landscape and nature while valuing fine detail. Assume her query is $Q_1 = \{(picture, 0.7), (resolution, 0.6)\}$ because she wants a camera with good image quality and resolution. Table 2 shows on the first two rows the sentiment values of the two cameras in the concepts appearing in the query. The second two rows show the satisfaction degree of the two cameras for each requirement and the overall $DS$ for the query. Notice that satisfaction is 1 when the sentiment value is higher than the required utility value for a concept.

The second example is query $Q_2 = \{(picture, 0.7), (resolution, 0.6), (video, 0.9)\}$ (second half of Table 2) is created by a user that, besides hiking, also loves recording video. Now, according to user reviews, Canon EOS70D has an outstanding video quality (1.0), while Nikon D7100 has an average quality video (0.64). Because of this new added requirement now the higher ranking camera is Canon EOS70D instead of Nikon D7100, the best ranking camera for $Q_1$.

## 5    Evaluation

In this section we compare and evaluate the different bundles of arguments with those of DPReview.com, a renowned website specialized in digital cameras. We are keen to study the differences between the sets of pros, cons and moots of the three different bundles of arguments, $B_G$, $B_\sigma$ and $B_F$, while assessing the impact that the number of reviews of a product has over the quality of the bundle of arguments. Therefore we evaluate the precision and recall of the product bundles by comparing them with the expert evaluations of products presented in DPReview. Finally, we present a ranking strategy for product bundles and compare the rankings of products obtained with each bundle type ($B_G$, $B_\sigma$, $B_F$) compared with two external product rankings (those of DPReview and Amazon).

**Table 3.** Average number of pros, cons and moot arguments for the 3 bundle types.

|  | Gini bundle $B_G$ | Agreement bundle $B_\sigma$ | Cardinality bundle $B_F$ |
|---|---|---|---|
| Avg. # pros | 9.42 | 12.44 | 12.65 |
| Avg. # cons | 0.54 | 3.42 | 2.60 |
| Avg. # moots | 26.04 | 20.14 | 20.75 |

The *Digital SLR Camera* dataset we use was extracted by us from Amazon during April 2014 [13] contained more than 20,000 user generated reviews over a set of 2,264 products. We pruned those products older than 1st January 2008 and with less than 15 user reviews, and merged any synonymous products, leaving us data on 50 products. Over the set of reviews of these products we extracted 251 different aspects, that were grouped by the hierarchical clustering algorithm presented in Sect. 2 into 36 concepts, that form the concept vocabulary $\mathcal{C}$. Using $\mathcal{C}$, we created three types of argument bundles for each of the 50 products as described in Sect. 3.

**Comparison between Argument Bundles $B_G$, $B_\sigma$ and $B_F$.** Here we study the differences among the pro, con and moot arguments of the three bundle types $B_G$, $B_\sigma$ and $B_F$. Since the criteria to select the arguments varies between the three bundle types, the quantity of pros, cons, and moot arguments obtained by each bundle type may differ. Table 3 presents a comparison between the average quantity of pros, cons and moot arguments of each bundle type.

The Agreement and Cardinality bundles have a similar average number of pros and cons, while Gini bundles are slightly smaller. The Gini average tends to move the argument sentiment value the towards 0 when there is dispersion in the distribution of sentiment values, and thus more arguments tend to be moots.

Next we study which concepts are considered pros in the different bundles. Figure 3 presents the quantity of pros shared between the three bundle types of each product, showing that most pros (almost 8 out of 10) are shared between two or three bundle types of a product, a good indicator of the consistency of our approach. This means that a pro concept in a $B_G$ is also likely form part of $B_\sigma$ pros and $B_F$ pros. Furthermore, the number of pros (and also cons, not included in this figure due to lack of space) of a bundle is directly related with the number of sentences in the reviews of that product: the more reviews the more richer the bundles are. Notice that we are only studying if a concept is categorized as a pro between the 3 bundle types of a product; we are not comparing the concrete positive sentiment values of the arguments.

**Bundle of Arguments Evaluation.** To evaluate the quality of bundles, we compared the bundles of arguments of the 15 products with more reviews with the product pros and cons textual descriptions from DPReview. The DPReview pros and cons of a product are separately formed by lists of sentences such as 'good detail and color in JPEGs at base ISO (pro)' or 'buggy Live View /

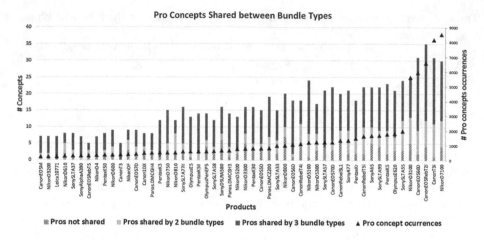

**Fig. 3.** Quantity of pros shared between the three bundles of arguments $B_G$, $B_\sigma$ and $B_F$, together with the number of occurrences of the pro concepts in the reviews of the product.

Movie Mode (con)'. In order to compare the DPReview pro and con items with our bundles of arguments, we first manually identify the concepts referenced in each item text and interpret that concept as one of the concepts in our concept vocabulary, if it exists. For instance, we consider that previous DPReview pro sentence 'good detail and color in JPEGs at base ISO' refers to the vocabulary concepts *jpeg*, *color* and *picture*, whilst 'buggy Live View / Movie Mode' refers negatively to concepts *live view* and *video*. Those sentences from DPReview that did not clearly refer to a concept in $\mathcal{C}$ were ignored. By grouping the vocabulary concepts present in the DPReview pro and con items of a product, we create the sets of DPReview pros $Pros_{dp}$ and cons $Cons_{dp}$ but without a sentiment value associated. We compare those DPReview sets with the pros and cons of the three different bundles of arguments of each product without taking into account the sentiment values, only whether the concept is selected as a pro or con.

Table 4 present the average precision, recall and $F_2$-score between the sets of pros and cons of the three bundle types and those of DPReview. We use the $F_2$-score to weight recall higher than precision, since we are keen to study whether the three different bundle types identify as pros and cons the same concepts listed in DPReview. Furthermore, we analyze the percentage of *contradictions*, which are those concepts selected as pros in our bundles of arguments but considered cons in DPReview and vice versa. A low percentage of contradictions is a good indicator of the quality of the bundles.

The bundle of arguments that performs best for the pro arguments is the cardinality bundle $B_F$, with an average recall of 0.822 and an $F_2$-score of 0.733. That means that the 82.2 % of the concepts listed as pros of product $p$ in DPReview also form part of the pros of the cardinality bundle $B_F(p)$. On the other hand, the sets of cons of all three bundles of arguments perform poorly. This is because

**Table 4.** Measures on precision, recall, $F_2$-score and contradictions between pros and cons of bundles $B_G$, $B_\sigma$ and $B_F$ with respect to DPReview pros and cons.

|      |           | Precision | Recall | $F_2$-score | Contradictions |
|------|-----------|-----------|--------|-------------|----------------|
| Pros | $B_G$     | **0.567** | 0.644  | 0.627       | **0.004**      |
|      | $B_\sigma$| 0.506     | 0.761  | 0.691       | 0.135          |
|      | $B_F$     | 0.513     | **0.822** | **0.733** | 0.065          |
| Cons | $B_G$     | 0.333     | 0.046  | 0.056       | 0.046          |
|      | $B_\sigma$| 0.285     | 0.558  | 0.468       | 0.132          |
|      | $B_F$     | 0.388     | 0.488  | 0.464       | 0.165          |

the granularity of the sentences is different between our concept vocabulary and DPReview. For us, the granularity level is given by our concept vocabulary, while DPReview sentences normally work at different levels of granularity. Furthermore, the granularity of DPReview sentences varies whether the sentence is a pro or a con. DPReview pro sentences tend to be more general: 'camera buttons and dials are useful and easily configurable', while con sentences tend to be more specific: 'the video dial is not easily accessible'. Although for us both sentences reference concept *button*, it is clear that the DPReview pro sentence better describes a general view of the buttons of the camera than the second one. Furthermore, note that the precision values of all bundles are lower than 0.6, suggesting that the sets of pros of the bundles of arguments are richer in concepts compared to those of DPReview summaries. This is not strange, since the sets of DPReview pros and cons are not exhaustive but a short list of the concepts that stand out from their point of view. The average size of the set of bundle pros is 12–14 arguments, while the average pro set size of DPReview identified concepts is 7–9. Finally, notice the low quantity of contradictions between the bundles of arguments and the DPReview sets. However low, we are interested in studying what are the most frequent concepts in contradictions.

The most common contradictions between the bundles and the set of pro and con concepts extracted from DPReview for the 15 selected products are: battery (10), viewfinder (5), recording (5) and button (3). In DPReview battery is often selected as a pro, however it is usually selected as a con in the bundles of arguments. That is because in the reviews people usually complain about the battery of a camera, while they do not seem to express positive opinions on cameras with a good battery (it would seem it is taken as a given). Other frequent contradictions are *viewfinder*, *recording* and *button*. This is because in DPReview those are commonly selected as cons for having not optimal behavior in certain types of situations (e.g. 'the video dial is not easily accessible') while the overall opinions about the rest of the buttons are positive. Therefore, our bundles will capture this average higher granularity sentiment of *button*. Similar situations are observed for *recording* and *jpeg* concepts.

Next we define the function $\Theta : B \times B \rightarrow [-1, 1]$ that estimates the degree in which a product bundle $B(p_i)$ is better or superior to another bundle $B(p_j)$:

$$\Theta(B(p_i), B(p_j)) = \frac{1}{2|\mathcal{C}|} \sum_{k=1}^{|\mathcal{C}|} \mathbf{s}_k^i - \mathbf{s}_k^j$$

where $\mathbf{s}_k^i$ and $\mathbf{s}_k^j$ are the sentiment values of respective arguments $\langle p_i, C_k, \mathbf{s}_k^i \rangle$ and $\langle p_j, C_k, \mathbf{s}_k^j \rangle$ in the bundles of products $p_i$ and $p_j$. $\Theta$ is the average of these differences over all concepts in $\mathcal{C}$, a value in $[-1, 1]$. If the value of $\Theta(B(p_i), B(p_j))$ is in $(0, 1]$, then $B(p_i)$ is superior than $B(p_j)$, while if this value is in $[-1, 0)$, then $B(p_i)$ is worse than $B(p_j)$.

Using $\Theta$, we take the 15 products with more reviews and we create a product ranking for each bundle type ($B_G$, $B_\sigma$ and $B_F$). Moreover, we create two more rankings over these 15 products: (1) *DPReview Ranking*, based on the DPReview overall product score, and (2) *Amazon Ranking*, based on Amazon's star rating score. Whenever two or more products had the same DPReview score, such as Olympus E620 and Nikon D3100 both with a score of 72 out of 100, we only kept the product with most reviews, in this example the Nikon D3100. This left us with 9 different products. Let us now compare these rankings. The top 3 products for the $B_G$ ranking are Nikon D7100, Pentax K-5 and SonySLT A-55. The top 3 products for $B_\sigma$ are Nikon D7100, SonySLT A-99 and SonySLT A-55, and the top 3 ranked products for $B_F$ are Nikon D7100, SonySLT A-99 and Pentax K-5. Notice that Nikon D7100 is the top product in all three bundles, and it is also ranked 1st (with a score of 85 points) in the DPReview ranking, followed by SonySLT A-99 and Pentax K-5. Table 5 shows the *Spearman Rank Correlation* of the 3 bundle rankings with the DPReview Ranking and the Amazon Ranking. We added a random ranking strategy to facilitate a baseline comparison. The random ranking correlation values were obtained by averaging the Spearman correlations of 1000 randomly generated product rankings with DPReview ranking and Amazon ranking.

The results show that $B_F$ ranking has the highest Spearman correlation with DPReview ranking (correlation of 0.904). This value tells there is a very strong

**Table 5.** Spearman rank correlation of the bundle rankings with DPReview product ranking and Amazon star ratings ranking.

| Rankings | Spearman rank correlation | |
|---|---|---|
| | DPReview ranking | Amazon ranking |
| $B_G$ ranking | 0.50 | −0.19 |
| $B_\sigma$ ranking | 0.57 | −0.33 |
| $B_F$ ranking | **0.90** | 0.09 |
| Random ranking | 0.34 | 0.34 |
| DPReview ranking | 1 | 0.33 |

correlation between the two rankings, a good indicator of the quality of the cardinality bundles $B_F$. The correlations $B_\sigma$ and $B_G$ are also strong, being notably higher than the random ranking correlations. Note that the Amazon star-based ranking does not correlate with any of the bundle rankings nor the DPReview score ranking. In fact, the random ranking obtains the highest Spearman rank correlation with the Amazon star ranking, showing no strong correlation between the star-rating ranking and the bundles extracted from the reviews. This may be understandable, since two people with similar arguments about a product can give different star-rating values. Nevertheless, the fact is that Amazon's star rating cannot be used as ground truth to test the quality of the bundles.

## 6   Related Work

There exist numerous applications that gather knowledge from user-generated reviews, usually oriented to help other users make more informed decisions in the area of recommendation systems and CBR. The most common approach consists in characterizing a set of products by considering product aspects (also called features) mentioned in the reviews [1, 2]. In this process, the set of aspects selected to characterize a product together with the sentiment analysis of the sentences have a crucial role in the final recommendation [3, 5, 12]. A related work on creating BLC is [6], but they have to mark by hand a corpus with the classes (concepts) to which words belong; then they use supervised learning while we discover th BLCs in an unsupervised way.

Another focus is identifying the sets of aspects with higher positive/negative polarity to give insights into the reason why items have been chosen [8]. Those approaches need previously to group the aspects to reduce the granularity in order to provide useful recommendations, often solved by clustering aspects using background knowledge to simplify the process. Our approach is different in a sense that we create basic level concepts [10] by exploring the usage of the aspects among the user-generated reviews in an unsupervised way.

Using these basic concepts, we build the bundles of arguments by identifying the pro and con concepts over the set of reviews of a product. Finally, we define a concept-wise satisfaction degree of a user query over a set of bundles using the notion of fuzzy implication [4]. User queries are the reason we define bundles: they allow to reuse experiences of other people to the needs a specific user.

## 7   Conclusions and Future Work

In this paper we extend our previous work on aspect extraction and sentiment analysis and propose a method to create a vocabulary of basic level concepts with the appropriate granularity to characterize a set of cameras. This concept vocabulary is useful to practically reuse other people's experiences with digital cameras because it abstracts the concrete terms used in the corpus as given by the aspect extraction approach. By analyzing the usage of the concepts over the reviews of a product, we find those concepts that have a clearly positive or negative polarity and create the argument bundles. We present three different types of argument bundles, each one defining the pros and cons of a product based on a different criteria. The argument bundles allow us to define a satisfaction degree, interpreted in fuzzy logic and modeled with a fuzzy implication operator, between products and a user query.

An evaluation of the three types of argument bundles is performed and compared with the expert descriptions of the DPReview website, showing that the bundles of arguments correctly identify the pro and con features listed in DPReview. Moreover, the cardinality bundle ranking proved to correlate with the overall DPReview score ranking over the subset of the most frequent products, while Amazon.com star rating ranking does not correlate with neither of them.

The characterization of products by means of the bundles of arguments and BLC is promising. We have observed that the quality of a product bundle is related to the quantity of reviews of that product: the products with more reviews have a richer vocabulary of pro and con arguments, while products with fewer reviews had more moots. This can be due to two reasons that open new lines for future work. First, improving the detection of aspects (for instance, considering also 3-gram aspects) could improve the argument bundles of those products with less reviews. And second, improving the sentiment analysis of reviews by developing a domain specific sentiment dictionary for digital cameras will enhance the accuracy of the arguments' sentiment.

**Acknowledgments.** This research has been partially supported by NASAID (CSIC Intramural 201550E022). We thank Lluís Godo and Pere García for their insightful comments.

# References

1. Chen, L., Chen, G., Wang, F.: Recommender systems based on user reviews: the state of the art. User Model. User Adap. Inter. **25**(2), 99–154 (2015)
2. Dong, R., Schaal, M., O'Mahony, M., McCarthy, K., Smyth, B.: Mining features and sentiment from review experiences. In: Case-Based Reasoning Research and Development. LNCS, vol. 7969, pp. 59–73. Springer, Heidelberg (2013)
3. Dong, R., Schaal, M., O'Mahony, M.P., McCarthy, K., Smyth, B.: Opinionated product recommendation. In: Delany, S.J., Ontañón, S. (eds.) ICCBR 2013. LNCS (LNAI), vol. 7969, pp. 44–58. Springer, Heidelberg (2013). doi:10.1007/978-3-642-39056-2_4
4. Hájek, P., Godo, L., Esteva, F.: A complete many-valued logic with product-conjunction. Arch. Math. Logic **35**(3), 191–208 (1996)
5. Hu, M., Liu, B.: Mining opinion features in customer reviews. In: Proceedings of the Nineteenth National Conference on Artificial Intelligence, pp. 755–760 (2004)
6. Izquierdo, R., Suárez, A., Rigau, G.: Word vs. class-based word sense disambiguation. J. Artif. Intell. Res. (JAIR) **54**, 83–122 (2015)
7. Meng, L., Huang, R., Gu, J.: A review of semantic similarity measures in wordnet. Int. J. Hybrid Inform. Technol. **6**(1), 1–12 (2013)

8. Muhammad, K., Lawlor, A., Rafter, R., Smyth, B.: Great explanations: opinionated explanations for recommendations. In: Hüllermeier, E., Minor, M. (eds.) ICCBR 2015. LNCS (LNAI), vol. 9343, pp. 244–258. Springer, Heidelberg (2015). doi:10.1007/978-3-319-24586-7_17
9. Plaza, E.: Semantics and experience in the future web. In: Althoff, K.-D., Bergmann, R., Minor, M., Hanft, A. (eds.) ECCBR 2008. LNCS (LNAI), vol. 5239, pp. 44–58. Springer, Heidelberg (2008). doi:10.1007/978-3-540-85502-6_3
10. Rosch, E.: Human categorization. Stud. Cross Cult. Psychol. **1**, 1–49 (1977)
11. Sani, S., Wiratunga, N., Massie, S., Lothian, R.: Term similarity and weighting framework for text representation. In: Ram, A., Wiratunga, N. (eds.) ICCBR 2011. LNCS (LNAI), vol. 6880, pp. 304–318. Springer, Heidelberg (2011). doi:10.1007/978-3-642-23291-6_23
12. Turney, P.: Thumbs up or thumbs down?: semantic orientation applied to unsupervised classification of reviews. In: Proceedings of the Annual Meeting on Association for Computational Linguistics, pp. 417–424 (2002)
13. Chen, Y.Y., Ferrer, X., Wiratunga, N., Plaza, E.: Sentiment and preference guided social recommendation. In: Lamontagne, L., Plaza, E. (eds.) ICCBR 2014. LNCS (LNAI), vol. 8765, pp. 79–94. Springer, Heidelberg (2014). doi:10.1007/978-3-319-11209-1_7
14. Chen, Y.Y., Ferrer, X., Wiratunga, N., Plaza, E.: Aspect selection for social recommender systems. In: Hüllermeier, E., Minor, M. (eds.) ICCBR 2015. LNCS (LNAI), vol. 9343, pp. 60–72. Springer, Heidelberg (2015). doi:10.1007/978-3-319-24586-7_5
15. Yitzhaki, S.: Relative deprivation and the Gini coefficient. Q. J. Econ. **93**, 321–324 (1979)

# Incorporating Transparency During Trust-Guided Behavior Adaptation

Michael W. Floyd[1](✉) and David W. Aha[2]

[1] Knexus Research Corporation, Springfield, VA, USA
michael.floyd@knexusresearch.com
[2] Navy Center for Applied Research in AI, Naval Research Laboratory
(Code 5514), Washington, DC, USA
david.aha@nrl.navy.mil

**Abstract.** An important consideration in human-robot teams is ensuring that the robot is trusted by its teammates. Without adequate trust, the robot may be underutilized or disused, potentially exposing human teammates to dangerous situations. We have previously investigated an agent that can assess its own trustworthiness and adapt its behavior accordingly. In this paper we extend our work by adding a transparency layer that allows the agent to explain why it adapted its behavior. The agent uses explanations based on explicit feedback received from an operator. This allows it to provide simple, concise, and understandable explanations. We evaluate our system on scenarios from a simulated robotics domain by demonstrating that the agent can provide explanations that closely align with an operator's feedback.

**Keywords:** Inverse trust · Behavior adaptation · Explanation · Transparency

## 1 Introduction

Robots can be valuable additions to human teams if they provide additional skills to the team, lessen the humans' workload, or can replace humans in dangerous situations. However, even if a robot provides such benefits, the humans may not utilize it to its full potential if they do not trust it. If the robot is underutilized, it may actually increase the humans' workload (e.g., spending extra time observing the robot's behavior) or exposure to risks (e.g., performing dangerous tasks instead of the robot).

One option would be to hard-code the robot's behavior to ensure trustworthiness. However, this may not be feasible as the type of behavior that is considered trustworthy can depend on the teammate (e.g., their amount of experience working with robots), time (e.g., how long the robot has been with the team), or context (e.g., a routine versus a dangerous situation). Alternatively, the humans could explicitly tell the robot whether it is trustworthy. Yet providing such feedback may not be feasible during run-time if the team is in a time-critical situation. Similarly, if the feedback is given at the end of a

© Springer International Publishing AG 2016
A. Goel et al. (Eds.): ICCBR 2016, LNAI 9969, pp. 124–138, 2016.
DOI: 10.1007/978-3-319-47096-2_9

mission (e.g., an after action trust survey) the robot may have performed the entire mission while acting in an untrustworthy manner.

Our previous work focused on an inverse trust estimate that allows a simulated robot[1] to estimate its own trustworthiness and adapt its behavior in situations where it believes it is untrustworthy. We investigated case-based methods that allow the agent to adapt its behavior in response to implicit [1] and explicit feedback [2]. In this paper, we extend our work by adding the ability for the agent to explain why it is adapting its behavior. Adding a level of transparency, wherein an automated system can explain the reasons for its actions, can increase the trustworthiness and reliance on automation [3]. By providing such explanations, even in situations where errors occur, it is possible to maintain trust at a higher level than if no explanations are provided.

In the remainder of this paper we will discuss how our case-based approach for behavior adaptation can also be used for explanation. Section 2 reviews how the agent estimates its trustworthiness using an inverse trust estimate and uses that estimate to guide behavior adaptation. In Sect. 3, we review the feedback model the agent uses to learn how to adapt its behavior in response to explicit feedback. Our approach for allowing the agent to use the same model for explanation is presented in Sect. 4. In Sect. 5, we use a military simulation to evaluate the ability of the agent to provide correct explanations to the user. Related work, with a specific focus on human-robot transparency and explanation, is discussed in Sect. 6, followed by conclusions and areas of future work in Sect. 7.

## 2    Trust-Guided Behavior Adaptation

We assume that the agent receives commands from a single teammate called the *operator*. The operator provides the agent with high-level commands (e.g., "*move to the flag*", "*patrol for threats*") and it performs the assigned tasks autonomously. The agent also has direct control over the *modifiable components* of its behavior. These could be parameter values (e.g., minimum and maximum speeds), algorithms (e.g., choosing among alternative path planning algorithms), or data sources (e.g., using alternative maps of the environment). For each modifiable component $i$, the agent is responsible for selecting a value $m_i$ from the partially ordered set $\mathcal{M}_i$ of possible values ($m_i \in \mathcal{M}_i$).

The agent's current behavior $B$ is represented by the tuple containing the currently selected value for each of the $n$ modifiable components: $B = \langle m_1, m_2, \ldots, m_n \rangle$. At any time, the agent can change the values of one or more modifiable components from its current behavior $B$ to a new behavior $B'$ (e.g., changing from $\langle m_1, m_2, \ldots, m_n \rangle$ to $\langle m_1', m_2', \ldots, m_n' \rangle$, where at least one $m_i \neq m_i'$). Although the agent can change its behavior for any reason, we will focus on trust-guided behavior adaptation (i.e., changing the robot's behavior in an attempt to make it more trustworthy).

---

[1] For the remainder of this paper, we use the term *robot* to refer to a physical (or simulated) robot and *agent* to refer to the intelligent agent controlling the robot.

Traditional trust metrics [4] allow an agent to measure its trust in other agents (e.g., teammates). However, for an agent to modify its behavior to be more trustworthy it must estimate another agent's trust in it. To perform such an estimate, we use an *inverse trust metric* [1]. Inverse trust is measured from the agent's perspective, so only observable indicators of human-robot trust can be used (i.e., none of the human's internal reasoning information can be used).

Our inverse trust metric is based on the strong correlation between the agent's performance and human-robot trust [5]. The operator's perception of the agent's performance is not limited to a mission-level evaluation and can be influenced in real-time by both suitable and poor performance [6]. Without any guarantees of explicit feedback from the operator (i.e., the operator may not always have time to provide feedback), our agent uses implicit feedback to estimate its trustworthiness. In particular, it uses three types of implicit feedback related to its performance: successful completion of an assigned task, failure to complete an assigned task, and interruption by the operator. This assumes that completing a task will be viewed as satisfactory performance, whereas failure or interruption will be viewed as poor performance.

The agent estimates the trustworthiness $Trust_B$ of its current behavior $B$ using the influence $inf_i$ of each of the $c$ commands it has completed. Successfully completed commands increase the trust estimate (i.e., $inf_i = 1$) whereas failed or interrupted commands decrease the trust estimate (i.e., $inf_i = -1$). Each command is also given a weight $w_i$ related to its relative importance (e.g., giving a higher weight to more recent commands, giving higher weight to commands that involve human safety).

$$Trust_B = \sum_{i=1}^{c} w_i \times inf_i$$

The trust estimate is recomputed after each command and compared to two threshold values: the trustworthy threshold $(\tau_T)$ and the untrustworthy threshold $(\tau_U)$. If the trust estimate reaches the trustworthy threshold $(Trust_B \geq \tau_T)$, the agent concludes it is behaving in a trustworthy manner and continues to monitor its trustworthiness in case of any changes (e.g., a change in operator or mission context). If the trust estimate reaches the untrustworthy threshold $(Trust_B \leq \tau_U)$, the agent concludes that its current behavior is untrustworthy and should be changed. Otherwise $(\tau_U < Trust_B < \tau_T)$, the agent continues to monitor the trust estimate until it is more confident about its trustworthiness.

In the event that the untrustworthy threshold is reached, the agent changes from its current behavior $B$ to a new behavior $B'$ and begins measuring the trustworthiness of that behavior (i.e., $Trust_{B'}$). The behavior $B$ along with the time $t$ it took to reach the untrustworthy threshold are stored as an *evaluated pair* $E$ $(E = \langle B, t \rangle)$. As the agent evaluates behaviors, it maintains a set $\mathcal{E}_{past}$ that contains all behaviors that have been found to be untrustworthy $(\mathcal{E}_{past} = \{E_1, E_2, \ldots\})$. This set represents behaviors encountered on the search path taken by the agent before it eventually finds a trustworthy behavior $B_{final}$. In a case-based reasoning context, the set of previously evaluated behaviors is the *problem* and the trustworthy behavior is the *solution*. A case is created and added to the case base each time the agent finds a trustworthy behavior. We use the following case representation:

$$C = \langle \mathcal{E}_{past}, B_{final} \rangle$$

This representation is motivated by the assumption that operators who find similar behaviors to be untrustworthy (i.e., similar problems) will also find similar behaviors to be trustworthy (i.e., similar solutions). A more detailed description of case acquisition, similarity calculation, case retrieval, and case-based behavior adaptation can be found in [1].

## 3   Feedback Model

Our inverse trust metric uses *implicit* operator feedback to estimate trust, but it is also possible for the agent to use *explicit* feedback. Although explicit feedback is provided at the operator's discretion (i.e., the agent does not know when or how often it will occur), it provides direct feedback on the agent's performance (e.g., *"go faster"*, *"slow down"*, *"watch out for obstacles"*). Initially, the agent has no knowledge about the type of feedback it will receive or what each piece of feedback means. As feedback is received, the agent learns a feedback model that contains information about how it should respond to operator feedback. For example, if the operator tells the agent *"go faster"*, the agent should learn that this means it should increase its speed.

The agent acquires its feedback model by learning the relationships between its behavior when feedback is received and a trustworthy behavior (i.e., how it was behaving when feedback was received and how it should behave). These relationships are stored in a *feedback base*, where each feedback relationship case *FR* is defined as:

$$FR = \langle f, R, cnt \rangle$$

Each feedback relationship case contains a piece of feedback $f$, a relationship $R$, and a frequency $cnt$. For any pair of behaviors $B_i$ and $B_j$, the relationship $R_{ij}$ encodes how the behaviors differ ($\mathcal{B} \times \mathcal{B} \rightarrow \mathcal{R}$, where $\mathcal{B}$ is the set of all behaviors and $\mathcal{R}$ is the set of all relationships). More specifically, a relationship encodes how each pair of modifiable component values differ ($rel : \mathcal{M}_i \times \mathcal{M}_i \rightarrow \mathcal{O}, \mathcal{O} = \{\prec, \succ, =\}$). The overall relationship $R_{ij}$ is a tuple containing each of the modifiable component relationships ($|B_i| = |B_j| = |R_{ij}| = n, R_{ij} = \langle rel(B_i.m_1, B_j.m_1), \ldots, rel(B_i.m_n, B_j.m_n) \rangle$).

The frequency $cnt$ measures how many times the relationship $R$ was found for feedback $f$. Since the cases in the feedback base are learned by the agent, it is possible for unnecessary or erroneous relationships to be learned (e.g., the operator gave incorrect feedback). The agent works under the assumption that unnecessary and erroneous relationships occur less frequently than correct ones, so preference is given to relationships with higher frequency values.

Consider an example where the agent has two modifiable components: its speed and its object padding (how far it attempts to stay away from obstacles when planning its movement). The agent receives the feedback *"go faster"* when using a behavior $B_1$

with a speed of 1.0 m/s and a padding of 0.5 m ($B_1 = \langle 1.0, 0.5 \rangle$). Eventually, the agent finds a trustworthy behavior $B_2$ with a speed of 5.0 m/s and a padding of 0.5 m ($B_2 = \langle 5.0, 0.5 \rangle$). The relationship $R_{12}$ between them would show the speed increased while the padding remained constant ($R_{12} = \langle \prec, = \rangle$). If this was the first time this relationship was learned for the feedback "*go faster*", the *cnt* value would be 1 ($FR_{example} = \langle$ "*go faster*", $\langle \prec, = \rangle$, 1$\rangle$). If the agent receives the feedback "*go faster*" again, it can retrieve this feedback relationship case and know to increase its speed. A full description of how the agent can learn a feedback model is described in [2]. For the remainder of this paper, we will assume that the agent already has a feedback model available to use.

## 4    Behavior Adaptation Explanation

As we explained in the previous sections, the agent has two methods for modifying its behavior: adapting in response to implicit feedback and adapting in response to explicit feedback. Adaptation in response to explicit feedback occurs directly after the feedback is received. This provides the operator with a direct connection between their feedback and the behavior change. However, adaptation in response to implicit feedback occurs over a longer period of time (i.e., the entire time the agent is measuring the trustworthiness of a behavior). Since there may not be any single event that caused the agent to change its behavior, it may not be clear to the operator why the behavior change occurred.

To obtain transparency between the agent and the operator, it can provide an explanation when it adapts its behavior in response to implicit feedback. The information contained in the agent's explanation could be in different forms and at varying levels of abstraction (e.g., a visual representation of the agent's trust estimate over time, a list of the commands that were failed or interrupted, an acknowledgement that a behavior change occurred). However, an explanation may not be useful to the operator if it is verbose or difficult to interpret. For example, if the agent provided an explanation that included a complete list of all assigned commands and their results (i.e., all the information it uses to compute the inverse trust estimate), the operator may ignore it.

We designed our agent to provide explanations using a method of communication that we believe will be appropriate for the operator. To achieve this, the agent uses the model of explicit feedback that it has learned from the operator, since the feedback is both understandable (i.e., the operator has used it to communicate with the agent) and succinct (i.e., the operator was able to provide the feedback under real-time constraints). By using the operator's own feedback in explanations, the agent relates that its behavior adaptation is motivated by the predicted actions of the operator. For example, if the agent adapts its behavior by increasing its speed, it can provide the explanation "*I adapted my behavior because I think you were going to tell me to **speed up***".

The agent generates an explanation (Algorithm 1) when it adapts its current behavior $B_{curr}$ to a new behavior $B_{adapt}$ (i.e., it performs case-based behavior adaptation when the inverse trust metric reaches the untrustworthy threshold). The relationship $R$ between the two behaviors is calculated (line 2) and compared to the relationship stored in each feedback relationship (i.e., *FR.R*) in the feedback base *FeedbackBase* (lines 3–4). Any feedback relationship that contains the relationship *R* is added to the set $\mathcal{F}$ (line 5). The *selectFeedback*(...) function selects a single piece of feedback from among the feedback items stored in $\mathcal{F}$ (line 6).

---

**Algorithm 1.** Adaptation Explanation

**Function:** *findExplanation($B_{curr}$, $B_{adapt}$) returns expectedFeedback*

---

1   $\mathcal{F} \leftarrow \emptyset$;
2   $R \leftarrow$ relationship($B_{curr}$, $B_{adapt}$);
3   **foreach** $FR \in FeedbackBase$ **do**
4      **if** $FR.R = R$ **then**
5         $\mathcal{F} \leftarrow \mathcal{F} \cup FR$;

6   **return** selectFeedback($\mathcal{F}$);

---

We propose four alternative implementations for the *selectFeedback*(...) function:

- **Highest Count:** The feedback relationship $FR_i$ with the largest frequency is selected $(FR_i \in \mathcal{F}, \forall FR_j \in \mathcal{F}, FR_i.cnt \geq FR_j.cnt)$ and its feedback $FR_i.f$ is returned. If multiple feedback relationships are tied for the largest frequency, one is selected at random according to a uniform distribution.
- **Highest Group Count:** The feedback relationships are partitioned into $k$ subsets such that all feedback relationships in a subset have the same associated feedback and there is only one subset per feedback type $(\mathcal{P}_1 \cup \mathcal{P}_2 \cup \ldots \cup \mathcal{P}_k = \mathcal{F}, \mathcal{P}_1 \cap \mathcal{P}_2 \cap \ldots \cap \mathcal{P}_k = \emptyset, \forall FR_i, FR_j \in \mathcal{P}_l, FR_i.f = FR_j.f)$. For each subset, the frequency of all feedback relationships in the subset are summed $\left( sum_l = \sum_{FR_i \in \mathcal{P}_l} FR_i.cnt \right)$ and the subset with the largest summed frequency has its feedback returned. If multiple subsets tie for the largest summed frequency, one is selected at random according to a uniform distribution.
- **Mean Group Count:** The feedback relationships are partitioned and the mean frequency count for each subset is calculated $\left( \mu_l = \frac{sum_l}{|\mathcal{P}_l|} \right)$. The subset with the largest mean frequency count has its feedback returned. If multiple subsets tie for the largest mean frequency count, then one is selected at random according to a uniform distribution.
- **Random:** One feedback relationship is randomly selected from $\mathcal{F}$ according to a uniform distribution. The feedback stored in the feedback relationship is returned.

The piece of feedback *expectedFeedback* that is returned from Algorithm 1 is used to produce a human-readable explanation for the operator. The explanation takes the following form:

> "*I adapted my behavior because I think you were going to tell me to*   *<expectedFeedback>*"

For example, if Algorithm 1 returned "*drive safely*", the generated explanation would be "*I adapted my behavior because I think you were going to tell me to **drive safely**".

## 5 Evaluation

In this section, we evaluate our claim that *the agent produces explanations for its behavior that align with the operator's evaluation of the agent*. The evaluation uses simulated operators so the agent's trustworthiness from a human's perspective cannot be directly measured. However, we can measure the agent's ability to perform actions that have been shown to positively influence trust (i.e., providing explanations). Our evaluation tests the following hypotheses:

**H1**: The explanations provided by the agent are consistent with the explicit feedback the operator would have provided had the opportunity arisen.
**H2**: The explanations provided by the agent outperform a random baseline.
**H3**: The agent provides better explanations using a manually authored feedback base compared to a learned feedback base.

### 5.1 Domain: eBotworks

We use the eBotworks simulator [7] for our evaluation. eBotworks allows autonomous agents to control simulated robots in a simulated urban environment. In our evaluation, the agent controls an unmanned ground vehicle (UGV) in an environment composed of other agents (e.g., humans, other simulated UGVs), obstacles (e.g., buildings, vehicles, traffic cones, boxes), and ground features (e.g., roads, grass). We chose to use eBotworks because it provides a built-in agent design framework, autonomy modules (e.g., natural language command interpretation, path planning), and allows for evaluation in a non-deterministic and noisy environment.

The scenario we use involves the agent-controlled robot receiving natural language commands from an operator. The commands instruct the agent to patrol between its current location and a goal location. While patrolling, it continuously scans for suspicious objects. If a suspicious object is found, the robot pauses its patrol, moves toward the object, and uses explosive-detection sensors to determine if it is *dangerous* or *harmless*. After classifying a suspicious object, the robot continues patrolling.

The robot has the following modifiable components (and possible values they can take):

- **Speed (meters per second)**: The maximum speed the simulated robot uses when moving through the environment. $\mathcal{M}_{speed} = \{0.5, 1.0, \ldots, 10.0\}$
- **Padding (meters):** How far the robot attempts to stay away from obstacles when planning its path. Higher paddings decrease the likelihood that it will collide with obstacles. $\mathcal{M}_{padding} = \{0.1, 0.2, \ldots, 2.0\}$

- **Scan Time (seconds)**: How much time the robot spends scanning each suspicious object. Higher scan times increase the probability that the robot will successfully classify the suspicious object. $\mathcal{M}_{scantime} = \{0.5, 1.0, \ldots, 5.0\}$
- **Scan Distance (meters)**: How close the robot gets to the suspicious object while scanning it. Lower scan distances increase the probability that the robot will successfully classify the suspicious object. $\mathcal{M}_{scandistance} = \{0.25, 0.5, \ldots, 1.0\}$

## 5.2    Experimental Setup

Our study uses simulated operators to issue commands to the agent and monitor its behavior. The simulated operators were selected to represent a subset of control strategies used by human operators (i.e., when to allow the agent to complete a task and when to interrupt). Two simulated operators are used: *speed-focused* and *detection-focused*. The speed-focused operator prefers the task to be performed quickly (i.e., 95 % probability of interrupting if the robot does not complete the task within 120 s) and correctly (i.e., 100 % probability of interrupting if the robot misses a suspicious object or incorrectly classifies it). The detection-focused operator prefers the task to be performed correctly but is less concerned about speed (i.e., 5 % probability of interrupting if the robot exceeds 120 s). Both operators place a relatively low emphasis on the robot's safety (i.e., 5 % probability of interrupting if the robot comes in contact with an obstacle).

At the start of each experimental *trial*, the robot is assigned a random initial behavior (i.e., a random value for each modifiable component from the set of possible values according to a uniform distribution). Each trial involves multiple *runs*. The robot is placed in an initial position before each run, and six suspicious objects are placed in the environment at random locations (uniformly distributed in predefined regions) and with random appearance (uniformly distributed from a set of small objects that the robot can detect). Each run starts when the operator issues a command to the robot and terminates when it successfully completes the task (i.e., reaches the destination and successfully classifies all suspicious objects) or is interrupted by the operator. At the end of each run, the agent updates its inverse trust estimate, compares the current trust estimate to the thresholds, and may adapt its behavior. The environment is then reset to its initial conditions before the next run.

Each time the robot is interrupted, the operator generates a piece of natural language feedback. The feedback comes in five categories: speed feedback (e.g., "*go faster*"), safety feedback (e.g., "*be careful*"), false positive feedback (e.g., "*that wasn't a threat*"), false negative feedback (e.g., "*that was a threat!*"), and missed object feedback (e.g., "*you missed one!*"). Although the operators provide multiple synonymous pieces of feedback for each category (e.g., "*speed up*", "*go faster*"), for this evaluation we treat all synonymous feedback as equivalent. The feedback is never actually provided to the robot (i.e., the robot cannot use it for adaptation). Instead, the feedback is logged and used to evaluate any explanations provided by the robot. For each trial, all feedback generated during that trial is recorded.

Each trial ends in one of two possible outcomes: the robot labels its behavior as trustworthy or as untrustworthy. If the behavior is found to be trustworthy, the data from the trial is discarded. This occurs because no behavior adaptation occurred and therefore no explanations were provided by the robot. However, if the behavior is found to be untrustworthy, case-based behavior adaptation is performed (i.e., adaptation in response to implicit feedback). The robot uses a case base that was learned during a previous study[2]. The robot's current behavior (i.e., the one randomly generated at the start of the trial) and the behavior returned by case-based behavior adaptation are used to generate an explanation. Eight variant methods for generating explanations are evaluated using Algorithm 1. The variants differ based on the feedback base that is used (i.e., *authored* by an expert or *learned*[3] by the robot) and the method for selecting feedback (i.e., *Highest Count, Highest Group Count, Mean Group Count, Random*). We also use a baseline approach that randomly selects an explanation according to a uniform distribution (labelled as *Baseline* to avoid confusion with the *Random* feedback selection method).

At the end of each trial, the robot's explanation is compared to the explicit feedback generated by the operator during the trial. The following metrics are computed:

- **Most Common**: The percentage of trials where the robot's explanation matched with the most common piece of feedback provided by the operator.
- **Matched One**: The percentage of trials where the robot's explanation matched at least one piece of feedback given by the operator.
- **Mean Rank**: The mean rank of the robot's explanation relative to a list that is ranked by the number of times each piece of feedback occurs during a trial. If the robot's explanation does not appear in the ranked list, it is given a value of the size of the ranked list plus one.

The average from 1000 trials was collected and the process was repeated 25 times (i.e., 25,000 total trials). The robot used a trustworthy threshold of $\tau_T = 5.0$, an untrustworthy threshold of $\tau_U = -5.0$, and the case-based adaptation approach described in [1].

### 5.3    Results

Figure 1 shows results for the *Most Common* and *Matched One* metrics, and Fig. 2 shows the *Mean Rank* results (error bars show 95 % confidence intervals). The results using the expert-authored feedback base are combined into a single entry, labelled as *Expert*. This was done for simplicity because the results were identical regardless of the explanation selection method used. The expert did not include redundancy so each relationship only appears once in the expert-authored feedback base. This causes the same explanation to be returned regardless of which explanation selection method is used.

---

[2] The case base described in [1] labelled *Patrol Random*. It contains cases learned from both the speed-focused and detection-focused operators (25 total cases).

[3] The learned feedback base is identical to the feedback base described in [2] where feedback is given by the operator 100 % of the time. It contains feedback from both operators.

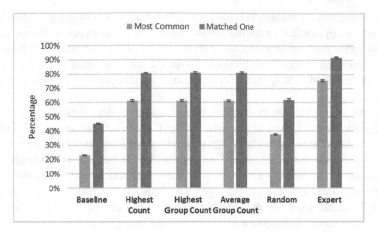

**Fig. 1.** The percentage of explanations that matched the most common feedback provided by the operator (Most Common) or any feedback provided by the operator (Matched One).

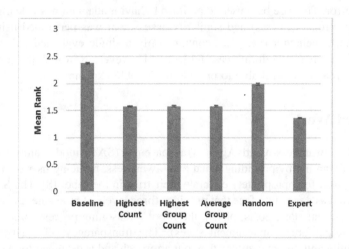

**Fig. 2.** The mean rank of the robot's explanation relative to a list that is ranked by the number of times each piece of feedback occurs during a trial.

All results that use our explanation approach (i.e., *Highest Count, Highest Group Count, Average Group Count, Random,* and *Expert*) were statistically significant improvements over *Baseline* (using a paired $t$-test with $p < 0.001$). These results provide support for **H2**. *Expert* outperformed the other approaches across all three metrics. The primary benefits of using the expert-authored feedback base are that there are no erroneous or redundant feedback relationships. Compared to the best results when using a learned feedback base (i.e., *Highest Count, Highest Group Count,* and *Average Group Count*), *Expert* can better provide an explanation that matches at least one piece of feedback given by the operator (91 % of the time vs. 81 % of the time), and regularly provides the most common piece of feedback (75 % of the time vs. 61 % of the time). These results support **H3**. However, the results are promising as they indicate that, while our approach for explanation works best with an expert-authored feedback base, reasonable performance can be achieved using a learned feedback base. Given that both the expert-authored and learned feedback bases resulted in explanations that closely matched feedback from the operator, the results provide evidence that **H1** is supported.

There are no significant differences between the results when using the *Highest Count*, *Highest Group Count*, or *Average Group Count* explanation selection methods. However, all three were significant improvements over *Random* explanation selection. This improvement occurred because some feedback relationships in the feedback base are erroneous. A higher count value in a feedback relationship indicates that the relationship has been observed more often (i.e., less likely to be the result of a single error), so the three methods that use the count value are better able to reduce the influence of erroneous relationships.

The primary reason why none of the approaches were able to achieve ideal performance (i.e., always providing an explanation that matched the most common piece of feedback given by the operator) is because the behavior adaptation process also introduced error. The case base used to perform behavior adaptation was learned (i.e., it may contain erroneous cases) and similarity assessment was performed using only a single evaluated behavior (i.e., $\mathcal{E}_{past}$ contains only a single evaluated pair so limited information was available during case retrieval). However, even with these sources of error our approach was still able to provide reasonable explanations.

# 6 Related Work

The Situation awareness-based Agent Transparency (SAT) model aims to improve human-robot teaming by providing situational awareness, reducing user overhead, and allowing the user to appropriately calibrate their trust in the robot [8]. The SAT model is implemented as a user interface that provides three levels of transparency: the robot's status (e.g., current state, goals, plans), the robot's reasoning process, and the robot's projections (e.g., future environment states). The transparency offered by the SAT model is significantly more complex than our approach and is designed for a user that is continuously monitoring the robot (i.e., constantly observing the robot through the interface). Instead, we focus on providing transparency for an operator who is performing their own tasks and may only monitor the robot sporadically.

Explanation in AI systems can be divided into *internal explanations* and *external explanations* [9]. Internal explanations are used by the system as part of its reasoning process. For example, DiscoverHistory [10] identifies discrepancies between observed environment states and expected environment states, and generates explanations for why those discrepancies occurred (e.g., the actions of other agents). These explanations provide the system with information about unobservable parts of the environment and allow it to respond more intelligently to unexpected events. The other category of explanation, external explanations, differs in that they aim to explain the system's reasoning process to a user. While many internal explanations can be used as external explanations, they are not intended to be understandable by a user (e.g., formatting, presentation, amount of content).

Storing concrete problem solving instances gives case-based reasoning an inherent advantage in providing explanations when compared to other learning approaches [11]. In some domains, providing the user with the cases themselves may be a sufficient explanation (e.g., a help desk system). Cunningham et al. [12] have shown that providing the retrieved case as an explanation improves user satisfaction compared to displaying a rule or giving no explanation. However, using the case as an explanation requires that the user can clearly understand why it is similar to the input problem and why the solution is appropriate [13]. In our system, explanations are a result of two separate case-based reasoning processes: one using the behavior adaptation case base and the other using the feedback base. Although the cases stored in the feedback base are relatively simple triples (i.e., feedback, relationship, and frequency), the behavior adaptation cases require a more complex similarity calculation. The complexity of retrieval, use of multiple interconnected case bases (i.e., the results of behavior adaptation are used as input when generating an explanation), and time constraints of the operator make it unsuitable to directly use the cases as explanations.

Sørmo et al. [13] identify five goals of explanation in CBR: *transparency* (i.e., how the answer was reached), *justification* (i.e., why the answer is good), *relevance* (i.e., why a question is relevant), *conceptualization* (i.e., clarify the meaning of a concept), and *learning* (i.e., teach the user). Our work falls under the transparency category since it is focused on how the robot made a decision. To a lesser extent, our explanations also provide justification by presenting the reason the agent thinks a behavior change was necessary (i.e., it is good because it aligns with the operator's preferences). Our work differs from the traditional use of explanations in CBR in that we are not explaining an answer that is given to a user, but instead explaining an agent's reasoning process. Our explanations are *cognitive explanations* since they explain the reasoning of an intelligent system [14]. Cognitive explanations have also been examined in ambient intelligence systems [15]. Similar to our work, they discuss how a system can explain to the user why a behavior was chosen. However, it differs in that it attempts to explain the reasoning that resulted in an incorrect or unexpected behavior being chosen, whereas we focus on explaining why a behavior was changed.

Issue-based prediction [16], like our own work, stores an explanation as the solution portion of a case. Case-based reasoning and a weak domain model are used to explain which side will win in a legal dispute. Our work uses much simpler explanations and does not require any domain knowledge during reasoning, instead learning knowledge about operator feedback. However, in a legal domain the explanations are much more complex and benefit from domain knowledge. FormuCaseViz [17] is similar to our own work in that the explanations are meant to reduce the cognitive burden on the user. The system visually displays the differences between the target problem and similar cases, helping the user to quickly understand the similarities and differences. This approach differs from our own in that it attempts to explain aspects of the CBR process to the user whereas we focus on explaining the agent's reasoning.

Explanation has been identified as an important feature of recommender systems, with numerous systems implementing explanation capabilities [18]. While many recommender systems explain why they gave a particular recommendation, explanations have also been used to explain why follow-up questions were asked in conversational recommender systems [19]. Muhammad et al. [20] have examined how the explain-ability of a case can be used to guide retrieval in a case-based recommender. Their approach is based on the idea that a useful case should be both similar and allow for informative explanations. Our work differs from recommender systems in that the agent is not providing alternatives for the operator to choose from, but is instead justifying a decision that has already been made.

It is not surprising that trust models have been examined in the context of case-based recommender systems [21], given the relationship between providing explanations and trust. Additionally, trust is an important factor in case-based agent collaboration [22] and case provenance [23]. Unlike our work, which focuses on inverse trust, these investigations examine traditional trust (e.g., trust in another agent or trust in the source of a case). Even outside of the CBR literature, most existing work focuses on traditional trust [4]. The two exceptions are the work of Kaniarasu et al. [24] and Saleh et al. [25]. Kaniarasu et al. use negative performance factors (e.g., how often the robot is warned of poor performance) and periodic performance feedback from the operator (e.g., whether the robot is currently performing well) to estimate the operator's trust. This differs from our own work in that their approach requires explicit perfor-mance feedback to estimate trust. Saleh et al. instead use a set of expert-authored rules to estimate operator trust. Unlike our inverse trust estimate, which can be used regardless of operator or mission context, their approach requires the rules to be redefined by an expert whenever a change in context occurs.

## 7 Conclusions

In this paper, we examined how an agent, which controls a simulated robot and performs trust-guided behavior adaptation, can provide explanations when it modifies its behavior. Introducing a layer of transparency should increase the agent's trust-worthiness by allowing it to present its operator with the motivation for any behavior changes. Our approach uses two case-based reasoning processes, both of which use cases that are learned while interacting with the operator. The first process, which was the focus of our prior work, involves evaluating the robot's trustworthiness and selecting a new behavior if the robot is behaving in an untrustworthy manner. When the robot adapts its behavior, a second case-based reasoning process is used to generate an explanation for why the change occurred. Given that the human-robot team may be in a time-sensitive situation, we designed our explanations to be simple, concise, and understandable.

Our evaluation involved an operator instructing a robot to patrol a simulated environment, identify suspicious objects, and classify them as threats or harmless. As the robot completed its tasks, it evaluated its trustworthiness and adapted its behavior if it determined it was untrustworthy. Our results indicate that the explanations provided by the robot for its behavior adaptations aligned closely with the explicit feedback provided by the operator. The primary area we wish to address in future work is to validate our results in a user study. While our system was based on the findings of existing research on human-robot trust and transparency, we plan to independently validate those findings using human operators in our robotics environment.

**Acknowledgements.** Thanks to ONR for sponsoring this research. Thanks also to Michael Drinkwater for his assistance in developing the eBotworks scenarios we used to evaluate our agent, and to the reviewers for their comments.

# References

1. Floyd, M.W., Drinkwater, M., Aha, D.W.: How much do you trust me? Learning a case-based model of inverse trust. In: Lamontagne, L., Plaza, E. (eds.) ICCBR 2014. LNCS, vol. 8765, pp. 125–139. Springer, Heidelberg (2014). doi:10.1007/978-3-319-11209-1_10
2. Floyd, M.W., Drinkwater, M., Aha, D.W.: Improving trust-guided behavior adaptation using operator feedback. In: Hüllermeier, E., Minor, M. (eds.) ICCBR 2015. LNCS, vol. 9343, pp. 134–148. Springer, Heidelberg (2015). doi:10.1007/978-3-319-24586-7_10
3. Dzindolet, M.T., Peterson, S.A., Pomranky, R.A., Pierce, L.G., Beck, H.P.: The role of trust in automation reliance. Int. J. Hum Comput Stud. **58**(6), 697–718 (2003)
4. Sabater, J., Sierra, C.: Review on computational trust and reputation models. Artif. Intell. Rev. **24**(1), 33–60 (2005)
5. Hancock, P.A., Billings, D.R., Schaefer, K.E., Chen, J.Y., De Visser, E.J., Parasuraman, R.: A meta-analysis of factors affecting trust in human-robot interaction. Hum. Factors J. Hum. Factors Ergon. Soc. **53**(5), 517–527 (2011)
6. Kaniarasu, P., Steinfeld, A., Desai, M., Yanco, H.A.: Potential measures for detecting trust changes. In: Proceedings of the Seventh International Conference on Human-Robot Interaction, pp. 241–242. ACM, Boston (2012)
7. Knexus Research Corporation: eBotworks (2016). Retrieved from http://www.knexus-research.com/products/ebotworks.php
8. Chen, J.Y.C., Barnes, M.J., Selkowitz, A.R., Stowers, K., Lakhmani, S.G., Kasdaglis, N.: Human-autonomy teaming and agent transparency. In: Proceedings of the Twenty-First International Conference on Intelligent User Interfaces, pp. 28–31. ACM, Sonoma (2016)
9. Aamodt, A.: Explanation-driven case-based reasoning. In: Wess, S., Richter, M., Althoff, K.-D. (eds.) EWCBR 1993. LNCS, vol. 837, pp. 274–288. Springer, Heidelberg (1994). doi:10.1007/3-540-58330-0_93
10. Molineaux, M., Kuter, U., Klenk, M.: Discover history: understanding the past in planning and execution. In: Proceedings of the Eleventh International Conference on Autonomous Agents and Multi-agent Systems, pp. 989–996. IFAAMAS, Valencia (2012)
11. Leake, D.B.: CBR in context: the present and future. In: Leake, D.B. (ed.) Case-Based Reasoning: Experiences, Lessons, and Future Directions. AAAI Press/MIT Press, Menlo Park (1996)
12. Cunningham, P., Doyle, D., Loughrey, J.: An evaluation of the usefulness of case-based explanation. In: Ashley, K.D., Bridge, D.G. (eds.) ICCBR 2003. LNCS, vol. 2689, pp. 122–130. Springer, Heidelberg (2003). doi:10.1007/3-540-45006-8_12
13. Sørmo, F., Cassens, J., Aamodt, A.: Explanation in case-based reasoning—perspectives and goals. Artif. Intell. Rev. **24**(2), 109–143 (2005)
14. Roth-Berghofer, T.R.: Explanations and case-based reasoning: foundational issues. In: Funk, P., González Calero, P.A. (eds.) ECCBR 2004. LNCS (LNAI), vol. 3155, pp. 389–403. Springer, Heidelberg (2004). doi:10.1007/978-3-540-28631-8_29

15. Kofod-Petersen, A., Cassens, J.: Explanations and context in ambient intelligent systems. In: Kokinov, B., Richardson, D.C., Roth-Berghofer, T.R., Vieu, L. (eds.) CONTEXT 2007. LNCS (LNAI), vol. 4635, pp. 303–316. Springer, Heidelberg (2007). doi:10.1007/978-3-540-28631-8_29

16. Brüninghaus, S., Ashley, K.D.: Combining case-based and model-based reasoning for predicting the outcome of legal cases. In: Ashley, K.D., Bridge, D.G. (eds.) ICCBR 2003. LNCS, vol. 2689, pp. 65–79. Springer, Heidelberg (2003). doi:10.1007/3-540-45006-8_8

17. Massie, S., Craw, S., Wiratunga, N.: Visualisation of case-case reasoning for explanation. In: Proceedings of the Seventh European Conference on Case-Based Reasoning Workshops, pp. 135–144. Madrid, Spain (2004)

18. Tintarev, N., Masthoff, J.: A survey of explanations in recommender systems. In: Proceedings of the Twenty-Third International Conference on Data Engineering Workshops, pp. 801–810. IEEE, Istanbul (2007)

19. McSherry, D.: Explanation in recommender systems. Artif. Intell. Rev. 24(2), 179–197 (2005)

20. Muhammad, K., Lawlor, A., Rafter, R., Smyth, B.: Great explanations: opinionated explanations for recommendations. In: Muhammad, K., Lawlor, A., Rafter, R., Smyth, B. (eds.) ICCBR 2015. LNCS, vol. 9343, pp. 244–258. Springer, Heidelberg (2015). doi:10.1007/978-3-319-24586-7_17

21. Tavakolifard, M., Herrmann, P., Öztürk, P.: Analogical trust reasoning. In: Ferrari, E., Li, N., Bertino, E., Karabulut, Y. (eds.) IFIPTM 2009. IFIP AICT, vol. 300, pp. 149–163. Springer, Heidelberg (2009). doi:10.1007/978-3-642-02056-8_10

22. Briggs, P., Smyth, B.: Provenance, trust, and sharing in peer-to-peer case-based web search. In: Althoff, K.-D., Bergmann, R., Minor, M., Hanft, A. (eds.) ECCBR 2008. LNCS, vol. 5239, pp. 89–103. Springer, Heidelberg (2008). doi:10.1007/978-3-540-85502-6_6

23. Leake, D.B., Whitehead, M.: Case provenance: The value of remembering case sources. In: Weber, R.O., Richter, M.M. (eds.) ICCBR 2007. LNCS, vol. 4626, pp. 194–208. Springer, Heidelberg (2007). doi:10.1007/978-3-540-74141-1_14

24. Kaniarasu, P., Steinfeld, A., Desai, M., Yanco, H.A.: Robot confidence and trust alignment. In: Proceedings of the Eighth International Conference on Human-Robot Interaction, pp. 155–156. ACM, Tokyo (2013)

25. Saleh, J.A., Karray, F., Morckos, M.: Modelling of robot attention demand in human-robot interaction using finite fuzzy state automata. In: Proceedings of the International Conference on Fuzzy Systems, pp. 1–8. IEEE, Brisbane (2012)

# Inferring Student Coding Goals Using Abstract Syntax Trees

Paul Freeman$^{(\boxtimes)}$, Ian Watson, and Paul Denny

Department of Computer Science, University of Auckland, Auckland, New Zealand
pfre484@aucklanduni.ac.nz, {id.watson,p.denny}@auckland.ac.nz

**Abstract.** The rapidly growing demand for programming skills has driven improvements in the technologies delivering programming education to students. Intelligent tutoring systems will potentially contribute to solving this problem, but development of effective systems has been slow to take hold in this area. We present a novel alternative, Abstract Syntax Tree Retrieval, which uses case-based reasoning to infer student goals from previous solutions to coding problems. Without requiring programmed expert knowledge, our system demonstrates that accurate retrieval is possible for basic problems. We expect that additional research will uncover more applications for this technology, including more effective intelligent tutoring systems.

## 1 Introduction

Across nearly all fields of study, students today are increasingly motivated to improve their skills in computer science, software engineering and, more specifically, computer programming [17]. As a result, programming education courses are being added to curricula at many universities [7]. Within the Science, Technology, Engineering, and Mathematics (STEM) fields, most students are expected to have skills in computer programming after completing their university studies [20], while many other faculties are also beginning to encourage education in this area [19].

This increased demand for computer programmers has pressured academic institutions to modify the delivery method of introductory computer science materials. In addition to textbooks and lectures, computer science courses now frequently include laboratory programming exercises. While programming exercises are nothing new, the access methods being used by students today have changed to keep up with improved technology.

Online services ease the delivery of educational materials to the students. Such services increase access to more students, regardless of platform choice, software configuration, or access to university computer labs. Additionally, this cloud-based method of instruction provides a valuable record of all student activities, making it easy for schools, universities, and education researchers to access data on student learning trends [14].

Our research addresses the application of Case-Based Reasoning (CBR) into the area of computer science education. Specifically, our work investigates the

© Springer International Publishing AG 2016
A. Goel et al. (Eds.): ICCBR 2016, LNAI 9969, pp. 139–153, 2016.
DOI: 10.1007/978-3-319-47096-2_10

potential for a system to intelligently infer the goal of a struggling student through naive analysis of both the student's current progress and existing solutions to the problem. Through inference of the goal, it is then possible to generate hints for the student.

This research targets students learning basic programming skills and completing short programming assignments. While some aspects of our research may be applicable to advanced courses, this is not the direct intention of this research.

## 2    Problem Description

Laboratory exercises are frequently used in programming courses, during which students develop solutions to simple programming problems. It is reasonable to assume that the data collected by these online learning environments is of some utility to future students solving the same problems. The data logs amount to hundreds or even thousands of prior code submissions for each programming problem. Could this data be used to generate hints for future students solving the same problem?

Our research uses a CBR approach, which we call Abstract Syntax Tree Retrieval (ASTR) to data mine prior solutions contained in a large dataset. Through analysis of retrieved solutions for specific code states, we attempt to answer the following questions:

- How can prior solutions be retained such that they are both readily accessible and easily compared to future submissions?
- What method of similarity can accurately approximate the work required by a student to move from the current state of a code string to the state of a solution string?
- Can our system frequently use the similarity method developed to select an appropriate existing solution that is not a drastically different approach to solving the problem?

The accuracy of the system is evaluated in two ways. Expert analysis is used to decide if the solution retrieved by ASTR is a reasonable solution to pursue, given the current code state. The percentage of retrievals determined to be *appropriate* provides the measure of success for this test. In a second test, we measure the system's success at retrieving a student's final solution when all of their prior attempts are submitted for ASTR. If the system selects the student's solution, this indicates that the system was able to *exactly* predict what the student would eventually do.

### 2.1    Human Tutor Emulation

ASTR attempts to emulate a human tutor, who is assumed to be an expert at solving the problem in question. The tutor observes the current failing state of the student's submission. If the student has used the correct approach, but has used incorrect code, the tutor should provide hints that lead to the correction of the code, but without changing the approach. Consider the following code strings:

```
def max_of_two_values(a, b):          def max_of_two_values(a, b):
    def max_of_two_values(a, b):          if (a <= b):
        if (a <= b):                          return b
            return b                      else:
        else:                                 return a
            return a
```

The student solved the problem correctly (shown on the left), but they were unaware that the system would add the function definition to the code they submitted. This is an artefact of the online platform and we would expect a tutor to guide the student toward an appropriate solution (shown on the right).

When the wrong approach has been taken by the student, the tutor should suggest modifications that result in a solution that uses as many parts of the current attempt as possible. The following submission and solution is considered a misunderstanding of the problem.

```
def max_of_two_values(a, b):          def max_of_two_values(a, b):
    if (a < b):                           if (a > b):
        return a                              return a
    else:                                 else:
        return b                              return b
```

On the left, the student calculated the incorrect return value. The tutor could help the student towards the solution on the right. Such a solution is certainly a logical step from the current state of the student's code, however, the tutor may also guide the student to a solution that swaps the return values, a and b, into the appropriate position, leaving the < operator alone. Both goals are considered equally appropriate.

The tutor should try to find a solution that is similar to the code the student has already created. In the previous example, it would be inappropriate for the tutor to move the student towards a radically different solution, such as the following:

```
def max_of_two_values(a, b):
    return max(a, b)
```

Ideally, an expert human tutor is able to provide appropriate guidance because they are able to consider many different ways of solving the problem and can accurately select a solution that seems most similar to what the student has written. ASTR provides this solution, which could then be used to generate any number of hints and guide the student in the appropriate direction.

## 2.2   A Case for Case-Based Reasoning

Online learning environments provide a tremendous opportunity to make use of CBR. Specifically, computer programming exercises lend themselves to CBR due to the fact that students are typically attempting to generate one of many (possibly infinite) solutions.

During many programming exercises, it is desirable to allow the student to explore the entire solution space rather than restricting them to finding one of only a few solutions. The use of CBR gives us the opportunity to allow this open exploration of solutions by using prior solutions to guide students back into the area of the total search space known to contain solutions.

## 3    Related Work

Our research builds off prior work in a number of areas. The most important of these areas include: Abstract Syntax Tree (ASTs), code clone detection, hint generation, and CBR. Here, we briefly introduce the current state of research in these topics as it relates to our work.

### 3.1    Abstract Syntax Tree

An AST is a hierarchical representation of a program into a branching sequence of operators. Each tree represents the hierarchy of a program. The leaves of the AST indicate which calculations are performed first. The results of these calculations become the next level of leaves. The process moves up the tree until the total program execution is calculated at the root of the trees. An example of a simple Python AST is provided in Fig. 1.

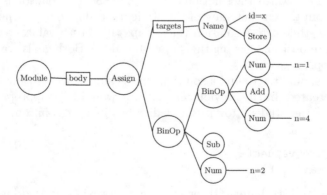

**Fig. 1.** The Python AST for x = 1 + 4 - 2.

ASTs provide a method for comparing source code fragments that focuses on the operations occurring within programs rather than the string labels used to identify the parts. They have a wide range of application areas. A tool developed by Falleri et al. [5] provides edit scripts (or *diff* files) for two versions of code by analyzing the AST. Generating edit scripts through a text-based process, while still correctly documenting the intended changes of the programmer, has an algorithmic complexity of $O(n^3)$ at a minimum. Using an AST method, the authors were able to develop an algorithm running, in worst case, in $O(n^2)$ time.

The use of ASTs in the educational domain has also been examined. Rivers and Koedinger [16] investigated the potential of using ASTs as part of an Intelligent Tutoring System (ITS). They highlight many of the advantages to using ASTs to compare student code to existing solutions. Their work attempts to establish a common framework upon which to build simple programming tutors using AST comparisons. They also explore possible methods of quickly generating hints for students from this existing data.

## 3.2   Code Clone Detection

One of the more successful applications of Abstract Syntax Tree has been in the areas of *code clone detection*, which is used to identify code segments that perform the same function, in the same way. Code clones are created through several processes and detecting these clones is useful. Many tools have been made that detect these clones [2] and they are often used to identify areas within a project for refactoring.

AST comparison methods cannot detect code clones in all situations. Inversion of `if-else` statements was identified as a problem area for AST-based code clone detection [4]. This clone generation process involves inverting the condition clause of an `if` statement and swapping the code contained within the `if` and `else` blocks. Other techniques, such as loop unrolling or inserting dummy methods, can also mask code clones from these detectors.

Tao et al. [18] researched methods by which ASTs can be adjusted to catch additional code obfuscation techniques. Their process provided procedural analysis for many common changes in logical structure. However, if the system translates the code into an AST so as to account for this, many common logical syntactic changes used to create code clones will be unable to mask their semantic similarity to the original.

Detecting code clones is useful for identifying potentially unwanted duplication of code. However, code clone detection can also be used in a more positive manner. In introductory programming courses, when students attempt to solve a programming exercise, they are attempting to create a code clone of one of the possible solutions. The measure to which a student's code is a clone of a given solution can be used as a measurement of their distance from that solution.

Leveraging this distance calculation allows the computation of a *nearest* solution when many are available. This observation is an important component of our ASTR system. Once a nearest solution is calculated, hint generation may be performed.

## 3.3   Hint Generation

During a problem solving task, especially one in which a tutor is involved, the ability to provide hints to the user is of interest.

Hints that have been preprogrammed into a system are known as *authored hints*. Authored hints are commonly used to provide feedback [12], but can only ever provide hints for a finite number of cases. It is therefore of interest to develop methods of automatically generating hints.

Arguably, some of the most robust hint generation systems fall into a category of systems known as *goal-directed hint generators*. Such systems possess the domain knowledge necessary to compute a path from any current state to the goal state. Goal-directed hint generation provides the greatest flexibility in generating hints, but at the expense of requiring large amounts of expert domain knowledge [14].

*State-based hint generation* is a broad category of hint generators that provide hints based on a known or computed path to a goal state, but lack the ability to provide hints for *all* possible states. State-based hint generators frequently make use of authored hints [1,13]. Because the system is waiting for specific state-driven events, the variety of hints required is limited. It is easy to give instructors the opportunity to improve the hint language used for a particular state.

Some tutoring systems have made use of annotations within authored solutions to improve the readability of generated hints [6]. Solutions are processed and the system indicates with a placeholder when annotations are needed. This allows the author to input custom annotations for hints, at the direction of the system.

### 3.4  Case-Based Reasoning

The availability of solutions leads to potential application of CBR. CBR aims to retrieve a prior solution, or case, with similarity to the current problem. The retrieved prior solution and the current problem can be used in conjunction to propose a solution to the current problem.

CBR has a range of applications in education. Kolodner et al. [10] wrote a commentary on the use of many different CBR-inspired approaches to learning. They concluded, "CBR-inspired educational approaches will be making their way more into the e-learning mainstream." Essentially, since CBR takes a similar problem solving approach to that of students, CBR technologies can be used to develop cognitive models of students.

Regan and Slater [15] developed a case-based hint generator for the virtual world DollarBay. The tutoring system was designed to teach users how to play the game effectively. An agent inside the virtual world would visit players who were struggling with certain aspects of the basic strategy. The agent would provide a message to the user, advising them how to improve their gameplay strategy. Inaction by the agent is also considered a valid action, and was the proposed "solution" to a subset of cases, as well. This event-driven approach is not a true CBR system, but does demonstrate the effective use of a case-base in an educational environment.

Although never explicitly referred to as a CBR system, the data mining hint generator, developed by Jin et al. [8], performs as such. Previous solutions and intermediate steps are mined by the system, with each solution being stored as a linkage graph, indicating relationships between variables. When a student requests a hint, the system transforms the code into a linkage graph and calculates the most similar "case". Since intermediate steps are saved by the system, the next step along a path to the goal is used to provide a hint to the student.

# 4    Abstract Syntax Tree Retrieval

Our system uses a process we refer to as Abstract Syntax Tree Retrieval (ASTR). It requires no prior knowledge of the problem being solved. It uses CBR and the grammar of the programming language to retrieve a prior solution with high similarity to a struggling student's failing submission.

Like many programming education systems, ASTR does not classify solutions as *more* or *less* correct. The system defers to the results of acceptance tests to determine correctness of a submission. *Every* student attempt at solving *any* programming problem is processed by the ASTR system. Attempts that pass the acceptance tests of a programming problem, and have not already been seen, are retained by the case-base as new solutions, while attempts that do not are matched to the most similar solutions from the case-base. Ideally, ASTR should perform in the same manner as the human tutor described in Sect. 2.1. During the retrieval process, the goal of the system is to retrieve a solution from the case-base that matches the intended solution of the student.

The case retrieved from the case-base is referred to as the *goal* of the student. If an expert believes the goal would be appropriate for the student to work towards, it is referred to as an *appropriate goal*. The challenge of ASTR is to correctly retrieve an appropriate goal from the case-base as often as possible.

## 4.1    Student Goal Assumptions

In the field of computer science, *edit distance* is a well-known method for determining the similarity of strings [11], trees [3,21,22], or graphs [9,23]. An underlying assumption is made that the student is attempting to reach the goal that will require the fewest number of edits to the current state of their program. By observing a student's initial failing attempt, and comparing it to their eventual solution, we can show this is frequently true.

In ASTR, edits are defined to be one of three possible modifications to the current state of the student submission:

1. Adding to the code.
2. Removing from the code.
3. Modifying an existing part of the code.

Additionally, edits are assumed to be made directly to the AST, despite the reality of students making edits to the source code text. However, since text modifications resulting in the same syntax tree are of no interest to our research, these types of modifications can be ignored.

It is also assumed that the case-base will always return one of the goals retained by the case-base. Although the goal will not always be in the case-base, assuming this would simply result in failing to return any case.

With these assumptions in place, we have reduced our problem to that for which CBR is most appropriate. Given a student submission and a set of successful cases, return the successful case that is the fewest number of edits from the state of the student submission.

## 4.2    Preprocessing and AST Generation

When a submission is received by the system, the Python string is parsed into an AST using the `ast` module. However, in addition to AST generation, a number of preprocessing steps are used to improve similarity calculations.

**Removing Unreachable Code.** Novice programmers frequently include unreachable code in their programs. Commercial tutoring systems would probably choose to notify the student about the unreachable code, but our research system would benefit from removing unreachable code as a preprocessing step before making any comparisons.

The following Python keywords are identified as being markers for potentially unreachable code: `break`, `continue`, and `return`. Code following any of these keyword arguments, provided it's at the same indentation level, will not execute. Consider the following example, which is a passing submission:

```python
def max_of_two_values(a, b):
    if a > b:
        return a
        print a
    if a < b:
        return b
        print b
```

The `print` statements in this code cannot execute. Functionally, this code would be the same as a submission without the print lines, however, the system would identify them each as a different solution, since the ASTs of the submissions would be different. It is for this reason that the preprocessing step removes any code following a `return` statement, or other block-terminating Python keyword.

In addition to these removals, the system removes code following `if`/`else` statements when one of these block-terminating keywords exists in both the `if` and `else` blocks, as in the following example:

```python
def is_odd(num):
    if num % 2 == 0:
        return False
    else:
```

```
    return True
is_odd(52)
```

These rules can be written into a tree pruning algorithm which traverses the AST recursively, removing unreachable code, resulting in fewer unique solutions for a given problem.

**Variable Standardization.** Online programming exercises rarely enforce naming conventions for the variables used in problem submissions. ASTs do not reach a level of abstraction that removes variable names. Therefore, submissions with different variable names will create different syntax trees.

A simple method for standardizing labels is to simply replace each variable name as it occurs with algorithmically generated values. In our system, each variable, as it is encountered, is replaced with an enumerated value. The associated variable names are stored in a table and future values are replaced with the same value as was used prior. This method is effective for problems with shorter length, but would need to be expanded to support problems with a larger number of variables. Additionally, there may be rare instances where the order of variable declaration makes the system unable to assign variables in a syntactically consistent way between submissions; however, such instances were not observed during our experimentation.

## 4.3 Case Retrieval

The preprocessed AST generated from the submission is now ready for ASTR. All submissions, whether passing for failing, are of some interest to the system. If a submission has successfully solved the problem, then solution retrieval is not needed. However, the submission should be added to the case-base if it is a new solution. The system does an equality check against existing solutions in the case-base. If the submission is unique, it is added to the case-base. Duplicate submissions are currently ignored, although retaining the duplication count for each submission could have future application. If the submission has not solved the problem (i.e. it is a failing submission), ASTR is used to calculate and return the most similar existing solution.

**Zhang Shasha Tree Edit Distance.** The similarity between ASTs is calculated with the Zhang Shasha (ZSS) tree edit distance algorithm [22]. The algorithm uses a dynamic programming approach to calculate the exact edit distance between the two input trees. The implementation used by the system is a Python implementation provided in the zss module. The implemented algorithm allows for customization of the cost weights for different edit operations. The algorithm returns a total cost value based on the cost summation of all edit operations necessary to transform one tree into the other.

For our experiments, we use two different weightings for the costs of edit operations. The first weighting assigns the cost to each edit operation as 1. We refer to this as the *metric* tree edit distance (TED) calculation in our results. The second weighting places a weight of 3 on *add* node operations, a weight of 2 on *change* node operations, and leaves the *remove* node weight as 1. We refer to this as the *weighted* TED calculation. The weighted calculation slightly favors smaller ASTs and is shown to perform better in many of our tests.

The ZSS algorithm is used to calculate a similarity value for all ASTs stored in the case-base. The AST with the smallest tree edit distance is the AST with the highest similarity and is the result of ASTR.

## 5  Dataset

Our research uses a large dataset to test the effectiveness of ASTR. The data consists of submissions created by students to solve a variety of programming problems. The submissions are written in the Python programming language. There are a total of 57,234 submissions, from 24 programming questions, in the dataset. The level of the exercises is targeted to beginning programmers.

### 5.1  Programming Questions

ASTR was evaluated against three problems from the dataset, which were uniquely identified as problems: 2593, 2594, and 2600.

Problem 2600 required students to determine the maximum value of two inputs. It had the greatest ratio of *total solutions* (154) to *unique solutions* (20), with an average of 7.7 repetitions of each solution. It was also the submission with the largest average number of users submitting the same solution, which was 6.35 users per solution. Almost half of the successful submissions, 74 out of 154, were reduced to the same unique solution. From a complexity standpoint, solutions to this problem were short, relative to the other problems in the dataset, and provided an acceptable baseline test for the system.

Problem 2593 required students to determine if an input was odd. Both problems 2593 and 2600 had a similar number of unique solutions, but this problem had a much greater number of unique failures. There could be many explanations for the large number of failures, but it would seem that a larger number of failures might correlate with a lower understanding of the problem by the students. However, student anomalies in submission patterns can also cause spikes in unique failures. For instance, some students will begin making arbitrary changes to their failing code in an effort to stumble upon a solution. This is especially likely when the student is given unlimited submissions and checking each submission is almost instantaneous. A sample solution to the problem 2593 can be seen here:

```python
def is_odd(num):
    if ((num % 2) == 0):
        return False
    else:
        return True
```

Problem 2594 was more complicated than the other problems in the experiment. It required students to calculate the largest divisor of the input value. Below is a sample solution to problem 2594:

```
def largest_divisor(num):
    largest = 0
    for i in range(1, num):
        if ((num % i) == 0):
            largest = i
    return largest
```

Students created 82 unique solutions to this problem and over 300 unique failing submissions. When examining the failing submissions, it was clear that many students either did not understand the definition of *largest divisor* or did not understand the algorithm used to calculate the largest divisor. This contrasted well with the other questions we used, as the definition of *odd* and *maximum* are typically well understood. Programming problem 2594 demonstrates our system's ability to handle submissions that correctly solve *a* problem, but the *wrong* problem.

## 6    Experiments

The intent of the experiments was to determine the performance of the case-base under ideal circumstances. To simulate ideal conditions, we would manually insert every solution into the case-base. With the well developed case-base, the experiment then presented each failing submission to the system. Identical failing submissions were removed. For each failing case, the case-base returned the most similar passing case. A log was created showing the source code of the failing case and the source code of the most similar matching case selected by the case-base. The log was reviewed by a human expert and marked for the appropriateness of the case returned by the case-base.

In assessing appropriateness, the expert is attempting to infer the goal of the student when the failing submission was written. Whether or not the retrieved case indicates the actual goal of the student is highly subjective. Therefore, the expert must decide if the goal proposed by the case-base is a natural progression from the current state of the failing code. Assessing the retrieved solutions was straightforward, but time intensive, requiring a couple hours to review logs containing the results of a couple hundred submissions.

The overall score of the case-base is the overall percentage of appropriate retrievals compared to the total number of failing submissions processed.

By providing the case-base with all solutions, there is an increased chance that the actual intended solution of a failed attempt will be contained. For example, if a student made a submission that failed the acceptance tests but then later made a second submission that passed the acceptance tests, it is likely that the second submission represents the original *goal* (see Sect. 4) of the student (although this is not always the case). By loading all solutions into the case-base prior to performing case retrieval, it is possible to match a student's failing submission with any future passing submission. If the case-base selects a student's own future submission as the most similar solution, this may be another indication that the ASTR system is performing accurately.

A separate experiment was performed to determine the frequency with which the student's own solution was retrieved as the goal solution by ASTR. For each failing submission, any solutions to the same problem, by the same user, were added to the case-base before any other cases were added. It was necessary to add the user's solutions first because the case-base only retains the first instance of each unique solution. Adding them first ensured that the user's solutions would be available for ASTR. The test performed the experiment and recorded the numbers of both accurate and inaccurate retrievals.

# 7    Results and Discussion

When ASTR was performed using a TED metric, the system was frequently able to retrieve an appropriate goal solution. Performance using a weighted TED calculation improved accuracy on problems 2594 and 2600 significantly while reducing the accuracy of problem 2593 only marginally. The results are shown in Table 1.

**Table 1.** ASTR appropriate retrieval performance

| Question | Metric | | | Weighted | | |
|---|---|---|---|---|---|---|
| | 2593 | 2594 | 2600 | 2593 | 2594 | 2600 |
| Unique solutions | 25 | 81 | 21 | 25 | 81 | 21 |
| Unique failures | 157 | 383 | 31 | 157 | 383 | 31 |
| Correct retrievals | 113 | 247 | 20 | 107 | 309 | 30 |
| Incorrect retrievals | 44 | 136 | 11 | 50 | 74 | 1 |
| Accuracy | 72 % | 64 % | 65 % | 68 % | 81 % | 97 % |

Question 2593 was the only programming problem that had a lowered accuracy using the weighted distance calculation. However, the number of inaccuracies only increased by 4 %. The cause for this change is not readily apparent. In any case, 2593 was the lowest performing test question.

Question 2594 contained the largest number of submissions, but still performed well. The initial performance of 64 % was the lowest recorded accuracy for the metric retrieval test, but this question responded very well to the weighted calculation, making 62 more correct retrievals than with metric calculation and putting the accuracy for this test at 81 %.

Question 2600 had the fewest examples of correct submissions and the fewest examples of unique failures. ASTR returned reasonable solutions for 30 of 31 submissions, or 97 % accuracy, using the weighted distance calculation. The metric

**Table 2.** User's solution identified as goal

| Question | Metric | | | Weighted | | |
|---|---|---|---|---|---|---|
| | 2593 | 2594 | 2600 | 2593 | 2594 | 2600 |
| Success | 111 | 66 | 30 | 76 | 64 | 34 |
| Failure | 142 | 415 | 22 | 177 | 417 | 18 |
| Accuracy | 44 % | 14 % | 58 % | 30 % | 13 % | 65 % |

calculation performed similarly to the other questions, with an accuracy of 65 % (Table 2).

ASTR accuracy rates were lower when we examined the rate at which a user's own solution was retrieved as the goal solution. Question 2600 performed the best, with over half the failing submissions being matched up with the correct solution. This could be a by-product of having many students submitting the same solution, however, question 2593 also had a high accuracy rate on this test (although the results dropped 14 % for the weighted distance test). Question 2593 underperformed on this test, with similar scores of 13 % and 14 % for the two similarity variations.

## 8   Conclusion

The results achieved by our system are encouraging. The ASTR system contains no information about the programming problem prior to observing successful submissions. Additionally, the system has no understanding of Python syntax. Despite these limitations, the system is usually able to correctly identify a solution that appears to an expert to be the student's intended goal.

The ASTR system is generally accurate for 2 out of 3 submissions at a minimum, with favorable problems potentially performing much better. Weighted TED calculations seem to result in improved retrieval accuracy over metric TED calculations, as was hypothesized. Retrieval performance was not improved enough to show this definitively, but the results are encouraging.

Student's "own solution retrieval" test had less favorable results. When taken together with the other results, this may indicate a need for more interaction with students during programming exercises. We have shown that ASTR frequently returns an appropriate goal, but this test shows that students may not have discovered this appropriate goal on their own. Although the exact reason for this is not known at this time, there is a potential application of ASTR to this problem in the future.

Prior solutions to introductory programming problems contain valuable knowledge for assisting students who are currently struggling. AST analysis seems to provide a strong naive data structure from which similarity values can be calculated between current state and goal state. Solutions with similar ASTs are often determined to be appropriate goals. Our ASTR system is able to leverage this knowledge with minimal effort, providing potential for the development of ITSs, hint generation systems, and working across a large of domain of problems.

# References

1. Antonucci, P., Estler, C., Nikolić, Đ., Piccioni, M., Meyer, B.: An incremental hint system for automated programming assignments. In: Proceedings of the 2015 ACM Conference on Innovation and Technology in Computer Science Education (ITiCSE 2015), pp. 320–325 (2015)
2. Baxter, I.D., Yahin, A., Moura, L., Sant'Anna, M., Bier, L.: Clone detection using abstract syntax trees. In: Proceedings of the International Conference on Software Maintenance (ICSM 1998), vol. 98, pp. 368–377 (1998)
3. Bille, P.: A survey on tree edit distance and related problems. Theoret. Comput. Sci. **337**(1–3), 217–239 (2005)
4. Cui, B., Li, J., Guo, T., Wang, J., Ma, D.: Code comparison system based on abstract syntax tree. In: 2010 3rd IEEE International Conference on Broadband Network and Multimedia Technology (IC-BNMT), pp. 668–673 (2010)
5. Falleri, J.R., Morandat, F., Blanc, X., Martinez, M., Montperrus, M.: Fine-grained and accurate source code differencing. In: Proceedings of the 29th ACM/IEEE International Conference on Automated Software Engineering (ASE 2014), pp. 313–324 (2014)
6. Gerdes, A., Heeren, B., Jeuring, J.: An interactive functional programming tutor. In: Proceedings of the 17th ACM Annual Conference on Innovation and Technology in Computer Science Education (ITiCSE 2012), pp. 250–255 (2012)
7. Guzdial, M.: A media computation course for non-majors. SIGCSE Bull. **35**(3), 104 (2003)
8. Jin, W., Barnes, T., Stamper, J., Eagle, M.J., Johnson, M.W., Lehmann, L.: Program representation for automatic hint generation for a data-driven novice programming tutor. In: Cerri, S.A., Clancey, W.J., Papadourakis, G., Panourgia, K. (eds.) ITS 2012. LNCS, vol. 7315, pp. 304–309. Springer, Heidelberg (2012). doi:10. 1007/978-3-642-30950-2_40
9. Kammer, M.L.: Plagiarism detection in Haskell programs using call graph matching. Master's thesis, Utrecht University (2011)
10. Kolodner, J.L., Cox, M.T., González-Calero, P.A.: Case-based reasoning-inspired approaches to education. Knowl. Eng. Rev. **20**(3), 1–4 (2005)
11. Lu, W., Du, X., Hadjieleftheriou, M., Ooi, C.: Efficiently supporting edit distance based string similarity search using B+-trees. IEEE Trans. Knowl. Data Eng. **26**(12), 2983–2996 (2014)
12. Mckendree, J.: Effective feedback content for tutoring complex skills. Hum. Comput. Interact. **5**(4), 381–413 (1990)
13. Paquette, L., Lebeau, J.-F., Beaulieu, G., Mayers, A.: Automating next-step hints generation using ASTUS. In: Cerri, S.A., Clancey, W.J., Papadourakis, G., Panourgia, K. (eds.) ITS 2012. LNCS, vol. 7315, pp. 201–211. Springer, Heidelberg (2012). doi:10.1007/978-3-642-30950-2_26
14. Piech, C., Sahami, M., Huang, J., Guibas, L.: Autonomously generating hints by inferring problem solving policies. In: Proceedings of the Second ACM Conference on Learning @ Scale (L@S 2015), pp. 195–204 (2015)

15. Regan, P.M., Slator, B.M.: Case-based tutoring in virtual education environments. In: Proceedings of the 4th International Conference on Collaborative Virtual Environments (CVE 2002), pp. 2–9 (2002)
16. Rivers, K., Koedinger, K.R.: A canonicalizing model for building programming tutors. In: Cerri, S.A., Clancey, W.J., Papadourakis, G., Panourgia, K. (eds.) ITS 2012. LNCS, vol. 7315, pp. 591–593. Springer, Heidelberg (2012). doi:10.1007/978-3-642-30950-2_80
17. Snyder, L.: Being Fluent with Information Technology. National Academy of Sciences, Washington, D.C. (1999)
18. Tao, G., Guowei, D., Hu, Q., Baojiang, C.: Improved plagiarism detection algorithm based on abstract syntax tree. In: 2013 Fourth International Conference on Emerging Intelligent Data and Web Technologies (EIDWT), pp. 714–719 (2013)
19. Vaidyanathan, S.: Fostering creativity and innovation through technology. Learn. Lead. Technol. **39**, 24–28 (2012)
20. Wilson, C., Sudol, L.A., Stephenson, C., Stehlik, M.: Running on empty: the failure to teach K-12 computer science in the digital age. Technical report, Association for Computing Machinery (ACM) (2010). http://runningonempty.acm.org/fullreport2.pdf
21. Yang, R., Kalnis, P., Tung, A.K.H.: Similarity evaluation on tree-structured data. In: Proceedings of the 2005 ACM SIGMOD International Conference on Management of Data (SIGMOD 2005), pp. 754–765 (2005)
22. Zhang, K., Shasha, D.: Simple fast algorithms for the editing distance between trees and related problems. SIAM J. Comput. **18**(6), 1245–1262 (1989)
23. Zheng, W., Zou, L., Lian, X., Wang, D., Zhao, D.: Efficient graph similarity search over large graph databases. IEEE Trans. Knowl. Data Eng. **27**(4), 964–978 (2015)

# Combining CBR and Deep Learning to Generate Surprising Recipe Designs

Kazjon Grace[✉], Mary Lou Maher, David C. Wilson, and Nadia A. Najjar

Department of Software and Information Systems,
University of North Carolina at Charlotte, Charlotte, NC, USA
{k.grace,m.maher,davils,nanajjar}@uncc.edu

**Abstract.** This paper presents a dual-cycle CBR model in the domain of recipe generation. The model combines the strengths of deep learning and similarity-based retrieval to generate recipes that are novel and valuable (i.e. they are creative). The first cycle generates abstract descriptions which we call "design concepts" by synthesizing expectations from the entire case base, while the second cycle uses those concepts to retrieve and adapt objects. We define these conceptual object representations as an abstraction over complete cases on which expectations can be formed, allowing objects to be evaluated for surprisingness (the peak level of unexpectedness in the object, given the case base) and plausibility (the overall similarity of the object to those in the case base). The paper presents a prototype implementation of the model, and demonstrates its ability to generate objects that are simultaneously plausible and surprising, in addition to fitting a user query. This prototype is then compared to a traditional single-cycle CBR system.

## 1 Introduction

A great challenge of applying AI techniques to generate creative designs is how known, successful objects can be the basis for new design that are both valuable and surprising [1]. We have developed a dual-cycle case based reasoning approach to creative design that uses known designs first as a basis for setting expectations and second as a basis for adaptation. The first cycle uses a deep learning based model of unexpectedness that is trained on our case base. This model is then used to search the space of possible "design concepts" (i.e. abstract ideas that have not yet been developed into complete objects) for those that are creative, which we define as being simultaneously surprising and valuable [2,3]. These concepts, which are known to be particularly unexpected given our extant case base, are then used as input into a second case-based reasoning cycle, which retrieves and adapts the closest matching known recipe to fit the new concepts. This is inspired by the division between idea-focused *conceptual design* and detail-focused *detailed design* common to many models of design [4]. We present a prototype implementation in the domain of recipes as a proof of concept.

The core contribution of this approach is that our first cycle generates a design concept that is unexpected given the known objects in the case base. This

© Springer International Publishing AG 2016
A. Goel et al. (Eds.): ICCBR 2016, LNAI 9969, pp. 154–169, 2016.
DOI: 10.1007/978-3-319-47096-2_11

is an inversion of the typical similarity-based SBR approach, in which generated objects are typically highly similar to exemplars from the case base. In our prototype in the domain of recipe generation, this concept is a set of ingredients without amounts or preparation steps. In other domains the concept could be a similar set of high-level object "features". This divergent new design is then used as input to the second cycle in which the most similar case is retrieved and adapted to integrate the surprising object's unexpectedness. The first cycle learns the latent structure of the domain from the case base in order to generate a new object that deliberately flouts that structure, while the second cycle re-integrates the new object with known cases through adaptation.

Our prototype system, called "Q-chef", short for the "curious chef", uses this approach to generate new recipes based on requirements given by a human user. Dietary diversity has been shown to correlate with overall health [5] and the greatest predictor of culinary preference is exposure [6]. Given these facts, our goal is to develop systems that can generate surprising recipes and make users curious, leading to broadened dietary preference over time. Q-chef generates new recipe designs that meet three objectives: they fit the requirements, they are "plausible" (i.e. as a whole they reflect the latent structure of the case base) and they are "surprising" (i.e. they contain a combination of ingredients judged as highly unexpected given the case base). We present several experiments with a set of queries matching three simulated "user personas".

## 2   Background

Three fields are germane to this research: computational design and creativity, novelty in case-based reasoning, and computational recipe generation.

### 2.1   Computational Design and Creativity

Computational creativity is the application of AI techniques to creative problems, including art, music, design, science and more [7]. Computational creativity has three purposes: to aid in our understanding of creative cognition by modeling it computationally, to aid the development of tools that support creative people, and to develop autonomously creative machines. Computational design is the study of systems that model, simulate or assist with parts of the design process. It has been a fundamentally transformative force in the design industry – from bridges to fashion – for the last sixty years. Computational creative design is the intersection of these two fields: it is the computational modeling of the creative designing in order to understand, aid and reproduce it [8].

Within computational creativity an enduring question has been the definition of a "creative" artefact or process, and how best to evaluate such computationally. There are a multitude of definitions ([9] lists more than 50) but perhaps the closest to consensus exists for the novelty/value duality: a creative product must be "novel and valuable for the thinker or [their] culture" [10]. The synthesis of the factors in this definition are challenging to operationalize. The value of an artefact (also referred to as "utility" or "appropriateness") is complex and subjective, and only becomes more so when considering it in a social and cultural context. In recent years novelty has been argued to be similarly complex and subjective, being the product of expectations grounded in experience rather than some objective measurement of difference from what has come before [2]. In this view, novel designs are surprising: they *violate confident expectations* formed about that design domain based on previous exposure. In our previous work we have used a variety of unsupervised machine learning techniques to model unexpectedness [11,12], and here we place those models in a case-based reasoning context. Expectation is fundamentally case-based, and searching for the unexpected is reasoning from the past in order to not repeat it.

## 2.2 Novelty, Diversity and Creativity in Case-Based Reasoning

Case-based reasoning (CBR) is a framework for problem solving that finds new solutions through the retrieval and adaptation of similar past cases. CBR approaches have been applied to solve creative design tasks in the domains of graphics [13], musical pieces [14], poems [15], and plots for story telling [16]. There are also many existing examples of CBR in recipe generation [17–19].

In a survey of CBR for design contexts Goel et al. [20] point out that creativity is one of the factors that make the design task very challenging. Research in this context had mainly focused on early explorations and innovative design in either architecture or engineering. It has also been suggested that creativity in design depends on the retrieval of cases that are either not literally similar or similar only in a specific and potentially non-obvious way, and that these "subtle" similarities can lead to more creative solutions [21].

Valitutti [13] presents a discussion on how to characterize a case-based generative system as "creative". He suggests that the case-based adaptive process can be viewed as a type of search in the space of possible artifacts, with the set of past examples as an "inspiring set". Gervás [15] discusses whether a CBR application that generates poetry versions of prose texts provided by a user should be considered creative (when surprising the user with results dissimilar from the query) and/or faithful (when keeping as close as possible to the query).

Diversity in CBR has been considered in the retrieval step. Smith et al. [22] argue that diversity is particularly important for case-based recommender systems. The similarity assumption in CBR can have an adverse effect on the diversity of the retrieved cases. In situations where multiple cases are retrieved for a given query we can explicitly consider the diversity among as well as their similarity to the query. Doing so, Smith argues, provides users with better coverage of the information space in the neighborhood of their query. By contrast McSherry [23] argues that increasing diversity at the expense of similarity may not always be acceptable. such as when items are available only for a limited period or sought in competition with other users. To address this problem, they have presented a retrieval algorithm (DCR-1) that increases diversity while preserving similarity.

## 2.3   Recipe Generation Systems

Creating new recipes has a long history in CBR. Early examples include CHEF [17], which applied case-based planning to the domain of Szechwan cooking recipes, and Julia [18], which incorporated recipe constraints (e.g., vegetarian, dairy allergy) as part of overall meal planning. More recently, entries to the Computer Cooking Contest series at ICCBR (2008–2015) have addressed a variety of recipe generation challenges (e.g., sandwich, mixology, adaptation). While there is a wide range of specific instantiations, systems typically have representation and domain knowledge conventions that serve a very similar case/domain knowledge role, accounting for notions of user taste preference, ingredient characteristics, recipes (typically the fundamental case representation), and meals. For example, ingredient hierarchies are prevalent [24,25], with hierarchy distance as a common enabling metric [24]. The main distinguishing factors are often the specific approaches to case similarity and adaptation. Case similarity is often a variation on aggregating local similarities for component ingredients, but almost universally with the goal of retrieving the cases with the most similar features to a given query. A number of approaches define explicit similarity metrics based on co-occurrence of ingredients in recipes, such as frequent itemset measures. For example, GoetheShaker [26] employs co-occurrence of ingredients as a measure of complementarity of cocktail ingredients, while Earl [19] employs an *a priori* algorithm based measure of coherence for sets of sandwich ingredients.

# 3   The Q-Chef CBR Model

The overall Q-chef CBR process consists of two complementary CBR cycles: *problem framing* and *problem solving*. Both CBR cycles employ the same case-base, but take different perspectives on how the cases and knowledge containers for similarity and adaptation are employed. During the problem framing cycle cases are viewed as "recipe concepts" — high-level sets of ingredients without regard for recipe detail, such as measure. These ingredient set combinations are synthesized and evaluated for their level of surprise and their plausibility. Based on this evaluation a set of ingredients is selected as the recipe concept and used as a revised query for the problem-solving cycle. During the problem-solving cycle cases are viewed as recipes and the selected set of ingredients from the problem framing cycle comprise an optimized query to retrieve similar recipes that are adapted to provide a more detailed recipe design. The system can also provide the query directly to the problem-solving cycle for comparative purposes.

Q-chef also uses the underlying case-base to learn the key enabling similarity knowledge employed in both cycles. A deep unsupervised neural network component is trained over the case-base of ingredient combinations, in order to model the latent ingredient associations present in the case knowledge. During the problem framing cycle, this model is used as synthesis knowledge; it provides a way to measure surprise and plausibility for generated recipe concepts. While we conceive of this as a design synthesis process, it could be analogously compared to a kind of mass adaptation from a set of past designs via a machine learning model and a genetic search. During the problem-solving cycle, the deep learning model is used as similarity knowledge to support recipe case retrieval. In this way, the problem-solving cycle focuses on case knowledge as the primary driver of the CBR process — the recipe cases best matching the design concept query from the problem framing cycle are retrieved, with limited adaptation sufficient to fit the retrieved case to the problem context.

The Q-chef model is illustrated in Fig. 1, in which the shaded boxes highlight how Q-chef differs from the characteristic CBR model: we include a deep learning model of expectation from existing cases. This model is trained to identify whether a new design concept is surprising and plausible, which are the objectives of our synthesis process. Q-chef starts with a request for a new recipe that has two parts: a set of ingredients as the requirements and a goal expressed as a desired level of surprise. If the desired level of surprise is 0, the problem framing cycle is bypassed (dotted line) and the problem solving cycle is activated using the set of ingredients specified in the requirements as the design concept. If the desired level of surprise is not 0, the first cycle of problem framing is activated. In Q-chef, problem framing has the goal of generating many sets of combinations of ingredients and evaluating their level of surprise and their plausibility. The generative process is a genetic algorithm that generates combinations of ingredients in the space of all possible ingredients and has a fitness function that expresses surprise and plausibility. The basis for evaluating surprise and plausibility is a model of expectation that is generated using a deep learning algorithm that allows Q-chef to associate a level of probability and surprise with combinations of ingredients.

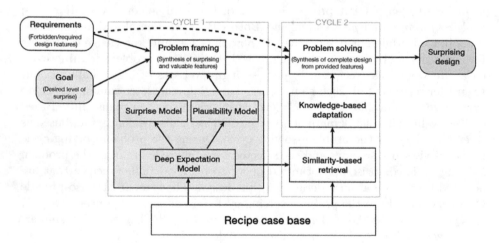

**Fig. 1.** Q-chef's dual-cycle CBR model, starting with the goal and requirements on the left. Grey shaded boxes indicate our deep learning model.

# 4    Experimenting with Q-Chef

We present a series of test cases designed as a proof of concept of the Q-chef dual-cycle CBR approach. In addition to demonstrating what is possible with our CBR approach to generating creative recipes, our goal is to distinguish between the generation of recipe designs that have a goal for surprising the user and the generation of recipe designs that bypass the problem framing cycle to generate expected recipe designs. We have designed the tests to compare our prototype's ability to synthesize plausible yet surprising designs with a modified version that retrieves only the closest match from the case base. Both models are presented with a list of required ingredients and a list of forbidden ones, and the dual-cycle model is additionally presented with a desired level of surprise in the interval [0–1], normalized to the most surprising discovery in the case base. Both models are configured to be non-learning for the purposes of simplicity during this test, with no results being added back to the case base.

## 4.1    The Dual-Cycle Q-Chef Prototype

We have developed a prototype implementation of the dual-cycle CBR model described in Sect. 3. The first CBR cycle uses a combination of a genetic algorithm and a deep neural network to synthesize sets of ingredients that are simultaneously plausible, surprising, and fit the requirements. The second CBR cycle takes this set of ingredients, which we refer to as a "recipe concept", and then retrieves the most similar recipe from the case base and adapts it to fit the concept and requirements. If the GA-based synthesis process of the first cycle is viewed as a kind of mass adaptation using knowledge from the whole case base, then both cycles can be considered complete CBR processes. From this perspective the first cycle focuses on adapting the query to produce a design concept that is both surprising and plausible, while the second cycle focusses on retrieving knowledge to flesh out that concept. Similarity and the evaluation of adaptations are also based on the neural network model of expectation.

The deep neural network we use to capture the latent structure of the case base is based on a model of expectation used in our previous work [2,11,12]. This approach uses unsupervised representation learning [27] to learn the latent structure of the case base. It uses a Variational Autoencoder (VAE [28,29]), a deep generative model, to learn a vector of random Gaussian variables. These variables can be sampled from to approximate the distribution of the case base. Currently this model is trained only on the set of ingredients in each recipe, which we call a "recipe concept", and ignores the amounts and preparation steps of those ingredients. This model captures the relationships between case features that an observer familiar with the case base would reasonably expect. We use the expectation model to estimate the plausibility and surprisingness of known cases and newly generated recipes, as well as in our similarity-based retrieval.

The expectation model is used to evaluate the surprisingness of a recipe concept – a set of ingredients that has either been synthesized or extracted from a known case. Our surprise model uses the "missing data imputation" method described in Rezende et al. [28] to estimate the likelihood of different ingredients given the context of a set of other ingredients. This is then expressed as the number of bits of information which would be provided by observing that ingredient in that context. This is then compared to the number of bits of information which would be gained by observing the same ingredient without that context. The number of *additional* bits given the context is referred to as the number of "wows" by Baldi and Itti [30]. For simplicity in the current prototype we only use those ingredients that appear in at least 1 % of recipes in the deep learning model. We also consider only the impact of all surprising combinations of up to length three for reasons of computational tractability. A full description of this process can be found in our previous work on the subject [31].

The same deep learning model used for estimating surprise can be used to evaluate the plausibility of a recipe. We define plausibility as the likelihood the expectation model assigns to that recipe concept based on the model of the case base's latent structure it has learnt. Plausibility counterbalances the system's drive to maximize surprise by forcing the search trajectory towards recipe concepts that are similar to those in the case base. These objectives are not the opposites they may at first seem: our surprise measure focuses on surprising combinations of ingredients, while the plausibility measure considers the recipe as a whole. A recipe can be of high surprise and plausibility if it is mostly mundane, with a few highly novel features. This pattern has been found to correlate with impact in scientific publications [32].

We use a genetic algorithm to generate the *recipe concepts* that are used as a query into the case base for the second cycle. We use the NSGA-2 multiobjective approach [33], in which Pareto dominance in objective space is used to determine which individuals are selected for the next generation. The effect of this is that the system can pursue both plausibility and surprise in its recipe concepts without having to combine the two into one objective. In the experiments presented in this paper the genetic algorithm was run for 50 generations with a population of 10,000. The population was initialized from a normal distribution with mean ($\sim$8) and standard deviation ($\sim$3) equal to the case base.

The deep learning model is also used to determine similarity in the retrieval process in the second CBR cycle. The network we are using is a kind of autoencoder, meaning that it learns a way to re-represent each case that captures its most salient features. This can be considered a form of nonlinear dimensionality reduction. We use Euclidean distance in the space thus produced as our similarity metric. This representation captures the latent structure of the case base, comparisons made using it will capture meaningful differences between cases.

The current recipe adaptation engine only operates at the recipe concept level – it adapts the set of ingredients suggested by the concept based on the set of ingredients in the closest matching case. Adapting the amounts and preparation steps to fit the recipe concept is intended as future work. The adaptation of recipe concepts is based on a simple knowledge base that assigns tags to each ingredient and permits only the substitution of ingredients matching a tag. Tags include "fruit", "protein", "vegetable", "fat", and "building block", with the latter being a special category for those ingredients whose role is based on their chemical properties, such as yeast or baking soda. We forbid the substitution of "building block" ingredients as to do so intelligently would require far more knowledge. This substitution approach is simplistic, but sufficient to demonstrate the strength of the dual-cycle model.

The system generates all valid substitutions given a source and a target recipe, and then evaluates the surprise, plausibility and fit to requirements of each. Recipes with a surprise further than 0.3 from the desired level specified in the query are discarded (an arbitrary threshold chosen to help demonstrate the diversity of "feasible" solutions), as are those that do not match the requirements. The remaining recipes are sorted by plausibility and returned to the user. This re-evaluation is required because while the set of recipe concepts provided to the second CBR cycle is known to be surprising, plausible and fit to the requirements, there is no guarantee that these traits survived adaptation. For the experiments presented here we follow a simplified version of this process, with the first cycle providing only 3 recipes to the second for each query rather than all recipes on the Pareto frontier that pass the requirements. We also find only the single highest scoring adapted case for each of those recipes. This results in output that can be effectively conveyed in a paper, although it does sacrifice some of the potential diversity generated by the first cycle.

## 4.2   The Single-Cycle Prototype

The single-cycle prototype bypasses the processes of problem framing and surprise modeling used in the dual-cycle prototype. In this system, which we have developed for comparison purposes, requirements are provided directly to the "problem solving" cycle of our model, which performs traditional similarity-based retrieval and case adaptation. The single-cycle prototype retains the deep-learning based similarity function used in the Q-chef prototype. Additionally, the results are ranked according to the same requirements-adjusted plausibility score derived from that deep learning network.

In this experiment the single-cycle prototype is used to retrieve the closest-matching recipe to the provided requirements. These closest-matching recipes are not guaranteed to fit the requirements (as the similarity function does not respect forbidden or required ingredients), so we iteratively discard the closest match until one which matches the requirements is found. We effectively treat any retrieved recipes with forbidden requirements as having a similarity of 0, and thus exclude them from the closest-matching recipes. All potential adaptations of the retrieved recipes are then evaluated for surprise, plausibility and fit to requirements in the same way in the dual-cycle prototype, and the Pareto-dominant front of adaptations is kept.

## 4.3   Test Queries

We have developed a set of three representative test queries, each reflecting a simulated user query, and each diverse enough from the others to show the diversity of our approach. These queries are deliberately simple as our user model is still under development, with each involving a single required ingredient and a set of forbidden ones. The test queries are shown in Table 1.

**Table 1.** The three test queries representing our user personas, each with their required and forbidden ingredients as well as desired level of surprise.

| Nickname | Required | Forbidden | Surp |
|----------|----------|-----------|------|
| Picky kid | Cheese | Hot sauce, chillies, turmeric, peanut butter, spinach, olives | 0.3 |
| Vegetarian | Beans | Fish, beef, chicken, prawns, bacon, sausage | 0.6 |
| Halal foodie | Chicken | Pork, sausage, bacon, wine, sherry | 0.9 |

### 4.4   Results

We present the results of our simulations in two steps. In the first we focus on the first cycle of our dual-cycle prototype, showing the ranges of surprise and plausibility it assigned to the recipe concepts it generated, as well as several exemplary such concepts. In the second we compare the output of the second cycle in the dual-cycle prototype with that of the single-cycle prototype.

#### 4.4.1 Recipe Concepts from the First Cycle of the Dual-Cycle Model.

Figure 2 shows a scatter plot of the surprise and plausibility evaluations for the best results of each of the three queries. This is the performance of the dual-cycle prototype only, as the single-cycle prototype does not generate a diverse set of concepts in this manner. This shows the top 100 recipes that fit the requirements while having the appropriate level of surprise and the highest possible plausibility. The multi-objective ranking uses Pareto dominance and crowding distance as per the NSGA-II algorithm (see [33]), which favors those recipes that are not worse at both objectives than any others, then those that are only worse at both objectives than one other recipe, and so on.

Note that each of the three result sets grow progressively further from the desired surprise level as plausibility increases, forming a leftward-pointing rough triangle in Fig. 2. This forms because of the search for Pareto-optimal combinations of maximal plausibility (the X axis) and proximity to the desired level of surprise (the Y axis). Designs at the leftmost point of the triangle have lower plausibility but are extremely close to the desired surprise, while designs further to the right diverge from the desired surprise in both directions. The well-covered Pareto frontier shows that the problem framing cycle is able to generate a diverse set of recipe concepts from the case base, each of which can then be used in retrieval. If combined with a compositional adaptation system (such as [34]) this diversity could have a significant positive impact.

After discarding the portion of the Pareto frontier with a surprise rating more than 0.3 from the desired level, we are left with a set of valid designs.

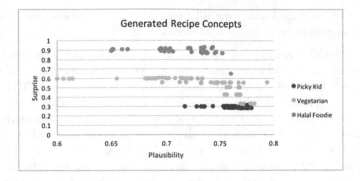

**Fig. 2.** Plot of the plausibility and surprise evaluations of recipe concepts generated for our three test queries: the "picky kid", the "vegetarian" and the "Halal foodie".

**Table 2.** Exemplary recipe concepts generated by the problem framing cycle of our dual-cycle model. For each query the first recipe has the highest plausibility of the result set, the second is a mix of plausibility and surprise, and the third is the recipe concept with the closest surprise to the desired level.

| Query | # | Ingredients | Plaus | Surp |
|---|---|---|---|---|
| Picky kid | 1 | Cheese, salt, eggs, butter, milk, black pepper, flour | 0.823 | 0.411 |
| | 2 | Cheese, lemon, water, flour, butter, sugar, salt | 0.804 | 0.237 |
| | 3 | Cheese, salt, vegetable oil, mayonnaise, black pepper, olive oil | 0.764 | 0.300 |
| Vegetarian | 1 | Beans, parmesan, garlic, salt, tomatoes, black pepper, onions | 0.756 | 0.405 |
| | 2 | Beans, salt, eggs, milk, onions, flour | 0.735 | 0.427 |
| | 3 | Beans, water, salt, baking powder, olive oil, flour sugar | 0.655 | 0.600 |
| Halal foodie | 1 | Chicken, salt, eggs, butter, icing sugar, flour | 0.769 | 0.749 |
| | 2 | Chicken, vanilla, salt, eggs, butter, milk | 0.765 | 0.981 |
| | 3 | Chicken, lemon, vanilla, salt, eggs, sugar | 0.740 | 0.949 |

Three exemplars from this set for each query, selected from the two extremes and the middle of the frontier, are shown in Table 2. Note that the first recipe for each query appears to be more well-formed than the third, although the latter best matches the desired surprise. For example, the third recipe for the "picky kid" query appears to be a collection of flavors without a base protein or starch.

Note that for the "Picky Kid" and "Vegetarian' queries in Table 2, the generative first-cycle process was able to exactly match the desired surprise level, while in the Halal Foodie query it was not. This is caused by the sparsity of the extreme high-end of the surprise distribution, as there are significantly fewer combinations of surprising ingredients around the 0.9 required by the query. The distribution of surprise levels over all possible feature combinations (of up to length 3) is approximately Poisson with mean ∼0.29.

**4.4.2 The Effectiveness of Dual-Cycle Retrieval.** The results from Sect. 4.4.1 form the dual-cycle model's input into the second cycle. We take the recipe concepts from Table 2 and use them as input to the second CBR cycle. For each of these the closest matching recipe is retrieved and substitution performed. We collate for each query persona the resultant list of all valid substitutions for each matched recipe, after removing all recipes that are not within 0.3 of the desired surprise level. Table 3 shows the best matching recipe for each concept in Table 2, along with the substitutions made by the second cycle of our system.

Table 3 shows that the dual-cycle model appears to be able to adapt recipes to increase their surprise while maintaining high plausibility. To put the plausibility

**Table 3.** Top retrieved and adapted recipe for each of the recipe concepts in Table 2. Ingredients added during adaptation have been bolded, ingredients substituted during adaptation are paired with the removed item struck out and the added item bolded.

| Query | Recipe | Plaus | Surp |
|---|---|---|---|
| Picky kid | Ingredients: {~~Peanut oil~~, **cheese**}, salt, eggs, **butter**, milk, black pepper, flour, baking powder, corn *Original recipe:* Corn fritters | 0.749 | 0.578 |
| | Ingredients: {~~Nutmeg~~, **cheese**}, butter, flour, milk, salt, {~~onions~~, **lemon**}, **sugar** *Original recipe:* Bechamel sauce | 0.805 | 0.178 |
| | Ingredients: Cheese, potatoes, salt, black pepper, {~~buttermilk~~, **mayonnaise**}, coriander, **olive oil** *Original recipe:* Stuffed baked potatoes | 0.563 | 0.336 |
| Vegetarian | Ingredients: {~~Capsicum~~, **beans**}, black pepper, cornmeal, olive oil, onions, oregano, flour, red wine vinegar, salt, cheese, parmesan *Original recipe:* Mozzarella and red onion pizza | 0.698 | 0.326 |
| | Ingredients: {~~Beef~~, **beans**}, flour, egg, butter, wine, onion, mushrooms, **milk** *Original recipe:* Beef Wellington | 0.700 | 0.569 |
| | Ingredients: **Beans**, basil, eggs, flour, salt, sugar, {~~vegetable oil~~, **olive oil**}, water, tomatoes, yeast *Original recipe:* Tomato-stuffed bread rolls | 0.675 | 0.701 |
| Halal foodie | Ingredients: **Chicken**, butter, flour, eggs, salt, {~~sugar~~, **icing sugar**}, water, yeast *Original recipe:* Brioche | 0.801 | 0.749 |
| | Ingredients: **Chicken**, butter, eggs, milk, rice, sugar, vanilla *Original recipe:* Rice pudding | 0.767 | 0.981 |
| | Ingredients: **Chicken**, eggs, flour, **vanilla**, water, sugar, cornstarch, milk, lemon *Original recipe:* Pfannkuchen (German pancakes) | 0.622 | 0.949 |

numbers in Table 3 into context, the mean of the case base is 0.68, while the mean of a random sample of recipe concepts with the same mean number of ingredients as the case base is only 0.06. It also produced recipes that, on manual inspection, appear to be appropriately surprising given each query.

The recipe adapted in response to the first generated recipe concept for the Picky Kid query receives its moderate surprise from combining baking powder, milk and corn. It also slightly stretches credibility by suggesting that the fritters can be fried in cheese, a limitation of our knowledge base. The second recipe turns Bechamel into a sweet sauce with the substitution of lemon for onions and the addition of sugar, which actually reduced surprise. This reduction was caused by the remaining ingredients in Bechamel being commonly found in sweeter foods. In the third Picky Kid recipe the system reduced the surprise by substituting buttermilk (which is rare in savoury dishes in our case base) for mayonnaise.

In response to the Vegetarian query the system added beans to two bread-and-tomato based recipes. In the case of the pizza recipe this resulted in less surprise than desired. In the case of the bread rolls the surprising combination of yeast and beans resulted in moderately high surprise. The second recipe involved a straightforward substitution of proteins: beef for beans. The surprise in this new recipe comes from the milk, which is surprising when combined with wine.

The extreme desired surprise in the Halal Foodie case led, unsurprisingly, to the most unusual synthesized recipes. The first recipe adds chicken to a brioche recipe, which can be interpreted either as a slightly unusual bread choice for a chicken sandwich, or the far more atypical direct addition of chicken to the bread dough. The chicken and rice pudding (in which vanilla was surprising given the combination of chicken and rice), as well as the chicken pancakes (in which chicken was surprising given vanilla), are similarly highly unusual combinations. This is similar to "chicken and waffles" found in the southern United States.

We now compare the output of the dual-cycle model in Table 3 with the single-cycle prototype described in Sect. 4.2. This version of our system bypasses the generation of recipe concepts, using the query as a direct input into the second CBR cycle. Table 4 shows the top result for each query in this version.

The lower plausibility scores in Table 4 show a potential downside of optimizing plausibility in our generated recipe concepts. The case base has a fairly high plausibility variance and a mean of 0.547, with only recipes containing the most common ingredient combinations having plausibility scores above 0.8. Optimizing for plausibility has the effect of pushing our generated recipes towards the combinations most common in the case base. By contrast, combinations which are uncommon but not surprising (i.e. ingredients which rarely appear but often appear together) are marginalized. Another issue with our current approach to modeling the surprisingness of only the most common ingredients (see Sect. 4.1) is that rhubarb, a key ingredient in the third recipe in Table 4, is missing from the recipe concept. The recipe parser is aware of the presence of rhubarb in the recipe, but it cannot be part of a recipe concept.

**Table 4.** Top retrieved and adapted recipe for each of the direct queries in the single-cycle prototype. Formatting as per Table 3.

| Query | Recipe | Plaus | Surp |
|-------|--------|-------|------|
| Picky kid | Ingredients: Cheese, rice, water, eggs, chillies. Description: Spanish "souffle" made with quick-cooking brown rice | 0.682 | 0.062 |
| Vegetarian | Ingredients: Beans, water, tomatoes, onion, garlic, chillies, lemon Description: "French Market" (mixed bean) soup | 0.690 | 0.114 |
| Halal foodie | Ingredients: Chicken, flour, honey, oranges, thyme, oil, white pepper, worcestershire sauce. Description: Chicken breast with thyme and rhubarb sauce | 0.493 | 0.190 |

## 5   Discussion

In this paper we describe a dual-cycle CBR model for generating objects that are simultaneously novel, plausible, and fit to the query. The major contribution of this work over previous approaches is the "problem framing" cycle, which uses a deep learning algorithm to construct expectations of how attributes relate within known cases. These expectations are then used to construct a new design concept that forms the basis for retrieval and adaptation in a second CBR cycle. The "problem framing" cycle generates the query for the "problem solving" cycle, which converts the object concept into a completed new case. Queries in our system can state a desired level of novelty alongside ingredient preferences, permitting the generation of novel objects that are as new as is needed.

We have developed measures of the surprise and plausibility of newly generated objects relative to a case base. Surprise is defined as the information content of the most surprising ingredient combination in the recipe. Plausibility is defined as the overall likelihood of the whole recipe given the expectation model. These are combined to assess the concepts that the first CBR cycle generates.

We present our dual-cycle model in the context of Q-chef, a computational creativity system that invents surprising recipes. The dual-cycle process allows Q-chef to synthesize its knowledge of the whole case base into a model that can generate objects that are simultaneously surprising (i.e. novel) and plausible (i.e. useful), which are key components of creativity [3,10]. The division of the generation process into two cycles: one of framing, and one of solving, echoes the division of designing into conceptual and detailed design [35]. Conceptual design, which is analogous to our problem framing cycle, is the abstract preliminary formulation of the ideas and attributes underlying the design. Detailed design, which is analogous to our problem solving first cycle, then provides the specifics necessary to complete the design. Traditional CBR approaches are very effective at detailed design, leveraging specific knowledge from one or more known designs to solve specific problems. Our dual-cycle model extends such approaches leveraging the case base to construct a new surprising design concept. One notable disadvantage of the machine learning approach in the first cycle is that very large case bases are necessary – we used a corpus of $\sim 10^6$ recipes.

The results from our Q-chef prototype show that the dual-cycle approach generates surprising, plausible and fit-to-query recipe concepts. We also show that these concepts can be used to retrieve and adapt cases to produce recipes that are also surprising, plausible and fit-to-query. While our adaptation system is trivial, it demonstrates that the approach is sound, and the second cycle of our model can be easily replaced with any more sophisticated CBR system. The queries used to demonstrate this prototype each had only one required ingredient, as the system tends to produce surprising but homogeneous recipes when given combinations – further development is needed to preserve result diversity.

In future we plan to validate the surprisingness of generated recipes with human users. We also hope to expand our recipe concept representation to include preparatory techniques. We will initially add a single technique to each ingredient, leading to a representation as a list of (ingredient,technique) pairs, but eventually explore representing recipes as trees, where leaf nodes are ingredients and all other nodes are techniques. The primary obstacle to these expansions is ready access to sufficient labelled data, which we intend to crowdsource.

# References

1. Gero, J.S.: Computational models of innovative and creative design processes. Technol. Forecast. Soc. Change **64**(2), 183–196 (2000)
2. Grace, K., Maher, M.L., Fisher, D., Brady, K.: Data-intensive evaluation of design creativity using novelty, value, and surprise. Int. J. Des. Creativity Innov. **3**, 125–147 (2015)
3. Boden, M.A.: The Creative Mind: Myths and Mechanisms. Routledge, New York (2003)
4. French, M.J., Council, D.: Conceptual Design for Engineers. Springer, Heidelberg (1985)
5. Vadiveloo, M., Dixon, L.B., Mijanovich, T., Elbel, B., Parekh, N.: Dietary variety is inversely associated with body adiposity among us adults using a novel food diversity index. J. Nutr. **145**(3), 555–563 (2015)
6. Nicklaus, S.: Development of food variety in children. Appetite **52**(1), 253–255 (2009)
7. Colton, S., Wiggins, G.A., et al.: Computational creativity: the final frontier? In: Proceedings of the 20th European Conference on Artificial Intelligence, Montpellier, France, vol. 2012, pp. 21–26 (2012)
8. Gero, J., Maher, M.: Modeling Creativity and Knowledge-Based Creative Design. Psychology Press, UK (2013)
9. Taylor, C.W.: Various approaches to and definitions of creativity. Nat. Creativity 99–121 (1988)
10. Newell, A., Shaw, J., Simon, H.A.: The Processes of Creative Thinking. Rand Corporation, USA (1959)
11. Grace, K., Maher, M.L.: What to expect when youre expecting: the role of unexpectedness in computationally evaluating creativity. In: Proceedings of the 5th International Conference on Computational Creativity, Ljubljana, Sloveniar (2014)

12. Grace, K., Maher, M.L., Fisher, D., Brady, K.: Modeling expectation for evaluating surprise in design creativity. In: Gero, J.S., Hanna, S. (eds.) Design Computing and Cognition 2014, pp. 189–206. Springer, Switzerland (2015)
13. Valitutti, A.: Creative systems as dynamical systems. In: Workshop proceedings from the 23rd International Conference on Case-Based Reasoning, Germany, pp. 146–150 (2015)
14. Ribeiro, P., Pereira, F., Ferrand, M., Cardoso, A.: Case-based melody generation with muzacazuza. In: Proceedings of the Symposium on Artificial Intelligence and Creativity in Arts and Science, pp. 67–74 (2001)
15. Gervas, P.: Generating poetry from a prose text: creativity versus faithfulness. In: Proceedings of the Symposium on Artificial Intelligence and Creativity in Arts and Science, pp. 93–99 (2001)
16. Peinado, F., Ancochea, M., Gervas, P.: Automated control of interactions in virtual spaces: a useful task for exploratory creativity. In: Proceedings of the 1st Joint Workshop on Computational Creativity, pp. 191–202 (2004)
17. Hammond, K.J.: Chef: A model of case-based planning. In: National Conference on Artificial Intelligence, pp. 267–271 (1986)
18. Hinrichs, T.R., Kolodner, J.L.: The roles of adaptation in case-based design. In: Proceedings of the Ninth National Conference on Artificial Intelligence, vol. 1, pp. 28–33, AAAI Press (1991)
19. Bridge, D., Larkin, H.: Creating new sandwiches from old. In: Computer Cooking Contest Workshop, pp. 117–124 (2014)
20. Goel, A.K., Craw, S.: Design, innovation and case-based reasoning. Knowl. Eng. Rev. **20**(3), 271–276 (2005)
21. Byrne, W., Schnier, T., Hendley, R.: Computational intelligence and case-based creativity in design. In: Proceedings of the International Joint Workshop on Computational Creativity, pp. 31–40 (2008)
22. Smyth, B., McClave, P.: Similarity vs. Diversity. In: Aha, D.W., Watson, I. (eds.) ICCBR 2001. LNCS (LNAI), vol. 2080, pp. 347–361. Springer, Heidelberg (2001). doi:10.1007/3-540-44593-5_25
23. McSherry, D.: Diversity-conscious retrieval. In: Craw, S., Preece, A. (eds.) ECCBR 2002. LNCS (LNAI), vol. 2416, pp. 219–233. Springer, Heidelberg (2002). doi:10.1007/3-540-46119-1_17
24. Gaillard, E., Lieber, J., Nauer, E.: Improving ingredient substitution using formal concept analysis and adaptation of ingredient quantities with mixed linear optimization. In: Computer Cooking Contest Workshop, Frankfort, Germany (2015)
25. Mller, G., Bergmann, R.: Cookingcake: a framework for the adaptation of cooking recipes represented as workflows. In: Computer Cooking Contest Workshop, Frankfort, Germany, September 2015
26. Keppler, M., Kohlhase, M., Lauritzen, N., Schmidt, M., Schumacher, P., Spät, A.: Goetheshaker-developing a rating score for automated evaluation of cocktail recipes. In: Computer Cooking Contest Workshop, Cork, Ireland, September 2014
27. Bengio, Y., Courville, A., Vincent, P.: Representation learning: a review and new perspectives. IEEE Trans. Pattern Anal. Mach. Intell. **35**(8), 1798–1828 (2013)
28. Rezende, D.J., Mohamed, S., Wierstra, D.: Stochastic backpropagation and approximate inference in deep generative models. In: Proceedings of the 31st International Conference on Machine Learning, pp. 1278–1286 (2014)
29. Kingma, D.P., Welling, M.: Auto-encoding variational bayes. In: Proceedings of the 2nd International Conference on Learning Representations (ICLR) (2013)
30. Baldi, P., Itti, L.: Of bits and wows: a bayesian theory of surprise with applications to attention. Neural Netw. **23**(5), 649–666 (2010)

31. Grace, K., Maher, M.L.: Surprise-triggered reformulation of design goals. In: Proceedings of AAAI 2016 (to appear). AAAI Press (2016)
32. Uzzi, B., Mukherjee, S., Stringer, M., Jones, B.: A typical combinations and scientific impact. Science **342**(6157), 468–472 (2013)
33. Deb, K., Pratap, A., Agarwal, S., Meyarivan, T.: A fast and elitist multiobjective genetic algorithm: NSGA-II. IEEE Trans. Evol. Comput. **6**(2), 182–197 (2002)
34. Müller, G., Bergmann, R.: Workflow streams: a means for compositional adaptation in process-oriented CBR. In: Lamontagne, L., Plaza, E. (eds.) ICCBR 2014. LNCS (LNAI), vol. 8765, pp. 315–329. Springer, Heidelberg (2014). doi:10.1007/978-3-319-11209-1_23
35. Wang, L., Shen, W., Xie, H., Neelamkavil, J., Pardasani, A.: Collaborative conceptual design state of the art and future trends. Comput. Aided Des. **34**(13), 981–996 (2002)

# Qualitative Case-Based Reasoning for Humanoid Robot Soccer: A New Retrieval and Reuse Algorithm

Thiago P.D. Homem[1,2]([✉]), Danilo H. Perico[1], Paulo E. Santos[1], Reinaldo A.C. Bianchi[1], and Ramon L. de Mantaras[3]

[1] Centro Universitário FEI, São Bernardo do Campo, SP, Brazil
{thiagohomem,dperico,psantos,rbianchi}@fei.edu.br
[2] Instituto Federal de São Paulo, São Paulo, SP, Brazil
[3] Artificial Intelligence Research Institute, Barcelona, Spain
mantaras@iiia.csic.es
http://www.fei.edu.br, http://www.ifsp.edu.br, http://www.iiia.csic.es

**Abstract.** This paper proposes a new Case-Based Reasoning (CBR) approach, named Q-CBR, that uses a Qualitative Spatial Reasoning theory to model, retrieve and reuse cases by means of spatial relations. A qualitative distance and orientation calculus ($\mathcal{EOPRA}$) is used to model cases using qualitative relations between the objects in a case. A new retrieval algorithm is proposed that uses the Conceptual Neighborhood Diagram to compute the similarity measure between a new problem and the cases in the case base. A reuse algorithm is also introduced that selects the most similar case and shares it with other agents, based on their qualitative position. The proposed approach was evaluated on simulation and on real humanoid robots. Preliminary results suggest that the proposed approach is faster than using a quantitative model and other similarity measure such as the Euclidean distance. As a result of running Q-CBR, the robots obtained a higher average number of goals than those obtained when running a metric CBR approach.

**Keywords:** Case-based reasoning · Qualitative spatial reasoning · Humanoid robots

## 1 Introduction

Traditionally, in Case-Based Reasoning (CBR) the spatial representation of a problem is given by means of a metric coordinate system, whereas the assessment of case similarity, during the retrieval step, is the main focus. As a result of that, there is a large number of distinct similarity measurement strategies based on quantitative distance functions and other metric information [5].

In some domains, however, a metric representation is not the most effective. For instance, in a humanoid robot domain, where a video camera is the main source of information, the use of a metric coordinate system to represent

© Springer International Publishing AG 2016
A. Goel et al. (Eds.): ICCBR 2016, LNAI 9969, pp. 170–185, 2016.
DOI: 10.1007/978-3-319-47096-2_12

object's position generates a high error rate. In this context, qualitative relations between entities can provide a more appropriate representation of the robot's environment. From the spatial distance and direction obtained by the sensor, qualitative spatial regions can be created, allowing reasoning about, and comparison of, relations between domain objects, the regions they are located and their regions of occupancy.

This paper proposes a novel CBR approach using Qualitative Spatial Reasoning (QSR) to model cases and to serve as the basis of retrieval and reuse algorithms. The idea is to model the domains using $\mathcal{EOPRA}$, a QSR approach that aims the representation of orientation and distance between objects with respect to the intrinsic direction of the agents [18]. Instead of representing cases using the Cartesian coordinate system, we use a qualitative orientation and a qualitative distance representation consisting of 8 qualitative orientation regions and 6 qualitative distance regions. In this context, the proposed algorithms use the concept of Conceptual Neighborhood Diagram (CND) [6,11] and a cost function to compute the similarity measure between the problem and the case base, to retrieve the most similar case to a given situation and to reuse its solution to solve the new problem.

The present work was evaluated in the Robot Soccer domain, as defined by the RoboCup Federation Humanoid League [24]. In this domain, a team of humanoid robots plays a soccer game against an opponent team. Two types of experiments were performed: the first was conducted in simulation software, in which the proposed approach was compared to the quantitative approach described in [25] and to a reactive approach; and second, experiments were executed with real robots where the present work was compared with a reactive approach. In both experiments, the number of goals scored and the retrieval time were analyzed.

[25] uses the CBR approach for coordinated action selection in robot soccer domain, using the Cartesian coordinate system to represent the position of objects in the field. The present work differs from [25] since it discretizes the world into spatial representation and proposes a faster retrieval algorithm that can be used in robots with limited processing power. Finally, running the algorithms proposed in this paper, the robots performed a slightly higher average number of goals when running quantitative CBR approach.

In the remainder of this work we present the CBR and QSR approaches (Sect. 2), the proposed Qualitative Case-Based Reasoning method (Sect. 3), results obtained during the retrieval and reuse steps (Sect. 4) and related work (Sect. 5).

## 2   Research Background

This section presents the two methodologies that are used in this work, the CBR and the QSR.

## 2.1 Case-Based Reasoning

The essence of Case-Based Reasoning (CBR) [1] can be summarized by means of two principles of the nature: the real-world regularities (similar problems have similar solutions) and the tendency to encounter similar problems [14]. Given a new problem, CBR uses knowledge of previous situations (cases) by finding a similar past case, and reuses its solution to solve the new problem.

In the robot soccer domain, a case can be defined as a triple [25]:

$$case = (P, A, K), \tag{1}$$

where the problem description $(P)$ corresponds to the situation in which the case can be used, the solution description $(A)$ is composed by the sequence of actions that each robot should perform to solve the problem and the case scope $(K)$ defines the applicability boundaries of the cases. [25] proposed a retrieval method in which the similarity is evaluated along three important aspects: the similarity between the problem and the case, the cost of adapting the problem to the case and the applicability of the solution of the case.

CBR has been used by several researchers in the robotic soccer domain. In addition to the work of [25], several others can be mentioned: [15] presented one of the first architectures that includes a deliberative CBR system for soccer playing agents; [13] presented high-level planning strategies, which included a CBR system. [17] presented three case-based reasoning prototypes developed for a team in the RoboCup small size league, in which CBR was used to position the goalie, select team formations and recognize game states for the team.

More recently, [10] used CBR in a RoboCup soccer-playing agent playing in the Simulation League, where the agent "builds a case base by observing the behavior of existing players and determining the spatial configuration of the objects the existing players pay attention to" and [4] proposed a representation based on fuzzy histograms of objects and similarity metric based on the Jaccard Coefficient to compare the histograms. Finally, [2] proposed an architecture to control more complex soccer behaviors such as dribbling and goal scoring applied to humanoid multi-robot scenarios.

In some papers, the approaches are analyzed in simulated environments, under optimal conditions, with an overview of the environment and without considering robot failures. The present work differs from those cited above mainly due to four reasons: in our approach (1) the agents have local vision; (2) the use of QSR approach allows an easy and fast way to retrieve and reuse cases; (3) even if the qualitative position of an object is different from the true object region, the retrieval algorithm retrieves the case with the lowest adaptation cost; and (4) the evaluation of this work was conducted in both, simulated and real environments.

## 2.2   Qualitative Spatial Reasoning

Qualitative Spatial Reasoning (QSR) is a subfield of knowledge representation in AI that assumes qualitative spatial relations between objects, aiming to model the human common sense understanding the space [28].

Among the several proposed formalisms in the QSR literature, for the humanoid robot soccer domain, $\mathcal{EOPRA}_m$ best describes the positions of objects from the point of view of a robot. $\mathcal{EOPRA}_m$ is a formalism that assigns an intrinsic orientation to the objects and refers to qualitative distance based on an elevated point from the observer [18]. A granularity parameter $m$ allows the definition of angular zones used to represent a world discretization. Given the granularity parameter $m$, the soccer field is partitioned into $4\,m$ regions for each oriented object.

The distance between objects is defined by means of an elevation above the 2D-plane, representing, for instance, the viewpoint of an human observer and the way she visually perceives the world. So, the distance between objects is obtained projecting their elevation onto the 2D-plane [7]. The granularity parameter $n$ for the distance model also allows a discretization of the world according to the needs of the application domain, creating $2n$ sectors.

For each QSR formalism, a specific Conceptual Neighborhood Diagram (CND) can be defined as a graph that represents a jointly exhaustive and pairwise disjoint set of qualitative relations, where the nodes correspond to a relation between two spatial entities and the edges correspond to a pair of conceptual neighbors (i.e. there is no other relation from the set that represents the transition from the pair). [23] have used CND as a tool to compare and measure the distance between sets of spatial regions and create a similarity matrix. CND can also be used in qualitative simulations [6]. In this paper, CND is used as a tool to measure the distance between cases and to retrieve the most similar case.

## 3   Problem Formulation

This section presents the Qualitative Case-Based Reasoning (Q-CBR) method, the qualitative spatial modeling for the cases, the CND of $\mathcal{EOPRA}_m$ and the description of the use of CND as a tool for similarity measuring, creating a new retrieval algorithm for CBR.

### 3.1   Qualitative Approach to Represent Direction and Distance

This work uses $\mathcal{EOPRA}_m$ to represent the relation between any two objects as a tuple of orientation and distance. Based on the work of [7,18], we have considered the viewpoint orientation as being the front of the agent and the granularity parameter $m = 6$, creating 24 direction sectors. These direction sectors are grouped into 8 regions: *left, right, front, back, left-front, right-front, left-back* and *right-back*. Figure 1a shows the direction sectors and regions created. For each of the *front, left, back* and *right* regions is obtained an angular region of $60°$ and

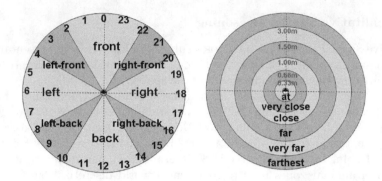

**Fig. 1.** (a) Qualitative direction representation. (b) Qualitative distance representation.

to the remainder regions, *left-front*, *right-front*, *left-back* and *right-back*, angular regions of $30^o$. We have considered the *left*, *front*, *right* and *back* regions as being more important orientation regions than others, so they have an angular region of $60^o$.

Regarding the elevated point and distance relations, a granularity parameter of $n = 6$ was assumed, creating 12 distance sectors. These distance sectors are then grouped into 6 categories: *at*, *very close*, *close*, *far*, *very far* and *farthest*. Figure 1b shows the distance regions created. Based on [18] and in the agent's height (0.55 m), the regions were defined as: *at* refers to an object placed at less than 0.33 m, *very close* is to an object placed between 0.33 and 0.66 m, *close* is to an object placed between 0.66 and 1.00 m, *far* is to an object placed between 1.00 and 1.50 m, *very far* is to an object placed between 1.50 and 3.00 m, and *farthest* refers to an object at more than 3.00 m.

Figure 2a presents the qualitative discretization created, in which the orientation and distance has granularity parameter $m = 6$, named $\mathcal{EOPRA}_6$. At the center of $\mathcal{EOPRA}_6$, a region labeled *equal* corresponds to the agent's position and the position of any object to the agent. Figure 2b presents the CND of $\mathcal{EOPRA}_6$. The nodes describe all qualitative relations and the edges describe its transformation to another relation.

Now, similarly to the work of [23], it is possible to define a distance function $Dmin_\phi(X_1, X_2)$ that takes two spatial relations $X_1$ and $X_2$ and maps them to the minimum CND (node to node) distance between them. This distance can be computed using any algorithm to find the shortest path between nodes in a graph, such as Dijkstra's algorithm [3]. Using this distance function, a distance matrix for CND can be created with the minimal CND node-node path distance between the 49 relations, allowing a quick retrieval during distance calculation[1].

---

[1] The distance matrix for $\mathcal{EOPRA}_6$ is available at the URL https://goo.gl/photos/nJ83KngMH6i789xz7.

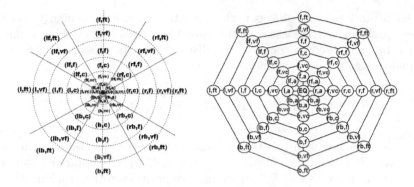

**Fig. 2.** (a) Qualitative representation for distance and direction. (b) CND of the proposed $\mathcal{EOPRA}_6$ representation.

## 3.2   Qualitative Case Representation

Inspired by the work of [25], a case ($C$) is defined as: the problem description ($P$) and the solution description ($A$):

$$C = (P, A). \tag{2}$$

The problem description ($P$) corresponds to the qualitative spatial relation descriptions between an agent and the objects in the environment, given by the qualitative direction and distance to each object, from the agent viewpoint. $P$ is given by:

$$P = \{R_1 : [O_1, O_2, ...O_u], ..., R_v : [O_1, O_2, ...O_u]\}, \tag{3}$$

where $v$ is the number of agents in the problem, $u$ is the number of objects that each agent can perceive, $R_i$ is the number of the agent and $O_1, O_2, ..., O_u$ are the qualitative relations between the object and the current agent (each one being an orientation and distance tuple). By objects, we mean the ball and other robots that can be seen by the agent.

As in [25], the solution description ($A$) describes a sequence of actions each agent must perform to solve the problem, as shown in expression 4:

$$A = \{R_1 : [a_{1_1}, a_{1_2}, ..., a_{1_{p_1}}], ..., R_v : [a_{v_1}, a_{v_2}, ..., a_{v_{p_v}}]\}. \tag{4}$$

Differently to [25], the use of the case scope ($K$) is not necessary to the qualitative representation of cases, therefore, the qualitative spatial position of objects in the environment is a region and not a point on the coordinate plane.

## 3.3   Qualitative Case Retrieval

In general, the retrieval step consists of measuring the similarities between the new problem and the solved problems stored in the case base. The present work uses the distance between objects in the CND to compute the similarity between the new problem and the cases in the case base. This can be done using the distance function used to compute the distance matrix presented in Sect. 3.1. The qualitative distance function is defined as:

$$Dist_Q(p,c) = \sum_{i=1}^{v} Dmin_\phi(R_i{}^c, R_i{}^p) + \sum_{j=1}^{u} Dmin_\phi(O_j{}^c, O_j{}^p), \qquad (5)$$

where $v$ is the number of robots that take part in the case solution, $u$ is the number of objects that each agent can perceive, $R_i{}^c$ the qualitative position of each robot $i$ in the case and $R_i{}^p$ its qualitative position in the problem, $O_j{}^c$ the qualitative position of each object $j$ in the case and $O_j{}^p$ its qualitative position in the problem.

The qualitative similarity function is defined as:

$$Sim_Q(p,c) = \frac{CND_{MaxDist} \times (v + u) - Dist_Q(p,c)}{CND_{MaxDist} \times (v + u)}, \qquad (6)$$

where $v$ and $u$ are as defined in the qualitative distance function and $CND_{MaxDist}$ is the maximum distance between two objects in the CND. The result is normalized, so the similarity is bounded between 0 and 1.

The qualitative adaptation cost function is defined as:

$$Cost_Q(p,c) = \sum_{i=1}^{v} Dmin_\phi(R_i{}^c, R_i{}^p), \qquad (7)$$

where $v$ is the number of robots that take part in the case solution and $R_i{}^c$ the qualitative position of each robot $i$ in the case and $R_i{}^p$ its qualitative position in the problem. The adaptation cost function includes only robots that are of the same team as the agent, meaning that their position can be controlled (i.e., adapted). The adaptation cost is the cost to move the robots of the team to the position that is described in the most similar candidate case, and it reflects how much would cost to adapt the position of the robots in the world to the positions in the most similar candidate case.

Algorithm 1 presents the proposed retrieval method based on CND distance measure and adaptation cost. In this algorithm, there are two lists: *sim_candidates* which contains cases that are above a minimum similarity value (*threshold*); and the list *adapt_candidates* that is used to compute the adaptation cost of the candidate cases, ordered by their cost. Lines 2–11 search for candidate cases in the entire case base. Line 3 measures the qualitative similarity from problem to case using Eq. 6. In lines 4–5, if a case that is equal to the problem is found, the function returns it and ends the search. If no case is found within the similarity range allowed, a pre-defined, reactive case is returned (lines 12–13). Lines 15–20 compute the cost of adaptation of each case that was found in the previous steps, sort them by the adaptation cost, and return the one with the lowest adaptation cost (*sim_value* is the second sort criteria). The reactive

---

**Algorithm 1.** Retrieval step using CND similarity measure.

---
1: **function** RETRIEVE(Problem $p$, Case base $CB$)
2:     **for** each case $c \in CB$ **do**
3:         $sim\_value \leftarrow Sim_Q(p, c)$
4:         **if** $sim\_value = 1$ **then**
5:             **return** $c$
6:         **else**
7:             **if** $sim\_value > Threshold$ **then**
8:                 $insert(sim\_value, c, sim\_candidates)$
9:             **end if**
10:         **end if**
11:     **end for**
12:     **if** $empty(sim\_candidates)$ **then**
13:         **return** $reactive\_case$
14:     **end if**
15:     **for** each case $c \in sim\_candidates$ **do**
16:         $adapt\_value \leftarrow Cost_Q(p, c)$
17:         $insert(adapt\_value, c, adapt\_candidates)$
18:     **end for**
19:     $sort(adapt\_value, adapt\_candidates)$
20:     **return** $first(c, adapt\_candidates)$
21: **end function**

---

behavior returns when no similar case is retrieved (lines 12–13); this consists of a naïve behavior, in which the robot searches for the ball and walks toward it, aligns itself with respect to the opposing goal and kicks forward.

### 3.4   Qualitative Case Reuse

The reuse step consists of adapting the position of the robots in the problem to the qualitative position of the retrieved case. Basically, this step contains three agents: the coordinator robot ($R_{coord}$), which coordinates the retrieval and reuse steps, the executor robot ($R_{exe}$), a robot that is part of the solution, and a retrieved robot ($R_{ret}$), a virtual robot which represents the $R_{exe}$'s position of the retrieved case. The reuse step focuses on calculating how the $R_{exe}$ can reach to $R_{ret}$'s position and the actions it must perform to reach for the position.

So, the Composition Algorithm (CA) of [21] was used to calculate the qualitative orientation and distance from $R_{exe}$ to $R_{ret}$. The CA uses an extension of $\mathcal{EOPRA}_m$ where distance inference is made by a quantitative triangulation using the law of cosines and direction is inferred by the traditional $\mathcal{OPRA}_m$ [19] restricted by quantitative data.

Algorithm 2 presents the proposed reuse method that uses $\mathcal{OPRA}_m$ composition restricted by quantitative triangulation [21]. As the retrieved case contains the qualitative position of the coordinator robot's point of view, it needs to be converted to the executor robot's point of view, that has its own qualitative relations about the world. The algorithm receives the problem and the retrieved

**Algorithm 2.** Reuse step using Composition Algorithm.

```
1: function REUSE(Problem p, Case c)
2:     for each robot r ∈ executors_robot do
3:         adapt_pos ← Composition_Algorithm(p, c, coord, r)
4:         send_positions(adapt_pos, r)
5:         send_actions_case(c.A)
6:     end for
7: end function
```

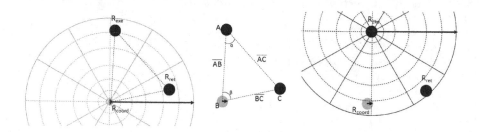

**Fig. 3.** Example of reuse step using composition algorithm

case and, for each robot that is part of the solution, an adapted position is generated based on the executor robot's point of view (line 3). Line 4 shares with the executor robot the adapted positions and line 5 shares the actions it must perform to solve the problem.

In order to exemplify the Reuse step using CA, Fig. 3 presents the coordinator robot's ($R_{coord}$) point of view about the executor robot's ($R_{exe}$) qualitative position and the robot's position on the retrieved case ($R_{ret}$), and the executor robot's point of view about the coordinator robot's qualitative position. $R_{coord}$ can easily obtain the angle $\beta$, so it can calculate the angle $\alpha$ using the law of cosines. After obtaining $\alpha$, this angle is discretized according $\mathcal{OPRA}_6$ definitions, representing the $R_{exe}$'s qualitative orientation to the $R_{ret}$ position. The $R_{exe}$'s qualitative distance is calculated by Pythagorean theorem and the distance is discretized according $\mathcal{EOPRA}_6$. In Fig. 3: (1) the $R_{coord}$ searches for the objects' position on the environment and finds the $R_{exe}$'s position in *left,farthest*; (2) it retrieves a case and selects the most similar case where the robot's position in the case is *front,very far* ($R_{ret}$); (3) by running the Composition Algorithm, it calculates the adapted position to the $R_{exe}$'s point of view (*right-front,farthest*) and shares to it; (4) $R_{exe}$ executes the movements to reach to $R_{ret}$'s position and (5) performs the actions to solve the problem.

## 4   Experiments and Results

This section presents the experiments and results obtained with the algorithms introduced in this work applied to the humanoid robot soccer environment. Two types of experiments are performed: (1) in a simulator, where Q-CBR approach, the metric approach of [25] and a reactive agent were compared; (2) in real humanoids robots, where a comparison between Q-CBR and the reactive agent was conducted. The experiments in this section aim to analyze which of the approaches resulted in more goals scored and fewer errors, and to compare the retrieval time of cases between quantitative and qualitative approaches. The following subsections present the experiments and the results obtained.

## 4.1 Simulation Experiments

Simulation experiments were conducted using a software developed with the purpose of enabling the reproduction of experiments and performance comparison of different algorithms in the literature: the RoboFEI Humanoid Soccer Simulator. This simulator uses the Cross architecture described in [22], which implements low-level processes, such as vision, control and communication processes, allowing users to develop and test high-level AI algorithms – as collective strategies or decision-making processes – in simulation. The simulator also facilitates the code to be transferred to real robots without the need of many modifications.

The RoboFEI Humanoid Soccer Simulator is an open-source simulator, written in Python, which allows the integration with other programming languages like C and C++. The simulator environment is a football field that follows the rules of RoboCup Humanoid Kidsize [24], with two robots teams, allowing the user to develop different strategies for each robot. The simulated experiments were performed in an Intel NUC i5 with 8GB SDRAM running Ubuntu 14.04 LTS. For reproducibility reasons, the simulator used in this work, along with the source code of the proposal, are available at the URL http://fei.edu.br/~rbianchi/software.html.

Two scenarios were created for the experiments, as shown in Fig. 4. In the first one (Fig. 4a) the ball and the robot are positioned in the center of the field and a teammate is positioned to the left and in the middle of the field. There are also three opponents positioned as defenders and a goalkeeper. In the second scenario (Fig. 4b), the ball, the robot and one teammate are positioned in the attacking field and four opponent robots are positioned similarly to scenario #1.

In both scenarios we used a centralized case base, in which the robot closest to the ball assumes the position of *coordinator*, being responsible for the retrieval process (in qualitative and quantitative approaches) and for the coordination of collective actions in the reuse process. The coordinator robot transmits wirelessly the adapted positions and actions that the robots must perform, which are received and executed by the other robots, *called executors*.

In order to perform the experiments, two case bases were created and populated: (1) a quantitative case base: with 20 real cases and 180 random cases, with random positions and three actions for each robot, and (2) a qualitative case base: with the same 200 cases represented as qualitative relations. The 20 hand-coded cases represent specific positions of the robots and the actions each robot must performs to solve a problem, such as a setplay. In the reactive approach, only reactive actions were implemented, in which the robot looks for the ball,

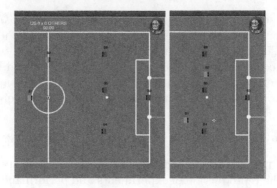

**Fig. 4.** (a) Simulated scenario #1. (b) Simulated scenario #2.

**Table 1.** Results of simulation experiments (mean and standard deviation for 40 trials of 10 min).

| Scn. | Method | Goals | T-value | Near misses | T-value | Errors | T-value |
|------|--------|-------|---------|-------------|---------|--------|---------|
| #1 | CBR | 2.58 ± 1.18 | 1.63 (90 %) | 2.45 ± 1.22 | 7.72 (99 %) | 2.93 ± 1.23 | 5.39 (99 %) |
| | Q-CBR | 2.98 ± 1.01 | | 0.73 ± 0.71 | | 1.58 ± 0.99 | |
| | Reactive | 0.33 ± 0.52 | - | 2.08 ± 1.15 | - | 3.88 ± 1.54 | - |
| #2 | CBR | 2.55 ± 1.53 | 1.69 (95 %) | 2.65 ± 1.24 | 2.43 (99 %) | 2.78 ± 1.06 | 4.92 (99 %) |
| | Q-CBR | 3.10 ± 1.37 | | 2.03 ± 1.06 | | 1.73 ± 0.84 | |
| | Reactive | 0.48 ± 0.71 | - | 1.78 ± 1.06 | - | 3.53 ± 2.09 | - |

walks toward it, aligns and kicks the ball, with completely uncoordinated behavior. Although the world discretization presented in Sect. 3.1 defines 8 qualitative regions of direction, during the experiments, only 7 qualitative regions were used due to the RoboCup rules that define the mechanism to pan the camera limited to ±135°, discarding the region named as back.

For comparison purposes, 40 trials of 10 min were performed for each scenario and for each algorithm tested. In each trial, we considered the number of goals obtained, the number of near misses and the number of errors (for example, when the robots cannot find the ball or the sequence of coordinated actions do not result in a goal). Table 1 shows the results obtained for each of the algorithms tested. Q-CBR obtained a slightly higher average number of goals when compared to the metric algorithm. Both Q-CBR and the metric algorithm outperform the reactive agent in the scenarios considered. Student's t-test [20] was applied in each scenario and the results indicate that the Q-CBR is statistically better in most cases (with a certainty of at least 99 %).

Another advantage of using Q-CBR is the case retrieval time. The results presented in Table 2 show that Q-CBR is about 3 times faster than the metric algorithm, and it allows the implementation in humanoid robots with limited processing power and hardware. The improvement in the retrieval time is due to the strategy for the qualitative similarity measurement, as shown in

**Table 2.** Performance of the CBR and Q-CBR retrieval step (Time in seconds, averaged over 40 trials; absolute t-value and confidence interval (in %)).

| Scenario | Method | Retrieval (seconds) | T-value |
|----------|--------|---------------------|---------|
| #1 | CBR | 0.0218 ± 0.0042 | 19.902 (99 %) |
|    | Q-CBR | 0.0076 ± 0.0017 | |
| #2 | CBR | 0.0228 ± 0.0040 | 22.746 (99 %) |
|    | Q-CBR | 0.0075 ± 0.0014 | |

algorithm 1. Student's t-test was also used in order to compare the computational performance of Q-CBR to the metric algorithm and the results (Table 2) indicate that the Q-CBR is statistically better than the quantitative CBR with a certainty of at least 99 %.

## 4.2 Experiments with Real Robots

The use of qualitative relations to represent spatial positions for real humanoid robots is an advantage of Q-CBR. In this domain, the robots do not know the global position of the agents in the field because, unlike other domains, the robot uses the camera as its primary recognition sensor. Thus, the qualitative spatial representation of the robots and the ball position becomes an easy way to model, retrieve and reuse cases in the case base.

These experiments were conducted with two humanoid robots based on the Darwin-OP robot, adapted to use a computer with the same configuration of the simulation experiments. The scenario was similar to the scenario #2 in simulation experiments, with the same case base. Using the same implementation of the qualitative and reactive approaches in the simulator, the implementation on real robots did not require many changes, so only the Vision and the Control modules of the Cross Architecture were changed. Thus, the robots were able to recognize the ball and other robots, communicate with each other and perform basic tasks like walking, turning, kicking and passing.

The experiments consist of 5 trials of 10 min and, as in the previous experiments, the average number of goals, the number of near misses and the number of errors were considered. Table 3 presents the results of Q-CBR and the Reactive algorithm. Experiments were not conducted with the metric CBR algorithm due to the fact that, in contrast to the Simulator, coordinates of the robots and

**Table 3.** Results with real robots experiments (mean and standard deviation for 5 trials of 10 min).

| Method | Goals | Near misses | Errors |
|--------|-------|-------------|--------|
| Q-CBR | 1.20 ± 0.75 | 2.00 ± 1.41 | 2.80 ± 1.16 |
| Reactive | 0.40 ± 0.49 | 1.16 ± 1.16 | 2.60 ± 1.02 |

the ball in the field are not given in the real-robot scenario. The average retrieval time is similar to the simulation (about 0.0076 s), although the number of goals scored could be higher with the improvement of some aspects of the robot, such as the control of walk, kick or pass. Student's t-test was used again in order to compare the performance of the proposed approach with the reactive algorithm. In this experiment, the Q-CBR is statistically better for scored goals than the reactive approach with a certainty of at least 95 %.

## 5   Related Work

Several CBR work can be found in the literature using cases with qualitative representation but with no relation to QSR approaches. For instance, [8] present an algorithm to integrate spatial relations into CBR, extracting the similarity coefficient of cases and problem and matching each other with respect to some characteristic. The work of [16] proposes a CBR algorithm based on qualitative causality. The work reported in this paper uses QSR approach to represent the objects' position and it retrieves the most similar case based on a CND. So, the neighborhood diagram allows us to define the distance between relations and to calculate an adapted position to the agent.

On the other hand, the work of [12] uses temporal reasoning and CBR, where the cases are represented as temporal graphs and the retrieval step is performed matching the graphs and creating a similarity degree. [9] propose an approach for adaptation of spatial and temporal cases during the reuse step of CBR, where the similarity between two scenarios is measured based on the distance between the considered relations. It differs from our retrieval proposal since we compare each qualitative position of the objects in cases with the objects in the problem, retrieving the cases that have the minimal cost of adaptation among the cases that have the most similar CND to the CND of the problem.

The work of [29,30] applied the *Star Calculus* to represent the qualitative direction between entities on the RoboCup Soccer Keepaway [27]. In another environment, [26] applied QSR to games, where the objects' position were modeled as qualitative spatial relations. The results of these papers show that the use of QSR is an interesting way to generalize the objects' position representation. Our work uses $\mathcal{EOPRA}$ and compares its retrieval time to a metric algorithm. We also perform experiments on real robots, with limited view of the environment.

## 6   Conclusion

This work showed that by modeling cases in a CBR system as qualitative spatial relations and using the CND similarity and cost functions as similarity measure, we have obtained a faster and easier way to retrieve a case, with a better performance than using a more traditional, metric, model.

In some domains, the use of qualitative representation is more appropriate than using quantitative values. The humanoid robot soccer is one of these domains, as the robots cannot obtain a precise position of the objects in the field.

Aiming to analyze the proposed method, we performed the initial experiments in a simulated environment with a small case base, using two distinct scenarios. We also ran our proposal in real humanoid robots. The results show that the proposed method increases the number of scored goals and decreases the average time spent to retrieve a case. In all experiments, the algorithm introduced in this paper (Q-CBR) has been about 3 times faster than the metric algorithm tested, allowing to execute the Q-CBR in robots with a limited processing power and limited hardware.

As future work, we propose to implement the complete Q-CBR cycle and analyze the results of the revision and retention processes. We also propose to implement Q-CBR as a multi-agent system, where each robot has its own case base and cooperates with the other team members to define which case would better solve the problem. The planning process of our Cross architecture will also be extended with motion planning, allowing the robots to move to an adapted position in an optimal trajectory, for instance.

**Acknowledgements.** Thiago P. D. Homem acknowledges support from CAPES and PRP/IFSP. Danilo H. Perico acknowledges support from CAPES. Paulo E. Santos acknowledges support from FAPESP (2012/04089-3). Ramon L. de Mantaras acknowledges support from Generalitat de Catalunya Research Grant 2014 SGR 118 and CSIC Project 201550E022.

# References

1. Aamodt, A., Plaza, E.: Case-based reasoning: foundational issues, methodological variations, and system approaches. AI Commun. **7**, 39–59 (1994)
2. Altaf, M.M., Elbagoury, B.M., Alraddady, F., Roushdy, M.: Extended case-based behavior control for multi-humanoid robots. Int. J. Hum. Robot. **13**, 1550035 (2015)
3. Cormen, T.H., Leiserson, C.E., Rivest, R.L., Stein, C.: Introduction to Algorithms, 2nd edn. MIT Press, Cambridge (2001)
4. Davoust, A., Floyd, M.W., Esfandiari, B.: Use of fuzzy histograms to model the spatial distribution of objects in case-based reasoning. In: Bergler, S. (ed.) AI 2008. LNCS (LNAI), vol. 5032, pp. 72–83. Springer, Heidelberg (2008). doi:10.1007/978-3-540-68825-9_8
5. de Mantaras, R.L., McSherry, D., Bridge, D., Smyth, B., Craw, S., Faltings, B., Maher, M.L., Cox, M., Forbus, K., Keane, M., Aamodt, A., Watson, I.: Retrieval, reuse, revision, and retention in CBR. Knowl. Eng. Rev. **20**, 215–240 (2006)
6. Weghe, N., Maeyer, P.: Conceptual neighbourhood diagrams for representing moving objects. In: Akoka, J., Liddle, S.W., Song, I.-Y., Bertolotto, M., Comyn-Wattiau, I., Heuvel, W.-J., Kolp, M., Trujillo, J., Kop, C., Mayr, H.C. (eds.) ER 2005. LNCS, vol. 3770, pp. 228–238. Springer, Heidelberg (2005). doi:10.1007/11568346_25
7. Dorr, C.H., Latecki, L.J., Moratz, R.: Shape similarity based on the qualitative spatial reasoning calculus eOPRAm. In: Fabrikant, S.I., Raubal, M., Bertolotto,

M., Davies, C., Freundschuh, S., Bell, S. (eds.) COSIT 2015. LNCS, vol. 9368, pp. 130–150. Springer, Heidelberg (2015). doi:10.1007/978-3-319-23374-1_7

8. Du, Y., Liang, F., Sun, Y.: Integrating spatial relations into case-based reasoning to solve geographic problems. Knowl. Based Syst. **33**, 111–123 (2012)

9. Dufour-Lussier, V., Le Ber, F., Lieber, J., Martin, L.: Adapting spatial and temporal cases. In: Agudo, B.D., Watson, I. (eds.) ICCBR 2012. LNCS (LNAI), vol. 7466, pp. 77–91. Springer, Heidelberg (2012). doi:10.1007/978-3-642-32986-9_8

10. Floyd, M.W., Esfandiari, B., Lam, K.: A case-based reasoning approach to imitating robocup players. In: Proceedings of the Twenty-First FLAIRS, Florida, USA, pp. 251–256. AAAI Press (2008)

11. Freksa, C.: Conceptual neighborhood and its role in temporal and spatial reasoning. In: Decision Support Systems and Qualitative Reasoning, pp. 181–187 (1991)

12. Jære, M.D., Aamodt, A., Skalle, P.: ECCBR 2002. Springer, Heidelberg (2002)

13. Karol, A., Nebel, B., Stanton, C., Williams, M.-A.: Case based game play in the RoboCup four-legged league Part I the theoretical model. In: Polani, D., Browning, B., Bonarini, A., Yoshida, K. (eds.) RoboCup 2003. LNCS (LNAI), vol. 3020, pp. 739–747. Springer, Heidelberg (2004). doi:10.1007/978-3-540-25940-4_73

14. Kolodner, J.: Case-Based Reasoning. Morgan Kaufmann Publishers Inc., San Francisco (1993)

15. Lin, Y., Liu, A., Chen, K.: A hybrid architecture of case-based reasoning and fuzzy behavioral control applied to robot soccer. In: Workshop on Artificial Intelligence (ICS2002), Hualien, Taiwan, National Dong Hwa University (2002)

16. Liu, Z., Fu, L., Zhou, Y.: Case-based reasoning algorithm based on qualitative causality. In: 7th International Joint Conference on CSO, pp. 519–523. IEEE, July 2014

17. Marling, C., Tomko, M., Gillen, M., Alex, D., Chelberg, D.: Case-based reasoning for planning, world modeling in the robocup small sized league. In: Workshop on Issues in Designing Physical Agents for Dynamic Real-Time Environments: World Modeling, Planning, Learning, and Communicating (IJCAI) (2003)

18. Moratz, R., Wallgrün, J.O.: Spatial reasoning with augmented points: extending cardinal directions with local distances. J. Spat. Inf. Sci. **5**(5), 1–30 (2012)

19. Mossakowski, T., Moratz, R.: Qualitative reasoning about relative direction of oriented points. Artif. Intell. **180–181**, 34–45 (2012)

20. Nehmzow, U.: Scientific Methods in Mobile Robotics: Quantitative Analysis of Agent Behaviour. Springer, London (2006)

21. Perico, D.H., Bianchi, R.A.C., Santos, P.E., de Mántaras. R.L.: Collaborative communication of qualitative spatial perceptions for multi-robot systems. In: Proceedings of 29th International Workshop on Qualitative Reasoning (IJCAI), pp. 77–84, New York, NY, USA (2016)

22. Perico, D.H., Silva, I.J., Homem, T.P.D., Destro, R.C., Bianchi, R.A.C.: Hardware, software aspects of the design, assembly of a new humanoid robot for robocup soccer. In: 2014 Joint Conference on Robotics (2011). Observation of strains. Infect Dis Ther. 3(1), 35–43.: SBR-LARS, pp. 73–78. IEEE, October 2014

23. Randell, D.A., Witkowski, M.: Tracking regions using conceptual neighbourhoods. In: Proceedings of the Workshop on Spatial and Temporal Reasoning ECAI 2004, pp. 63–71 (2004)

24. RoboCup 2015 robocup soccer humanoid league rules and setup (2015). http://www.robocuphumanoid.org

25. Ros, R., Arcos, J.L., de Mantaras, R.L., Veloso, M.: A case-based approach for coordinated action selection in robot soccer. Artif. Intell. **173**(9–10), 1014–1039 (2009)

26. Southey, T., Little, J.J.: Relations, learning qualitative spatial for object classification. In: IROS: From Sensors to Human Spatial Concepts (2007)
27. Stone, P., Sutton, R.S., Kuhlmann, G.: Reinforcement learning for RoboCup soccer keepaway. Adap. Behav. **13**(3), 165–188 (2005)
28. Wolter, D., Wallgrün, J.: Qualitative spatial reasoning for applications: new challenges and the SparQ toolbox. In: Qualitative Spatio-temporal Representation and Reasoning: Trends and Future Directions, pp. 336–362 (2012)
29. Young, J., Hawes, N.: Predicting situated behaviour using sequences of abstract spatial relations. In: AAAI 2013 Proceedings, Fall Symposium Series (2013)
30. Young, J., Hawes, N.: Learning by observation using qualitative spatial relations. In: Proceedings of the 2015 AAMAS, pp. 745–751, Richland, SC (2015)

# Ensemble of Adaptations for Classification: Learning Adaptation Rules for Categorical Features

Vahid Jalali[1]([✉]), David Leake[1], and Najmeh Forouzandehmehr[2]

[1] School of Informatics and Computing, Indiana University, Bloomington, IN, USA
{vjalalib,leake}@indiana.edu
[2] Electrical and Computer Engineering Department,
University of Houston, Houston, TX, USA
nforouzandehmehr2@uh.edu

**Abstract.** Acquiring knowledge for case adaptation is a classic challenge for case-based reasoning (CBR). To provide CBR systems with adaptation knowledge, machine learning methods have been developed for automatically generating adaptation rules. An influential approach uses the *case difference heuristic* (CDH) to generate rules by comparing pairs of cases in the case base. The CDH method has been studied for case-based prediction of numeric values (regression) from inputs with primarily numeric features, and has proven effective in that context. However, previous work has not attempted to apply the CDH method to classification tasks, to generate rules for adapting categorical solutions. This paper introduces an approach to applying the CDH to cases with categorical features and target values, based on the *generalized case value difference heuristic* (GCVDH). It also proposes a classification method using ensembles of GCVDH-generated rules, *ensemble of adaptations for classification* (EAC), an extension to our previous work on ensembles of adaptations for regression (EAR). It reports on an evaluation comparing the accuracy of EAC to three baseline methods on four standard domains, as well as comparing EAC to an ablation relying on single adaptation rules, and assesses the effect of training/test size on accuracy. Results are encouraging for the effectiveness of the GCVDH approach and for the value of applying ensembles of learned adaptation rules for classification.

**Keywords:** Case adaptation learning · Case difference heuristic · Classification · Value difference metric

## 1 Introduction

The difficulty of acquiring case adaptation knowledge is a well-known challenge for case-based reasoning. In response, much research has focused on methods for learning adaptation knowledge. A popular approach is to learn adaptation rules from cases in the case base. The *case difference heuristic* (CDH) method,

© Springer International Publishing AG 2016
A. Goel et al. (Eds.): ICCBR 2016, LNAI 9969, pp. 186–202, 2016.
DOI: 10.1007/978-3-319-47096-2_13

first proposed by Hanney and Keane [4], generates adaptation rules by comparing pairs of stored cases. Given two stored cases, it computes the difference between the problems the two cases solve and between their two solutions. It ascribes the difference between the solutions to the difference between the two problems, and formulates a rule capturing that the observed difference in problem descriptions between an input query and source case would require adapting the source solution to achieve the corresponding solution difference. That rule is then applied whenever an input problem and the problem part of a retrieved case have a similar difference. The CDH approach has been applied to many regression tasks (i.e., prediction tasks with numeric target values) (e.g., [8,13,14]). Some difference-based approaches have also been applied to adaptation for for structured cases (e.g., [1,16]).

Existing CDH regression methods have focused on domains with primarily numeric problem descriptions, and have relied on a very simple approach—exact matching—for categorical features. To apply CDH-based adaptation rule learning to classification, for which problem features are often categorical, methods are needed for comparing the categorical features in problem descriptions. Such methods are also needed for adaptation rule selection.

The paper makes two primary contributions. First, it extends the case difference heuristic approach by proposing the *Generalized Case Difference Heuristic* (GCDH), a version of the case difference heuristic for domains in which problem and solution descriptions can include both categorical and numeric feature/target values. To alleviate the need for knowledge acquisition, the GCDH determines differences using the *heterogeneous value difference metric*, a knowledge-light method which we define based on the classic *value distance metric* [20]. The GCDH approach enables applying the case difference heuristic to domains with categorical feature values and to classification tasks, without the overhead required for knowledge-rich categorical similarity metrics (e.g., such as an external ontology or thesaurus). Second, it investigates the benefit of combining adaptation rule learning for classification with the use of ensembles of adaptation rules. The rationale for applying ensemble methods is to exploit the availability of many automatically generated adaptation rules, to neutralize the unavoidable inaccuracies of automatic adaptation rule generation. Our ensemble method, *Ensembles of Adaptations for Classification* (EAC), extends our previous work on *Ensembles of Adaptations for Regression* (EAR) to develop an ensemble method for predicting categorical target values using the Generalized Case Difference Heuristic.

To assess the first contribution, the paper presents an evaluation of a specific instantiation of the Generalized Case Difference Heuristic approach, which we call GCDH1, in a standard CBR context, in which solutions are adapted using a single automatically-generated adaptation rule. To assess the second, it evaluates EAC1, a specific version of EAC. Results show improvement in estimation accuracy for EAC over other methods in the test domains. The evaluation also assesses the comparative data requirements of the methods, examining the effect of different training to test data size ratios on classification accuracy. Results support that EAC1 generally provides better performance improvement with additional training data, compared to baseline methods.

The paper begins by discussing related work. It then discusses the roles of a categorical distance measure in CDH adaptation for categorical features, and describes the Generalized Case Difference Heuristic approach. It then defines our case value difference heuristic metric, describes the EAC approach, and presents evaluations.

## 2   Related Work

Much CBR research studies learning case adaptation knowledge for case-based regression, but not for classification (knowledge-light classification methods have tended to rely simply on kNN and majority voting to generate new solutions). Here we briefly illustrate some adaptation learning approaches for regression and then focus specifically on methods using the case difference heuristic.

Patterson et al. [17] present a method for learning adaptations by training a linear regression method on the differences between the top nearest neighbors of an input query. Specifically, the top nearest neighbors are used to build a generalized case whose differences with the input query are used in training the linear regression model. Policastro et al. [18] propose an adaptation learning method based on training a set of regression learners for estimating the initial solutions and adjusting/combining their values. They use a multi-layer neural network, an M5 regression tree, and a support vector machine both for generating the initial estimations and combining them. Craw et al. [3], Jarmulak et al. [11], and Wiratunga et al. [22] propose acquiring adaptation knowledge from a case base by partitioning it to smaller subsets called probe cases. The top $k$ similar cases to each probe case are retrieved and a set of adaptation rules generated, each addressing the differences in an individual input feature between the pairs of cases considered. Resulting rules are filtered to retain rules expected to have above average accuracy.

Most relevant to this paper are difference-based methods, often based on the *Case Difference Heuristic* [4]. As discussed, the CDH generates adaptation rules by comparing pairs of cases in the case base. Based on the differences in their problem descriptions, it forms the antecedent of the new rule; based on differences in their solutions, it forms the consequent of the new rule. As a very simple example, consider the task domain of predicting apartment rental price. If two cases record apartments whose descriptions differ only in that one has an additional bedroom, and their rates differ by $200, a rule might be generated to add $200 per month to the estimated rent when an input apartment has one bedroom more than the retrieved apartment case. (Many other rules could be generated as well; the example is simply an illustration.)

McSherry [14] introduces a difference-based rule learning method that, given an input problem, generates rules by retrieving a stored case that differs from the input in a single feature, called the distinguishing attribute, and generates an adaptation by retrieving a pair of cases also differing only in that attribute, from which the difference is used to generate a rule addressing that specific difference. When multiple rules must be applied, to address multiple differences, the final solution is generated by averaging the individual rules' proposed adjustments.

McDonnell and Cunningham [13] extend the case difference heuristic by taking into account how variations in feature values and the values of other features of a case affect the solution. Their method generates adaptation rules by comparing the input query to the top $k$ neighbors whose gradient is similar to the selected source case, where gradient is approximated by using local linear regression.

Jalali and Leake [8,10] introduce EAR, a family of methods for learning and applying adaptation rules for regression. EAR combines adaptation rule learning, using a case difference approach, with the use of rule ensembles to perform adaptations. Different EAR variants can be generated based on choices for source case selection and rule generation. EAR has been extended to consider the role of context [6] and adaptation rule confidence [9] in the rule retrieval process, as well as adapted to use methods for big data platforms for scaleup [5].

# 3   Generalizing the Case Difference Heuristic

## 3.1   Motivations

The case difference heuristic approach has proven effective for a range of regression tasks. The knowledge-light nature of the case difference heuristic makes it more desirable than knowledge-intensive alternatives when generating knowledge-based adaptation rules would be time-consuming or costly. However, prior CDH research does not provide a comprehensive solution for domains with categorical input features, instead relying on simple exact matching for categorical features, and has not attempted to apply the CDH to classification.

*The Need for Finer-Grained Treatment of Categorical Features During Rule Generation:* To develop rules that appropriately reflect categorical features, the CDH requires methods beyond exact match for comparing categorical features. For example, in an automobile price estimation domain, car body condition might be described by a categorical input feature, with values such as poor, fair, good, and excellent. If the system relies on exact matching for determining the similarity of categorical features, an automobile with poor body condition will be considered equally similar to two other automobiles with (respectively) fair and excellent condition (assuming other features are identical), even though the difference between poor and excellent is intuitively greater than between poor and fair, requiring generation of a rule effecting a bigger price adjustment.

*The Need for Finer-Grained Treatment of Categorical Features During Rule Selection:* Richer categorical feature similarity judgments are equally important for selecting adaptation rules to apply. Suppose that a system is adapting the price estimate from a retrieved case for a car in excellent condition to apply to a car in *good* condition, and two adaptation rules are available, neither of which is an exact match:

**Rule1:** If input body condition is *poor* and the retrieved case's body condition is *excellent*, then adapt the retrieved case's price by subtracting $5000.

**Rule2:** If input body condition is *fair* and the retrieved case's body condition is *excellent*, then adapt price by subtracting $3000.

Intuitively *rule2* is more appropriate to address the differences between the input query and the retrieved case compared to *rule1*, and should be selected.

### 3.2 The Generalized Case Difference Heuristic Approach

The *Generalized Case Difference Heuristic* (GCDH) approach extends the CDH to handle categorical features and target values. This section describes the basic approach; the following section describes the distance metric applied in the specific instantiation we call GCDH1. The distance metric compares is applied in comparisons at two points: When two cases are compared to generate adaptation rules, and when the difference between a source and target case is compared to the differences encoded by adaptation rules, in order to select the most applicable rule.

*GCDH Rule Representation:* Following other case difference heuristic research, we assume that input problem descriptions are represented as feature vectors, with the differences between problems represented as vectors of component-wise differences. However, it is not always possible to define a full ordering on categories, enabling them to be compared out of context. An important difference between the case difference heuristic and GCDH is that, to provide the needed context in domains with categorical target values, the GCDH rule selection process also considers the target value of the source case to adapt. Consequently, GCDH considers the target value of the source cases for rule generation, to filter out irrelevant adaptations in the rule selection process. Equation 1 shows the generic format of rules generated by GCDH where each pair of cases is turned into a single rule:

$$(\Delta_{f_1}, ..., \Delta_{f_k}) \Rightarrow \Delta_t \tag{1}$$

In Eq. 1, $f_1, ..., f_k$, and $t$ represent the input and target features in the underlying domain respectively, and $\Delta_x$ represents the distance between the given pair of values for feature $x$, if $x$ is numeric, and the ordered pair of given values of feature $x$ (e.g. $(t_i, t_j)$) if $x$ is a categorical feature. For numeric features, distance may be calculated by subtraction, percent difference, or any other function appropriate to the domain.

*GCDH Rule Selection and Application:* For GCDH, as for the standard case difference heuristic approach, adaptation rules are considered relevant if the difference between the input problem and problem of the source case is similar to the difference captured in the antecedent of the generated adaptation rule. Given an input problem as described above, and a retrieved source case $c_1$ with $t_1$ as its target value, GCDH filters rules by target value, only considering rules in the rule base generated to adapt the same target value, *i.e.*, rules with consequents of the form $(t_1, t_i)$, where $t_i$ is any arbitrary value. After the rule filtering step, the distance/similarity between different components (i.e. input feature differences) of two adaptation rules generated by GCDH can be measured by the same method used for case retrieval. This enables ranking the remaining rules according to their (e.g., Euclidean) distance to the pair of the source case and input query.

*An Illustration:* The following example illustrates the GCDH's rule selection and rule application process for a classification task (the steps for the specific GCDH1 instantiation are illustrated in Fig. 1). The example concerns a highly simplified credit approval domain, with the input query a customer with a high school diploma, and the source case the case of a customer with master's degree whose credit request was approved. We assume that the system's adaptation rules are:

**Rule1:** If input education degree is *bachelor* and the retrieved case's education degree is *diploma*, then adapt the retrieved case's outcome from *approve* to *reject.*

**Rule2:** If input education degree is *less than high school* and the retrieved case's education degree is *diploma*, then the case's outcome remains *reject.*

**Rule3:** If input education degree is *bachelor* and the retrieved case's education degree is *master*, then the case's outcome remains *approve.*

During rule selection, GCDH only considers rules generated to adapt the same categorical target value as in the source case. Consequently, in this case, only rules whose consequent starts with "approve" are considered, leaving rules 1 and 3. These remaining rules are ranked according to their distance to the pair of source case and the input query. Assuming that the problem similarity/distance measure considers the pair (master, diploma) (generated from problem description part of source case and input query) to be more similar to *rule1*'s pair (bachelor, diploma) than to *rule3*'s pair (bachelor, master), the most similar rule to the differences between the source case and the input query will be *rule1*. When the value of the source case is adjusted by applying *rule1*, then adapted decision is "reject".

## 3.3 Instantiating GCDH Using the Case Value Difference Heuristic Metric

Different methods of measuring the distance between categorical features' values provide different instantiations of GCDH. For example, distances could be based on a semantic similarity measure (e.g. [19]) or an external ontology or thesaurus. Here we apply a probabilistic method, the *Value Difference Metric* VDM, [20], which calculates the distance between values "a" and "b" of the categorical feature $f$ by:

$$vdm_f(a, b) \equiv \sum_{c=1}^{C} \mid \frac{N_{f,a,c}}{N_{f,a}} - \frac{N_{f,b,c}}{N_{f,b}} \mid \qquad (2)$$

where $N_f, a$ is the number of cases that have value "a" for feature $f$, $N_{f,a,c}$ is the number of cases that have value "a" for feature $f$ and target value "c", and C is the number of distinct target classes in the domain. To extend this approach with the ability also to handle numeric features, we use Wilson and Martinez's *Heterogeneous Value Difference Metric* (HVDM) [21], which applies the VDM for categorical features and calculates differences between numeric features by:

$$\text{diff}_f(a, b) \equiv \frac{|a - b|}{4\sigma_f} \tag{3}$$

where $\sigma_f$ is the standard deviation of feature "f". If the values of feature "f" follow a normal distribution then the expectation is that 95 % of them fall within a $4\delta$ distance range of each other.

We define the *Case Value Distance Heuristic Metric*, CVDHM, as the metric that calculates the difference between values "a" and "b" of feature $f$ as follows (the feature difference is limited to 1 to eliminate extreme changes for outlier values):

$$CVDHM_f(a, b) \equiv \begin{cases} 1, & \text{if a or b is unknown} \\ 1, & \text{if f is categorical and} \\ & \text{a or b is not observed in the training data} \\ vdm_f(a, b), & \text{if f is categorical} \\ min(1, \text{diff}_f(a, b)), & \text{if f is numeric} \end{cases} \tag{4}$$

We then define GCDH1 as the instantiation of GCDH that uses Case Value Distance Heuristic Metric for both case and rule retrieval. Case retrieval using Heterogeneous Value Difference Metric is examined by Wilson and Martinez [21]; in this paper we discuss the rule retrieval process only.

As described by Eq. 1, the generic form of GCDH1 rules is a vector of rule differences $\Delta_x$, where $\Delta_x$ is a numeric value for numeric features and is a pair of categorical values for categorical features, whose distance to other pairs of categorical values in other rules is calculated using CVDHM, with Heterogeneous Value Difference Metric trained on the rule base to determine the VDM values. After training, CVDHM can use the results to find the distance between individual features of two adaptation rules by their Minkowski distance as follows:

$$\text{rulediff}_f(r_1, r_2) \equiv \left(\sum_{i=1}^{n} | CVDHM_i(r_{1,i}, r_{2,i}) |^p\right)^{\frac{1}{p}} \tag{5}$$

where $n$ is the number of features, and $r_{1,i}$, $r_{2,i}$ are the values of the $i^{th}$ feature of rules $r_1$ and $r_2$ respectively. As previously mentioned, $r_{1,i}$ and $r_{2,i}$ will be numeric values for numeric features and will be pairs of categorical values for categorical features. $p$ is a real value that is greater or equal to 1. For GCDH1 $p$ is set to 2, for Euclidean distance. GCDH1 applies rulediff to rule retrieval, to select the rule most applicable to adapting a source case to an input query.

It treats the pair (input query, source case) as determining the antecedent of an adaptation rule with no consequent, and uses rulediff to select the adaptation rule with the most similar antecedent.

## 4  Ensemble of Adaptations for Classification

Ensemble of Adaptations for Classification is an extension to Ensemble of Adaptations for Regression (EAR) [8,10], applying the EAR approach to domains including categorical features in problem descriptions and target values. EAR combines automatic generation of adaptation rules with adaptation by ensembles of adaptations. EAR was developed to exploit the ability of CDH to generate many adaptation rules by applying rule ensembles, and experiments supported EAR's ability to outperform baselines.

EAR can apply a range of approaches for selecting the source cases to adapt to a new problem, as well as for selecting the cases from which adaptations will be generated. EAC supports the same range of approaches for source case selection and adaptation rule generation, summarized briefly below and described in detail by Jalali and Leake [10]. EAC must also include approaches for selecting cases for training HVDM.

*Selecting Source Cases and Cases for Training HVDM:* Selecting source cases to adapt in EAC is very similar to EAR [8,10]. As for EAR, EAC can be applied with any of the following three alternatives for selecting a source case or cases to adapt:

1. Nearest: choosing the most similar case to the input query
2. Local: choosing the top $k$ nearest neighbors to the input query
3. Global: using all cases in the case base.

However, EAC differs from EAR in using Heterogeneous Value Difference Metric as the metric for determining the nearest neighbors to the input query. Because Heterogeneous Value Difference Metric must learn the distance between categorical feature values, a training process must be defined as well. For such training, EAC uses all cases in the case base. Note that because this training can be done offline, this exhaustive training is feasible for moderate size case bases. Efficiency could be increased by using subsets of the case base to train Heterogeneous Value Difference Metric for source case selection, and is a potential topic for future work (see Sect. 6).

*Selecting Cases for Rule Generation:* As for selecting source cases to adapt, various criteria could be applied to select the pairs of cases from which adaptation rules are generated. The strategies can be described by how each element in a rule generation pair is selected. For example, one strategy is to select a local case—a case in the same neighborhood as the source case—and generate a rule from it and one of its local neighbors. Another is to select a global case—from anywhere in the case base—and generate a rule by comparing that case to one of its local neighbors. The local-global choices give rise to three strategies: 1- Local cases - Local neighbors; 2- Global cases - Local neighbors; 3- Global cases - Global cases. Note that with alternative 1, Local cases - Local neighbors, EAC enables on demand generation of adaptations: adaptations do not necessarily need to be generated in advance and the system can learn them as needed for specific queries. Once cases for generating rules are selected, rules are generated according to Eq. 1. As mentioned, $\Delta_x$ represents the distance between the given pair of values for feature $x$, if $x$ is numeric, and the ordered pair of given values of feature $x$ (e.g. $(t_i, t_j)$) if $x$ is a categorical feature. Rules are generated without applying any retention or generalization mechanism. Details can be found in Jalali and Leake [7].

*Selecting Cases for Training HVDM for Rule Features:* After adaptation rules have been generated, another round of HVDM training is required to learn the distance between the pairs of categorical feature values in the adaptation rules, to use in the rule retrieval step. Again, there are three options:

1. Local cases - Local neighbors: training HVDM based on the local neighborhood of the input query
2. Global cases - Local neighbors: training HVDM based on the local neighborhoods of cases throughout the case base
3. Global cases - Global cases: training HVDM on all pairs of cases in the case base.

By combining the above alternatives for selecting cases to generate adaptations, and the alternatives for training HVDM, nine different methods could be generated for rule generation and retrieval in EAC. When source case selection methods are combined with these nine alternatives, there are 27 possible variations of EAC.

## 4.1  Estimating the Target Value with EAC

Algorithm 1 summarizes EAC's value estimation process. *CaseHVDM* input contains the categorical value distances for every symbolic feature (this can be calculated in advance and reused for processing any incoming query), *NeighborhoodSelector* and *RankRules* are the case and rule retrieval methods respectively. *HVDMRuleTrain* is one of the three alternatives for learning the distance between pair of values of categorical features in the case base, and *FilterRules* filters out the non-applicable rules from the set of generated rules by removing those whose first element of the consequent pair does not match the value of the selected source case. *MajorityRuleVote* returns the majority value of the second element of the consequent pair of the retrieved rules and *MajorityVote* returns the majority of the adjusted source cases' values.

---

**Algorithm 1.** EAC's basic algorithm

---

**Input:**

$Q$: query

$n$: number of source cases to adapt to solve query

$r$: number of rules to be applied per source case

$CB$: case base

$CaseHVDM$: categorical value similarities

**Output:** Estimated solution value for Q

$CasesToAdapt \leftarrow$ NeighborhoodSelector($Q,n,CB,CaseHVDM$)

$RuleHVDM \leftarrow$ HVDMRuleTrain($Q,CB$)

$NewRules: \leftarrow$ RuleGenerator($Q,CasesToAdapt,CB$)

**for** $c$ in $CasesToAdapt$ **do**

    ApplicableRules $\leftarrow$ Filter($NewRules$)

    RankedRules $\leftarrow$ RankRules($ApplicableRules,c,Q,RuleHVDM$)

    $SolutionEstimate(c) \leftarrow$ MajorityRuleVote($RankedRules, c, r$)

**end for**

return MajorityVote($\cup_{c \in CasesToAdapt}SolutionEstimate(c)$)

---

### 4.2  EAC1: An Instantiation of EAC

As described, EAC offers a wide range of customizations, ranging from methods for source case selection and selecting cases, to building adaptation rules, and to the choice of similarity measure used for categorical feature values and training the similarity measure used. In this section we introduce EAC1, a specific instantiation of EAC which we will use as the basis for evaluation.

The choice of source case selection used by EAC1 is "Local," meaning that EAC1 uses the top $k$ nearest neighbors of the input query to build its estimations. GCDH1 is used as the underlying method for learning and selecting adaptations (without applying any retention or generalization mechanism). EAC1 uses CVDHM to retrieve these top nearest neighbors and trains HVDM on all cases in the case base as discussed in Sect. 4. Adaptation rules are learned by applying the "Global cases - Local neighbors" approach in EAC1 and for adaptation rule similarity, HVDM is trained on "Global cases - Local neighbors" as well.

Figure 1 illustrates EAC1's process. EAC1 first trains HVDM to learn the distance between categorical feature values. Next, it uses CVDHM (which uses the HVDM results) to select cases from which adaptation rules will be generated. It also uses CVDHM to find the source cases (nearest neighbors of the input query). After generating the adaptation rules, EAC1 trains a new instance of HVDM on the rule base to learn the distance between pairs of pairs of categorical values. The newly trained HVDM along with the differences between the input query and the source cases are used by CVDHM to select the rules that should be applied. The value of each source case will be adapted by applying an ensemble of adaptation rules and the final estimation will be generated by combining the adjusted values of source cases using majority voting.

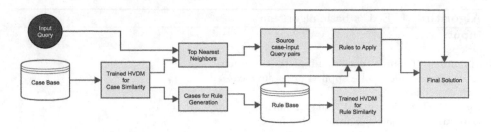

**Fig. 1.** Illustration of the generic process of EAC1

## 5   Experiments

We hypothesize that the GCDH, in conjunction with a general-purpose, knowledge-light distance metric, is sufficient to enable more accurate case-based classification than standard classification baselines, and that the ensemble-based approach of EAC will improve accuracy compared to base line of kNN and EAC1. Our experiments address these through the following questions:

1. Comparative accuracy: How does classification accuracy of EAC1 compare to the accuracy of standard classification methods Random Forest, Naive Bayes, and kNN?
2. Effect of applying ensembles of adaptations: How does classification accuracy of EAC1, which applies an ensemble of adaptation rules, compare to the accuracy of an ablated version for which only one adaptation is applied per source case?
3. Effectiveness exploiting training: How does the ratio of training to test data affect the performance of EAC1 compared to other alternative classification methods?

### 5.1   Experimental Design

We implemented EAC1 using Spark MLlib [15], Apache Spark's scalable machine learning library. Spark MLlib provides a set of built in classification methods, grid search for tuning the model parameters and cross validation for assessing the performance of the tested methods. The accuracy experiments measure the accuracy in five classification domains from UCI repository [12]:

1. Balance Scale: Predict whether scale will tip right, left, or stay balanced
2. Qualitative Bankruptcy: Predict bankruptcy or non-bankruptcy from qualitative parameters from experts
3. Car Evaluation: Predict the acceptability of an automobile (unaccepted, accepted, good, and very good)
4. Credit Approval: Predict whether a credit card application will be approved

**Table 1.** Characteristics of the test domains

| Domain name | # categorical features | # numeric features | # cases | # unique combination of categorical features | # unique solutions |
|---|---|---|---|---|---|
| Balance | 4 | 0 | 625 | 625 | 3 |
| Bankruptcy | 6 | 0 | 250 | 729 | 2 |
| Car | 6 | 0 | 1728 | 1728 | 4 |
| Credit | 9 | 6 | 653 | 54432 | 2 |

We refer to these domains respectively as Balance, Bankruptcy, Car, and Credit. Balance, Bankruptcy, and Car have only categorical features; Credit has a mix of categorical and numeric features. We note that in some cases, categorical features could have been ordered and converted to numerical ranges. However, to test the method's performance as a knowledge-light method without additional knowledge, no ordering information was provided to the system. Because the credit approval data includes no human-comprehensible values, for that domain only knowledge-light symbolic methods can be applied. Data were cleaned by removing cases from each domain with unknown feature values (EAC can be applied to cases with unknown feature values by combining it with existing methods for handling missing feature values (e.g., [2])). Balance, Bankruptcy and Car had no missing values, but cleaning changed the number of Credit cases from 690 to 653.

The experiments compare the percent accuracy of EAC1 and three baseline classification methods: Random Forest (RF), Naive Bayes (NB), and kNN. The tests used Apache Spark implementations of Random Forrest and Naive Bayes.[1] The kNN version used, which we designate *kNN-HVDM* is an implementation of k Nearest Neighbors with HVDM used as the underlying similarity measure. kNN-HVDM and EAC1 are both implemented using Spark MLlib; where kNN-HVDM is a special case of EAC1 in which no adaptation rule is applied. EAC1 and our implementation of kNN consequently extend the *ClassificationModel* class in Spark MLlib. In addition, we implemented EAC1-a, an ablated version of EAC1 using single adaptations rather than ensemble of adaptations. EAC1-a's tuning is exactly the same as EAC1's except that the number of tested adaptations is always one.

We used grid search to tune the parameters using ten fold cross validation on the training data. Unless mentioned otherwise we split the data sets into 70 % training and 30 % test data. Table 2 summarizes the parameters tuned for each classifier and their candidate values. All experimental results are averages over 100 runs.

---

[1] *RandomForestClassifier* and *NaiveBayes* from the *org.apache.spark.ml.classification* package.

**Table 2.** Parameters and their corresponding values used to tune each classifier

| Method name | parameters | values |
|---|---|---|
| RF | [# trees, max depth, impurity] | [[2, 5, 10, 20, 100], [2, 4, 6, 8], [entropy, gini]] |
| NB | [lambda] | [1.0] |
| kNN-HVDM | [# source cases] | [1, 2, 3, 5, 10] |
| EAC1 | [# source cases, # rules to apply, # neighbor cases for rule learning] | [[1, 2, 3, 5, 10], [1, 2, 3, 5, 10], [5, 10, 20]] |

**Table 3.** Estimation error of EAC1, kNN-HVDM, RF, and NB methods in four sample domains

| Domain name | EAC1 | RF | kNN-HVDM | NB |
|---|---|---|---|---|
| Balance | **15.98 %** | 16.76 % | 25.26 % | 44.51 % |
| Bankruptcy | **1.35 %** | 2.05 % | 3.43 % | 13.51 % |
| Car | **3.95 %** | 6.25 % | 6.50 % | 30.71 % |
| Credit | **15.64 %** | 17.10 % | 18.11 % | 30.81 % |

## 5.2  Experimental Results

**Q1: Comparative Accuracy.** To compare the performance of EAC1 with other classification methods we conducted experiments on the four sample domains from Sect. 5.1. Table 3 shows the performance of each method.

In all domains, EAC1 achieves the best accuracy. Random Forest achieves second best performance in all domains. EAC1 shows 5 %, 34 %, 37 %, and 8 % decreases in estimation error over Random Forest in the Balance, Bankruptcy, Car, and Credit domains respectively.

EAC1 shows 37 %, 60 %, 38 %, and 14 % improvement in decreasing estimation error over kNN-HVDM in Balance, Bankruptcy, Car, and Credit domains respectively. For domains with only categorical features (i.e. all domains except Credit), we note EAC1's gain over kNN-HVDM is higher when curse of dimensionality seems to be more severe (we assessed dimensionality by dividing the number of cases in a domain by the number of unique combinations of the input features; both values are shown in Table 1). For Balance and Car domains, on average there is one case per n-dimensional unit of the problem specification space while this number decreases to 0.3 for Bankruptcy domain. However, the possible effect of dimensionality needs more investigation before a general conclusion will be possible.

**Q2: Effect of Applying Ensembles of Adaptations.** In order to assess the effect of using ensembles of adaptations, we implemented an ablated version of EAC1 named EAC1-a, which always applies a single adaptation to adjust the source case values. Note that EAC1-a only limits the number of adaptations

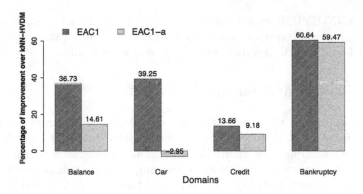

**Fig. 2.** Percentage of improvement of EAC1 and EAC1-a compared to kNN-HVDM

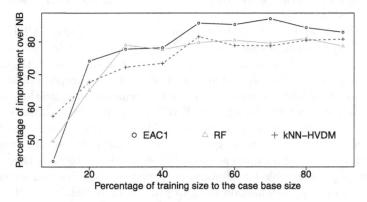

**Fig. 3.** Percentage of improvement of EAC1, RF, and kNN-HVDM over NB at different ratios of training size to the case base size in the car evaluation domain

to be applied per source case and not the number of source cases to be used in building the solution. Figure 2 depicts the percentage decrease in estimation error for EAC1 and EAC1-a over kNN-HVDM. In all test domains EAC1 outperforms EAC1-a. For the Car domain applying a single adaptation (EAC1-a), rather than an ensemble of adaptations (EAC1), results in average accuracy slightly lower than that of kNN-HVDM, which uses no adaptations. Also, as seen in the results for Question 1, EAC1-a's best results compared to kNN-HVDM are achieved in the Bankruptcy domain. In this case even applying a single adaptation can improve the estimation accuracy by a large margin, and the use of ensembles has little effect. We conjecture that this may reflect greater uniformity in the rules generated for the Bankruptcy domain, but more study is needed.

**Q3: The Effect of Test/Train Ratio on EAC Performance** Fig. 3 shows the percentage of improvement of EAC1, kNN-HVDM, and RF over NB in the car evaluation domain for different ratios of training vs test data size. As the training versus test size ratio increases, EAC1 best benefits from the training data, with kNN-HVDM second, but fairly close to Random Forest. For smaller sizes of training set, Random Forest can outperform EAC1 (when training size ratio is 10 % and 30 %). Trends in other domains were similar.

# 6    Conclusions and Future Research

This paper addressed the problem of automatic adaptation rule learning for cases with categorical features and for classification tasks. It introduced the Generalized Case Difference Heuristic approach, an extension to the case difference heuristic approach, that enables learning and applying adaptation rules in domains with categorical input features and target values. It also introduced Ensemble of Adaptations for Classification, a generic approach to predicting categorical target values that works by applying an ensemble of GDHC adaptations to adjust classifications.

To instantiate and test the GCDH approach, we introduced GCDH1, a specific instantiation using a difference metric based on the heterogeneous value difference metric, and EAC1, an instantiation of EAC using that GCDH1. Experiments conducted on five test domains showed promising improvement in estimation accuracy of EAC1 compared to other alternative classification methods. Likewise, they showed the benefits of EAC1 compared to an ablated version of EAC1 that used single adaptation rules. These results support the feasibility of automatically generating adaptations for classification tasks, as well as for value of including adaptation in case-based classification. Likewise, they support the benefit of the ensemble-based adaptation approach, which can leverage the availability of multiple adaptation rules to provide more reliable results.

Possible extensions include exploring how categorical similarity measures other than HVDM affect performance and examining the effects of additional alternatives for the 27 possible variations of EAC. Because EAC involves additional processing costs compared to traditional methods, are also interested in employing the same ideas we developed in BEAR [5] for increased scalability of EAC for large case bases. As part of this goal, we intend to introduce Locality Sensitive Hashing methods tailored for domains with categorical input features and target values.

# References

1. Badra, F., Cordier, A., Lieber, J.: Opportunistic adaptation knowledge discovery. In: McGinty, L., Wilson, D.C. (eds.) ICCBR 2009. LNCS (LNAI), vol. 5650, pp. 60–74. Springer, Heidelberg (2009). doi:10.1007/978-3-642-02998-1_6
2. Bogaerts, S., Leake, D.: Facilitating CBR for incompletely-described cases: distance metrics for partial problem descriptions. In: Funk, P., González Calero, P.A. (eds.) ECCBR 2004. LNCS (LNAI), vol. 3155, pp. 62–76. Springer, Heidelberg (2004). doi:10.1007/978-3-540-28631-8_6
3. Craw, S., Jarmulak, J., Rowe, R.: Learning and applying case-based adaptation knowledge. In: Aha, D.W., Watson, I. (eds.) ICCBR 2001. LNCS (LNAI), vol. 2080, pp. 131–145. Springer, Heidelberg (2001). doi:10.1007/3-540-44593-5_10

4. Hanney, K., Keane, M.T.: Learning adaptation rules from a case-base. In: Smith, I., Faltings, B. (eds.) EWCBR 1996. LNCS, vol. 1168, pp. 179–192. Springer, Heidelberg (1996). doi:10.1007/BFb0020610

5. Jalali, V., Leake, D.: CBR meets big data: a case study of large-scale adaptation rule generation. In: Hüllermeier, E., Minor, M. (eds.) ICCBR 2015. LNCS (LNAI), vol. 9343, pp. 181–196. Springer, Heidelberg (2015). doi:10.1007/978-3-319-24586-7_13

6. Jalali, V., Leake, D.: A context-aware approach to selecting adaptations for case-based reasoning. In: Brézillon, P., Blackburn, P., Dapoigny, R. (eds.) CONTEXT 2013. LNCS (LNAI), vol. 8175, pp. 101–114. Springer, Heidelberg (2013). doi:10.1007/978-3-642-40972-1_8

7. Jalali, V., Leake, D.: An ensemble approach to instance-based regression using stretched neighborhoods. In: Proceedings of the 2013 Florida AI Research Symposium, pp. 381–386. AAAI Press (2013)

8. Jalali, V., Leake, D.: Extending case adaptation with automatically-generated ensembles of adaptation rules. In: Delany, S.J., Ontañón, S. (eds.) ICCBR 2013. LNCS (LNAI), vol. 7969, pp. 188–202. Springer, Heidelberg (2013). doi:10.1007/978-3-642-39056-2_14

9. Jalali, V., Leake, D.: On deriving adaptation rule confidence from the rule generation process. In: Delany, S.J., Ontañón, S. (eds.) ICCBR 2013. LNCS (LNAI), vol. 7969, pp. 179–187. Springer, Heidelberg (2013). doi:10.1007/978-3-642-39056-2_13

10. Jalali, V., Leake, D.: Enhancing case-based regression with automatically-generated ensembles of adaptations. J. Intell. Inform. Syst. **5**, 1–22 (2015)

11. Jarmulak, J., Craw, S., Rowe, R.: Using case-base data to learn adaptation knowledge for design. In: Proceedings of the 17th International Joint Conference on Artificial Intelligence, vol. 2, IJCAI 2001, pp. 1011–1016. Morgan Kaufmann, San Francisco (2001)

12. Lichman, M.: UCI machine learning repository (2013). http://archive.ics.uci.edu/ml

13. McDonnell, N., Cunningham, P.: A knowledge-light approach to regression using case-based reasoning. In: Roth-Berghofer, T.R., Göker, M.H., Güvenir, H.A. (eds.) ECCBR 2006. LNCS (LNAI), vol. 4106, pp. 91–105. Springer, Heidelberg (2006). doi:10.1007/11805816_9

14. McSherry, D.: An adaptation heuristic for case-based estimation. In: Smyth, B., Cunningham, P. (eds.) EWCBR 1998. LNCS, vol. 1488, pp. 184–195. Springer, Heidelberg (1998). doi:10.1007/BFb0056332

15. Meng, X., Bradley, J.K., Yavuz, B., Sparks, E.R., Venkataraman, S., Liu, D., Freeman, J., Tsai, D.B., Amde, M., Owen, S., Xin, D., Xin, R., Franklin, M.J., Zadeh, R., Zaharia, M., Talwalkar, A.: Mllib: Machine learning in apache spark. CoRR abs/1505.06807 (2015). http://arxiv.org/abs/1505.06807

16. Müller, G., Bergmann, R.: Learning and applying adaptation operators in process-oriented case-based reasoning. In: Hüllermeier, E., Minor, M. (eds.) ICCBR 2015. LNCS (LNAI), vol. 9343, pp. 259–274. Springer, Heidelberg (2015). doi:10.1007/978-3-319-24586-7_18

17. Patterson, D., Rooney, N., Galushka, M.: A regression based adaptation strategy for case-based reasoning. In: Proceedings of the Eighteenth Annual National Conference on Artificial Intelligence, pp. 87–92. AAAI Press (2002)

18. Policastro, C.A., Carvalho, A.C., Delbem, A.C.: A hybrid case adaptation approach for case-based reasoning. Appl. Intell. **28**(2), 101–119 (2008)

19. Resnik, P.: Semantic similarity in a taxonomy: an information-based measure and its application to problems of ambiguity in natural language. CoRR abs/1105.5444 (2011). http://arxiv.org/abs/1105.5444
20. Stanfill, C., Waltz, D.L.: Toward memory-based reasoning. Commun. ACM **29**(12), 1213–1228 (1986)
21. Wilson, D.R., Martinez, T.R.: Improved heterogeneous distance functions. J. Artif. Int. Res. **6**(1), 1–34 (1997)
22. Wiratunga, N., Craw, S., Rowe, R.: Learning to adapt for case-based design. In: Craw, S., Preece, A. (eds.) ECCBR 2002. LNCS (LNAI), vol. 2416, pp. 421–435. Springer, Heidelberg (2002). doi:10.1007/3-540-46119-1_31

# Similarity Metrics from Social Network Analysis for Content Recommender Systems

Guillermo Jimenez-Diaz[✉], Pedro Pablo Gómez Martín[✉],
Marco Antonio Gómez Martín[✉], and Antonio A. Sánchez-Ruiz[✉]

Department of Software Engineering and Artificial Intelligence,
Universidad Complutense de Madrid, Madrid, Spain
{gjimenez,antsanch}@ucm.es, {pedrop,marcoa}@fdi.ucm.es

**Abstract.** Online judges are online systems that test programs in programming contests and practice sessions. They tend to become big problem *live archives*, with hundreds, or even thousands, of problems. This wide problem statement availability becomes a challenge for new users who want to choose the next problem to solve depending on their knowledge. This is due to the fact that online judges usually lack of meta information about the problems and the users do not express their own preferences either. Nevertheless, online judges collect a rich information about which problems have been attempted, and solved, by which users. In this paper we consider all this information as a *social network*, and use social network analysis techniques for creating similarity metrics between problems that can be then used for recommendation.

## 1 Introduction

Online judges are online systems that test programs in programming contests and practice sessions. They are able to compile and execute the code sent by users and deliver a verdict regarding its correctness according to the problem statement. Usually, execution time and memory consumption are restricted in order to force the users to look for efficient solutions. Examples of such systems are the UVa Online Judge (https://uva.onlinejudge.org) and Codeforces (http://codeforces.com/), to mention just two of them.

These systems are usually used for training on-site programming contests such as ACM-ICPC International Collegiate Programming Contest. Some of them are not mere problem archives with automatic judges, but they also hold on-line contests that coincide with the publication of new problems. This way, users may practice with fresh problems and their final standing is reflected in different rankings managed by the system.

Moreover, most of them can even emulate past online and on-site contests for those users that could not participate on them. The user manifests his intention of joining to a past contest and the system starts a virtual contest for him. System displays problem statements and a ranking where the real submissions

Supported by UCM (Group 910494) and Spanish Committee of Economy and Competitiveness (TIN2014-55006-R).

sent by the user are merged with submissions that were made by participants in the original contest.

For that reason, online judges are valuable resources for expert users who want to increase their performance on competitive programming. Unfortunately, these systems pay little or no attention to newbies who are not biased by programming contests but just want to practice algorithms or data structures. Usually they are overwhelmed by the great amount of problems in the archive and they have no idea about which one should try to solve next. The main reason is that online judges do not usually have any recommendation mechanism that guide those users. However, when they have one, they usually use the *Global Ranking Method* that just recommend the problem with more correct solutions in the system that the user has not resolved yet.

Having a recommender system on online judges is not easy, though. The users hardly ever rate problems, they do not express their preferences on their profiles and the information about the problem is nonexistent or, at most, they are just tagged with the programming concepts that should be used in their solutions.

This paper introduces a technique to incorporate recommendation methods into these online judges. In order to do that, we characterize the user-problem interaction (which users have solved which problems) as an implicit social network. Then, we apply Social Network Analysis (SNA) techniques to extract information about the problems that will be used for a recommendation. In particular, we suggest the use of similarity-based link prediction techniques over the implicit user-problem interaction network to predict new links that may be seen as problem recommendation to users.

The experimental evaluation uses *Acepta el reto*[1] (Spanish translation of *Take on the challenge*) as dataset, an online judge developed by some of the authors. Our evaluation will point out that using link prediction similarity-based methods we are able to build recommender systems whose performance is comparable to those based on classical recommendation methods without the need of additional knowledge about the content recommended.

The reminder of this work runs as follows. Next section describes our online judge used as dataset. It is followed by the description of how to characterize the user interaction of online judges as a social network. Section 4 describes the link prediction problem and enumerates the similarity metrics that we have used in our recommendation algorithms. Section 5 describes the experimental setup and the evaluation results. Then the related work is present (Sect. 6) and finally a section with conclusions and future work closes this work.

## 2    *Acepta el reto* Online Judge

*Acepta el reto* (ACR) is an online judge created by two of the authors in 2014. It focuses on Spanish students, who find hard to use other judges (with English statements) because of the language barrier. Problems are tagged according to

---

[1] https://www.aceptaelreto.com (Spanish only).

the programming concepts needed to solve them, the kind of data structures required and some other aspects.

Users select the next problem to confront with and then try to solve it submitting code solutions in one of the accepted languages (currently C, C++ and Java). The system compiles the source code and runs it against many *test cases* whose solutions are known by the judge. The output generated by the submitted solution is compared with the official solution and a verdict is provided.

From the system point of view, a submission can be seen as a tuple $(d, p, u, c, v)$ where $d$ is the submission date, $p$ and $u$ are the problem and user respectively, $c$ is the source code sent by the user, and $v$ is the verdict emitted by the judge. As in many other online judges, the verdicts and their meanings are the following:

AC (*Accepted*): The solution sent was correct because it produces the right answer and it did not exceed the time and memory usage thresholds.

PE (*Presentation Error*): The solution was almost correct, though it failed to write the output in the exact required format (having an excess of blanks or line endings, for example).

CE (*Compile Error*): The solution did not compile.

WA (*Wrong Answer*): The program failed to write the correct answer for one or more test cases.

RTE (*Runtime Error*): The program crashed during the execution (because of segmentation fault, floating point exception...).

TLE (*Time Limit Exceeded*): The execution took too much time and was cancelled.

MLE (*Memory Limit Exceeded*): The solution consumed too much memory and was aborted.

OLE (*Output Limit Exceeded*): The program tried to write too much information. This usually occurs if it goes into an infinite loop.

Generally, users suffering a negative verdict try to fix their code and they resubmit it. It is not unusual to receive resubmissions of users making changes to their *accepted* code, in order to improve their ranking position creating optimized code. It could happen that those assumed improvements lead into a negative verdict. Nevertheless, from the system point of view, the user will have the problem still accepted, despite the non-AC verdict in his last submission.

We carried out an exploratory analysis of the ACR database in order to familiarize with the data contained in it and to find relevant information for our recommendation purposes. Although ACR system do allow all these resubmissions described above with no restrictions, for the sake of simplicity we filter the submissions in order to make easier to model the relationship between users and problems. After all, from a *user's* point of view, a problem can be:

- *Unattempted*: the user did not submit any solution to that problem yet.
- *Attempted*: the user submitted one or more solutions to the problem, but all of them were invalid.

– *Solved*: the user submitted one ore more solutions to the problem, and at least one of them was judged as *Accepted*.

Note that the solutions evaluated as *Presentation Error* are closed to be correct, usually just deleting extra blanks at the end of the lines. Although a bad token separator can ruin, for example, a network protocol, this is not the case for the kind of problems that online judges focus on. At the same time, we have found that 90 % of the users obtaining a PE adapt their solutions and get an AC verdict. Instead of categorizing the other 10 % as users that did not solve the problem yet, we considered PE and AC as synonyms in our analysis.

At the time of this writing (April 2016), ACR has more than 3,000 registered users, 241 problems and more than 68,000 submissions (including resubmissions). During the exploratory analysis, initially we group together runtime errors (RTE) and all the verdicts related to limits exceeded (TLE, MLE and OLE). Later, we ignore the repeated attempts made by the users in order to get their solution accepted and therefore removed all the submission prior their AC. In those cases where a user finally gives up the problem without having their solutions accepted, we just keep their last attempt. The final number of submissions considered in the data analysis drops from the original 68,000 to 13,863.

Figure 1 shows the number of submissions per month after the filtering since the ACR deployment. The high number of submissions in March 2014-2016 is due to a local contest organized by the authors on that months. The online judge contains the problems of past editions and contestants use the judge for training purposes. Figure 2 shows the cumulative number of submissions. It is

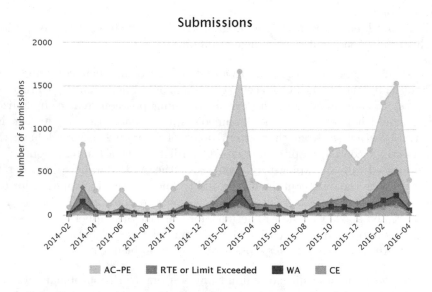

**Fig. 1.** Submissions in *Acepta el reto* online judge between 2014/02 and 2016/04, categorized by verdict.

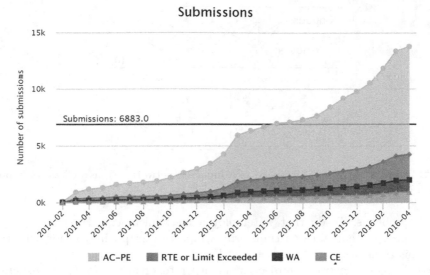

**Fig. 2.** Cumulative number of submissions in *Acepta el reto* online judge between 2014/02 and 2016/04, categorized by verdict.

worth noting that June 2015 was the point where the judge reached the midpoint of the number of submission that it currently has. As we will see later, this point in time will be chosen as the split timestamp in our evaluation.

## 3   User Interactions as a Social Network

A social network is a group of individuals or entities that are related to each other, traditionally represented as a graph. In an online judge system, user-problem interactions can be abstracted and represented into a user-problem non-weighted bipartite network $G$, where nodes belong exclusively to one of two disjoint sets, the problem-set $P = \{p_1, \ldots, p_m\}$ or the user-set $U = \{u_1, \ldots, u_n\}$. Therefore, we define an adjacency matrix $A = \{a_{ij}\}$, where $a_{ij} = 1$, if the user $u_i$ attempted to solve (or correctly solved, depending on the use of the matrix) the problem $p_j$.

The network representation allows us to analyse the user-problem interactions that occurred in the online judge system using the methods and metrics defined by the Social Network Analysis field [3]. Link prediction is a technique used in social network analysis that aims to predict new links that might be formed between nodes in a future time or to predict missed links in the current network [7]. Then, problem recommendation in an online judge system can be viewed as a link prediction problem: *given a graph that represents the implicit user interactions with the problems, how can we predict a link between a user and a problems?*

Instead of using a bipartite graph we will define a non-bipartite graph where the user-problem interactions will be transformed into implicit relationships

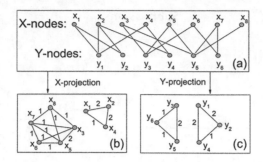

**Fig. 3.** Illustration (from Wikipedia) of a bipartite network (a), as well as its X projection (b) and Y projection (c)

among problems. To do that we employ a mechanism called *network projection*, a process that aims to transform a bipartite graph into a non-bipartite one. This transformation is depicted in Fig. 3.

We employ a network projection in order to create a *Problem-projection* graph, a network containing only *Problem nodes* where two nodes are connected if they have at least one common user that attempted to solve (or solved, depending on the matrix) both problems. To avoid losing information from the original network about user interactions we use a simple weighting method where an edge $(p_i, p_j)$ is weighted with the number of different users that attempted to solve both $p_i$ and $p_j$ problems. This way, the problem recommendation task can be reformulated as: *given the problems resolved by a user, how can we predict a link between these problems and future interesting problems?*. Our proposal consists on reusing the metrics employed by similarity-based link prediction methods for recommending purposes, as we will describe in next section.

## 4 Link Prediction and Similarity Metrics

Link prediction algorithms are a family of graph-based algorithms that aims to predict new links that might be formed between nodes in a future time or to predict missed links in the current network [7]. The alternative approaches to predict these links are the following [12]:

- *Similarity-based methods*, which compute the proximity or *similarity* between pairs of unconnected nodes in order to predict new links.
- *Learning-based methods*, which use machine learning algorithms to classify non-connected pair of nodes as positive or negative potential links.

Similarity metrics employed in the former approach can be also classified into:

- *Node-based metrics*, which compute the similarity using node attributes.
- *Neighbour-based metrics*, which compute the similarity using each node neighbourhood.

– *Path-based metrics*, which do not only use node neighbours but also the paths between two nodes.
– *Random walk-based metrics*, which use transition probabilities from a node to its neighbours and to non-connected nodes in order to simulate social interactions.

We will focus on similarity-based methods and we will employ a variation of these metrics in order to calculate a similarity score that expresses how similar two problems are in the graph according to the users' interactions with these problems. These scores will be employed in the recommendation process for suggesting similar problems to the ones that a user successfully solved.

These metrics must be considered as a *score* for a pair of nodes $(x, y)$ instead of a classic similarity metric because, in general, the value performed by these metrics does not lie in $[0,1]$ range.

For clarity of the descriptions of the similarity metrics, we give some notation:

– $N(x)$ represents the neighbours of node $x$.
– $|N(x)|$ represents the number of neighbours (or *node degree*) of node $x$.
– $WD(x)$ represents the *weighted node degree* of node $x$, which means the sum of the weights in the edges directly connected with node $x$.
– $A_{xy}$ represents the weight of the edge that links node $x$ and node $y$; in our context this weight expresses the number of users that have resolved both problem $x$ and $y$.

Most of these metrics are detailed in [5, 12] and some of them are defined in two different flavours: unweighted and weighted metrics [8].

Using this notation, now we can describe the similarity metrics used in our study.

**Edge Weight.** This simple metric measures the similarity between two nodes as the weight of the edge that links them. Two problems are more similar if there are more users that solved both of them. No connection between two nodes is represented by $A_{xy} = 0$.

$$EW(x,y) = A_{xy} \tag{1}$$

Although an unweighted version of this metric exists ($A_{xy} = 1$ if the edge exists; 0 otherwise), we have not used it because it cannot be employed as a similarity metric.

**Common Neighbours.** This metric measures the similarity between two nodes as the number of neighbours they have in common. The rationale behind this metric is that the greater the intersection of the neighbour sets of any two nodes, the greater the chance of future association between them.

$$CN(x,y) = |N(x) \cap N(y)| \tag{2}$$

Its weighted version is defined as:

$$WCN(x,y) = \sum_{z \in N(x) \cap N(y)} A_{xz} + A_{yz} \qquad (3)$$

**Jaccard Neighbours.** This is an improvement of $CN(x,y)$ as it measures the number of common neighbours of $x$ and $y$ compared with the number of total neighbours of $x$ and $y$.

$$JN(x,y) = \frac{|N(x) \cap N(y)|}{|N(x) \cup N(y)|} \qquad (4)$$

This metric does not have an equivalent weighted metric.

**Adar/Adamic.** This metric considers to evaluate the likelihood that a node $x$ is linked to a node $y$ as the sum of the number of neighbours they have in common. This metric also measures the intersection of neighbour-sets of two nodes in the graph, but emphasizing in the smaller overlap.

$$AA(x,y) = \sum_{z \in N(x) \cap N(y)} \frac{1}{log|N(z)|} \qquad (5)$$

Its weighted version is defined as:

$$WAA(x,y) = \sum_{z \in N(x) \cap N(y)} \frac{A_{xz} + A_{yz}}{log(1 + WD(z))} \qquad (6)$$

**Preferential Attachment.** This metric is based on the consideration that nodes create links, with higher probability, with those nodes that already have a larger number of links. The similarity between nodes $x$ and $y$ is calculated as the product of the degree of the nodes $x$ and $y$, so the higher the degree of both nodes, the higher is the similarity between them. This metric has the drawback of leading to high similarity values for highly connected nodes to the detriment of the less connected ones in the network.

$$PA(x,y) = |N(x)| \cdot |N(y)| \qquad (7)$$

The weighted version is an improvement of the previous one, where the edge weights are taken into account when computing the degree of nodes $x$ and $y$.

$$WPA(x,y) = WD(x) \cdot WD(y) \qquad (8)$$

## 5    Experimental Setup and Evaluation Results

In this section we evaluate the performance of the previous similarity metrics to recommend new problems to the ACR users in comparison with a classical recommendation method. ACR users cannot rate the problems so we cannot use collaborative filtering methods. Instead, the problems are catalogued into different problem categories represented by *labels*. A problem can belongs to several categories (or is defined by more than one label). Apart from just watching what other users are submitting, these programming categories are the only tool available to the ACR users to select new problems. For this reason, our baseline recommendation method will be a content-based recommender that uses Jaccard as similarity metric. We name this metric as *Jaccard Label*:

$$JL(x, y) = \frac{|L(x) \cap L(y)|}{|L(x) \cup L(y)|} \tag{9}$$

where $L(x)$ and $L(y)$ are the set of labels or problem categories that problems $x$ and $y$ belongs to, respectively.

The first step to perform the evaluation was to select a particular timestamp $t$ to split the database of submissions into two sets for training and evaluation. The former set contained the submissions made *before* time $t$ and was used to create the graph of problems and compute the different similarity metrics described above. The latter set contained the submissions made *after* time $t$ and was used to evaluate their performance in the context of a $k$ Nearest Neighbour (k-NN) recommender. The timestamp selected to split the database was 2015/06/30, when ACR approximately reached the midpoint of the current number of submissions.

Using the submissions made before time $t$ we built a projection graph where nodes represent problems, edges between nodes represent that at least one user achieved an accepted (AC) or a presentation error (PE) verdict in both problems, and edge weights correspond to the number of users that achieved AC or PE in both problems. The graph had 169 nodes[2] and 14,041 edges, which corresponds to an edge density of 98.9 %. This high level of density can be explained because there are a few very active ACR users that have resolved a lot of problems. In fact, if we filter the edges with *weight* < 5, i.e. we only connect problems solved simultaneously by at least 5 users, the number of edges drops to 6,343 and a density of 45.17 %. In the following experiments we will use this last filtered graph, shown in Fig. 4.

From the projection graph, we computed the different similarity measures between problems detailed in Sect. 4 and used a k-NN recommender to generate a ranked list of problems for each user. Then, we compared the list of recommendations with the problems each user actually attempted to solve after timestap $t$ using the following standard evaluation metrics:

– *Precision, Recall* and *F-Score* in top $k$ recommendations [9].
– *At least one hit* (1-hit): ratio of recommendations in which at least one recommended problem was attempted by the user. It corresponds to the metric *Success@k* with a success condition of guessing right at least one problem.

---

[2] Although at the time of this writing ACR has 241 problems, it only had 169 at time $t$.

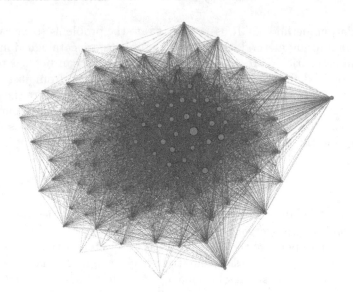

**Fig. 4.** Problem-problem graph. Edges with *weight* < 5 are filtered. Node size is proportional to the node weighted degree and edges are coloured with a gradient from yellow to red, with edges with more weight in red colour. (Color figure online)

– *Mean Reciprocal Rank* (MRR): it evaluates the quality of a ranked list of recommendations based on the position of the first correct item [11]. Since we only provide one list of recommendations per user, the MRR can be computed as follows:

$$MRR = \frac{1}{rank_i}$$

where $rank_i$ is the position of the first correct item.

It is important to note that all those metrics require that the users interact with the system after the time $t$: we cannot evaluate a recommendation to a user if that user does not submit any problem after that. Although ACR had 1,733 registered users at time $t$, many of them showed a transience behaviour, working with the online judge during 3–4 months and then leaving the system. For our analysis, we filtered the users in order to consider only those ones who have solved at least 5 different problems before the timestamp $t$ and submitted at least 5 different problems after $t$ (no matter the status that they achieved). Only 37 users fulfilled these constraints.

Table 1 summarizes the results of the evaluation using $k = 3$ and computing the average values for all those users. We can extract some interesting conclusions from these results. First, the Precision, 1-hit and MRR values of Jaccard Labels are not as good as we had anticipated, which reflects that most of the users do not use the programming categories to choose the next problem to solve.

**Table 1.** Evaluation of the different similarity metrics with $k = 3$.

| Similarity | Precision | Recall | F-Score | 1-hit | MRR |
|---|---|---|---|---|---|
| Jaccard Labels (JL) | 0.1982 | 0.0267 | 0.0429 | 0.4054 | 0.4414 |
| Jaccard Neighbours (JN) | 0.1171 | 0.0111 | 0.0190 | 0.2432 | 0.3873 |
| Common Neighbours (CN) | 0.1081 | 0.0104 | 0.0177 | 0.2162 | 0.3626 |
| Adar/Adamic (AA) | 0.1081 | 0.0104 | 0.0177 | 0.2162 | 0.3626 |
| Preferential Attachment (PA) | 0.1171 | 0.0118 | 0.0200 | 0.2432 | 0.3738 |
| Edge Weights (EW) | **0.3153** | **0.0447** | **0.0731** | **0.5676** | **0.5540** |
| Weighted Common Neighbours (WCN) | 0.1351 | 0.0132 | 0.0225 | 0.2703 | 0.3986 |
| Weighted Adar/Adamic (WAA) | 0.1261 | 0.0128 | 0.0218 | 0.2703 | 0.3986 |
| Weighted Preferential Attachment (WPA) | 0.1171 | 0.0121 | 0.0205 | 0.2432 | 0.3738 |

Moreover, all the other similarities, except Edge Weights, obtain very similar results and are comparable to Jaccard Labels. Unsurprisingly the weighted versions of the similarities seem to work better that the ones without weights because they can take advantage of the extra domain information. Finally Edge Weights is the clear winner in the comparison and dominates all the other similarities in every evaluation metric. Edge Weights obtains a precision value of 31.53 %, 1-hit of 56.76 % and MRR of 55.40 % recommending only 3 problems to the user, quite far from the next best similarity metric.

Recall (and therefore F-Score) values are really small, but this fact can be explained because we compare a list of 3 recommendations with all the problems the user attempted to solve after 2015/06/30. For example, if we increase the number of recommendations to $k = 50$ the recall of Edge Weights increases significantly to 47.55 %. However, we think that it does not make sense to recommend so many problems in a system like ACR. Our intuition is that 3–5 problems is a good number of recommendations, but we will have to implement and evaluate the recommender module in ACR in order to prove these hypotheses.

Finally, Fig. 5 shows the evolution of the precision, 1-hit and MRR metrics when we increase the number of recommendations (parameter $k$) from 1 to 10. For the sake of simplicity, we have removed the non-weighted similarity metrics CN, AA and PA from the figure because they always achieve worse results than the weighted versions WCN, WAA and WPA. As we expected, the precision and MRR values decrease slowly because the last recommended problems are usually not as good as the first ones (their similarity is smaller). The trade off between the quality of recommendations and the number of choices available to the users is not easy to decide, and we will have to perform some tests with real users to adjust it. On the other hand, the probability of guessing right at least one problem increases as we make more recommendations. For every value of $k$, Edge Weights dominates all the other similarities.

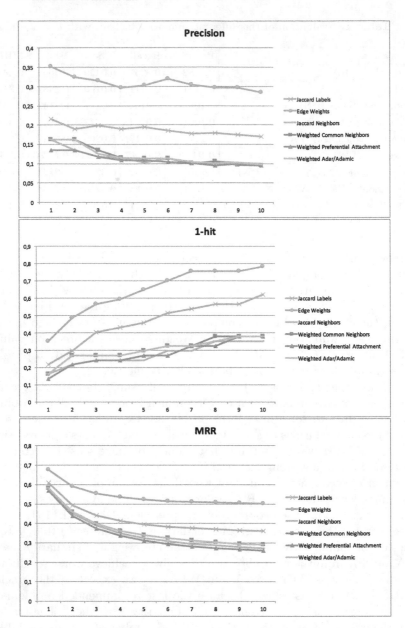

**Fig. 5.** Precision, 1-HIT and MRR evolution when we increase the number of recommended problems.

## 6    Related Work

Most of the recommendation systems, like collaborative filtering, content-based or case-based recommenders, try to recommend items *similar* to those a user has liked in the past [10]. *Similarity* is therefore one of the most important metrics in these systems. The process of recommending items to users can be considered as a link prediction problem in the user-item bipartite networks [2]. For these reason, link prediction has been widely applied in recommender systems.

As we detailed in Sect. 4, similarity-based methods employ different similarity metrics in order to predict new links. Liben-Nowell and Kleinberg [5] systematically compared some neighbour, path and random walk based node similarity indices for link prediction problem in co-authorship networks. These algorithms have been also applied in the user-product bipartite graphs in recommender systems and its performance has been evaluated with a Flickr dataset, outperforming collaborative filtering methods in some cases [2]. The work in [4] also proposes bipartite graph-based algorithms to analyse user-item interactions in order to alleviate the data sparsity problem in collaborative filtering recommenders. They compare CN, JN, AA, PA and two path-based similarity metrics using a book sales dataset and the results show that both path-based and neighbour-based approaches can significantly outperform the standard user-based and item-based algorithms.

Most of the studies on link prediction focused on unweighted networks but ignored the naturally existed link weights. Proximity between nodes can be estimated better by using both graph proximity measures and the weights of existing links. The work in [8] proposed a simple way to extend similarity metrics for binary networks to weighted metrics. However, the latter performed even worse in several real networks, opposed to our experimental results. For this reason they introduced a free parameter to control the relative contributions of weak ties to the similarity measure. The experimental study highlighted that the contributions of weak ties can enhance the prediction accuracy in some networks, suggesting that weak ties are not as weak as their weights show.

Other research works transform bipartite graphs into non-bipartite graphs using projections. Our work projects the graph and weigh the edges in a straightforward way using the number of users that resolved a pair of problems. However, this is not the only way to weigh the edges. The work in [13] details a two-step weighting method and a recommendation algorithm that outperforms classical collaborative filtering algorithms in Movielens dataset.

Finally, several works aggregate similarity based methods with another approaches in order to enhance the recommendation. [1] proposes an hybrid similarity measure that combines network similarity with node profile similarity. Profile similarity compares personal data stored in the profile items associated with two social network users, in a different manner as our Jaccard-Label metric works. Other works describe alternative link prediction methods based on social theory based metrics that enhance well-known neighbour-based metrics with node centrality measures like betweenness, closeness or node degree [6].

# 7   Conclusions and Future Work

Online judges contain hundreds or thousands of problems to be solved by myriads of users. Unfortunately, these systems usually lack of recommenders, so users are on their own in the challenging task of choosing the next problem to solve. Some online judges, as ACR, have the problems categorized with labels, but we have shown that users seem to ignore them.

Incorporating a recommender is not easy. Online judge users are not used to rate problems, and even in the case they do it, rating would be too dependent on the user knowledge. A point in favour of online judges is that they collect many information about users and problem interactions and this information cannot be dismissed. In this paper, we have considered those user-problem relationships as a social network, and used social network analysis techniques, such as the similarity-based link prediction, for creating similarity metrics between problems. These metrics have become the base for our desired recommender.

We have used different similarity metrics and we have compared them with what the users are currently doing in ACR. The analysis has shown that the *edge weight* metric provides the best results. However, this similarity metric relies on the existence of an edge between two problem nodes and it is not feasible in order to find new relationships between nodes.

The research trend in the link prediction problem is growing and the number of measures and methods employed in this field increases quickly. New similarity measures and approaches should be incorporated and evaluated in our recommendation problem in order to achieve better results.

Nevertheless, this work only describes some theoretical recommendation results. The recommender has not been used by real users yet so we plan to incorporate it into the system and carry out A/B evaluations with users in order to test if these results are aligned with reality.

# References

1. Akcora, C.G., Carminati, B., Ferrari, E.: User similarities on social networks. Soc. Netw. Anal. Min. **3**(3), 475–495 (2013)
2. Chiluka, N., Andrade, N., Pouwelse, J.: A link prediction approach to recommendations in large-scale user-generated content systems. In: Clough, P., Foley, C., Gurrin, C., Jones, G.J.F., Kraaij, W., Lee, H., Mudoch, V. (eds.) ECIR 2011. LNCS, vol. 6611, pp. 189–200. Springer, Heidelberg (2011). doi:10.1007/978-3-642-20161-5_19
3. Furht, B.: Handbook of Social Network Technologies and Applications. Springer Science & Business Media, New York (2010)
4. Huang, Z., Li, X., Chen, H.: Link prediction approach to collaborative filtering. In: Proceedings of the 5th ACM/IEEE-CS Joint Conference on Digital Libraries, pp. 141–142. ACM (2005)
5. Liben-Nowell, D., Kleinberg, J.: The link-prediction problem for social networks. J. Am. Soc. Inf. Sci. Technol. **58**(7), 1019–1031 (2007)
6. Liu, H., Hu, Z., Haddadi, H., Tian, H.: Hidden link prediction based on node centrality and weak ties. EPL (Europhys. Lett.) **101**(1), 18004 (2013)
7. Lu, L., Zhou, T.: Link prediction in complex networks: a survey. Phys. A Stat. Mech. Appl. **390**(6), 1150–1170 (2010)
8. Lü, L., Zhou, T.: Link prediction in weighted networks: the role of weak ties. Europhys. Lett. **89**(1), 18001 (2010)

9. Pazzani, M.J.: A framework for collaborative, content-based and demographic filtering. Artif. Intell. Rev. **13**(5–6), 393–408 (1999)
10. Ricci, F., Rokach, L., Shapira, B. (eds.): Recommender Systems Handbook. Springer US, Boston (2015)
11. Said, A., Bellogín, A.: Comparative recommender system evaluation. In: Proceedings of the 8th ACM Conference on Recommender systems - RecSys 2014, pp. 129–136 (2014)
12. Wang, P., Xu, B.W., Wu, Y.R., Zhou, X.Y.: Link prediction in social networks: the state-of-the-art. Sci. China Inf. Sci. **58**(1), 1–38 (2014)
13. Zhou, T., Ren, J., Medo, M., Zhang, Y.C.: Bipartite network projection and personal recommendation. Phys. Rev. E Stat. Nonlinear Soft Matter Phys. **76**(4), 1–7 (2007)

# Analogical Transfer in RDFS, Application to Cocktail Name Adaptation

Nadia Kiani[1], Jean Lieber[2,3,4(✉)], Emmanuel Nauer[2,3,4],
and Jordan Schneider[1]

[1] Université de Lorraine, SCA Master (Cognitive Sciences and Their Applications),
54000 Nancy, France
{Nadia.Kiani4,Jordan.Schneider1}@etu.univ-lorraine.fr
[2] Université de Lorraine, LORIA, 54506 Vandœuvre-lès-Nancy, France
{Jean.Lieber,Emmanuel.Nauer}@loria.fr
[3] CNRS, 54506 Vandœuvre-lès-Nancy, France
[4] Inria, 54602 Villers-lès-Nancy, France

**Abstract.** This paper deals with analogical transfer in the framework of
the representation language RDFS. The application of analogical transfer
to case-based reasoning consists in reusing the problem-solution dependency to the context of the target problem; thus it is a general approach
to adaptation. RDFS is a representation language that is a standard of
the semantic Web; it is based on RDF, a graphical representation of data,
completed by an entailment relation. A dependency is therefore represented as a graph representing complex links between a problem and a
solution, and analogical transfer uses, in particular, RDFS entailment.
This research work is applied (and inspired from) the issue of cocktail
name adaptation: given a cocktail and a way this cocktail is adapted by
changing its ingredient list, how can the cocktail name be modified?

**Keywords:** Adaptation · Analogical transfer · RDFS · Cocktail name
adaptation

## 1 Introduction

This paper presents an approach to analogical transfer in RDFS, with an application to cocktail name adaptation.

Adaptation is a research issue of case-based reasoning (CBR [11]) that has
received some attention during the last years in the CBR community (see,
e.g., [2,7,9]). In particular, this has been an issue for the competitors of the
Computer Cooking Contests (CCCs). Such a CCC competitor is meant to answer
cooking query problems, such as Q = "I want a dessert with pear but without
orange", using a recipe book as a case base. TAAABLE is one of these competitors [4]. In TAAABLE, several adaptation issues have been tackled:

– Adaptation of ingredients stating, e.g., that the substitution apple ⤳ pear
  (consisting in replacing apples with pears) applied to an apple pie recipe gives
  an answer to the query Q;

© Springer International Publishing AG 2016
A. Goel et al. (Eds.): ICCBR 2016, LNAI 9969, pp. 218–233, 2016.
DOI: 10.1007/978-3-319-47096-2_15

- Adaptation of quantities (e.g., modifying the mass of granulated sugar in the recipe);
- Adaptation of the preparation: add, remove and/or re-order the preparation steps when needed.

These three adaptation issues have been addressed thanks to the principle of revision-based adaptation, i.e., adaptation based on belief revision [3] (the first one has also been addressed using other techniques).

In the 2014 edition of the CCC, the jury has suggested the issue of adapting recipe names. The CookingCAKE system [10] has addressed this challenge for the CCC-2015, using a few rules.[1] The application motivation of this paper is to address this issue for TAAABLE, with a recipe base restricted to cocktails.

Unlike the other adaptation issues addressed in TAAABLE, revision-based adaptation has not appeared as a useful guideline for adapting cocktail names. Indeed, revision-based adaptation can be understood as modifying the source case so that it becomes consistent with the target problem, given the domain knowledge, and, for many examples, the inconsistency of a cocktail name wrt a cocktail recipe was—at least—difficult to capture. By contrast, analogical transfer (AT) has appeared as a useful guideline for this issue. In works about AT, graph-based representations, such as semantic networks [12], are often used. RDFS is such a representation formalism and is the standard of the semantic Web that has been chosen for this work. Actually, it is also used by TUUURBINE, which is a generic retrieval engine that is used in TAAABLE [5]. If some adaptation strategies proposed in this paper are domain-dependent, it is hoped that other ones cover a broader range of applications, yet this work contributes to adaptation and, more specifically, to AT, in the representation framework of RDFS.

The paper starts with preliminaries recalling notions related to AT and to RDFS (Sect. 2), and describes its application issue (Sect. 3). Then, it follows the steps of the study. First, a collection of cocktail name adaptations has been gathered; Sect. 4 describes this gathering and gives a few representative examples. These adaptations have been analyzed in details and Sect. 5 exemplifies such an analysis. From that, several approaches to cocktail name adaptation are proposed that cover the majority of the examples (Sect. 6). This work is discussed afterwards (Sect. 7). Section 8 concludes and presents some directions for future work.

## 2  Preliminaries

### 2.1  Adaptation by Analogical Transfer

A *case* in a particular application is a chunk of experience that is frequently represented by a problem pb and a solution sol(pb) of pb, where the notions

---

[1] For instance, the adjective "cheesy" is added to the recipe name if the adapted recipe of a sandwich contains some cheese. This is not a published material, though. In the paper [10], the accent is put on other issues. The authors wish to thank Gilbert Müller and Ralph Bergmann for having given them some hints about the recipe adaptation in CookingCAKE.

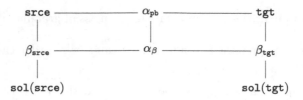

**Fig. 1.** Notations used to describe analogical transfer. tgt is the target problem. (srce, sol(srce)) is the retrieved case. $\beta_{srce}$ is a dependency between srce and sol(srce). $\alpha_{pb}$ is a matching from srce to tgt. From that, $\alpha_\beta$, $\beta_{tgt}$ and sol(tgt) are inferred.

of problem and solution are application-dependent. A case from the case base is called a *source case*, denoted by a pair (srce, sol(srce)). The problem to be solved is called the *target problem*, denoted by tgt. A classical way to perform the case-based inference consists in choosing a source case (srce, sol(srce)) judged similar to tgt (retrieval step) and in modifying sol(srce) into a solution sol(tgt) of tgt (adaptation step).

One approach to adaptation is analogical transfer (AT) that has been studied within the analogical reasoning community (see, e.g., [6,14]). Figure 1 presents notations related to AT.

A *dependency* $\beta_{pb}$ between a problem pb and a solution sol(pb) of pb is constituted by pieces of information relating pb to sol(pb): it can be seen as a partial explanation of why sol(pb) solves pb. Given a source case (srce, sol(srce)), the dependency $\beta_{srce}$ can either be stored with the case, at case authoring time, or inferred.

The *matching* $\alpha_{pb}$ from a problem srce to a problem tgt is constituted by pieces of information about the differences and/or the similarities between the two problems. It is often inferred during retrieval time, the retrieved case being in general the one that best matches the target problem.

Given srce, sol(srce), $\beta_{srce}$, $\alpha_{pb}$ and tgt, the adaptation by AT can be described by the following steps:

AT1 From $\beta_{srce}$ and $\alpha_{pb}$, infer a dependency $\beta_{tgt}$ between tgt and the (future) solution of tgt, sol(tgt). This inference consists in using the differences represented in $\alpha_{pb}$ to modify $\beta_{srce}$ into $\beta_{tgt}$. The matching between $\beta_{srce}$ and $\beta_{tgt}$ is denoted by $\alpha_\beta$.

AT2 sol(srce) is modified into sol(tgt), using $\beta_{tgt}$ and $\alpha_\beta$. The principle is to modify sol(srce) using $\alpha_\beta$ so that the result sol(tgt) respects the constraints given by $\beta_{tgt}$.

AT can be compared with derivational analogy (DA [1]) since in DA, a problem-solution link is also used. The difference in AT is that this link, the dependency, can be incomplete whereas it is supposed to be a complete "proof" in DA.

AT is an abstract approach for performing adaptation. To make it operational it is necessary to make some choices on the way the matchings and the dependencies are represented. For the latters, they are often represented in graph structures (e.g., in [6,14]). Therefore, this justifies the use of RDFS formalism for representing and handling dependencies, as a way to implement AT.

## 2.2  RDFS

RDFS is a representation formalism based on RDF.

*RDF (Resource Description Framework[2])* is a language that can be used to encode assertions using triples, e.g. "Romeo loves Juliet and knows someone whose age is 40" can be encoded by:

⟨romeo loves juliet⟩        ⟨romeo knows ?x⟩        ⟨?x age 40⟩

In a *triple* ⟨s p o⟩, s (the subject) is a *resource*, p (the predicate) is a *property*, and o (the object) is either a resource or a *literal*. A resource is either a constant or a variable (generally called identified resource and blank node, respectively). By naming convention, variables start with the symbol ? whereas constants do not. So, `juliet` is a constant and `?x` is a variable. A property is a particular type of resource, intended to represent a binary relation. A set of simple types (including `integer`, `float` and `string`) is fixed and a literal is a value of one of these types. For the sake of simplicity, triples ⟨s p o⟩ where s is a literal are also accepted in this paper, though this is not compliant with the RDFS standard: this will make some explanations simpler, avoiding some useless technicalities. An *RDF base* is a set of triples. An *RDF graph* is the graphical representation of an RDF base by a graph whose nodes are resources and literals, and whose edges are labeled by properties. For example, the RDF graph of Fig. 2 represents a cocktail recipe. Given an RDF base $\mathcal{B}$, the set of nodes of the corresponding RDF graph is denoted by `Nodes`($\mathcal{B}$). Given $n_1, n_2 \in$ `Nodes`($\mathcal{B}$), $n_1$ and $n_2$ are *connected* in $\mathcal{B}$ if there exists a non-directed path relating them in the graph corresponding to $\mathcal{B}$.

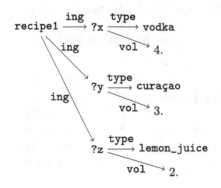

The "Blue Lagoon" recipe is identified by the resource `recipe1`. Its ingredients are 4 cl of vodka, 3 cl of curaçao, and 2 cl of lemon juice. The preparation is not represented here. `ing` relates a recipe to one of its ingredients. `type` (abbreviation of `rdf:type`) is an RDF property relating a class to its instance (for example, the triple ⟨?x type vodka⟩ means that ?x is an instance of vodka. The variables are existentially quantified (there exist ?x, ?y and ?z such that...). The property `vol` relates an ingredient to its volume in centiliters.

**Fig. 2.** An RDF representation of the "Blue Lagoon" recipe.

[2] https://www.w3.org/RDF/.

*RDFS* (*RDF Schema*[3]) is a representation formalism whose syntax is RDF and semantics is defined by a set of inference rules. Only a few rules are used in this paper:

$$\frac{\langle \text{a type C}\rangle \qquad \langle \text{C subc D}\rangle}{\langle \text{a type D}\rangle}r_1 \qquad \frac{\langle \text{a p b}\rangle \qquad \langle \text{p subp q}\rangle}{\langle \text{a q b}\rangle}r_2$$

$$\frac{\langle \text{A subc B}\rangle \qquad \langle \text{B subc C}\rangle}{\langle \text{A subc C}\rangle}r_3 \qquad \frac{\langle p \text{ subc } q\rangle \qquad \langle q \text{ subc } r\rangle}{\langle p \text{ subc } r\rangle}r_4$$

type, subc and subp are abbreviations for rdf:type, rdfs:subClassOf and rdfs:subPropertyOf. type is the membership relation between an instance and a class. subc (resp., subp) is the relation between a class and a superclass (resp., a property and a superproperty). $r_1$ means that if a is an instance of a class it is also an instance of its superclasses. $r_2$ means that if a and b are related by a property, they are also related by any of its superproperties. $r_3$ and $r_4$ state that subc and subp are transitive. For example, the following inference can be drawn:

$$\left\{ \begin{array}{c} \langle \text{?x type vodka}\rangle, \\ \langle \text{vodka subc alcoholicBeverage}\rangle \end{array} \right\} \vdash \langle \text{?x type alcoholicBeverage}\rangle$$

RDFS does not include negation, thus only positive facts can be entailed. However, an inference with closed world assumption (CWA) can be drawn, stating that if $\mathcal{B} \nvdash t$ then $t$ is considered to be false (given the RDFS base $\mathcal{B}$), denoted by $\mathcal{B} \vdash_{\text{cwa}} \neg t$.

*SPARQL* (*SPARQL Protocol and RDF Query Language*[4]) enables to write queries to RDF or RDFS bases. If a SPARQL engine uses RDFS entailment, this means that the query is done on the RDF base completed by RDFS entailment. For example, the following SPARQL query addressed to a base describing recipes such as the one of Fig. 2 returns the set of recipes ?r containing some alcohol:[5]

$$\mathtt{Q_{alcohol}} = \text{SELECT ?r WHERE } \{\text{?r ing ?a . ?a type alcoholicBeverage}\} \quad (1)$$

Given a SPARQL query Q and an RDFS base $\mathcal{B}$, the result of the execution of Q on $\mathcal{B}$ is denoted by $\text{exec}(\mathtt{Q}, \mathcal{B})$.

## 3   The Cocktail Name Adaptation Issue

In this application, a problem pb is a representation of a cocktail recipe by an RDFS graph. For the first version of this application, only ingredient types are considered, neither the quantities, nor the preparation steps. Therefore,

---

[3] https://www.w3.org/2001/sw/wiki/RDFS.

[4] https://www.w3.org/2001/sw/wiki/SPARQL.

[5] The CWA is assumed: if it cannot be entailed that a recipe contains some alcohol, then it is concluded that it does not.

a problem is an RDFS base $\text{pb} = \bigcup_{k=1}^{n}\{\langle id \ \text{ing} \ ?v_k\rangle, \langle ?v_k \ \text{type} \ f_k\rangle\}$ where $id$ is a constant (a resource identifying the recipe), $?v_1, \ldots, ?v_n$ are $n$ variables, and $f_1, \ldots, f_n$ are food classes.

A solution $\text{sol(pb)}$ of $\text{pb}$ is a literal of type string that gives a name to $\text{pb}$. It is assumed to be in lower case for the sake of simplicity, e.g., $\text{sol(pb)} = $ "blue  lagoon" solves the problem $\text{pb}$ represented in Fig. 2. The following operations on strings are used: concatenation (denoted by $+$, e.g., "ab" $+$ "cd" $=$ "abcd"), substring checking (denoted by subStringOf, e.g., subStringOf("bc", "abcd") $=$ true), and string replacement (e.g., replace("ab", "cd", "bababa") $=$ "bcdcda").

A dependency $\beta_{\text{pb}}$ between $\text{pb}$ and $\text{sol(pb)}$ is an RDFS base. Usually, at least one food class $f_k$ of $\text{pb}$ and the literal $\text{sol(pb)}$ occur in $\beta_{\text{pb}}$: when it is not the case, $\beta_{\text{pb}}$ does not relate $\text{pb}$ to $\text{sol(pb)}$ (which is possible, e.g., when $\beta_{\text{pb}} = \emptyset$, i.e., there is no known dependency between $\text{pb}$ and $\text{sol(pb)}$). For each case $(\text{srce}, \text{sol(srce)})$, $\beta_{\text{srce}}$ is assumed to be given.

A matching $\alpha_{\text{pb}}$ from srce to tgt is either simple or complex. A simple matching has the form $f \rightsquigarrow g$ where $f$ is a food class of srce and $g$ is a food class of tgt; it represents the substitution of $f$ by $g$. The removal of a food class $f$ will be denoted by $f \rightsquigarrow \emptyset$. A complex matching is a composition $\alpha_{\text{pb}} = \alpha_{\text{pb}}^{q} \circ \alpha_{\text{pb}}^{q-1} \circ \ldots \circ \alpha_{\text{pb}}^{1}$ of simple matchings. $\alpha_{\text{pb}}$ is built during the adaptation of ingredients process of TAAABLE.

The matching $\alpha_{\beta}$ from $\beta_{\text{srce}}$ to $\beta_{\text{tgt}}$ is built during the cocktail name adaptation. It consists of a set of ordered pairs $(d, d')$ where $d$ is a descriptor of $\beta_{\text{srce}}$ and $d'$ is a descriptor of $\beta_{\text{tgt}}$, a descriptor being either a resource (that can be a property) or a literal.

Finally, the domain knowledge is represented by an RDFS base DK.

## 4    Collecting Examples of Cocktail Name Variations

The first step of this work has been to collect 20 examples of cocktail name adaptations, with an attempt to have diverse types of adaptation. 10 of them have been taken from variants of classical cocktails.[6] The other 10 have been imagined for this study.

Here is a selection of examples, knowing that for some of them, only the relevant part of the information has been given (hence the "etc."):

**ex. 1** srce contains the ingredient classes vodka, curaçao and lemonJuice, $\text{sol(srce)} = $ "blue  lagoon", $\alpha_{\text{pb}} = $ curaçao $\rightsquigarrow$ appleJuice , $\text{sol(tgt)} = $ "yellow  lagoon". $\text{sol(srce)}$ depends on the color of curaçao, which is substituted by a yellow beverage.

**ex. 2** srce contains rhum, mintLeave, lime, brownSugar and ice, $\text{sol(srce)} = $ "mojito", $\alpha_{\text{pb}} = $ rhum $\rightsquigarrow \emptyset$, $\text{sol(tgt)} = $ "virgin  mojito". For some reason (that we do not wish to justify), transforming a recipe with alcohol into a recipe without alcohol makes it virgin.

---

[6] In particular, http://www.1001cocktails.com/ has proven to be useful, since it contains descriptions of cocktails with some named variants.

**ex. 3** `srce` contains `scotchWhisky` and `amaretto`, `sol(srce)` = "godfather", $\alpha_{pb}$ = `scotchWhisky` ⤳ `irishWhisky`, `sol(tgt)` = "the new godfather". No explanation of the name is known, so a default rule proposes a variant name (it could have been "godfather 2").

**ex. 4** `srce` contains `irishWhisky`, `coffee`, etc., `sol(srce)` = "irish coffee", $\alpha_{pb}$ = `irishWhisky` ⤳ `tequila`, `sol(tgt)` = "mexican coffee". An Irish ingredient is replaced with a Mexican one.

**ex. 5** (`srce`, `sol(srce)`) is the case of example 1, $\alpha_{pb}$ = `curaçao` ⤳ `indianTonic`, two solutions are proposed: `sol(tgt)` = "bitter lagoon" and `sol(tgt)` = "sparkling lagoon". Indeed, blue is an organoleptic property of curaçao, whereas bitter and sparkling are organoleptic properties of Indian tonic.

## 5 From Blue Lagoon to Yellow Lagoon: Analysis of an Example

This section models the adaptation example 1 following the two steps of AT introduced in Sect. 2.1.

**AT1.** A partial explanation of the name `sol(srce)` = "blue lagoon" is that the color of curaçao is blue (this is partial, since it does not explain the term "lagoon"), which can be modeled by

$$\beta_{srce} = \{\langle \text{curaçao color blue}\rangle, \langle \text{blue inEnglish "blue"}\rangle,$$
$$\langle \text{"blue" subStringOf "blue lagoon"}\rangle\}$$

Since $\alpha_{pb}$ = `curaçao` ⤳ `appleJuice` , in order to build $\beta_{tgt}$, the idea is to apply $\alpha_{pb}$ on $\beta_{srce}$ and then to make some modifications on the resources and literals to make it consistent with DK. This consistency test must be considered wrt CWA because there is no way to have $\langle$appleJuice color blue$\rangle$ inconsistent with DK in the classical semantics. It is assumed that DK $\vdash_{cwa}$ $\neg\langle$appleJuice color blue$\rangle$, thus the mere substitution $\alpha_{pb}$ on $\beta_{srce}$ gives an inconsistent result wrt DK under CWA. So, the idea is to relax this triple. One way to do it is to replace blue with a variable ?x. More generally, the strategy consists in replacing the descriptors of $\beta_{srce}$ by variables, with the exception of the predicates (that are higher order resources) and of the descriptors occurring in `tgt`. The variable that replaces `sol(srce)` is ?solTgt: solving `tgt` consists in giving a value `sol(tgt)` to this variable. This gives the following dependency (obtained by applying $\alpha_{pb}$ and turning some constants into variables):

$$\beta_{gen} = \{\langle \text{appleJuice color ?x}\rangle, \langle \text{?x inEnglish ?y}\rangle,$$
$$\langle \text{?y subStringOf ?solTgt}\rangle\}$$

$\beta_{gen}$ is so-called, since it generalizes $\alpha_{pb}(\beta_{srce})$ (in the sense $\alpha_{pb}(\beta_{srce}) \vdash \beta_{gen}$), where $\alpha_{pb}(\beta_{srce})$ is the result of applying the substitution $\alpha_{pb}$ on $\beta_{srce}$.

Now, in order to get $\beta_{tgt}$, the idea is to unify the variables ?x and ?y with some constants, using the domain knowledge. Therefore DK is interrogated with the following SPARQL query:

> SELECT ?x ?y WHERE {appleJuice color ?x . ?x inEnglish ?y}

Assuming the only result is the pair {?x ← yellow, ?y ← "yellow"}, it comes:

$$\beta_{\text{tgt}} = \{\langle \text{appleJuice color yellow}\rangle, \langle \text{yellow inEnglish "yellow"}\rangle,$$
$$\langle \text{"yellow" subStringOf ?solTgt}\rangle\}$$
$$\text{and } \alpha_\beta = \{(\text{curaçao}, \text{appleJuice}), (\text{blue}, \text{yellow}), (\text{"blue"}, \text{"yellow"})\}$$

**AT2.** Therefore, $\beta_{\text{tgt}}$ involves that sol(tgt) has to respect the following constraint:

$$\text{sol(tgt)} \in \{s : \text{string} \mid \text{"yellow" is a substring of } s\} \qquad (2)$$

Now, sol(srce) must be modified using $\alpha_\beta$ into sol(tgt) that respects (2). Here, a domain-dependent choice has to be made: it concerns the way the solution space is structured, i.e., how can modifications be applied on solutions. It is assumed that in this application, the only modification operation is based on the replace operation on the set of strings (which is the solution space). Hence, since ("blue", "yellow") $\in \alpha_\beta$, the following cocktail name that is consistent with (2) is proposed:

> sol(tgt) = replace("blue", "yellow", sol(srce)) = "yellow  lagoon"

## 6   Cocktail Name Adaptation Strategies

An adaptation strategy is a function with the following signature:

**input** srce, sol(srce), tgt, $\beta_{\text{srce}}$, $\alpha_{\text{pb}}$, DK;
**output** a set of strings, each of them being a proposed solution sol(tgt) for tgt.

When the output is empty, this means that the strategy has failed. Each element of the output have to be different from sol(srce).

The two first strategies presented below (Sects. 6.1 and 6.2) are application-dependent, whereas the last ones should be adaptable to other applications. Strategies presented in Sects. 6.3 and 6.4 are designed for simple matchings whereas the strategy of Sect. 6.5 combines strategies for dealing with complex matchings.

### 6.1   Strategy "Removing Alcohol Makes the Cocktail Virgin"

A simple strategy consists in generalizing the example 2. It is presented by the algorithm of Fig. 3(a). Note that the condition of the test can be performed by executing the SPARQL query $Q_{\text{alcohol}}$ (cf. Eq. (1)) twice:

- "srce contains some alcohol" is encoded by $\text{exec}(Q_{\text{alcohol}}, \text{DK} \cup \text{srce}) \neq \emptyset$ and
- "tgt contains no alcohol" is encoded by $\text{exec}(Q_{\text{alcohol}}, \text{DK} \cup \text{tgt}) = \emptyset$.

```
function adaptNameWhenRemovingAlcohol(...)        function adaptDefault(...)
begin                                             begin
    if srce contains some alcohol and tgt does not then      sol(tgt) ← "the new "
        sol(tgt) ← "virgin " + sol(srce)                              + sol(srce)
        return {sol(tgt)}                             return {sol(tgt)}
    else                                          end
        return ∅
    end
end
```

(a) Adaptation strategy of §6.1.                  (b) Adaptation strategy of §6.2.

**Fig. 3.** Two basic cocktail name adaptation strategies.

## 6.2 Default Strategy

The default strategy is applied when all other strategies fail. It is presented by the algorithm of Fig. 3(b). Example 3 is an application of this strategy.

## 6.3 Strategy "Turn Constants into Variables"

This strategy has been generalized from the analysis of the example 1 that is described in Sect. 5. Its algorithm is presented in Figs. 4 and 5. The two first tests of the algorithm define conditions under which the strategy applies (∅ is returned otherwise): $\alpha_{pb}$ is a substitution of ingredient $f \rightsquigarrow g$ and $\beta_{srce}$ relates $f$ to sol(srce).

Then the AT1 step of analogical transfer is implemented. First, $\widehat{\alpha_{\beta}}$ is computed: it corresponds to pairs $(d, d')$ where $d$ is a descriptor of $\beta_{srce}$ and $d'$ is either a variable or a value (constant or literal). New variables $d'$ are generated that correspond to nodes of $\beta_{srce}$ that are connected to the substituted ingredient class $f$. Second, $\beta_{gen}$ is computed by replacing in $\beta_{srce}$ $d$ by $d'$ for each $(d, d') \in \widehat{\alpha_{\beta}}$. Third, $\widehat{\alpha_{\beta}}$ and $\beta_{gen}$ are instantiated by $\alpha_{\beta}$ and $\beta_{tgt}$; since there may be several instantiations, pairs $(\alpha_{\beta}, \beta_{tgt})$ are generated. To find these instantiations of variables, the domain knowledge is queried: a SPARQL query is built that enables to find variable instantiations respecting the constraints of $\beta_{tgt}$.

Then, for each pair $(\alpha_{\beta}, \beta_{tgt})$ so generated, the AT2 step is applied. It consists mainly in applying the function modifyUnderMappingAndConstraints that is described by the algorithm of Fig. 5: substrings of sol(srce) occurring in $\beta_{srce}$ are replaced by the corresponding strings in $\beta_{tgt}$. In theory, this algorithm is underspecified, since the order of the replacements matters for the result. In practice, however, in all the examples we have met, this has had no influence.

It is worth noticing that the only part of this strategy that is domain-dependent lies in the function modifyUnderMappingAndConstraints and that this latter depends essentially on information on how to "travel" in the solution space, that is, for the application presented here, the substring checking and the replace function.

**function** turnConstantsIntoVariables(srce, sol(srce), tgt, $\beta_{\text{srce}}$, $\alpha_{\text{pb}}$, DK)
**begin**
    ▷ **testing whether the strategy applies**
    **if** $\alpha_{\text{pb}}$ *is not a simple matching* **then**
       | **return** $\emptyset$                                    ▷ adaptation strategy failure
    **end**
    Let $f, g$ be resources such that $\alpha_{\text{pb}} = f \rightsquigarrow g$ ($f$ is necessarily a food class of srce).
    **if** $f$ *is not connected to* sol(srce) *in* $\beta_{\text{srce}}$ **then**
       | **return** $\emptyset$
    **end**
    ▷ **AT1: computing** $\widehat{\alpha_\beta}$
    $\widehat{\alpha_\beta} \leftarrow \{(f, g), (\text{sol(srce)}, ?\text{solTgt})\}$
    Add to $\widehat{\alpha_\beta}$ every $(a, a)$ such that $a \in \text{Nodes}(\beta_{\text{srce}})$ and $a$ is not connected to $f$ in $\beta_{\text{srce}}$.
    **for** $n \in \text{Nodes(srce)}$ **do**
       | **if** *there is no image of $n$ by $\widehat{\alpha_\beta}$* **then**
          | $n' \leftarrow$ new variable
          | $\widehat{\alpha_\beta} \leftarrow \widehat{\alpha_\beta} \cup \{(n, n')\}$
       | **end**
    **end**
    ▷ **AT1: computing** $\beta_{\text{gen}}$
    $\beta_{\text{gen}} \leftarrow \emptyset$
    **for** $\langle s \; p \; o \rangle \in \beta_{\text{srce}}$ **do**
       | $s' \leftarrow \widehat{\alpha_\beta}(s)$                          ▷ i.e., $s'$ is such that $(s, s') \in \widehat{\alpha_\beta}$
       | $o' \leftarrow \widehat{\alpha_\beta}(o)$
       | $\beta_{\text{gen}} \leftarrow \beta_{\text{gen}} \cup \{\langle s' \; p \; o' \rangle\}$
    **end**
    ▷ **AT1: computing the set** $P_{\alpha_\beta \beta_{\text{tgt}}}$ **of ordered pairs** $(\alpha_\beta, \beta_{\text{tgt}})$
    $P_{\alpha_\beta \beta_{\text{tgt}}} \leftarrow \emptyset$
    Let Q be the SPARQL query such that the selected variables of Q are the variables occurring
    in $\widehat{\alpha_\beta}$ different from ?solTgt and the body of Q (i.e., what follows WHERE) is constituted
    by triples of $\beta_{\text{gen}}$, except the ones with ?solTgt
    R $\leftarrow$ exec(Q, DK)                  ▷ R is a set of assignments of the variables
    **for** $A \in$ R **do**
       | $\alpha_\beta \leftarrow$ result of applying the assignment $A$ on $\widehat{\alpha_\beta}$
       | $\beta_{\text{tgt}} \leftarrow$ result of applying the assignment $A$ on $\beta_{\text{gen}}$
       | $P_{\alpha_\beta \beta_{\text{tgt}}} \leftarrow P_{\alpha_\beta \beta_{\text{tgt}}} \cup \{(\alpha_\beta, \beta_{\text{tgt}})\}$
    **end**
    ▷ **AT2: computing the set** SOLs(tgt) **of candidate values for** sol(tgt)
    SOLs(tgt) $\leftarrow \emptyset$
    **for** $(\alpha_\beta, \beta_{\text{tgt}}) \in P_{\alpha_\beta \beta_{\text{tgt}}}$ **do**
       | sol(tgt) $\leftarrow$ modifyUnderMappingAndConstraints(sol(srce), $\beta_{\text{srce}}$, $\alpha_\beta$, $\beta_{\text{tgt}}$)
       | **if** sol(tgt) $\neq$ sol(srce) **then**
          | SOLs(tgt) $\leftarrow$ SOLs(tgt) $\cup \{\text{sol(tgt)}\}$
       | **end**
    **end**
    **return** SOLs(tgt)
**end**

Fig. 4. The turn constants into variables strategy.

```
function modifyUnderMappingAndConstraints(sol(srce), βsrce, αβ, βtgt)
begin
    sol(tgt) ← sol(srce)
    for (d, d') ∈ αβ such that βsrce ⊢ ⟨d subStringOf sol(srce)⟩ do
    |   sol(tgt) ← replace(d, d', sol(tgt))
    end
    return sol(tgt)
end
```

**Fig. 5.** An implementation of analogical transfer step AT2.

Another illustration of this algorithm is given below. Consider the example 4, with the following dependency:

$$\beta_{srce} = \{\langle \texttt{irishWhisky origin ireland}\rangle,$$
$$\langle \texttt{ireland englishAdjective "irish"}\rangle,$$
$$\langle \texttt{"irish" subStringOf "irish  coffee"}\rangle,$$
$$\langle \texttt{coffee nameInEnglish "coffee"}\rangle,$$
$$\langle \texttt{"coffee" subStringOf "irish  coffee"}\rangle\}$$

recalling that $\texttt{sol(srce)} = \texttt{"irish  whisky"}$. Applying the algorithm of Fig. 4, the computation of $\widehat{\alpha_\beta}$ and $\beta_{gen}$ gives:

$$\widehat{\alpha_\beta} = \{(\texttt{irishWhisky}, \texttt{tequila}), (\texttt{ireland}, ?x), (\texttt{"irish"}, ?y),$$
$$(\texttt{"irish  coffee"}, ?\texttt{solTgt}), (\texttt{coffee}, \texttt{coffee}), (\texttt{"coffee"}, \texttt{"coffee"})\}$$
$$\beta_{gen} = \{\langle \texttt{tequila origin } ?x\rangle, \langle ?x \texttt{ englishAdjective } ?y\rangle,$$
$$\langle ?y \texttt{ subStringOf } ?\texttt{solTgt}\rangle, \langle \texttt{coffee nameInEnglish "coffee"}\rangle,$$
$$\langle \texttt{"coffee" subStringOf } ?\texttt{solTgt}\rangle\}$$

From this, the following SPARQL query is built and executed on DK:

SELECT ?x ?y WHERE {tequila origin ?x . ?x englishAdjective ?y}

Assuming the result R contains the only assignment $A = \{?x \leftarrow \texttt{mexico},$ $?y \leftarrow \texttt{"mexican"}\}$, it comes that $P_{\alpha_\beta \beta_{tgt}} = \{(\alpha_\beta, \beta_{tgt})\}$ with

$$\alpha_\beta = \{(\texttt{irishWhisky}, \texttt{tequila}), (\texttt{ireland}, \texttt{mexico}), (\texttt{"irish"}, \texttt{"mexican"}),$$
$$(\texttt{"irish  coffee"}, ?\texttt{solTgt}), (\texttt{coffee}, \texttt{coffee}), (\texttt{"coffee"}, \texttt{"coffee"})\}$$
$$\beta_{tgt} = \{\langle \texttt{tequila origin mexico}\rangle, \langle \texttt{mexico englishAdjective "mexican"}\rangle,$$
$$\langle \texttt{"mexican" subStringOf } ?\texttt{solTgt}\rangle,$$
$$\langle \texttt{coffee nameInEnglish "coffee"}\rangle,$$
$$\langle \texttt{"coffee" subStringOf } ?\texttt{solTgt}\rangle\}$$

And, finally, the solution proposed is obtained by replacing successfully (a) "irish" by "mexican" and (b) "coffee" by "coffee" in sol(srce) = "irish coffee":

$$\text{"irish coffee"} \overset{(a)}{\longmapsto} \text{"mexican coffee"} \overset{(b)}{\longmapsto} \text{"mexican coffee"} = \text{sol(tgt)}$$

## 6.4 Strategy "Generalization-Specialization of Dependencies"

Now, consider the example 5, of the adaptation of sol(srce) = "blue lagoon" when $\alpha_\beta$ = curaçao $\rightsquigarrow$ indianTonic with the same $\beta_{\text{srce}}$ as in example 4 (cf. Sect. 5), and assuming that DK gives no color to Indian tonic (i.e., there is no triple of the form $t = \langle$indianTonic color c$\rangle$ such that DK $\vdash t$), the adaptation strategy of Sect. 6.3 fails. However, it is assumed that

$$\text{DK} \vdash \left\{ \begin{array}{l} \langle\text{indianTonic taste bitter}\rangle, \langle\text{indianTonic texture sparkling}\rangle, \\ \langle\text{bitter inEnglish "bitter"}\rangle, \langle\text{sparkling inEnglish "sparkling"}\rangle, \\ \langle\text{color subp hOP}\rangle, \langle\text{taste subp hOP}\rangle, \langle\text{texture subp hOP}\rangle \end{array} \right\}$$

meaning that Indian tonic is bitter and sparkling, and that color, taste and texture are organoleptic properties (hOP is an abbreviation for hasOrganolepticProperty). Therefore, the adaptation strategy described in Sect. 6.3 can be applied with a slight modification: it is sufficient to replace in $\beta_{\text{gen}}$ the triple $\langle$indianTonic color ?x$\rangle$ by $\langle$indianTonic hOP ?x$\rangle$, which is more general according to DK.

One way to address this problem is to replace all the resources and literals of $\beta_{\text{srce}}$—including the predicates—by variables, with the exception of the ones matched by $\alpha_{\text{pb}}$ (i.e., indianTonic in the example). This would lead in the example to

$$\alpha_{\text{pb}}(\beta_{\text{srce}}) = \{\langle\text{indianTonic color blue}\rangle, \langle\text{color inEnglish "blue"}\rangle,$$
$$\langle\text{"blue" subStringOf sol(srce)}\rangle\}$$

generalized into

$$\beta_{\text{gen}} = \{\langle\text{indianTonic ?p1 ?x}\rangle, \langle\text{?x ?p2 ?y}\rangle, \langle\text{?y ?p3 ?solTgt}\rangle\}$$

However, we choose to discard this approach because it may give too many results and since it is based on a too shallow semantics. For example, sol(tgt) = "food lagoon" would be justified by the assignment $\{?p1 \leftarrow$ subc, $?x \leftarrow$ food, $?p2 \leftarrow$ inEnglish, $?y \leftarrow$ "food"$\}$.

Another way to address this problem is to search in the domain knowledge for triples for building $\beta_{\text{gen}}$ that are *similar* to $\alpha_\beta(\beta_{\text{srce}})$. This can be likened to the retrieval issue in CBR, which can be implemented by a least generalization of the query (see, e.g., [5]). A similar idea is proposed here. It consists in making a best-first search (e.g., an A* search) in a space of dependencies $\beta$ such that:

– The initial state $\beta_0$ corresponds to the $\beta_{\text{gen}}$ as it is computed in the strategy of Sect. 6.3.

- The successors of a state consists in making a generalization of one of its triples. The following generalization operators can be considered: replace a class (resp., a property) by a direct superclass (resp., direct superproperty) in DK, replace a resource or a literal by a variable, etc. A cost function must be associated to generalization operators, in order to choose the least costly generalization.
- A final state $\beta$ is such that the SPARQL query associated with it gives a nonempty set of results.

Once a final state $\beta$ is found, the rest of the algorithm of Sect. 6.3 can be applied with $\beta_{\text{gen}} = \beta$.

Back to the example, it comes:

$$\beta_0 = \{\langle \texttt{indianTonic color ?x} \rangle, \langle \texttt{?x inEnglish ?y} \rangle,$$
$$\langle \texttt{?y subStringOf ?solTgt} \rangle\}$$

In the first triple, `color` can be generalized into `hOP` (since DK $\vdash$ $\langle \texttt{color subp hOP} \rangle$), giving

$$\beta = \{\langle \texttt{indianTonic hOP ?x} \rangle, \langle \texttt{?x inEnglish ?y} \rangle,$$
$$\langle \texttt{?y subStringOf ?solTgt} \rangle\}$$

$\beta$ is a final state since $\texttt{exec}(\texttt{Q}, \texttt{DK}) \neq \emptyset$ for

$$\texttt{Q} = \text{SELECT } \texttt{?x ?y } \text{WHERE } \{\texttt{indianTonic hOP ?x . ?x inEnglish ?y}\}$$

Indeed, $\texttt{exec}(\texttt{Q}, \texttt{DK}) = \{A_1, A_2\}$ where $A_1 = \{\texttt{?x} \leftarrow \texttt{bitter}, \texttt{?y} \leftarrow \texttt{"bitter"}\}$ and $A_2 = \{\texttt{?x} \leftarrow \texttt{sparkling}, \texttt{?y} \leftarrow \texttt{"sparkling"}\}$, leading to the two expected solutions: `"bitter lagoon"` and `"sparkling lagoon"`.

Therefore this strategy consists in finding the minimal generalization $\beta$ of the initial dependency $\beta_0$ and then in specializing $\beta$ into $\beta_{\text{tgt}}$'s thanks to SPARQL querying on DK, hence the name of the strategy.

### 6.5 Composing Strategies When the Matching is Complex

When the matching $\alpha_{\text{pb}}$ is complex, it can be written $\alpha_{\text{pb}} = \alpha_{\text{pb}}^q \circ \alpha_{\text{pb}}^{q-1} \circ \ldots \circ \alpha_{\text{pb}}^1$, with $q \geq 2$. The idea is then to apply in sequence the strategies associated with simple matchings. For example, for $\texttt{sol(srce)} = \texttt{"irish coffee"}$, $\alpha_{\text{pb}}^1 = \texttt{irishWhisky} \rightsquigarrow \texttt{tequila}$, $\alpha_{\text{pb}}^2 = \texttt{coffee} \rightsquigarrow \texttt{hotChocolate}$, the strategy of Sect. 6.3 can be applied twice to give the name $\texttt{sol(tgt)} = \texttt{"mexican hot chocolate"}$. This adaptation is an application of the adaptation based on reformulations and similarity paths (see e.g. [8]).

Another example consists in substituting in the "Blue Lagoon" recipe all the ingredients by sparkling water, in the order curaçao, vodka and lemon juice, giving birth to the name `"the new virgin sparkling lagoon"` for a glass of sparkling water, which can arguably be considered as the result of a creative naming process!

# 7   Discussion

Among the 20 examples listed in the first phase of this study (cf. the sample presented in Sect. 4), 13 are modeled in the strategies above: 1 in Sect. 6.1, 1 in Sect. 6.2, 9 in Sect. 6.3 and 2 in Sect. 6.4[7] (0 in Sect. 6.5). 5 of the remaining ones corresponds to a strategy consisting in adding or substituting a qualifier to the source recipe name when a new ingredient is added or replaces an ingredient that has no connection with the source recipe name. For example, `"gin fizz"` becomes `"silver fizz"` when an egg white is added to the recipe. Finally, 2 examples are not covered because they would require a more complex case representation, for example, `"tequila sunrise"` becomes `"tequila sunset"` partially because of the change in the order of the preparation steps. These figures do not constitute statistically significant information but give some ideas on how the examples has led to strategies and how they can be used to guide future work.

Apart from the ad hoc strategies, the analogical transfer strategies presented in the previous section corresponds to a scheme of modifying (by generalization) a dependency so that it becomes consistent with the target context (under CWA). Modifying a case until it reaches consistency with the target problem wrt the domain knowledge is what revision-based adaptation (RBA) does [3]. Therefore, though RBA has not been a useful guideline for starting this research, it could be used to re-describe this contribution and to go one step further, in order to examine formal properties of the analogical transfer and to propose new strategies. Actually, in previous studies, RBA was used in order to modify the *solution* `sol(srce)` of the retrieved case (which is uneasy to formalize when solutions are strings), whereas RBA could be used as a tool to modify the *dependency* $\beta_{\texttt{srce}}$ within the AT process.

Following this idea, the analogical transfer amounts to travel in a dependency space structured by modifications (only generalizations in the examples given in this paper), with a good choice of the travel costs. The strategy described in Sects. 5 and 6.3 works with a constant modification that turns the edges of the RDFS graph into variables but does not modify the properties that label the edges of this graph. This could be understood as the fact that the cost of the former generalizations is much lower than the cost of the latter ones. To justify this, it is considered that the properties (e.g., `color`) are more abstract descriptors than other resources (e.g., `blue`, `yellow`). According to the heuristic saying that it is better to modify a concrete descriptor than an abstract one, this is justified. This heuristic principle has been defended for a long time in the analogical reasoning community [6].

# 8   Conclusion and Future Work

Starting from the application problem of cocktail name adaptation, this paper describes a research work on strategies for analogical transfer in the context of

---

[7] Actually, the 9 of Sect. 6.3 could also be counted as modeled by Sect. 6.4: the second strategy generalizes the latter.

the representation language RDFS. If some proposed strategies are application-dependent, it is claimed that other ones can be applied—or adapted—to a larger framework. Indeed, they match the principles described in some related work about analogical transfer (e.g., [6,14]) while proposing an approach having profit of the standard RDFS as well as on associated tools (RDFS SPARQL engines, RDF stores).

The operationality of this work is demonstrated by a first prototype in Python that covers some of the strategies. However, some work remains to be done to cover all of them, in particular the one based on generalization-specialization of dependencies, which constitutes an ongoing work. Furthermore, new strategies have to be developed (the strategies presented here covers the majority of the examples but not all of them: 13 on 20) and a way to control the application of strategies should be designed.

A first direction of future work aims at addressing two current limitations of the approach. First, there is an important workload for acquiring dependencies $\beta_{\texttt{srce}}$, which is currently done manually. Second, in order to get more relevant results, it is important to have more triples in the domain knowledge. In order to address these issues, it is planned to interrogate the Linked Open Data (LOD), i.e., a huge cloud of RDF and RDFS bases freely accessible on the Web (DBPedia, a base of the LOD, contains about 3 billions triples). For example, there are tools that enable to find paths in the LOD from a resource to another one (see, e.g., [13]), and such a tool could be used to find a link from an ingredient name of a cocktail recipe to a word occurring in the name of the cocktail or—more generally—from a problem $\texttt{srce}$ to a solution $\texttt{sol(srce)}$ of this problem. The union of such paths would constitute the dependency $\beta_{\texttt{srce}}$ and the domain knowledge should contain at least the union of all the $\beta_{\texttt{srce}}$'s.

# References

1. Carbonell, J.G.: Derivational analogy: a theory of reconstructive problem solving and expertise acquisition. In: Machine Learning, vol. 2, pp. 371–392. Morgan Kaufmann Inc. (1986)
2. Ceci, F., Weber, R.O., Gonçalves, A.L., Pacheco, R.C.S.: Adapting sentiments with context. In: Hüllermeier, E., Minor, M. (eds.) ICCBR 2015. LNCS (LNAI), vol. 9343, pp. 44–59. Springer, Heidelberg (2015). doi:10.1007/978-3-319-24586-7_4
3. Cojan, J., Lieber, J.: Applying belief revision to case-based reasoning. In: Prade, H., Richard, G. (eds.) Computational Approaches to Analogical Reasoning: Current Trends. Studies in Computational Intelligence, vol. 548, pp. 133–161. Springer, Heidelberg (2014)
4. Cordier, A., Dufour-Lussier, V., Lieber, J., Nauer, E., Badra, F., Cojan, J., Gaillard, E., Infante-Blanco, L., Molli, P., Napoli, A., Skaf-Molli, H.: Taaable: a case-based system for personalized cooking. In: Montani, S., Jain, L.C. (eds.) Successful Case-based Reasoning Applications-2. Studies in Computational Intelligence, vol. 494, pp. 121–162. Springer, Heidelberg (2014)
5. Gaillard, E., Infante-Blanco, L., Lieber, J., Nauer, E.: Tuuurbine: a generic CBR engine over RDFS. In: Case-Based Reasoning Research and Development, Cork, Ireland, vol. 8765, pp. 140–154, September 2014

6. Gentner, D.: Structure-mapping: a theoretical framework for analogy. Cogn. Sci. **7**(2), 155–170 (1983)
7. Jalali, V., Leake, D.: CBR meets big data: a case study of large-scale adaptation rule generation. In: Hüllermeier, E., Minor, M. (eds.) ICCBR 2015. LNCS (LNAI), vol. 9343, pp. 181–196. Springer, Heidelberg (2015). doi:10.1007/978-3-319-24586-7_13
8. Melis, E., Lieber, J., Napoli, A.: Reformulation in case-based reasoning. In: Smyth, B., Cunningham, P. (eds.) Fourth European Workshop on Case-Based Reasoning, EWCBR-98. LNCS, vol. 1488, pp. 172–183. Springer, Heidelberg (1998)
9. Müller, G., Bergmann, R.: Workflow streams: a means for compositional adaptation in process-oriented CBR. In: Lamontagne, L., Plaza, E. (eds.) Case-Based Reasoning Research and Development, ICCBR-2014. Lecture Notes in Artifichal Intelligence, vol. 8765, pp. 315–329. Springer, Cork, Ireland (2014)
10. Müller, G., Bergmann, R.: CookingCAKE: a framework for the adaptation of cooking recipes represented as workflows (2015)
11. Riesbeck, C.K., Schank, R.C.: Inside Case-Based Reasoning. Lawrence Erlbaum Associates Inc., Hillsdale (1989)
12. Sowa, J.F.: Conceptual Structures: Information Processing in Mind and Machine. Addison Wesley, Reading (1984)
13. Tiddi, I., d'Aquin, M., Motta, E.: Learning to assess linked data relationships using genetic programming. In: Proceedings of the 15th International Semantic Web Conference (ISWC). Springer (2016)
14. Winston, P.H.: Learning and reasoning by analogy. Commun. ACM **23**(12), 689–703 (1980)

# Adaptation-Guided Feature Deletion: Testing Recoverability to Guide Case Compression

David Leake$^{(\boxtimes)}$ and Brian Schack

School of Informatics and Computing, Indiana University,
Bloomington, IN 47408, USA
{leake,schackb}@indiana.edu

**Abstract.** Extensive case-based reasoning research has studied methods for generating compact, competent case bases. This work has focused primarily on compressing the case base by deleting entire cases, based solely on their competence contributions. Recent work proposed an alternative which compressed individual cases by selectively deleting their internal contents. Early studies of this approach, termed *flexible feature deletion* (FFD), demonstrated that for suitable domains, such as domains with cases of varying sizes for which case usefulness can be retained despite internal deletions, even very simple FFD approaches may outperform standard per-case methods. However, more sophisticated methods are needed. Because FFD's internal changes to cases can be seen as a form of case adaptation, this paper investigates whether the adaptation knowledge of a system can be harnessed to improve FFD. This paper proposes tying FFD choices directly to adaptation knowledge and presents results on a competence-preserving FFD method which prioritizes feature deletions by the *recoverability* of deleted features through case adaptation. Evaluation of recoverability-based FFD in a path-finding domain supports that it provides superior competence retention compared to standard flexible feature deletion at the same level of compression.

**Keywords:** Case-base maintenance · Competence-guided deletion · Flexible feature deletion

## 1 Introduction

Case-base maintenance has received extensive attention in CBR research (e.g., [1,2]). A particular focus of this work has been on developing compact, competent case bases (e.g., [3]). This work has focused on retention decisions at the case level, aimed at guiding case retention or deletion decisions based on the overall competence contributions of the cases. Per-case strategies are appropriate for the task domains to which they have been applied, which generally share two characteristics: (1) That cases are of fairly uniform size, and (2) that preserving the usefulness of a case depends on retaining its entire contents. However, in previous work [4], we observed that these assumptions do not always hold and proposed that in some circumstances, it may be useful to apply a finer-grained approach, focusing on compacting the contents of cases themselves by selectively deleting

© Springer International Publishing AG 2016
A. Goel et al. (Eds.): ICCBR 2016, LNAI 9969, pp. 234–248, 2016.
DOI: 10.1007/978-3-319-47096-2_16

components. We called this approach *flexible feature deletion* (FFD). Flexible feature deletion generalizes per-case maintenance by dropping the assumptions of uniform case size and indivisible cases.

Flexible feature deletion applies when information can be removed from a case while retaining some usefulness. For example, for cases capturing medical images, it may be possible to compact cases while retaining usefulness by adjusting resolution; for traces in trace-based reasoning (e.g., [5]) or plans in case-based planning, it may be possible to compact while retaining usefulness by deleting routine portions of the steps in a case that are easily re-generated; for large and rich cases capturing recommendation information (e.g., movie recommendations), it may be possible compact while retaining usefulness by selectively deleting features likely to hold less interest. In each of these instances, some information is lost—just as information is lost when deleting entire cases. However, our results showed that for suitable case bases, the FFD approach provided better competence retention for a given case base size than conventional per-case deletion approaches [4].

The key question for FFD is how to determine which features to remove. Our initial tests of FFD selected deletion targets by simple knowledge-light methods based on statistical feature properties. The tests demonstrated that for cases with varying sizes and for which not all information was essential to case usefulness, even such simple approaches can be sufficient to provide improved competence retention. However, a natural question is how to integrate richer knowledge into the FFD process.

Because FFD is revising internal case contents, the operations of FFD can be viewed as performing a form of case adaptation, though with the goal of reducing case size rather than of solving a particular problem. The competence loss from FFD can be mitigated if adaptation knowledge can recover the original case from the changed case. Consequently, this paper proposes using adaptation knowledge to guide the FFD process, by focusing deletion on case components that can be recovered by adaptation. From the perspective of Richter's CBR "knowledge containers" of CBR [6], this approach aims to delete case knowledge overlapping with knowledge contained in the adaptation knowledge container. The perspective of storing and recovering partial cases can also be seen as related to reconstructive models such as Dynamic Memory Theory [7] and Constructive Similarity assessment [8].

The paper begins with a discussion of potential roles of adaptation in flexible feature deletion. It next describes a sample domain and case study of the use of adaptation knowledge to guide choices during FFD and presents an evaluation of the approach. As expected, the evaluation shows that adding recoverability considerations can enable FFD to improve competence retention for given levels of compression. It also shows, surprisingly, that in some situations case-base compression by FFD may actually improve case-base competence, a phenomenon we call *creative destruction*.

# 2   Adaptation Knowledge in Flexible Feature Deletion

Flexible feature deletion removes components of cases. For flat feature representations, its function may be as simple as deleting particular features from a feature vector. However, FFD for structured cases may include a wider space of possible operations, not restricted to deleting individual features, or even limited to deletion per se. For example, FFD could include compressing cases through:

- **Substructure deletion:** FFD deletion operation removes components of any size, ranging from individual features to larger feature collections such as subplans of a plan.
- **Substructure substitution:** FFD substructure substitution replaces components with more compact components.
- **Substructure abstraction:** FFD substructure abstraction is a knowledge-guided form of substructure substitution. Rather than deleting a substructure entirely, it replaces the substructure with a more compact abstraction. For example, in case-based image recognition, abstraction could be applied to decrease the resolution of some or all of the image, saving space. To reuse a case, it may sometimes be necessary to do an inverse adaptation, to replace the abstraction with a more specific instantiation.

FFD may result in storing incomplete or unelaborated cases. For example, if a case records a path between points A and B, and some internal segments of the path are deleted, the path case would no longer be intact. However, such a deletion could still be allowed, with an annotation recording the gap. In that situation, the case would require adaptation before use, to recover from the deletion.

FFD need not preserve the original usefulness of a case, or may transform its usefulness, making it less useful for the original problem but more useful for a different problem. For example, after substructure deletion the case might require adaptation to solve the original problem, but might no longer require the adaptation it once required to solve some different problem. For example, for a case containing a route plan, FFD might delete some initial segments. In that case, the plan would no longer be directly applicable to the same starting point, but could be directly applied to generate a shorter path with the new start.

## 2.1   Applying Case Adaptation Knowledge for FFD

All of the FFD operators correspond to common operations for case adaptation. Any such operations can be applied successively in an adaptation chain [9], to provide varying levels of adaptation-based compression. If procedures for these are already available in the adaptation component, they can be applied directly for FFD. If the adaptation knowledge includes specific guidance on applicability, or on circumstances when a particular adaptation is suitable, the use of the adaptation knowledge for maintenance automatically makes that guidance available to the maintenance phase.

Even if particular adaptation knowledge is not framed in terms of deletion, it may be useful for compression. For example, consider a path planning system able to adapt plans to avoid roads that are closed. If the adaptation results in a route that can be described more compactly, that adaptation contributes to compression of the case, and in principle could be applied as an FFD operation.

Exploiting adaptation knowledge for FFD raises the key question of when a particular adaptation should be applied for FFD. This depends crucially on four factors: compression benefit, case/feature recoverability, quality retention, and recovery cost.

- **Compression benefit:** Compression refers to the reduction in case base size effected by FFD. Because FFD can change sizes of individual cases by feature deletion, compression is measured not in terms of the number of cases in the case base, but in terms of finer-grained subunits directly related to storage requirements. For cases represented by feature vectors, a natural unit is a feature-value pair. Compression benefit reflects the advantage to the CBR system of having a smaller case base. Often, this benefit is judged in terms of retrieval speed or overall processing cost [10]. However, this could also reflect factors such as hard limits on case base size (e.g., in a legacy or high-reliability system with limited storage) or transmission cost, if the case base will be provided to other agents.

- **Case/feature recoverability:** Case/feature recoverability refers to the ability of the system to regenerate its knowledge state prior to compression from adaptation knowledge and the remaining case base. (Note that regenerating the knowledge may not require regenerating identical solutions, if multiple solutions are satisfactory, and that the knowledge state might include knowledge not directly connected to competence, such as features used in indexing to increase retrieval speed.) We refer to FFD operations that are always recoverable as lossless; those that are not necessarily recoverable are lossy. Whether a particular strategy is lossy or lossless depends on the entire set of adaptations available to the system, on the length of adaptation chains allowed, and on all the cases in the case base.

- **Quality retention:** Quality retention refers to the quality of the solutions the system is able to generate, beyond simply generating a correct solution. For example, in a path planning domain, a deletion from a path would be recoverable if the system were still able to generate *some* path between the same endpoints. Quality retention might be measured by the ratio of the costs of old and new paths.

- **Recovery cost:** Recovery cost refers to the resources required to generate a new solution to the problem. For example, in a case-based planner able to draw on a generative planner when needed, all deletions might be recoverable, but some might be computationally expensive when done by reasoning from scratch. In those instances, FFD deletions might be more appropriate. Likewise, some domains, a complete case that is deleted may be unrecoverable, while internal deletions can be recovered. For example, consider a system whose cases are medical X-ray images. If an image is deleted, there may be no recourse but re-taking the X-ray, at considerable expense. However, if portions are stored at lower resolution, it may be possible to recover needed information by image processing algorithms at much lower cost.

Feature-centric recoverability is closely related to the notion of *reachability*, defined by Smyth and Keane [11]. However, there is an important difference. Reachability refers to the ability to adapt other cases in the case base to cover the problem addressed by a case; feature-centric recoverability refers to the ability to adapt *either* other cases in the case base or the FFD-revised original case to cover the competence contributions of that case (which may require adapting an internal subpart of the case). In the following, we will apply a restricted variant of recoverability, which we term *local recoverability*. When FFD is applied to a case, the case is locally recoverable if the system can still solve the problem that the case originally solved. Local recoverability is a weaker approximation of recoverability, because it does not test whether the system is still able to solve all the problems it could originally, but it is more efficient to calculate.

## 3   A Case Study of Recoverability-Based FFD

We are studying recoverability-based flexible feature deletion in the context of a path planning task. The recoverability-based approach prioritizes FFD deletion targets according to local recoverability. This section describes the underlying domain and system. The following section describes our experimental questions and results.

### 3.1   Testbed Domain

The path planning task is carried out on a road network represented by a weighted graph with labeled vertices. The vertices represent neighborhoods, and they are collected into groups representing boroughs, with each group having an equal number of vertices. Each vertex in a group is intra-connected via an edge to another vertex in the same group. The groups are also inter-connected with one or more vertices from each group having an edge to another vertex in a different group. This design abstracts characteristics such as streets intra-connecting neighborhoods in a borough and bridges inter-connecting boroughs in a city. To generate a road network, connections are randomly selected, with the constraint that at least one path must exist between any two vertices. Figure 1 illustrates a sample graph satisfying these constraints. Each edge has a random integer weight in [0, 100], representing the cost (e.g., distance or time) to travel across that edge. Problems to be solved by the system are described as lists of vertices which the solution path must include in order, starting with the source and ending with the destination.

### 3.2   Adaptation Strategies

The testbed system has five adaptation strategies, summarized in Table 1. No attempt was made to optimize the adaptation process, which is done by the

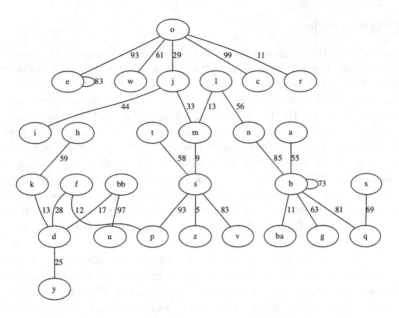

**Fig. 1.** Sample graph for path planning task

**Table 1.** Testbed system adaptation strategies

| Reuse strategy | Description |
|---|---|
| Reverse | Reverse the given solution so the source swaps with the destination, the destination swaps with the source, and the intermediate points reverse |
| Drop vertices | Drop the vertices in the given solution before the source of the given problem |
| Reverse drop vertices | Drop the vertices in the given solution after the destination of the given problem |
| Cons vertex | Append the source of the given problem to the front of the given solution |
| Reverse cons vertex | Append the destination of the given problem to the back of the given solution |
| Compose | Fill in a gap in the given solution with the solution from another case |

exhaustive application of adaptations. When adaptation must be done to generate solutions, all adaptations are tried; when the system assesses recoverability, the system attempts to adapt all cases until a solution is found. The system does not combine adaptations, except for the special case of the compose strategy, which can chain with one other strategy.

## 3.3   FFD Strategies

Leake and Schack [4] present a set of knowledge-light FFD strategies, categorized according to how they prioritize items for deletion and the type of item on which they operate (whether they delete cases, collections of features, or a mixture). These are illustrated in Table 2. For reasons of space, we do not describe them in detail here, but refer the reader to that work for the full descriptions.

Given any flexible feature deletion strategy, it is possible to develop a recoverability-based version by guiding deletion decisions according to recoverability. Specifically in our recoverability-based FFD approach, the system applies one of the original FFD strategies to rank items for deletion. It then tests deletion candidates in order to determine whether the deletion is recoverable. This is done by attempting to solve the problems from the cases that would be modified, either (1) by adapting the modified cases or (2) by CBR applied to cases from the remainder of the case base. If these problems are solvable (and even if the solutions are different but satisfactory), then the modifications are accepted as recoverable. Otherwise, the process continues through the rest of the deletions according to the ordering specified by the given strategy. The process stops after the first successful recovery (in which case the modification is accepted), or after a maximum number of trials or when no untried cases remain, in which case the strategy fails and no further compression can be done.

The testbed system applies the five deletion strategies in Table 3, selected to include high-performing FFD strategies from our previous tests. These include two lossy strategies, Largest Case (which first deletes the largest cases, measured by the number of vertices in each solution) and Random Case-Feature (which deletes a randomly-selected vertex from the solution to a randomly-selected case and marks the gap for potential future recovery). Adding recoverability considerations leads to the strategies Reachability-Based Largest Case

**Table 2.** Strategies for selecting the next item to delete. From Leake and Schack [4].

| Strategy | Type of Bundling | Hybrid or Non-Hybrid |
|---|---|---|
| Random case-features | Unbundled | Non-hybrid |
| Random cases | Case-bundled | Non-hybrid |
| Large cases | Case-bundled | Non-hybrid |
| Least coverage | Case-bundled | Non-hybrid |
| Most reachability | Case-bundled | Non-hybrid |
| Random features | Feature-bundled | Non-hybrid |
| Rarest features | Feature-bundled | Non-hybrid |
| Most common features | Feature-bundled | Non-hybrid |
| Rarest cases/Least coverage | Case-bundled | Hybrid |
| Rarest features/Least coverage | Unbundled | Hybrid |
| Rarest features/Large cases | Unbundled | Hybrid |

**Table 3.** Sample FFD retention strategies, including recoverability-based strategies.

| Deletion Target | Lossiness | Description |
|---|---|---|
| Shared Component | Lossless | Extract components shared by the solutions of multiple cases into separate cases. Mark gaps for completion during recovery |
| Reachability-Based Largest Case | Lossy | Delete cases in order from largest to smallest number of case-features, deleting only recoverable cases |
| Largest Case | Lossy | Delete cases in order from largest to smallest number of case-features regardless of recoverability |
| Recoverability-Based Random Vertex | Lossy | Delete randomly-chosen case-features from the solutions to cases, deleting only recoverable features |
| Random Vertex | Lossy | Delete randomly-chosen case-features from the solutions to cases regardless of recoverability |

and Recoverability-Based Random Vertex. The recoverability-based variants help alleviate the lossiness of the strategies by using recoverability to filter their deletion recommendations.

## 4 Evaluation

### 4.1 Experimental Questions

Our evaluation focuses on three questions:

1. **Competence retention:** How does ability to solve problems compare for given levels of compression compare for adaptation-guided and non-adaptation-guided FFD strategies?
2. **Solution quality retention:** For problems that can be solved for a given level of compression, how does solution quality compare for adaptation-guided and non-adaptation-guided FFD strategies?
3. **Processing time:** How does the choice of FFD strategy affect the processing time of the case-based reasoning cycle, for different levels of compression?

### 4.2 Experimental Design

The case base was seeded with a set of training problems with randomly chosen vertices, with beginning and ending vertices chosen from different groups which are not neighbors. This ensures that solving a problem requires exiting the source group, traversing one or more other groups, and then entering the target group. The Bellman-Ford path-finding algorithm was used to generate optimal solutions to the training problems, minimizing the sum of the weights of the edges along the path.

Experiments averaged results of 12 trials, each one using a different randomly generated route graph with 28 vertices, initial case base of 33 randomly-generated seed cases and 17 test problems. Tests were run for each retention strategy in Table 3, with three-fold cross-validation, averaging the results by strategy and level of compression. Reported processing times reflect processing on a MacBook Pro with a 2.5 GHz Intel Core i5 processor and 8 GB of RAM.

For lossy strategies, compression was continued until only a single case remained. Sometimes a strategy can no longer compress a case base, either because the strategy is a lossless strategy and cannot compress beyond full competence, or because it exceeded a pre-set limit of 100 trials to find a recoverable deletion.

### 4.3   Question 1: Competence Retention

To evaluate competence retention, we measured how many problems the system could solve at increasing compression levels (decreasing numbers of case-feature pairs) for five deletion strategies: Shared Component, Recoverability-Based Largest Case, Largest Case, Recoverability-Based Random Vertex, and Random Vertex. Figure 2 shows the percent of competence retained from the uncompressed case base to the compressed case base, as a function of the percent of case-feature pairs retained from the uncompressed case base to the compressed case base, ranging from the full case base (100 %) to 50 % compression.

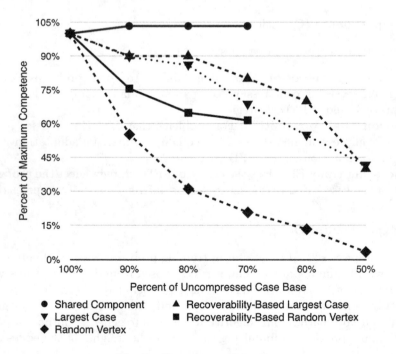

**Fig. 2.** Competence retention

The best performing strategy for competence was Shared Component. Because it is lossless, its high performance is expected. However, its ability to compress the case base stops at 70 % size when it cannot find any more shared components. To achieve more compression, one of the lossy strategies must be used.

Recoverability-based largest case does next best, and enables compression to 50 %. Comparison to the non-recoverability-based version shows that the recoverability-based approach improves competence retention.

The worst-performing strategy was the simple unguided strategy Random Vertex. Considering recoverability, in the Recoverability-Based Random Vertex strategy, markedly improves competence over Random Vertex, but Recoverability-Based Random Vertex can only compress the case base to 70 %, after which it can no longer find recoverable vertices.

We expected that as compression increases, competence would remain stable or decrease. That was true for four of the five strategies. However, surprisingly, competence *increased* slightly (103 % of original competence) for the Shared Component strategy at 70 % of the original case base. We discuss this in Sect. 4.6.

### 4.4   Question 2: Solution Quality Retention

We measured the quality of the solutions by the sum of their edge weights such that lower aggregate weights were preferred. Figure 3 shows the relative average sum (percent of maximum) of the weights of the solutions generated at different levels of compression with the five retention strategies, as a function of the percent of case-feature pairs retained from the uncompressed case base. Here no strategies are clearly the best or worst. This suggests that more knowledge would be needed to reliably ensure high-quality solutions.

### 4.5   Question 3: Processing Time

Figure 4 shows the average total processing time for case-based problem-solving for the test problems at each stage of compression for each of the five retention strategies, as a function of the percent of case-feature pairs retained from the uncompressed case base. The Largest Case strategy was most efficient, and the Recoverability-Based Random Vertex strategy was least efficient, with very rapid growth, due to checking many alternatives before finding vertices to delete. The line for this strategy continues beyond the top edge of the plot; the graph is cropped in order to show the performance on the other strategies in more detail.

### 4.6   Creative Destruction

In the competence retention experiment reported in Sect. 4.3, we expected that competence would always decrease with increased compression, as is normally expected for any case-base compression method. We were surprised to find that occasionally adaptation-guided flexible feature deletion applied to the case base

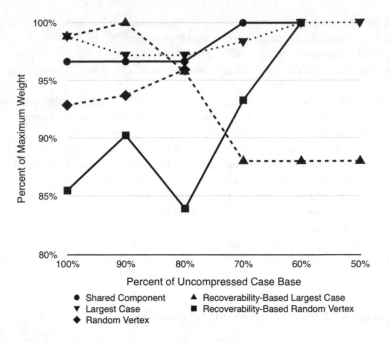

**Fig. 3.** Relative average solution quality as a function of compression

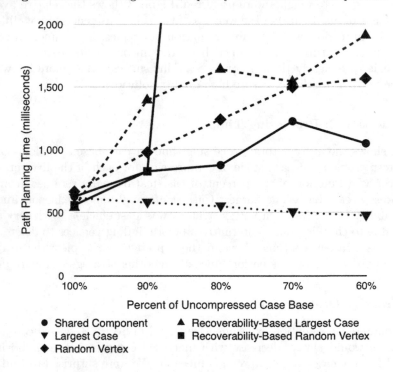

**Fig. 4.** Average total CBR path planning time

could slightly improve the competence of the system. Our explanation is that certain FFD strategies, by reorganizing contents of cases, can sometimes make certain case contents more accessible, enabling adaptations of limited power to exploit them more effectively.

For example, if the Shared Component FFD strategy finds a component of a solution shared between several cases, it moves this shared component into a separate case, leaving a marker in each of the cases from which it was removed. Later, the CBR process can manipulate the shared component independently of the rest of the components of the case. Normally the Drop Vertices reuse strategy can only remove vertices at the ends of a path, not within the path. However, after extraction, the middle of the case is "exposed" as its own case, and is therefore available to the Drop Vertices strategy.

Similarly, any retention strategy that removes a component creates a gap which could be filled by a different component which may be useful for further adaptation. Typically, the benefit of this effect is small, but as shown in Fig. 5, taken from a single example, the effect can be large in some circumstances. This suggests that it could be worthwhile, when designing FFD strategies, to consider creative destruction opportunities they might provide. However, more study is needed to corroborate these results more generally and understand the characteristics and potential for creative destruction for different domains.

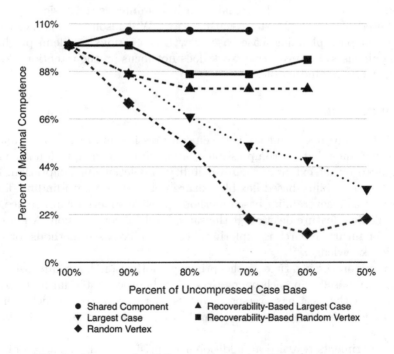

**Fig. 5.** An example of the creative destruction phenomenon

## 5   Related Work

Flexible feature deletion relates to the many approaches in CBR which address the construction of compact competent case bases (a number of which are surveyed in Wilson and Leake [1]), including case retention and forgetting strategies [12–15], diversity-preserving deletion strategies [16], making a trade-off between accuracy and case base size [17], taking into account local complexity in order to consider class boundaries [18], and competence-based deletion of cases [11]. Unlike flexible feature deletion, however, all these methods assume that cases are indivisible.

When CBR maintenance research has considered internal contents of cases, its goal has generally been to improve the quality of the contents (e.g., [19,20]), whereas the goal of FFD is to reduce case size. Some research on case-based abstraction has replaced concrete cases with abstractions [21] which is similar to our more fine-grained substructure abstraction operation. Most similar to FFD is recent work by Abdel-Aziz and Hüllermeier, who consider the removal of parts of cases in the context of maintenance to control case-base size for preference-based CBR [22].

The notion of connecting adaptation more directly to retention can be seen as in the same spirit as Smyth and Keane's adaptation-guided retrieval, which connects adaptation to similarity assessment [23].

Related to the case study domain, path planning is a classic application of case-based reasoning [24–26]. Kruusmaa and Willemson [27] observe that in mobile robot path planning, case base growth is a serious problem, precluding retaining all cases. However, that work does not focus on the retention strategy and considers cases indivisible.

## 6   Conclusion

This paper proposed a symmetry between compression and reuse for flexible feature deletion, such that the compression strategy draws on adaptation knowledge and is guided by the extent of the reversibility of deletions by adaptation. Evaluation of recoverability-based flexible feature deletion in a path-finding domain supported that recoverability-based methods provide superior competence compared to flexible feature deletion at the same levels of compression. An interesting area for future research is exploring recoverability-based methods for richer adaptation knowledge.

If overall processing time is the primary motivation for compression, the recoverability-based approach offers a trade-off between potential reductions in retrieval from the case base versus increases in recovery cost, which could be used to guide maintenance decisions. This is another interesting topic for further study.

The experiments revealed an additional surprising result: That compression by potentially lossy feature deletion strategies can sometimes actually improve case base competence. This creative destruction phenomenon suggests two interesting avenues for research. The first, in the context of case-base compression, is how to prioritize FFD to maximize the chance of creative destruction occurring. The second arises because creative destruction, by improving competence, would be valuable even if compression was not needed. This suggests opportunities for maintenance aimed at improving competence by revising or restructuring cases to make them more amenable to adaptation, given characteristics of the adaptation knowledge of the system.

# References

1. Wilson, D., Leake, D.: Maintaining case-based reasoners: dimensions and directions. Comput. Intell. **17**(2), 196–213 (2001)
2. Leake, D., Smyth, B., Wilson, D., Yang, Q. (eds.): Maintaining Case-Based Reasoning Systems. Blackwell, Oxford (2001). Special issue of Computational Intelligence, 17(2)
3. Smyth, B., McKenna, E.: Building compact competent case-bases. In: Althoff, K.-D., Bergmann, R., Branting, L.K. (eds.) ICCBR 1999. LNCS, vol. 1650, pp. 329–342. Springer, Heidelberg (1999). doi:10.1007/3-540-48508-2_24
4. Leake, D., Schack, B.: Flexible feature deletion: compacting case bases by selectively compressing case contents. In: Hüllermeier, E., Minor, M. (eds.) ICCBR 2015. LNCS (LNAI), vol. 9343, pp. 212–227. Springer, Heidelberg (2015). doi:10.1007/978-3-319-24586-7_15
5. Cordier, A., Lefevre, M., Champin, P.A., Georgeon, O., Mille, A.: Trace-based reasoning - modeling interaction traces for reasoning on experiences. In: Proceedings of the 2014 Florida AI Research Symposium. AAAI Press, pp. 363–368 (2014)
6. Richter, M.: Introduction. In: Lenz, M., Bartsch-Spörl, B., Burkhard, H.D., Wess, S. (eds.) CBR Technology: From Foundations to Applications, pp. 1–15. Springer, Berlin (1998)
7. Schank, R.: Dynamic Memory: A Theory of Learning in Computers and People. Cambridge University Press, Cambridge (1982)
8. Leake, D.: Constructive similarity assessment: Using stored cases to define new situations. In: Proceedings of the Fourteenth Annual Conference of the Cognitive Science Society, Hillsdale, NJ, pp. 313–318. Lawrence Erlbaum (1992)
9. Fuchs, B., Lieber, J., Mille, A., Napoli, A.: Differential adaptation: an operational approach to adaptation for solving numerical problems with CBR. In: Knowledge-Based Systems (2014)
10. Smyth, B., Cunningham, P.: The utility problem analysed. In: Smith, I., Faltings, B. (eds.) EWCBR 1996. LNCS, vol. 1168, pp. 392–399. Springer, Heidelberg (1996). doi:10.1007/BFb0020625
11. Smyth, B., Keane, M.: Remembering to forget: a competence-preserving case deletion policy for case-based reasoning systems. In: Proceedings of the Thirteenth International Joint Conference on Artificial Intelligence, pp. 377–382. Morgan Kaufmann, San Mateo (1995)
12. Muñoz-Avila, H.: A case retention policy based on detrimental retrieval. In: Proceedings of ICCBR-1999 (1999)
13. Ontañón, S., Plaza, E.: Collaborative case retention strategies for CBR agents. In: Ashley, K.D., Bridge, D.G. (eds.) ICCBR 2003. LNCS (LNAI), vol. 2689, pp. 392–406. Springer, Heidelberg (2003). doi:10.1007/3-540-45006-8_31
14. Romdhane, H., Lamontagne, L.: Forgetting reinforced cases. In: 9th European Conference on Advances in Case-Based Reasoning, ECCBR 2008, Trier, Germany, pp. 474–486, 1–4 September 2008

15. Salamó, M., López-Sánchez, M.: Adaptive case-based reasoning using retention and forgetting strategies. Know. Based Syst. **24**(2), 230–247 (2011)
16. Lieber, J.: A criterion of comparison between two case bases. In: Haton, J.-P., Keane, M., Manago, M. (eds.) EWCBR 1994. LNCS, vol. 984, pp. 87–100. Springer, Heidelberg (1995). doi:10.1007/3-540-60364-6_29
17. Lupiani, E., Craw, S., Massie, S., Juarez, J.M., Palma, J.T.: A multi-objective evolutionary algorithm fitness function for case-base maintenance. In: Delany, S.J., Ontañón, S. (eds.) ICCBR 2013. LNCS (LNAI), vol. 7969, pp. 218–232. Springer, Heidelberg (2013). doi:10.1007/978-3-642-39056-2_16
18. Craw, S., Massie, S., Wiratunga, N.: Informed case base maintenance: a complexity profiling approach. In: Proceedings of the Twenty-Second National Conference on Artificial Intelligence, pp. 1618–1621. AAAI Press (2007)
19. Racine, K., Yang, Q.: Maintaining unstructured case bases. In: Leake, D.B., Plaza, E. (eds.) ICCBR 1997. LNCS, vol. 1266, pp. 553–564. Springer, Heidelberg (1997). doi:10.1007/3-540-63233-6_524
20. Salamó, M., López-Sánchez, M.: Rough set based approaches to feature selection for case-based reasoning classifiers. Pattern Recogn. Lett. **32**, 280–292 (2011)
21. Bergmann, R., Wilke, W.: On the role of abstraction in case-based reasoning. In: Smith, I., Faltings, B. (eds.) EWCBR 1996. LNCS, vol. 1168, pp. 28–43. Springer, Heidelberg (1996). doi:10.1007/BFb0020600
22. Abdel-Aziz, A., Hüllermeier, E.: Case base maintenance in preference-based CBR. In: Hüllermeier, E., Minor, M. (eds.) ICCBR 2015. LNCS (LNAI), vol. 9343, pp. 1–14. Springer, Heidelberg (2015). doi:10.1007/978-3-319-24586-7_1
23. Smyth, B., Keane, M.: Adaptation-guided retrieval: questioning the similarity assumption in reasoning. Artif. Intell. **102**(2), 249–293 (1998)
24. Haigh, K.Z., Veloso, M.: Route planning by analogy. In: Veloso, M., Aamodt, A. (eds.) ICCBR 1995. LNCS, vol. 1010, pp. 169–180. Springer, Heidelberg (1995). doi:10.1007/3-540-60598-3_16
25. Anwar, M.A., Yoshida, T.: Integrating OO road network database, cases and knowledge for route finding. In: Proceedings of the 2001 ACM Symposium on Applied Computing, pp. 215–219 ACM (2001)
26. Goel, A., Ali, K., de Silva Garza, A.G.: Computational tradeoffs in experience-based reasoning. In: Proceedings of the AAAI-94 workshop on Case-Based Reasoning, Seattle, WA, pp. 55–61(1994)
27. Kruusmaa, M., Willemson, J.: Covering the path space: a casebase analysis for mobile robot path planning. Knowl. Based Syst. **16**(5), 235–242 (2003)

# Applicability of Case-Based Reasoning for Selection of Cyanide-Free Gold Leaching Methods

Maria Leikola[1(✉)], Lotta Rintala[1], Christian Sauer[2],
Thomas Roth-Berghofer[2], and Mari Lundström[1]

[1] School of Chemical Technology, Aalto University, Helsinki, Finland
{maria.leikola,lotta.rintala,mari.lundstrom}@aalto.fi
[2] School of Computing and Technology,
University of West London, London, UK
{chriatian.sauer,thomas.roth-berghofer}@uwl.ac.uk

**Abstract.** Designing hydrometallurgical experimental work, not to mention entire processes, is a complex task involving various ore properties and their combined effects on the available treatment methods. Gold leaching is one hydrometallurgical process, cyanide being the predominantly utilized leaching agent since late 1800s. Case-based reasoning (CBR) has previously been applied for selecting established process chains for a given gold ore, but with this paper, we are taking this previous research of gold processing towards cyanide-free leaching methods that are currently in development stage and not yet industrially applied. The utilization of CBR for cyanide-fee gold extraction experiment design is tested by building a preliminary CBR knowledge model to recommend treatments for gold extraction. Publications on cyanide-free leaching were analyzed and metallurgical researchers were interviewed in order to define the necessary attributes and their value ranges to be included in the model. We report the challenges encountered while building the CBR knowledge model, discuss its functionality and make suggestions for future research on the topic.

**Keywords:** Case-based reasoning · Gold leaching · Cyanide-free gold extraction

## 1 Introduction

When a new ore deposit is found, selecting a processing method for that particular ore depends on various characteristics of the deposit. Experts, such as geologists and metallurgists, analyze the whole ore body with great detail and make the process decisions based on the mineral type, deposit size, gold content, gold grain size, impurity content and many other attributes. Two ore deposits are never identical, so the process always needs to be tailored to serve the deposit in question. This makes the process selection and plant design a complicated task for professionals, who are trying to maximize the financial profitability of the plant project.

The utilization of Case-based reasoning (CBR) as a selection tool for gold processing has been investigated previously [30, 32], but the focus has been on established

© Springer International Publishing AG 2016
A. Goel et al. (Eds.): ICCBR 2016, LNAI 9969, pp. 249–264, 2016.
DOI: 10.1007/978-3-319-47096-2_17

technologies that use cyanide as a leaching agent. Cyanide, however, is highly toxic and due to recent environmental catastrophes [36] there is an increasing pressure for finding environmentally sustainable ways for gold extraction. Alternative methods are being developed [19], but only one, thiosulfate, is industrially utilized [7]. Furthermore, there is not just one universally valid cyanide-free leaching method applicable for all raw material types. At the moment, there is research information available on the new-coming processes, but due to the novel nature of cyanide-free leaching, many published articles lack an industrial point of view, focusing more on chemical phenomenon.

When a new gold leaching experiment series is designed for a given raw material, the researchers need to gather information about previous results acquired for similar materials or with similar solutions of interest. This can be a time consuming task due to the variety of material options and possible similarities between raw materials. If the designer of an experimental series of treatments could formalize the literature in a way that experiments done on similar raw materials could be found easily and their results compared, the efficiency of experiment design and even the speed of the process design project could be improved. For example, a researcher that has access to a waste product (tailings) from a gold mine that still contains gold, could conduct chemical and mineralogical analysis for the material and compare it to previous materials that have been experimented on by using a case base of examined treatments. If the expert finds out that a certain ore, with similar qualities to the raw material of interest, has been successfully leached with thiosulfate, the experiments could be designed for thiosulfate instead of experimenting on a variety of different leaching methods first.

The purpose of this article is to consider, if CBR is a potential tool for comparing previous cyanide-free leaching results on gold containing raw material in order to improve experiment design time and effort. We will investigate this concept by constructing a preliminary CBR knowledge model and then assessing the challenges related to its construction and the quality and functionality of the model. Finally, we will draw our conclusions and make suggestions for future research on this topic.

## 2 Background

### 2.1 Case-Based Reasoning for Metallurgical Process Selection

The methodology of CBR is based on a four-step process; Retrieve, Reuse, Revise and Retain that are applied upon a case-base of previous cases that consist of problem-solution pairs. A new problem is entered into the system as a query-case and similar problems are retrieved from the case base. Then the solution to the most similar case retrieved is examined and reused in order to solve the query-case problem. Evaluation of the solution usually leads to revision of the solution to serve the query-case as efficiently as possible. The revision phase is adapting any non-matching aspects of the retrieved solution to the query-case problem. Thus, eventually, if no full match was retrieved, a new problem-solution pair emerges and it can be added to the case base in the retain step [1].

CBR tools can be developed for several industrial applications, for example, malfunction diagnosis of complex equipment [40]. Decision rules and fuzzy sets have been studied for selecting gold ore treatments and calculating related cost estimates [34, 35]. However, CBR has been shown to be more suitable for metallurgical process selection due to, for example, its flexibility regarding incomplete input data [31, 32].

The previous work of Rintala et al. [30–32] concentrated on developing a process selection tool for established gold processing methods. By using ore properties such as gold content and information on other materials present in the ore, the user would receive suggestions of process chains that have been utilized for processing similar gold ores. This could aid mineralogical and metallurgical experts, who are designing a process for an ore body of industrial interest. After experiments on the new ore, based on the suggestions by the CBR tool, a new process could be developed and added into the case base. Rintala et al. performed knowledge formalization and evaluation by using myCBR 3.0[1] and the same open source software will be utilized in this work. The system constructed by them was able to retrieve similar ores with similar processing methods and they concluded that the system was suitable for modelling knowledge of established hydrometallurgical processes for gold extraction.

Where Rintala et al. [30–32] developed a CBR tool for selecting whole process chains for gold processing, from ore to final product utilizing cyanide chemistry, this project aims for a tool comparing an individual treatment i.e. gold leaching stage.

## 2.2 Cyanide-Free Gold Leaching Methods

There are several cyanide-free gold leaching methods under development. The new-coming methods are based on, for example, chloride [22], thiosulfate [21], thiourea [8], or bromine-bromide [37] as the cyanide replacing reagents. These methods are considered as challenging regarding profitability and the stage of technological development. However, great advances could be made, if an efficient way of recycling the reagent could be generated because often the reagent costs are one of the notable disadvantages of these new processes [19]. Stabilizing the processes and achieving a reliable performance level are also under development as some of the leaching processes are complex and difficult to control [17].

We have chosen two leaching methods for the construction of our preliminary model. Thiosulfate and chloride were selected, because the former is already industrially established and the latter has historically been the predominant process in the form of gold chlorination, before the appearance of cyanidation in 1888 [23]. Therefore, these two methods were seen as the most promising for further development and eventually having potential in becoming the new norm in the gold processing industry.

---

[1] http://mycbr-project.net/.

## 3    Attribute Determination

The selection of attributes to be formalized into the case base for similarity calculations is essential in building a functional knowledge model. The selection was performed in two parts using two different methods. First, we analyzed scientific review-articles about cyanide-free gold leaching methods. Second, we interviewed hydrometallurgical researchers on their opinion about important attributes for a system comparing knowledge on cyanide-free gold leaching.

### 3.1    Review Article Analysis

In the field of hydrometallurgy, treatment review articles are an important part of the scientific writing scheme as they often compile recent developments in a particular area and compare results acquired by other researchers. In other words, they gather knowledge from various sources and formalize it for the convenience and benefit of the reader. Therefore, gold treatment review-articles can be seen to function partly in the same manner as the knowledge model that is the objective of this research. We saw this as a way to extract the same type of knowledge from treatment review-articles as was elicited from researchers through interviews, as discussed in the next section.

Three review articles from three different peer-reviewed scientific journals were selected for thorough analysis from the 'what was seen as important information' – perspective. The selected articles along with their respective authors, journals and years of publishing are presented in Table 1.

**Table 1.**  Selected three review articles of cyanide-free gold leaching methods [17, 19, 33]

| Article | Authors | Journal | Year |
|---|---|---|---|
| Gold leaching in non-cyanide lixiviant systems: critical issues on fundamentals and applications | Senanayake, G. | Minerals Engineering | 2004 |
| Alternatives to cyanide in the gold mining industry: what prospects for the future? | Hilson, G. Monhemius, A. J. | Journal of Cleaner Production | 2006 |
| Non-cyanide Leaching Processes in Gold Hydrometallurgy and Iodine-Iodide Applications: A Review | Konyratbekova, S. Baikonurova, A. Akcil, A. | Mineral Processing and Extractive Metallurgy Review | 2015 |

The articles were analyzed systematically for information patterns that might occur between them. Six questions were taken into consideration in order to define the focus of the papers and the similarity of their way of discussing the topic of cyanide-free gold leaching methods. This information could then be used in a similar manner as the results from the researcher interviews – as a base for defining the objectives for the preliminary knowledge model. The questions were as follows:

- What cyanide-free gold leaching methods are mentioned in the article?
- What leaching methods are discussed further than a single mention? For example, the reaction mechanism is discussed.
- What method attributes, such as temperature, are mentioned in the article?
- What method attributes are discussed further? For example, an exact value is mentioned or the effect of the attribute on the leaching process is specified.
- What ore attributes are mentioned in the article?
- What ore attributes are discussed further?

The results of the analysis were tabulated for effective comparison. The amount of methods, method attributes and ore attributes mentioned or discussed further in at least one or all of the review articles is presented in Table 2.

**Table 2.** Quantitative results of the cyanide-free gold leaching review article analysis

|  | Mentioned in at least one article | Mentioned in all three articles | Discussed further in at least one article | Discussed further in all three articles |
|---|---|---|---|---|
| Methods | 27 | 7 | 11 | 6 |
| Method attributes | 29 | 20 | 25 | 10 |
| Ore attributes | 13 | 7 | 10 | 4 |

The concepts that were discussed further in all three review articles were concluded as being the most auspicious and influential due to the fact that they were seen as of high importance by all the review article authors. The 6 leaching methods, 10 method attributes and 4 ore attributes discussed further in the articles are presented in Table 3.

The leaching methods we chose earlier for the preliminary knowledge model were thiosulfate and chloride leaching. Both are discussed further in all review articles and this was interpreted as a confirmation of our selection. Other methods that were discussed further in all articles were bromine-bromide, iodine-iodide, thiocyanite and

**Table 3.** Methods and attributes discussed further in all three review articles

| Methods | Method attributes | Ore attributes |
|---|---|---|
| Bromine-bromide | Gold solubility | Gold content |
| Chloride | Gold extraction % | Ore type |
| Iodine-iodide | Leaching selectivity | Copper present |
| Thiocyanate | Reaction rate | Sulfur present |
| Thiosulfate | pH | |
| Thiourea | Oxidant concentr. | |
| | Ligand concentr. | |
| | Redox potential | |
| | Retention time | |
| | Additives | |

thiourea. The incorporation of these leaching methods for the second version of our knowledge model should be taken into consideration.

The method attributes concerning precise process conditions are not included in our preliminary knowledge model, but they could be included in a second version, if the user has a reason to predetermine some process conditions of interest. The ore attributes will form the basis of the query function in our model as the objective is to compare similar raw materials. Gold content, ore type and the presence of copper and sulfur were the four ore attributes discussed in all articles. Gold content is straightforward to transfer into a case attribute expressed as gold content in grams per ton of ore. Ore type refers to the mineralogical composition of the ore as do copper present and sulfur present attributes. Copper and sulfur are present in some minerals and if the minerals that are present in the ore are known, then the presence of copper and sulfur along with the ore type are known. Therefore, ore type was seen as an obsolete attribute as the minerals present and gold content would define the ore type. Furthermore, also copper and sulfur presence is known if the governing mineral types are stated. The interest towards mineralogical attributes is increased by the fact that ore mineralogy highly determines the feasibility of industrial gold extraction processes. It was noted during the manual case acquisition that generally ores contain one or two predominant minerals while the other minerals were of lower importance. Therefore, modelling two major minerals present in the raw material was seen as sufficient for the preliminary knowledge model.

Next to the manual analysis of the treatment review-articles we also investigated the possibility to use natural language extraction techniques to semi-automatically extract cases of treatments from the existing literature. We did so as the abstracts of the articles indicated an almost consistent format and thus are offering the opportunity to perform case extraction from a semi-structured text format. We initially built an ANNIE (A nearly new information extraction) application within the GATE[2] natural language processing architecture. Based on frequent term analysis we identified categories of terms like: "amount", "ore type", "parameters", "results", "substances" and "treatments". Initial experiments, see Fig. 1, in annotating such terms in abstracts from treatment articles show a promising structure in these abstracts. However, further refinement of the extraction application is still required.

### 3.2 Interviews of Researchers

Interviewing techniques can be divided into three categories based on the predefined questions and their control over the course of the interview; structured i.e. a form, semi-structured or theme interview and unstructured or open interview [28]. In this study we used a semi-structured interviewing technique where the questions were predefined, asked in the same order and no additional questions were asked. The questions were open questions [5], that the interviewee could answer with their own

---

[2] https://gate.ac.uk/.

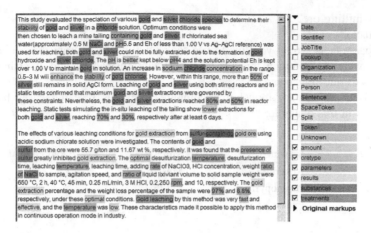

**Fig. 1.** Initial annotation of relevant term types in treatment article abstracts [6, 25]

words. The answers were analyzed for themes, concepts and ideas while constructing a summary of all the interviewing results.

The sample size, i.e. the number of interviewed researchers, can be determined based on the concept of data saturation, which is widely utilized in health sciences to confirm content validity [12]. Data saturation means conducting new interviews until no new ideas or concepts emerge from increasing the number of interviews [11]. One proposed method is a minimum of 10 interviews after which interviews are conducted until 3 consecutive interviews do not introduce any new information [10]. We could not adopt this method directly due to the fact that the availability of researchers for personal interviewing was limited. Therefore, the interviewing sample size could not be defined based on a stopping criterion, but the data saturation could be estimated.

We chose to interview both junior and senior level researchers that were specialized in hydrometallurgy. We defined "junior" as having conducted hydrometallurgical research for under two years and "senior" as at least ten years. A preliminary questionnaire was drafted aiming to find out what attributes the interviewees found most important and useful. The questions also included the question "Did we ask the right questions?" in order to gather feedback for developing the questionnaire itself. After the first round, we revised the original questions based on our observations during the interview and the received feedback.

The first interview was conducted with one junior level researcher. The questions were then revised, for example, the first interviewing round did not produce any suggestions of ore attributes when the essential purpose of the model should be to find similar materials. Therefore, we decided to include three ore attributes and ask the interviewees what attributes they would add to better describe the ore. The questions for the second round were as follows:

- Imagine you were designing an experiment series for chloride leaching. If all information from previous research articles was thoroughly organized, what knowledge and parameters would you compare for

- free-milling ore?
- refractory ore?
- What parameters would you like to use for excluding cases from the comparison?
- If you were designing a thiosulfate experiment instead, would it change your answers to question 1 and 2?
- There is a preliminary model that compares previous research cases based on attributes in this example:

| Method | Mineral 1 | Mineral 2 | Gold content |
|---|---|---|---|
| Chloride | Ankerite | Muscovite | 1.5 |
| Thiosulfate | Arsenopyrite | Pyrite | 56 |
| Thiosulfate | Pyrite | | 94.63 |

- What attributes would you add to the list, in order to better describe the ore?

These questions were presented to another junior level researcher and the third interview round included both a junior and a senior level researcher. There appeared no reason to revise the questions after the first round, so the questions stated previously were used throughout the rest of the interviews. The fourth round was conducted with three senior level researchers and the first junior level researchers that participated in round 1, resulting in three junior and four senior level interviewees combined.

Attributes mentioned by the interviewees were assigned an importance rating equal to the amount of interviewees who mentioned that particular attribute. There were only one method and one ore attribute (extraction rate and refractoriness) that were mentioned by five interviewees, as shown in Table 4. Naturally, these attributes were considered more important or useful than the attributes that were mentioned only once (3 method and 11 ore attributes). Table 4 presents the amount of method and ore attributes in each importance rating class. It needs to be noted that three ore attributes were mentioned in the final version of the questionnaire and therefore they were not mentioned by the interviewees.

**Table 4.** Amount of method and ore attributes mentioned by interviewees in each importance rating class. Importance class (1–5) referring to the amount of interviewees that mentioned the attribute.

| Importance rating class | Amount of method attributes | Amount of ore attributes |
|---|---|---|
| 5 | 1 | 1 |
| 4 | 3 | 1 |
| 3 | 6 | 2 |
| 2 | 6 | 3 |
| 1 | 3 | 11 |
| Combined amount of attributes | 19 | 18 |

Data saturation was not reached, because the last interviewee still mentioned attributes that had not been mentioned before. None of the attributes were mentioned by six or seven interviewees and therefore the maximum importance rating achieved (attributes extraction rate and refractoriness) was five.

Extraction rate indicates how quickly gold is dissolved into the solution. However, extraction rate is seldom stated outright in scientific articles, but the common practice is to report the overall extraction as percentage of gold (importance rating of 4). Therefore, it was decided that extraction percentage would be included in the model instead of extraction rate and be referred to as simply *Extraction*.

By definition, refractoriness describes the ore resistance towards cyanide leaching due to phenomenon such as (i) gold being locked inside insoluble minerals e.g. pyrite, (ii) mineral containing organic carbon causing gold back-precipitation (preg-robbing), (iii) mineral containing high cyanide consuming elements making gold dissolution unfeasible or (iv) any combination thereof. However, in practice the term refractoriness is most often related to explanation (i). These kind of minerals (e.g. pyrite) are often considered refractory without emphasizing the term. This inconsistency in scientific writing makes it challenging to use the attribute refractoriness in knowledge modelling. For this reason, we decided that the mineralogical attributes would suffice in implying the refractoriness level of the ore.

## 4   Preliminary Knowledge Model

We utilized myCBR Workbench, an open-source retrieval tool for knowledge formalization, in building our preliminary knowledge model. This section discusses the construction phases of the model from attribute formalization to acquiring cases and defining local similarities for the found attribute values.

### 4.1   Selected Attributes

We chose five attributes to be compared for each case, based on the treatment review article analysis and researcher interviews. This was seen being the minimal amount of attributes necessary in order to build a functional model. The attributes and their respective types are presented in Table 5.

**Table 5.**   Selected attributes for the preliminary knowledge model

| Attribute | Type of attribute |
|---|---|
| Method | Symbol |
| Mineral 1 | Symbol |
| Mineral 2 | Symbol |
| Gold content [g/t] | Floating point number |
| Extraction [%] | Floating point number |

*Method* implies the leaching method; thiosulfate vs. chloride. *Mineral 1* and *Mineral 2* are the two minerals most abundant in the ore, excluding barren quartz, because practically all ores contain quartz [23]. If the gold was reported occurring within the quartz minerals, then it would be included. *Gold content* expresses the amount of gold in grams per ton of ore. *Extraction* is a term used to describe the percentage of gold that has been successfully leached. These attributes were seen as vital for comparing cases at a sufficient level of scientific relevance.

## 4.2   Composing the Case Base

After the attributes were selected, we collected cases and formalized the information within them. All cases were extracted form peer-reviewed scientific articles published by respected journals in the field of hydrometallurgy. Some articles contained more than one test series, producing several cases. Altogether 24 cases were extracted from 20 articles [2–4, 6, 9, 13–16, 18, 20, 21, 24–27, 29, 38, 39, 41]. Chloride was the leaching agent used in 10 cases and thiosulfate was used in 14 of them. Most of the articles, 13, defined two main minerals, but 11 articles did not specify more than one dominant mineral. Combined, 11 different minerals and 2 rock species (combinations of several minerals) were mentioned in the articles.

The cases were named based on the method and the gold grade. Thiosulfate and chloride leaching were represented with $t$ and $c$ respectively. For example, if the leaching experiment was conducted with thiosulfate and the gold grade of the material was 1.5, the name of the case was X_$t$_1.5, X representing a running number.

## 4.3   Local Similarity Values

Local similarities were assigned between different values within the attributes. *Method* had only two possible values, "*Thiosulfate*" or "*Chloride*". These methods cannot be technically combined opposite to e.g. chloride and bromide methods that can be used in combined leaching. Therefore, the similarity between "*Thiosulfate*" and "*Chloride*" was set at 0.

The data type for the attributes *Gold content* and *Extraction* was chosen as floating point number and the values within the case base varied from 1.5 to 94.6 g/t and from 42.7 to 100 % respectively. A linear distance function was chosen to determine the local similarity, which in practice converts the difference between compared values into percentage of the entire range and subtracts it from full similarity of 100 %.

13 different minerals or rock species were present in the case base. Similarities between different minerals were defined based on rules concerning the elements within the minerals. The elements were divided into non-metallic (e.g. sulfur and oxygen) and metallic elements (e.g. iron and copper). The rules were constructed as follows:

- If all the same elements are present, but in different ratios, the similarity is 0.9.
- If the minerals share both non-metallic and metallic elements, but one or both include element(s) the other one does not, the similarity is 0.8.

- If the minerals share a nonmetallic element, the similarity is 0.6. This is justified by the fact that most minerals can be classified, for example, as sulfidic (containing sulfur) or oxidic (containing oxygen), defining largely its leaching behavior.
- If the minerals only share a metallic element the similarity is 0.3. This is justified by the fact that metallic elements do not usually affect the leaching behavior as significantly as the pre-mentioned non-metallic fraction.
- If the minerals share no elements, the similarity is 0.0.

Two rock species, andesite and rhyolite, were mentioned in one of the cases with no specification of the mineral ratios they encompassed. However, both usually share the minerals plagioclase, hornblende, and biotite and include from two to four other minerals. Therefore, the similarity between rhyolite and andesite was determined as 0.7, because they share several of the minerals they are composed of. Their similarities between rock species and single minerals was set at 0.

It is obvious that the similarity definitions are and cannot be perfect due to the unique characteristics of each ore body and changes in the chemical activity, even within the same mineral class. For example, the nature of gold containing pyrite ($FeS_2$) minerals varies, some being much easier to leach compared to others. However, the values listed above can give a good indication of the similarities of the *Mineral 1* and *Mineral 2* attribute values that are not accurate, but rather fuzzy in nature.

Initially, the global similarity measure chosen for the comparison of entire problem descriptions is a weighted sum with a (yet) even weight distribution between all attributes. The similarity of two problem descriptions is defined as a value in the interval [0,1], with 0 being non-similar at all and 1 being entirely similar.

# 5 Model Validation and Results

After the preliminary model had been constructed its functionality was assessed through test queries. This section discusses the configuration of the test queries and then the results and their analysis.

## 5.1 Test Queries

The preliminary knowledge model was validated by conducting test queries. The ores that were used for queries were selected from current and past industrial mining projects that utilize cyanide. Furthermore, the focus was on ores that are refractory i.e. difficult to treat by conventional methods, enabling more economic benefits if an alternative cyanide-free gold leaching method could be found.

Table 6 shows the ore attributes of the used query ores. In principal, the optimum *Extraction* is 100 %, though full extraction is not realistic. The value of this attribute was set at 100 % favoring results having higher *Extraction* values. The mining projects utilizing the test ores are called Joutel, Canada (T1), Flimston Gidji, Australia (T2), Giant Yellowknife, Canada (T3) and Grasberg-Ertsberg, Indonesia (T4) [23].

After testing with only the ore attributes in Table 6, the leaching methods chloride and thiosulfate were added to the queries individually, resulting in 12 test queries in

**Table 6.** Ore attributes of selected query cases [23]

| Test case | T1 | T2 | T3 | T4 |
|---|---|---|---|---|
| Gold content | 5.8 | 3.8 | 7.0 | 1.5 |
| Mineral 1 | Pyrite | Pyrite | Arsenopyrite | Chalcopyrite |
| Mineral 2 | Arsenopyrite | | Pyrite | |

**Table 7.** Results of test queries

| Query case | 1st case | Sim | 2nd case | Sim | 3rd case | Sim |
|---|---|---|---|---|---|---|
| T1 | 16_t_32 | 0.88 | 2_t_4.3 | 0.75 | 9_c_5.17 | 0.72 |
| T1 thiosulfate | 16_t_32 | 0.91 | 2_t_4.3 | 0.80 | 11_t_45.5 | 0.66 |
| T1 chloride | 9_c_5.17 | 0.77 | 8_c_27 | 0.75 | 15_c_20.45 | 0.72 |
| T2 | 2_t_4.3 | 1.00 | 9_c_5.17 | 0.97 | 8_c_27 | 0.94 |
| T2 thiosulfate | 2_t_4.3 | 1.00 | 5_t_94.63 | 0.79 | 1_t_33.89 | 0.79 |
| T2 chloride | 9_c_5.17 | 0.97 | 8_c_27 | 0.95 | 15_c_20.45 | 0.92 |
| T3 | 19_t_56 | 0.84 | 24_c_55.7 | 0.61 | 7_t_46 | 0.53 |
| T3 thiosulfate | 19_t_56 | 0.87 | 7_t_46 | 0.62 | 4_t_90 | 0.60 |
| T3 chloride | 24_c_55.7 | 0.68 | 19_t_56 | 0.67 | 9_c_5.17 | 0.57 |
| T4 | 3_c_11 | 0.93 | 2_t_4.3 | 0.74 | 18_c_11 | 0.72 |
| T4 thiosulfate | 2_t_4.3 | 0.79 | 20_t_1.646 | 0.76 | 3_c_11 | 0.74 |
| T4 chloride | 24_c_55.7 | 0.68 | 19_t_56 | 0.67 | 9_c_5.17 | 0.57 |

total. The three most similar cases of these test queries are shown in Table 7. Best similarity values for T2 and T2 Thiosulfate are rounded instead of exact matches.

## 5.2    Result Analysis

**Mineral Data Modelling.** Some shortcomings were noted in the way the mineralogy is modelled. Firstly, in some cases, there is only one major mineral and in these cases the value for *Mineral 2* is "*unknown*". The operational principle of myCBR leads to a situation where the "*unknown*" value of *Mineral 2* produces a local similarity of 1 with cases that also have only one dominant mineral. The ores having only one major mineral is not enough to justify such an increase in the global similarity value. Secondly, it was noted that *Mineral 1* and *Mineral 2* should be cross compared. Currently, if the queried value for *Mineral 1* is "*pyrite*" and for *Mineral 2* "*arsenopyrite*" and a potential case has the same minerals, but in reverse order, the similarity calculations will reflect them having different values in both attributes. However, the model should reflect the total similarity, which in this case is almost identical with respect to the behavior of the material in leaching processes.

Furthermore, it was evident that the case base did not include all the possible minerals. In addition, it seems that it is not reasonable to try to model all hundreds of minerals and their similarities in order to improve the model further. The next target in the research is to conduct a literature survey and interview senior professionals to define minerals, mineral groups and mineral classes that are known to have chemically similar leaching behavior and/or are of the highest importance during gold leaching.

**Leaching Method Modelling.** When the leaching methods Thiosulfate and Chloride were included in the queries, the most similar cases did not consist of solely the queried method. In all cases, the most similar case was shown to correspond the *Leaching method* in query. However, already the second case was differed from the query's *Leaching method* in tests "T3 chloride" and "T4 chloride". Therefore, it was concluded that the weight factor of the attribute *Leaching method* should be increased to a point where it becomes clearly dominant over the other attributes.

**Model Performance.** myCBR calculated the similarities according to the given local similarity values as planned. However, the shortcomings of the knowledge modelling of similarity measures of minerals were seen as significant enough to conclude that the model needs to be reconfigured. Especially the fact that the mineral attributes are not cross compared with each other, renders the model insufficient at comparing ores, concentrates and other materials based on natural minerals.

# 6 Conclusions and Future Work

The constructed knowledge model built in this work for cyanide-free gold leaching was shown to work in principle. However, it is evident that the knowledge modelling of minerals needs to be reconfigured. A different type of attribute set-up needs to be developed and possibly the calculation method needs to be reconsidered. There needs to be numerous material attributes and they all need to be compared with each other in a way that best describes the overall behavior of the material in a metallurgical process. There are several issues that need to be considered when refining the mineral knowledge modelling:

- mineral attributes are parallel, all affecting each other
- the amount of minerals present in the ores vary
- the amount of minerals available for querying cannot be restricted by the minerals present in the case base
- two minerals that behave similarly in one leaching process might act differently from each other in another leaching process.

The incorporation of the mentioned issues into the knowledge model will be challenging due to certain characteristics of the myCBR software. For example, at the moment myCBR compares attribute values with values of the same attribute and not values of other attributes. However, this cross comparing functionality would be essential for comparing raw materials with each other.

In addition to the reconfiguration of the mineral knowledge modelling, the case base needs to be expanded significantly and also incorporate other leaching methods. This way the model will serve the needs of researchers better. More attributes need to be modelled, especially method attributes such as pressure, temperature and solid-liquid ratio. These will enable researchers to target, for example, process conditions that are industrially feasible. Future work will aim to provide a knowledge model serving both academic researchers and industrial experts in designing experiments and advancing the scientific knowledge on environmentally safe gold processes.

# References

1. Aamodt, A., Plaza, E.: Case-based reasoning: foundational issues, methodological variations, and system approaches. AI Commun. **7**(1), 39–59 (1994)
2. Abbruzzese, C., Fornari, P., Massidda, R., Vegliò, F., Ubaldini, S.: Thiosulphate leaching for gold hydrometallurgy. Hydrometallurgy **39**, 265–276 (1995)
3. Aylmore, M.G.: Treatment of a refractory gold-copper sulfide concentrate by copper ammoniacal thiosulfate leaching. Miner. Eng. **14**(6), 615–637 (2001)
4. Baghalha, M.: Leaching of an oxide gold ore with chloride/hypochlorite solutions. Int. J. Miner. Process. **82**, 178–186 (2007)
5. Burgess, T.F.: Guide to the design of questionnaires: a general introduction to the design of questionnaires for survey research. University of Leeds (2001)
6. Cheng, Y., Shen, S., Zhang, J., Chen, S., Xiong, L., Liu, J.: Fast and effective gold leaching from a desulfurized gold ore using acidic sodium chlorate at low temperature. Ind. Eng. Chem. Res. **52**, 16622–16629 (2013)
7. Choi, Y., Baron, J.Y., Wang, Q., Langhans, J., Kondos, P.: Thiosulfate processing – from lab curiosity to commercial application. In: Proceedings of the World Gold 2013, pp. 45–50. The Australian Institute of Mining and Metallurgy, Melbourne, Australia (2013)
8. Deschênes, G., Ghali, E.: Leaching of gold from a chalcopyrite concentrate by thiourea. Hydrometallurgy **20**(2), 179–202 (1988)
9. Feng, D., van Deventer, J.S.J.: The role of oxygen in thiosulphate leaching of gold. Hydrometallurgy **85**, 193–202 (2007)
10. Francis, J.J., Johnston, M., Robertson, C., Glidewell, L., Entwistle, V., Eccles, M.P., Grimshaw, J.M.: What is adequate sample size? Operationalising data saturation for theory-based interview studies. Psychol. Health **25**(10), 1229–1245 (2010)
11. Glaser, B.G., Strauss, A.L.: The Discovery of Grounded Theory: Strategies for Qualitative Research. Aldine Transaction, New Brunswick (1967)
12. Guest, G., Bunce, A., Johnson, L.: How many interviews are enough? An experiment with data saturation and variability. Field Methods **18**(1), 59–82 (2006)
13. Hasab, M.G., Rashchi, F., Raygan, S.: Simultaneous sulfide oxidation and gold leaching of a refractory gold concentrate by chloride-hypochlorite solution. Miner. Eng. **50–51**, 140–142 (2013)
14. Hasab, M.G., Rashchi, F., Raygan, S.: Chloride-hypochlorite leaching and hydrochloric acid washing in multi-stages for extraction of gold from a refractory concentrate. Hydrometallurgy **142**, 56–59 (2014)
15. Hasab, M.G., Raygan, S., Rashchi, F.: Chloride-hypochlorite leaching of gold from a mechanically activated refractory sulfide concentrate. Hydrometallurgy **138**, 59–64 (2013)
16. Hashemzadehfini, M., Ficoriová, J., Abkhoshk, E., Shahraki, B.K.: Effect of mechanical activation on thiosulfate leaching of gold from complex sulfide concentrate. Trans. Nonferrous Met. Soc. China **21**, 2744–2751 (2011)
17. Hilson, G., Monhemius, A.J.: Alternatives to cyanide in the gold mining industry: what prospects for the future? J. Clean. Prod. **14**, 1158–1167 (2006)

18. Kononova, O.N., Kholmogorov, A.G., Kononov, Y.S., Pashkov, G.L., Kachin, S.V., Zotova, S.V.: Sorption recovery of gold from thiosulphate solutions after leaching of products of chemical preparation of hard concentrates. Hydrometallurgy 59, 115–123 (2001)
19. Konyratbekova, S.S., Baikonurova, A., Akcil, A.: Non-cyanide leaching processes in gold hydrometallurgy and iodine-iodide applications: a review. Min. Process. Extr. Metall. Rev. 36, 198–212 (2015)
20. Lampinen, M., Laari, A., Turunen, I.: Ammoniacal thiosulfate leaching of pressure oxidized sulfide gold concentrate with low reagent consumption. Hydrometallurgy 151, 1–9 (2015)
21. Langhans Jr., J.W., Lei, K.P.V., Carnahan, T.G.: Copper-catalyzed thiosulfate leaching of low-grade gold ores. Hydrometallurgy 29, 191–203 (1992)
22. Lundström, M., Ahtiainen, R., Haakana, T., O'Callaghan, J.: Techno-economical observations related to Outotec gold chloride processes. In: Proceedings of ALTA 2014 Gold-Precious Metals Sessions, pp. 89–104. ALTA Metallurgical Services Publications, Melbourne (2014)
23. Marsden, J.O., House, C.I.: The Chemistry of Gold Extraction, 2nd edn. Society for Mining, Metallurgy, and Exploration, Colorado (2009)
24. Molleman, E., Dreisinger, D.: The treatment of copper-gold ores by ammonium thiosulfate leaching. Hydrometallurgy 66, 1–21 (2002)
25. Nam, K.S., Jung, B.H., An, J.W., Ha, T.J., Tran, T., Kim, M.J.: Use of chloride-hypochlorite leachants to recover gold from tailing. Int. J. Miner. Process. 86, 131–140 (2008)
26. Navarro, P., Vargas, C., Villarroel, A., Alguacil, F.J.: On the use of ammoniacal/ammonium thiosulphate for gold extraction from a concentrate. Hydrometallurgy 65, 37–42 (2002)
27. Pangum, L.S., Browner, R.E.: Pressure chloride leaching of a refractory gold ore. Miner. Eng. 9(5), 547–556 (1997)
28. Preece, J., Rogers, Y., Sharp, H.: Interaction Design: Beyond Human-Computer Interaction. Wiley annotated edition, New York (2002)
29. Puvvada, G.V.K., Murthy, D.S.R.: Selective precious metals leaching from a chalcopyrite concentrate using chloride/hypochlorite media. Hydrometallurgy 58, 185–191 (2000)
30. Rintala, L.: Development of a process selection method for gold ores using case-based reasoning. Aalto University Publication Series: Doctoral dissertations, Finland (2015)
31. Rintala, L., Lillkung, K., Aromaa, J.: The use of decision and optimization methods in selection of hydrometallurgical unit process alternatives. Physicochem. Probl. Miner. Process. 46(1), 229–242 (2011)
32. Sauer, C.S., Rintala, L., Roth-Berghofer, T.: Two-phased knowledge formalization for hydrometallurgical gold ore process recommendation and validation. Künstliche Intelligenz 28(4), 283–292 (2014)
33. Senanayake, G.: Gold leaching in non-cyanide lixiviant systems: critical issues on fundamentals and applications. Miner. Eng. 17, 785–801 (2004)
34. Torres, V.M., Chaves, A.P., Meech, J.A.: Intelligold – a fuzzy expert system for gold plant process design. In: Proceedings of the 18th International Conference of the North American Fuzzy Information Processing Society, NAFIPS 1999, pp. 899–9014. IEEE (1999)
35. Torres, V.M., Chaves, A.P., Meech, J.A.: Intelligold – an expert system for gold plant process design. Cybern. Syst. 31(5), 591–610 (2000)
36. United Nations Environmental Programme Report: Cyanide Spill at Baia Mare Romania: UNEP/OCHA Assessment Mission, Geneva, p. 56 (2000)
37. Vukcevic, S.: A comparison of alkali and acid methods for the extraction of gold from low grade ores. Miner. Eng. 9(10), 1033–1047 (1996)
38. Xia, C., Yen, W.T.: Improvement of thiosulfate stability in gold leaching. Miner. Metall. Process. 20(2), 68–72 (2003)

39. Xu, B., Yang, Y., Jiang, T., Li, Q., Zhang, X., Wang, D.: Improved thiosulfate leaching of a refractory gold concentrate calcine with additives. Hydrometallurgy **152**, 214–222 (2015)
40. Yousuf, A., Cheetham, W.: Case-based reasoning for turbine trip diagnostics. In: Agudo, B. D., Watson, I. (eds.) ICCBR 2012. LNCS, vol. 7466, pp. 458–468. Springer, Heidelberg (2012)
41. Zipperian, D., Raghavan, S.: Gold and silver extraction by ammoniacal thiosulfate leaching from a rhyolite ore. Hydrometallurgy **19**, 361–375 (1988)

# Competence Guided Casebase Maintenance for Compositional Adaptation Applications

Ditty Mathew$^{(\boxtimes)}$ and Sutanu Chakraborti

Department of Computer Science and Engineering,
Indian Institute of Technology Madras, Chennai 600036, India
{ditty,sutanuc}@cse.iitm.ac.in

**Abstract.** A competence guided casebase maintenance algorithm retains a case in the casebase if it is useful to solve many problems and ensures that the casebase is highly competent in the global sense. In this paper, we address the compositional adaptation process (of which single case adaptation is a special case) during casebase maintenance by proposing a case competence model for which we propose a measure called retention score to estimate the retention quality of a case. We also propose a revised algorithm based on the retention score to estimate the competent subset of the casebase. We used regression datasets to test the effectiveness of the competent subset obtained from the proposed model. We also applied this model in a tutoring application and analyzed the competent subset of concepts in tutoring resources. Empirical results show that the proposed model is effective and overcomes the limitation of footprint based competence model in compositional adaptation applications.

**Keywords:** Casebase maintenance · Case competence · Footprint based competence model · Compositional adaptation

## 1 Introduction

Case Based Reasoning(CBR) systems solve new problems by retrieving similar past problems from a casebase and adapting their solutions. The adaptation process can be done in two ways - single case adaptation and compositional adaptation. In single case adaptation, the solution of a single case can be adapted to solve the target problem whereas in compositional adaptation the solutions from multiple cases are combined to produce a new composite solution [16]. Casebase Maintenance is a branch of CBR, which aims at looking into the quality of cases that should be retained in the casebase; the goal is often to maintain a compressed casebase that can solve new problems effectively [12]. We need to ensure that the cases in the compressed casebase would be able to be retrieved and adapted for a wide range of problems in the casebase. Thus, the competence of a casebase can be determined by the ability of the cases in the casebase to solve a large number of problems. A competence guided casebase maintenance

© Springer International Publishing AG 2016
A. Goel et al. (Eds.): ICCBR 2016, LNAI 9969, pp. 265–280, 2016.
DOI: 10.1007/978-3-319-47096-2_18

algorithm retains a case in the casebase if it is useful to solve many problems and ensures that the casebase is highly competent in the global sense [13]. For this, it is important to mark the cases that are involved in both the single case and compositional adaptation process in the past so that we can use this knowledge to measure coverage.

Footprint-based retrieval [15] is an efficient retrieval approach in CBR, which guides the search procedure using a case competence model [14]. This approach identifies a compact competent subset of the casebase called footprint set, using the case competence model. However, the competence model used in the footprint-based approach covers only the situation where a single case is adapted to solve a problem. It turns out that many CBR applications require compositional adaptation for their adaptation process. In such cases, the dependency between the cases has to be taken into consideration when we estimate the competence of each case in the casebase. To the best of our knowledge, no previous work has attempted to address the maintenance of casebase which requires compositional adaptation. So, we are motivated by the research question, *"How can we model a competence guided casebase maintenance model where the adaptation process involves compositional adaptation?"*

In this paper, we propose a new competence model which can be applied in an application that involves compositional adaptation (of which the single case adaptation is a special case). This model is based on a measure called retention score which estimates the retention quality of a case in the casebase. We also propose a revised approach to identify the footprint set where compositional adaptation is required. Section 2 reviews the literature on case competence model and footprint-based approach in particular. Section 3 summarizes the research in compositional adaptation applications and illustrates the flaw of footprint-based approach when used compositional adaptation applications. Our approach to measure the retention quality and the revised footprint approach are described in Sect. 4 using examples based on synthetic casebases. Section 5 presents the empirical results obtained on synthetically generated datasets. In Sect. 6, we demonstrate the proposed approach in a tutoring application and show the importance of the retention score measure with the support of experimental results.

## 2    Footprint-Based Approach

In Case Based Reasoning, the impact of utility depends on the size and growth of the casebase. Since efficiency (and on occasions effectiveness) is adversely affected in the presence of large number of not-so-useful cases, it is desirable to weed out such cases. Markovitch and Scott [5] have characterized an information filtering approach based on selective utilization and selective retention strategies to deal with the utility problem. This selective utilization and selective retention strategies ensure that stored knowledge is genuinely useful, and the performance will not be affected by the deletion of any information. In [13], Smyth and Keane introduced a case competence model to guide the learning and deletion of cases.

The competence of a CBR system is the range of target problems that the given system can solve. The global competence of a system also relies on the local problem-solving properties such as the coverage and reachability of each case. For the purpose of defining these properties, Smyth and McKenna [15] defined a relation *solves* between a case $c$ and a target problem $t$ as $c$ solves $t$ ($solves(c, t)$) and this relation is defined as in Definition 1.

**Definition 1.** $solves(c, t)$ iff $c$ is retrieved and $c$ can be adapted for $t$

Using this relation the competence properties such as coverage and reachability of each individual case is defined as in the Definitions 2 and 3 respectively.

**Definition 2.** $Coverage(c) = \{c' \in \mathbb{C} : solves(c, c')\}$

**Definition 3.** $Reachability(c) = \{c' \in \mathbb{C} : solves(c', c)\}$

Global competence of a casebase is a function of how the local competences of the cases interact when they are combined. When there is any overlap between the coverage of cases in the casebase, its individual contribution may not contribute globally [14]. The unique competence contribution of an individual case to solve a target problem depends on the presence of alternate solutions for the target problem. Smyth and McKnenna [15] defined a measure called relative coverage based on the idea that if a case $c$ can be solved by $n$ other cases then each of the $n$ cases will get a contribution of $1/n$ from $c$ to their relative coverage measures. Thus, relative coverage provides a mechanism to estimate the contribution of each case to global competence.

$$RelativeCoverage(c) = \sum_{c' \in Coverage(c)} \frac{1}{|Reachability(c')|} \qquad (1)$$

The maintenance strategy of the casebase becomes more and more critical in real-world situations. Competence directed casebase maintenance should delete irrelevant cases that guide the casebase to maximizes its competence [13]. Smyth and McKenna [15] estimated the set of cases that is to be retained using the relative coverage measure where the final set (i.e. footprint set) contains cases with large competence contributions and this set covers the rest of the cases in the casebase. For the construction of footprint set, first the cases are sorted in the descending order of the relative coverage values and then each case is added to the footprint set in this order only if the current footprint does not already cover it. As the cases are sorted based on the relative coverage, the larger competent cases will get added before the smaller competent cases and thus keep the footprint size to a minimum. The retrieval strategy based on this footprint set is not only a simple and novel approach but also it directs the use of competence model to guide the retrieval process. However, the relation *solves* considers only a single case for adaptation while estimating the footprint set.

## 3   Compositional Adaptation

In compositional adaptation, solutions from multiple similar cases are combined to obtain a new solution for a query problem. For example, in a regression setting where the data instances are the cases, the solutions from the k-nearest neighbor cases can be adapted to predict the target value of the corresponding case [10]. In the Airquap CBR system for predicting pollution levels, the solution to the target problem is the mean value of the solutions of the most similar cases [3]. Arshadi et al. [1] proposed an approach for designing a tutoring library by applying compositional adaptation. This method identifies the books or parts of the book for the user's search topic in the library by combining the solutions of the similar past requests by other users. Atzmueller et al. [2] examined the compositional case adaptation approach in the multiple disorder situation during medical diagnosis. The proposed approach identifies the solution based on the solutions of the $k$ most similar cases. In [8], Muller et al. attempted the compositional adaptation of cooking recipes by decomposing the cooking recipe cases into reusable streams [9]. The adaptation process compensates the deficiencies of the retrieved recipe by replacing the retrieved one with the streams of appropriate cooking recipes.

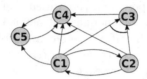

**Fig. 1.** An example of casebase where compositional adaptation is involved

We illustrate the drawback of the competence model in footprint-based approach when the adaptation process involves compositional adaptation. Figure 1 shows a network of cases where each node represents cases and an edge from one case (say $c_1$) to other case (say $c_2$) indicates that the case $c_1$ can be retrieved and its solution can be adapted to solve $c_2$. As per the definition of *solves* in Definition 1, the edge $c_1 \rightarrow c_2$ implies $c_1$ *solves* $c_2$. The arc (*AND* arc) between the edges represents compositional adaptation. For example, the arc between the edges $c_1 \rightarrow c_3$ and $c_2 \rightarrow c_3$ in the network indicates that the composite solution of the case $c_1$ and $c_2$ can solve the problem $c_3$. It is to be noted that neither case $c_1$ nor $c_2$ can solve $c_3$ in isolation. The footprint-based approach discussed in Sect. 2 cannot have the *AND* arcs between incoming edges, and outputs a footprint set $\{c_1\}$ corresponding to this network. Though Smyth et al. [15] proposed the footprint approach such that the footprint set covers the entire casebase, the footprint set identified for the casebase in Fig. 1 solves all the cases in the network only when compositional adaptation is not taken into consideration. For example, case $c_3$ cannot be solved by this footprint set as the case $c_3$ needs case $c_2$ which is not present in the footprint set, apart from $c_1$ to solve it. The current competence model has to be enhanced to include compositional adaptation.

# 4     Approach

In this section, we present a case competence model which covers the compositional adaptation (CA) process. Compositional adaptation composes a new solution by combining the solutions of multiple cases; cases which are used for adapting the new solution form an *AND* relation. It is possible to have multiple adapted solutions (either single case or compositional) for a target problem. These multiple solutions for a target problem shape an *OR* relation. The *AND* relation implies all the cases that are part of this relation are required to adapt a new solution and the *OR* relation indicates any of the cases can solve the target problem. The casebase is comprised of *AND-OR* relations between cases (or a disjunction over conjunctions). We assume that the compositional adaptation operator is a disjunction over conjunctions.

We define the relation $solves_{CA}$ in the context of compositional adaptation corresponding to the relation *solves* in Definition 1. For a casebase $\mathbb{C}$, $solves_{CA}$ is defined in Definition 4.

**Definition 4.** A set of cases $\mathbb{C}' \subset \mathbb{C}$ $solves_{CA}$ a target problem $t$ if and only if all the cases in $\mathbb{C}'$ are retrievable for $t$ and the solutions of the cases in $\mathbb{C}'$ can be adapted to solve $t$.

For example, in Fig. 1 the combined solution of the cases $c_1$ and $c_2$ solves the problem $c_3$ i.e.,. $solves_{CA}(\mathbb{C}', c_3)$ where $\mathbb{C}' = \{c_1, c_2\}$. As compared to the Smyth's competence model [14] which considers $c_1$ solving $c_3$ independent of $c_2$, here we need to model the fact that the cases $c_1$ and $c_2$ cannot individually solve the target problem. We exploit $solves_{CA}$ in the competence model and redefine the $coverage_{CA}$ and $reachability_{CA}$ as in Definitions 5 and 6. The Coverage$_{CA}$ is defined for a set of cases and Reachability$_{CA}$ is defined for each case. Each element in Reachability$_{CA}$ of a case $c$ is a set of cases which can be used for either single case or compositional adaptation to solve the target $c$.

**Definition 5.** Coverage$_{CA}(\mathbb{C}' \subset \mathbb{C}) = \{c \in \mathbb{C} : solves_{CA}(\mathbb{C}', c)\}$

**Definition 6.** Reachability$_{CA}(c) = \{\mathbb{C}' \subset \mathbb{C} : solves_{CA}(\mathbb{C}', c)\}$

For example, in Fig. 1 $Coverage_{CA}(c_1, c_2) = \{c_3\}$ and $Reachability_{CA}(c_4) = \{\{c_1, c_2, c_5\}, \{c_3\}\}$. The dependency between the cases in solving the problems has to be considered when we estimate the competence of each case in the casebase. Finally, this should reflect in the footprint set.

We propose a measure called *retention score* which orders the cases by considering compositional adaptation based on the extent to which a case is to be retained in the casebase. This measure quantifies the competence of a case in the casebase. Then, we propose a modified algorithm of Smyth's footprint [15] identification called footprint$_{CA}$ algorithm which identifies the footprint$_{CA}$ which reflects compositional adaptation.

## 4.1   Retention Score

The retention score is a measure which quantifies the importance of a case in terms of whether it is required to be retained in the casebase or not. To illustrate the idea of retention score, consider the graphs constructed out of synthetic casebases in Figs. 2 and 3. In the first one, the cases $c_1$ and $c_2$ are essential to retain as both are required to cover the other cases $c_3$ and $c_4$. However, in the second one the case $c_1$ requires $c_2$ to solve $c_3$, and both $c_2$ and $c_5$ to solve $c_4$. The factors that determine the retention quality of a case are the range of problems that it solves and the number of cases that are required to solve those problems. In a casebase, we would like to retain fewer good retention quality cases that cover more useful cases. To estimate the retention score, we define two terms - covered cases and support cases.

The covered cases of a case $c$ ($CoveredCases(c)$) include all the cases that $c$ can be used to solve either on its on, or in conjunction with other cases. For example, $CoveredCases(c_1)$ in the network shown in Fig. 2 is $\{c_3, c_4\}$.

**Fig. 2.** Synthetic network 1

The support cases of a case $c_i$ to solve the problem $c_j$ ($SupportCases(c_i, c_j)$) is the set of cases that the case $c_i$ requires to solve $c_j$. For example, in Fig. 3 the $SupportCases(c_1, c_3)$ is $\{c_2\}$ and the $SupportCases(c_1, c_4)$ is $\{c_2, c_5\}$.

The proposed measure for retention score is based on these two sets and it is based on the idea that *a case has high retention score if it can solve several cases that have high retention*

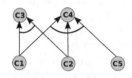

**Fig. 3.** Synthetic network 2

*score with as few cases that have high retention score*. More precisely, the retention score of a case is high if there are more covered cases that have high retention score with less number of support cases that have high retention score. Using this idea we came across the recursive formulation as given in Eq. 2.

$$RetentionScore_{k+1}(c) = \sum_{c_i \in CoveredCases(c)} \frac{RetentionScore_k(c_i)}{\sum\limits_{c_j \in SupportCases(c, c_i)} RetentionScore_k(c_j) + 1} \quad (2)$$

where $RetentionScore_{k+1}(c)$ is the retention score of a case $c$ at $k+1^{th}$ iteration. Each covered case contributes to the estimation of the retention score based on its retention score and the retention score of the support cases that solve this covered case. The addition of 1 in the denominator is to handle the situation when a case does not need any support case to solve the corresponding covered case. For the first iteration of the retention score estimation, the retention score of a case $c$ can be estimated as,

$$RetentionScore_0(c) = \sum_{c_i \in CoveredCases(c)} \frac{\frac{1}{1 + |\{\mathbb{C}' \, : \, \mathbb{C}' \in Reachability_{CA}(c_i) \text{ and } c \notin \mathbb{C}'\}|}}{1 + |SupportCases(c, c_i)|} \quad (3)$$

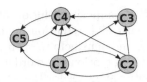

**Fig. 4.** A sample casebase graph

**Table 1.** RetentionScore and Relative-Coverage of cases

| Case ($c$) | $c_1$ | $c_2$ | $c_3$ | $c_4$ | $c_5$ |
|---|---|---|---|---|---|
| RetentionScore($c$) | 2 | 1.75 | 1.29 | 1.23 | 1 |
| RelativeCoverage($c$) | 2.25 | 1.75 | 0.25 | 0.5 | 0.25 |

The numerator part for each covered case $c_i$ in Eq. 3 captures the individual contribution of $c$ in solving $c_i$. The contribution of $c$ in solving $c_i$ is high if $c$ is involved in all the solutions of $c_i$. Thus, the individual contribution of $c$ to solve $c_i$ decreases with increase in the number of alternate solutions which do not contain $c$. The denominator of Eq. 3 ensures that the retention score increases with decrease in the number of support cases that $c$ requires to solve $c_i$ and vice versa. The addition of 1 in the denominator handles the situation when there are no supporting cases.

The retention score recursively measures the global competence of each case in the casebase. The recursive formulation of retention score captures the transitive solving property of cases. For example, if a case $c_1$ solves $c_2$ and $c_2$ solves $c_3$, then $c_1$'s contribution in solving $c_3$ will also be captured. But, relative coverage measure used in the footprint-based approach [15] cannot reveal the transitive coverage of a case. The relative coverage measure express only the individual contribution of each case irrespective of the requirements of other cases in solving a target problem.

In Fig. 4, the graph of the casebase example has been reproduced from Fig. 1. The retention score of the cases in the network become stable after 15 iterations. The scores are given in Table 1. We normalize the retention score values to range 1 to 2 after each iterations. The ordering based on retention score is obtained as $c_1, c_2, c_3, c_4, c_5$. The case $c_1$ secured highest retention score as it covers two cases without any supporting cases and two other cases with support cases. Though $c_3$ needs no support cases, its score is less due to the lack of its coverage. However, the case $c_4$ has even lesser score than $c_3$ although it covers same number of covered cases with no support cases. This is because, the covered case of $c_4$ i.e., $c_5$ can be alternatively solved by $c_1$ which has more coverage. However, the relative coverage values of the cases shown in Table 1 shows that the values are estimated only based on the participation of solving a target case. For example, the case $c_5$ secures a relative coverage value as it helps in solving $c_4$ irrespective of the requirement of $c_1$ and $c_2$ in solving $c_4$. This notion has been captured by the retention score.

## 4.2 Footprint$_{CA}$ Algorithm

The footprint algorithm proposed by Smyth et al. [15] does not consider compositional adaptation while constructing the footprint set. We modified the Smyth's footprint algorithm to obtain the footprint$_{CA}$ set and the algorithm is described in Algorithm 1. This algorithm estimates the footprint by adding the cases in the decreasing order of retention score if none of the composite solution of a case is present in the footprint set. Thus the cases with high retention quality are added before the cases with less retention quality, and thus help to keep the good quality cases in the footprint set. We preserve the retention score ordering of cases in the final footprint set. In this way, the footprint$_{CA}$ set for the example in Fig. 1 is obtained as $\{c_1, c_3\}$. It may be noted that this set can cover all concepts in the given network whereas the Smyth's footprint set $\{c_1\}$ which is based on relative coverage cannot cover all the cases in the network.

---

**Algorithm 1.** Footprint$_{CA}$ algorithm

---

**Input:** Cases sorted based on retention score, **Output:** Footprint$_{CA}$ (FP)
Cases ← Sorted cases according to their retention score
FP ← {}
*Changes* ← true
**while** *Changes* **do**
    *Changes* ← false
    **for** *each $c \in Cases$* **do**
        **if** *none of the composite solution of c is a subset of FP* **then**
            *Changes* ← true
            Add $c$ to FP

---

## 5    Evaluation

We empirically tested the proposed competence model by using synthetic regression datasets. The datasets are generated based on the factors like dimensions, the number of data points, the distance between neighbors, and non-linearity. The generation process of datasets used for analysis are illustrated below.

1. *Synthetic data 1:* $y = x_1 + x_2 + x_3 + x_4 + x_5 + x_6 + x_7 + x_8 + x_9 + x_{10}$ +noise
2. *Synthetic data 2:* $y = x_1^4 + x_2^3 + x_3^2 + x_4 + \cos^2(x_5)$ +noise
3. *Synthetic data 3:* $y = \sin(x_1 x_2) + \sqrt{x_3 x_4} + \cos^2(x_5) + x_6 x_7 + x_8 + x_9 + x_{10}$ +noise

The data points across each dimension of all the datasets are sampled uniformly with values between 0 and 10; we added a random gaussian noise with mean 0 and standard deviation 10. The structure of the datasets are - *Synthetic data 1* is linear and high dimensional; *Synthetic data 2* is nonlinear and low dimensional; *Synthetic data 3* is nonlinear and high dimensional.

### 5.1    Experimental Setup

Each data instance is considered as a case in the casebase and each case is assumed to be solved by the compositional adaptation solution of its k-nearest neighbor cases. Thus the casebase graph contains cases as nodes, and edges from the k-nearest neighbors of each case which are connected to it by an AND arc. Then the footprint$_{CA}$ set is estimated using this graph and is compared with the footprint$_{OR}$[1] set which is obtained from the same graph by removing the composition (AND) condition. The experiments are done with k = 1,2 and 4 and by varying the number of instances (casebase size) from 10 to 100. At k = 1, the adaptation process uses a single case; multiple cases are used when k> 1.

### 5.2    Evaluation Criteria

The analysis of the footprint size is one of the common criteria for evaluation. However, the size of both the footprint sets are not strictly comparable as the footprint$_{CA}$ is expected to have more cases than the footprint$_{OR}$ set due to composition condition in the former set. Figure 5 illustrates that the footprint$_{OR}$ size is less compared to footprint$_{CA}$. The size of footprint$_{OR}$ decreases with increase in the value of k where as the size of footprint$_{CA}$ increases with increase in the value of k. For a high value of k, more cases are involved in compositional adaptation

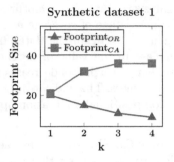

**Fig. 5.** Footprint size analysis

during which the footprint$_{OR}$ size compresses more and thereby loses composition knowledge of adaptation. Hence, we propose two measures to estimate the effectiveness of footprint$_{CA}$ obtained based on the retention score in a compositional adaptation application - casebase coverage and footprint sanity measure. We compare the results obtained over the footprint$_{CA}$ with the footprint$_{OR}$ set computed using the relative coverage measure in the same application.

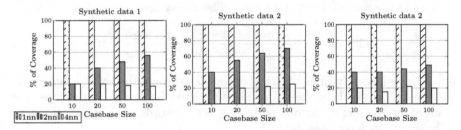

**Fig. 6.** Casebase coverage by footprint$_{OR}$

**Casebase Coverage.** The essential idea of the footprint set is that the footprint cases solve all the cases in the casebase. The casebase coverage of a footprint set $fp$ is measured as follows,

$$\text{Casebase Coverage}(fp) = \frac{|\text{Cases that are solved by } fp|}{\text{Casebase Size}} \tag{4}$$

---

[1] We refer the Smyth's footprint set [15] as the footprint$_{OR}$ set.

The main aim of this evaluation measure is to examine the effectiveness of footprint$_{CA}$ and footprint$_{OR}$ in the compositional adaptation application. As footprint$_{CA}$ is formulated for compositional adaptation; this set is expected to cover the entire casebase. However, the usefulness of footprint$_{CA}$ set can be observed by analyzing the casebase coverage of footprint$_{OR}$ set.

We analyzed that footprint$_{CA}$ has full casebase coverage all the dataset. However, the footprint$_{OR}$ set covers the entire casebase only when k = 1. The analysis of coverage on footprint set is illustrated in Fig. 6. We can observe that the percentage of coverage increases with increase in the number of data points when k = 2 in all the datasets. Also, the coverage percentage decreases with increase in the value of k. The reason behind this is that the increase in the number of neighbors decreases the size of footprint set and there by reduces its effectiveness. This indicates the ineffectiveness of the footprint$_{OR}$ set to apply it in compositional adaptation applications.

**Sanity Check.** To measure the sanity of the footprint set, we found a method to identify a set of cases that can cover the entire casebase using a graph-theoretic approach. We estimate the footprint set from the case network that is constructed using the relation *solves$_{CA}$*. In the same network, if we repeatedly remove the cases that do not solve any other cases until there are no such cases, the final network turns out to be a compressed set of cases that can solve all the cases in the casebase transitively. This final network is called the *kernel* of the case network. The algorithm for computing the kernel is given in Algorithm 2. Though there is no ordering of cases provided within the kernel, the cases in the kernel are the potential cases that can be presented in a footprint set. So, we compare the cases in the footprint set and kernel. The sanity measure is defined as,

$$\text{Sanity rate} = \frac{|\text{footprint cases} \cap \text{kernel cases}|}{|\text{kernel cases}|} \times 100 \qquad (5)$$

This idea is adapted from [6] where Masse et al. estimate the grounding kernel of a dictionary graph where the graph is constructed from word definitions. Here the grounding kernel turns out to be the set of words from which the entire dictionary words have been defined.

---

**Algorithm 2.** Computing the Kernel of the Case Network

---

**Input:** Case Network $\mathbb{G}$, **Output:** Kernel $\mathbb{K}$
$\mathbb{K} \leftarrow \mathbb{G}$ **do**
    Let $\mathbb{C}$ be the set of cases (vertices) in $\mathbb{K}$
    $\mathbb{U} \leftarrow \{v \in \mathbb{C} : \text{out-degree of v in } \mathbb{K} = 0\}$
    Remove all elements in $\mathbb{U}$ from $\mathbb{K}$
**while** $\mathbb{U} == \emptyset$;

---

In Fig. 7, the sanity rate of footprint$_{CA}$ and footprint$_{OR}$ are compared in all the three datasets for 1 nn, 2 nn, 4 nn and various casebase sizes. We can observe that footprint$_{CA}$ has high sanity rate for all the results with k = 2,4, and there

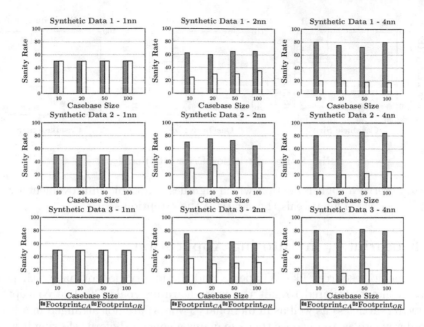

**Fig. 7.** Sanity rate of footprint cases in synthetic datasets

is a significant difference in the sanity rate between footprint$_{CA}$ and footprint$_{OR}$ sets. At k = 1 (single case adaptation), both the methods are performing similar which indicates that footprint$_{CA}$ is as good as footprint$_{OR}$ in the single case adaptation process.

We also check the sanity of the footprint sets by performing a reconstruction of noisy compression of the regression data using the footprint sets as a set of representative cases. In order to test the quality of the reconstruction of footprint sets, we performed a regression analysis where we used each footprint set as the the training data. The test data are the cases that belong to neither the footprint$_{CA}$ set nor the footprint$_{OR}$ set. The reconstruction error (RE) is evaluated by the mean square error and we compared the reconstruction error obtained by both the training data. The comparison of results for all the three synthetic datasets are shown in Fig. 8. The comparison is done based on the percentage of the difference between the reconstruction error received by the two training sets, with respect to the footprint$_{OR}$ set error. The comparison measure is given in Eq. 6. As we compute the reduction with respect to the footprint set, a high error percentage indicates a significant improvement by the footprint$_{CA}$ set.

$$\text{Reduction w.r.t footprint}_{OR}\ \text{RE} = \frac{\text{footprint}_{OR}\ \text{RE} - \text{footprint}_{CA}\ \text{RE}}{\text{footprint}_{OR}\ \text{RE}} \times 100 \tag{6}$$

The $k$ value for finding the neighbors is varied from 1 to 4. At $k = 1$, the reduction is close to zero which indicates footprint$_{CA}$ and footprint$_{OR}$ perform similarly in single case adaptation. For $k > 1$ and casebase size =10,

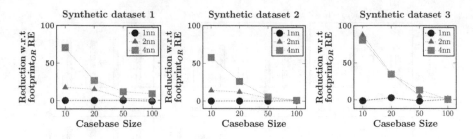

**Fig. 8.** Reconstruction Error (RE) analysis

we can observe a high reduction which signifies a notable improvement by the footprint$_{CA}$ set. This reveals the sanity of the retention score.

## 6   Footprint$_{CA}$ in Tutoring Application

Encyclopedic resources like Wikipedia and dictionary do not have rich peda-gogical content, tailored to suit the users learning goals [7]. The concepts in Wikipedia (articles) as well as in dictionary (words) are not arranged in a learn-ing order where as an ideal textbook explains a concept before referring it which results in a sequential order for learning [11]. So, sequencing the concepts in Wikipedia like resources may help the online learners to fulfill their learning goal. Each article in Wikipedia is explained in terms of other articles which, in turn explained using other articles. These articles are interconnected using hyperlinks. In the CBR perspective, Wikipedia articles are the cases and the concepts in Wikipedia that help in understanding a target concept are com-posed together to explain the target [7]. Hence, those set of cases acts as a composite solution of the target. The definition of a Wikipedia article can be approximated as the first sentence in the article [17]. So, the articles pointed to, by hyperlinks in the first sentence can be assumed as the concepts or cases that help in understanding the corresponding concept. We can construct a graph of Wikipedia casebase by marking these set of concepts as the cases that provide one composite solution for a Wikipedia article. In such graph, it is possible to adapt many composite solutions to explain a concept by using the transitivity property of the graph. This is because every Wikipedia concept is explained in terms of other concepts. Figure 9 illustrates an example of casebase graph con-structed from English Wikipedia. Each node corresponds to Wikipedia articles. The Edges are drawn from the concepts in the first sentence of each article. For example, the concept *atom* is explained in terms of *chemical element* and *matter*. Hence, the arc between the edges from *chemical element* and *matter* to *atom* which forms an *AND* relation indicates that the cases *chemical element* and *matter* are composed together to explain *atom*.

We can construct a casebase graph for a given topic, and our case competence model can identify a competent subset of concepts which covers the rest of the concepts in that topic. The retention score ordering implies the importance of each concept based on the extent to which the concept to be retained. A concept

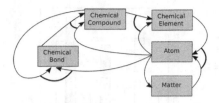

**Fig. 9.** An example of Casebase network from Wikipedia

**Table 2.** Retention score values

| Concepts | Retention score |
|---|---|
| Atom | 2.0 |
| Matter | 1.19 |
| Chemical element | 1.18 |
| Chemical compound | 1.12 |
| Chemical bond | 1 |

with high retention value is likely to be a basic concept as its coverage will be high due to its repetitive usage in defining other concepts. Thus, the ordering based on retention score provides an order in which one can learn the entire set of concepts under a specific topic.

The retention scores obtained for the Wikipedia concepts in the network shown in Fig. 9 are given in Table 2. The footprint$_{CA}$ set for this example is obtained as *{atom, chemical element, chemical compound}*. This set can cover the entire casebase. The ordering of elements in the footprint$_{CA}$ set indicates the learning ordering where the position in the order implies the level of completion of learning. For example, let the learning goal be *Chemical Compound*. To satisfy the learning goal, one can learn the concepts in footprint$_{CA}$ in the retention score ordering. While learning each concept in the footprint$_{CA}$, the concepts that are solved by the elements in footprint$_{CA}$ can be learnt. Note that these concepts may not be present in the footprint$_{CA}$ set. A learner who is familiar with any of the concept in footprint$_{CA}$ can skip all the concepts that are positioned before this concept in the footprint$_{CA}$. This is because a concept subsumes all the previously present concepts in footprint$_{CA}$. Thus, footprint$_{CA}$ and the retention score ordering helps a learner to satisfy his/her goal.

### 6.1   Empirical Results

The effectiveness of retention score and footprint$_{CA}$ set is analyzed on the casebase extracted from the Wikipedia and dictionary. We extracted the articles in Wikipedia Artificial Intelligence (AI) category[2] and sub-categories up to three levels. The composed solution cases of each article are marked from the hyperlink articles that are present in the first sentence. This casebase (wikiAI) contains 6,536 cases. In the dictionary, concepts (cases) are the words that are defined in it and the content words in the definition are marked as the cases that are used for compositional adaptation to define a word. We make simplifying assumptions that the words in the dictionary are sense disambiguated. So, the content words present in the first definition of the first sense is considered as the composed solution of each word. Thus, we have taken definitions from the Longman dictionary of contemporary English (ldoce) and WordNet (wn). The graph constructed from this casebase results in an *AND-OR* graph due to the presence of multiple compositional solutions. Thus, we have 81,653 cases in the casebase.

---

[2] https://en.wikipedia.org/wiki/Category:Artificial_intelligence.

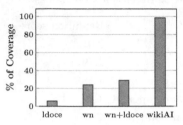

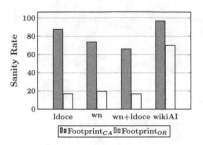

**Fig. 10.** Casebase coverage of footprint$_{OR}$

**Fig. 11.** Sanity rate analysis

Similarly, other casebases are constructed using only WordNet (wn) and only Longman dictionary (ldoce). The wn casebase includes 79,582 cases and ldoce casebase contains 26,984 cases. All these four casebases are used for the analysis of retention score and footprint$_{CA}$ in tutoring application.

**Casebase Coverage.** We analyzed the casebase coverage by the footprint$_{CA}$ set and footprint$_{OR}$ set in all the casebases. The footprint$_{CA}$ is observed as covering the full casebase whereas the footprint$_{OR}$ set does not cover the entire casebase due to the presence of *AND* composition. Thus, the entire dictionary words can be defined using the words in the footprint$_{CA}$. We analyzed the casebase coverage by the footprint$_{OR}$ and this is shown in Fig. 10. In all the casebases except wikiAI, the footprint$_{OR}$ set solves only less than 30 % of the cases in the casebase. The higher coverage of footprint$_{OR}$ in wikiAI can be because of the less number of hyperlinks in the first sentence of each article which is considered as the cases in the composed solution.

**Sanity Check.** The sanity of the footprint$_{CA}$ and footprint$_{OR}$ are analyzed using the sanity rate formulated in Sect. 5.2. The results are given in Fig. 11. We can observe that the sanity rate of footprint$_{CA}$ cases in all the casebases are more than 65 % and that of footprint$_{OR}$ cases are less than 20 % except the wikiAI dataset which might be due to the lack of compositional information in the dataset. This indicates that the footprint$_{CA}$ set is useful for compositional adaptation applications.

## 7    Conclusion and Future Work

We start with the observation that the Smyth's footprint-based approach [15] is not designed for compositional adaptation applications. We proposed a measure called retention score to estimate the retention quality of a case that involves compositional adaptation. Using the retention score, we proposed a revised approach to identify the footprint$_{CA}$ set where compositional adaptation is required. We tested the effectiveness of the footprint$_{CA}$ using regression datasets and compared it with the Smyth's footprint set. The empirical results demonstrated the improved performance of our model when compositional adaptation is required; the proposed model performs equally well as Smyth's model during single case adaptation process. We also illustrated and tested the effectiveness of our method in a tutoring application which uses compositional adaptation.

The proposed retention score measure assumes that the compositional adaptation operator is a disjunction over conjunctions which makes a *hard-AND* relation between the cases that solves a problem using compositional adaptation. In some applications, the *soft-AND* relation might solve the problem. For example, the mean value of the solutions of the similar cases is taken as the composed solution for the target problem in applications such as pollution prediction in Aiquap CBR system [3]. The dropping of any of the similar cases might not affect the resulting solution. It would be interesting to introduce the softness in the retention score.

# References

1. Arshadi, N., Badie, K.: A compositional approach to solution adaptation in case-based reasoning and its application to tutoring library. In: Proceedings of 8th German Workshop on Case-Based Reasoning (2000)
2. Atzmueller, M., Baumeister, J., Puppe, F., Shi, W., Barnden, J.A.: Case-based approaches for diagnosing multiple disorders. In: FLAIRS, pp. 154–159 (2004)
3. Lekkas, G.P., Avouris, N.M., Viras, L.G.: Case-based reasoning in environmental monitoring applications. Appl. Artif. Intell. Int. J. **8**(3), 359–376 (1994)
4. Lieber, J.: A criterion of comparison between two case bases. In: Haton, J.-P., Keane, M., Manago, M. (eds.) EWCBR 1994. LNCS, vol. 984, pp. 87–100. Springer, Heidelberg (1995). doi:10.1007/3-540-60364-6_29
5. Markovitch, S., Scott, P.D.: Information filtering: selection mechanisms in learning systems. Mach. Learn. **10**(2), 113–151 (1993)
6. Massé, A.B., Chicoisne, G., Gargouri, Y., Harnad, S., Picard, O., Marcotte, O.: How is meaning grounded in dictionary definitions? In: Proceedings of the 3rd Textgraphs Workshop on Graph-Based Algorithms for Natural Language Processing, pp. 17–24 (2008)
7. Mathew, D., Eswaran, D., Chakraborti, S.: Towards creating pedagogic views from encyclopedic resources. In: Proceedings of the Tenth Workshop on Innovative Use of NLP for Building Educational Applications, pp. 190–195 (2015)
8. Müller, G., Bergmann, R.: Compositional adaptation of cooking recipes using workflow streams. In: Computer Cooking Contest, Workshop Proceedings ICCBR (2014)
9. Müller, G., Bergmann, R.: Workflow streams: a means for compositional adaptation in process-oriented CBR. In: Lamontagne, L., Plaza, E. (eds.) ICCBR 2014. LNCS (LNAI), vol. 8765, pp. 315–329. Springer, Heidelberg (2014). doi:10.1007/978-3-319-11209-1_23
10. Patterson, D., Rooney, N., Galushka, M.: A regression based adaptation strategy for case-based reasoning. In: Proceedings of the Eighteenth National Conference on Artificial Intelligence and Fourteenth Conference on Innovative Applications of Artificial Intelligence, pp. 87–92 (2012)
11. Agrawal, R., Chakraborty, S., Gollapudi, S., Kannan, A., Kenthapadi, K.: Quality of textbooks: an empirical study. In: ACM Symposium on Computing for Development (2012)

12. Reinartz, T., Ioannis, I., Thomas, R.: Review and restore for case base mainte-
    nance. Comput. Intell. **17**(2), 214–234 (2001)
13. Smyth, B., Keane, M.T.: Remembering to forget. In: Proceedings of the 14th Inter-
    national Joint Conference on Artificial Intelligence (IJCAI), pp. 377–382 (1995)
14. Smyth, B., McKenna, E.: Modelling the competence of case-bases. In: Smyth, B.,
    Cunningham, P. (eds.) EWCBR 1998. LNCS, vol. 1488, pp. 208–220. Springer,
    Heidelberg (1998). doi:10.1007/BFb0056334
15. Smyt, B., McKenna, E.: Footprint-based retrieval. In: Althoff, K.-D., Bergmann,
    R., Branting, L.K. (eds.) ICCBR 1999. LNCS, vol. 1650, pp. 343–357. Springer,
    Heidelberg (1999). doi:10.1007/3-540-48508-2_25
16. Wilke, W., Bergmann, R.: Techniques and knowledge used for adaptation during
    case-based problem solving. In: Pasqual del Pobil, A., Mira, J., Ali, M. (eds.)
    IEA/AIE 1998. LNCS, vol. 1416, pp. 497–506. Springer, Heidelberg (1998). doi:10.
    1007/3-540-64574-8_435
17. Ye, S., Chua, T., Lu, J.: Summarizing definition from Wikipedia. In: Proceedings
    of the Joint Conference of the 47th Annual Meeting of the ACL and the 4th
    International Joint Conference on NLP of the AFNLP, pp. 199–207 (2009)

# On the Transferability of Process-Oriented Cases

Mirjam Minor[1(✉)], Ralph Bergmann[2], Jan-Martin Müller[1],
and Alexander Spät[1]

[1] Business Information Systems,
Goethe University, Robert-Mayer-Str. 10, 60629 Frankfurt, Germany
minor@cs.uni-frankfurt.de, jan-martin.mueller@t-online.de,
spaet@stud.uni-frankfurt.de
[2] Business Information Systems II, University of Trier, 54286 Trier, Germany
bergmann@uni-trier.de

**Abstract.** This paper studies the feasibility of using transfer learning for process-oriented case-based reasoning. The work introduces a novel approach to transfer workflow cases from a loosely related source domain to a target domain. The idea is to develop a representation mapper based on workflow generalization, workflow abstraction, and structural analogy between the domain vocabularies. The approach is illustrated by a pair of sample domains in two sub-fields of customer relationship management that have similar process objectives but different tasks and data to fulfill them. An experiment with expert ratings of transferred cases is conducted to test the feasibility of the approach with promising results for workflow modeling support.

**Keywords:** Process-oriented case-based reasoning · Transfer learning · CRM application

## 1 Introduction

*Transfer learning* (TL) addresses the "question of how the things that have been learned in one context can be re-used and adapted in related contexts" [14, p. 5]. TL has a long tradition in diverse research disciplines, ranging from psychology and education [23,32] to cognitive science [8] and artificial intelligence (AI) [11,14,22,30]. In the context of *case-based reasoning* (CBR), TL approaches use knowledge from a source domain "to enhance an agent's ability to learn to solve tasks from a target domain" [11, p. 54]. The *source domain* denotes the problem solving context in which knowledge is available at a mature level. The *target domain* is the problem solving context where the knowledge is sparse.

*Process-oriented case-based reasoning* (POCBR) is a recent research area of CBR that aims at applying and extending CBR methods for process and workflow management [16]. *Workflows* are "the automation of a business process, in whole or part, during which documents, information or tasks are passed from one participant to another for action, according to a set of procedural rules" [1]. The control flow of a workflow specifies the order of tasks to be executed.

© Springer International Publishing AG 2016
A. Goel et al. (Eds.): ICCBR 2016, LNAI 9969, pp. 281–294, 2016.
DOI: 10.1007/978-3-319-47096-2_19

The data flow specifies the interaction of tasks with data items (documents or information). In POCBR a case is usually a workflow or process description expressing procedural experiential knowledge. There are many application domains for POCBR where procedural experiential knowledge is sparse. It requires time-consuming efforts to populate a case base for a POCBR system from scratch, also referred to as the cold-start problem. In certain cases, there is a related application domain where workflows are available at a mature level, either resulting from previous modeling activities, from process mining [17] or extracted from other information sources, such as Internet Communities [28]. The transfer of procedural knowledge provides an approach to solve the cold start problem of POCBR systems. In addition, it might strengthen mature process-oriented case bases by introducing a larger variety of cases to be reused.

TL has been successfully applied in several CBR application fields, such as games [2] or physics [12]. However, TL has not yet been studied in the context of POCBR and workflows. The aim of this paper is to investigate transferability of knowledge from a POCBR system in a source domain to a POCBR system in a target domain. In particular, we will propose a novel approach on TL for POCBR that claims that generalization and abstraction of workflows, as well as structural analogies between the vocabularies of the source and target domain support the transfer of process-oriented cases. We will use two related domains of customer relationship management (CRM) as a running sample to illustrate our approach and to test it in a lab experiment.

## 2    Related Work

A large amount of work on TL in machine learning, especially for reinforcement learning, has been reported; see the 2009 survey [30] and the 2014 special issue of the German "KI" journal [14] for a review. A good overview on TL in data mining for classification, regression, and clustering is given by the 2010 survey of Pan & Yang [22]. The approaches from these research lines transfer a general concept that has been achieved by "eager learning" from training data. This means that a model has been learned from a data collection in a first phase to be used in a second phase. In contrast, there is a research line on *Case-based transfer learning* that is mainly addressing "lazy learning". CBR collects the examples in a case base [27, p. 280], learning from recording problem solving episodes. This means that the learning phase continues while the knowledge that has been learned so far is already in use. While TL has proven a significant benefit in several learning scenarios [14,22,30], it has not yet been studied in the context of POCBR.

A topic that has already been studied in case-based TL is the use of *models of analogy*. Sample analogy models that have been used for TL are structure-mapping engine [7,12], graph isomorphism [15], cognitive modeling [25], or goal-driven analogical mapping [13].

As CBR can be viewed as a kind of analogical problem solving, existing approaches to adaptation in CBR can already be considered to perform a kind of transfer learning. Klenk et al. [11] call this approach "CBR as transfer learning method". It requires a certain amount of overlap between the source and the target domain such that certain pieces of knowledge (for example, cases or adaptation knowledge) learned or acquired in the source domain can be directly used in the target domain. In case of hierarchical case representations, such a transfer can also occur on a higher level of abstraction that is a proper common abstraction of both domains. In the context of POCBR, recently developed adaptation methods can be considered in this respect. In compositional adaptation, workflow are decomposed into meaningful sub-workflows called *workflow streams*, which immediately provide a means for case abstraction [3]. An abstract case is a structurally simplified workflow, using more abstract terms as descriptions of task and data items. During problem solving, abstract cases can be retrieved and reused by refining the occurring abstract items. This refinement step can then transfer an abstract case towards a specific case in the target domain. Also, adaptation by generalization and specialization can be used for transfer learning in POCBR [19]. A generalized workflow is structurally identical to the base workflow but the semantic descriptions of task and data items are generalized. If this generalization is performed to a level that covers the source and the target domain, the generalized cases from the source domain can be immediately be used in the target domain to solve problems by being appropriately specialized.

This approach to CBR as transfer learning is clearly limited to source and target domains in which there is a significant overlap between the domain ontologies. For transfer learning between two more distant domains, analogical mapping approaches are required that enable the alignment of the two ontologies and thereby support the mapping of abstract and generalized cases from the source to the target domain. This paper presents a first step towards the development of such a transfer learning method.

## 3    A Typical Example of a Pair of POCBR Domains

A usage scenario for TL on POCBR is modeling support to alleviate the cold-start problem when a company starts a new process repository and a set of workflows is to be created. The following running sample uses two typical business application areas for POCBR. Both areas are sub-fields of customer relationship management (CRM). We have chosen the domain of opportunity management as a sample target domain. Workflows for opportunity management, for example, comprise activities to identify and nurture sales opportunities. The related documents and data items interact with the company's CRM system [9]. Second, we have chosen churn[1] management as a sample source domain. Churn management is a domain that aims at predicting customers with a high probability for churn [31]. The domain shares some commonalities with opportunity

---

[1] The meaning of *churn*, according to the Cambridge English Dictionary, is: "If customers churn between different companies that provide a particular service, they change repeatedly from one to another.".

management since in both domains customer data is analysed. Churn management involves tasks that are related to the identification of leads in opportunity management. A lead is a person that is likely to become a customer [9]. Sales persons aim at transforming leads into opportunities, i.e. to create new sales opportunities by nurturing leads with marketing activities [9]. Figure 1 depicts a typical churn management workflow [31] on churn analysis in Business Process Modeling Notation (BPMN) [4]. It starts with a task *"Measure cause variables"* which stands for a sub-workflow to measure different cause variables for churn from CRM data. The dotted box in the lower part of the figure contains the sub-workflow with the particular measuring tasks in parallel, such as for the customer complaint behavior and for the duration of the customer relationship. The resulting indices are further processed by a subsequent data mining task *"Logistic regression"*, which creates a model to divide customers into groups by their likelihood for churn. Churn management and opportunity management are a pair of typical domains for POCBR systems. We will study the transfer of knowledge from one POCBR system to another POCBR system illustrated by churn and opportunity management processes.

## 4   The Transfer Setting

The goal of our novel TL approach is to transfer parts of a case base $CB_S$ from a source domain $D_S$ to a target domain $D_T$ in order to extend a sparsely populated case base $CB_T$ to a richer case base $CB'_T$. The transfer setting is characterized by the transfer distance between $D_S$ and $D_T$ and by the means that are used to bridge the gap between the two domains.

The *transfer distance* can be delineated by the differences between the source and target problems [11]. Sample transfer distances consider the proportion of vocabulary that is shared across source and target or whether the transfer includes restructuring or composing of source knowledge (compare [11]). It has been stated in the literature [22] that the transfer distance may help to provide a measure for the transferability. Without providing a formal measure for the transfer distance yet, we make the assumption that both domains in our transfer setting are loosely related, i.e. $D_S$ and $D_T$ share little vocabulary to describe the process-oriented cases and the processes from both domains address slightly related objectives. We assume that ontologies $O_S$ and $O_T$ are available (or can be created) as vocabulary for both domains covering the workflow tasks and the data items of the workflows in $CB_S$ and $CB_T$. In addition, we assume that the ontologies contain some concepts which both have in common, i.e. there is an overlap $O_S \cap O_T \neq \emptyset$. Please note that this includes concepts at a higher hierarchical level. For instance, the workflow task *"Behavioral scoring"* for leads in our running sample opportunity management is a *"Customer scoring"* task, as specified in the ontology. It provides a scoring of a lead who has shown interest based on patterns observed in interacting with the company, such as responding to an email, registering for a Webinar, or attending that Webinar. In churn management, there is a typical workflow task *"Customer complaint*

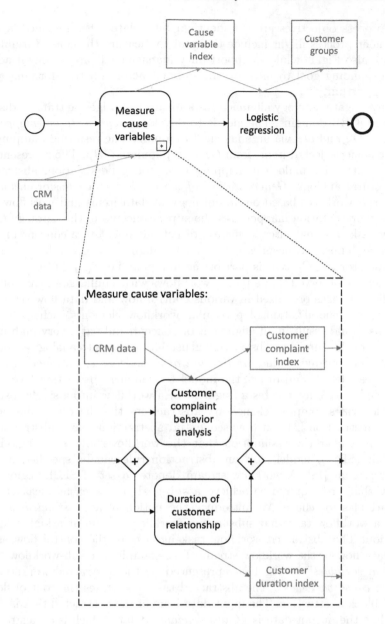

**Fig. 1.** Simplified sample workflow on churn analysis.

*behavior analysis*" (compare the sample workflow in Fig. 1) that is obviously different from "*Behavioral scoring*" for leads but has the super-concept "*Customer scoring*" in the churn management ontology. This means that both ontologies share the common concept "*Customer scoring*". Further, we assume that there are workflows in $CB_S$ and $CB_T$ addressing corresponding goals or sub-goals, i.e. the process objectives are not identical but related. For instance, a churn management process might include the goal to measure the cause variables for churn (see also Fig. 1) while an opportunity management process might address the corresponding goal to measure the impact factors on transforming a lead into an opportunity.

We investigate ontology alignment as a means to bridge the transfer distance between POCBR domains. The idea is to develop a *representation mapper* [13] that aligns $O_S$ and $O_T$ via an analogical mapping. The resulting mapping $f$ is used to transfer selected items from $CB_S$ to populate $CB'_T$. The representation mapper creates the analogical mapping based on generalization, abstraction, and structural analogy. *Generalization of a workflow* is a transformation into an isomorph workflow based on an ontology of data items and workflow tasks [19]. The representation mapper uses the super-concepts in the ontology $O_S$ for workflow tasks and data items where a direct alignment to a concept in $O_T$ is not feasible. Thus, a concept $x \in O_S$ can be aligned via generalization to the closest ancestor $\hat{x} \in O_S$ that is part of the mapping, i.e. $f(\hat{x}) \in O_T$.

Further, we observed that a pair of workflows with similar goals can comprise quite different tasks organized in various control flow and data flow structures. In such cases generalization of particular workflow elements such as tasks or data items would result in an alignment of concepts only at a very high hierarchical level of the ontology. Abstraction is used as a means to analogize workflow fragments. Polyvyanyy defines *abstraction of a workflow* as "a function that ... hides process details and brings the model to a higher abstraction level." [24]. An *abstraction rule* aggregates a fragment of a workflow into a single task [24]. Abstraction rules comprise elementary abstractions that have been introduced for BPM abstraction [24], such as sequential abstraction, block abstraction, or elimination. Under the assumption that the workflows follow a Single-Entry, Single-Exit (SESE) model [26]² an abstraction rule can be specified for each workflow stream [18]. A workflow stream denotes a set of SESE regions of a workflow that are required to achieve a sub-goal [18], such as measuring the cause variables for churn. We introduce the notion of an *abstracted workflow task* for a workflow task that subsumes a workflow stream at a higher level of abstraction. The abstracted workflow task aggregates the control flow as well as the data flow of the workflow stream. For example, the sub-workflow *"Measure cause variables"* in Fig. 1 is represented by the abstracted workflow task *"Measure cause variables"*. The abstracted task aggregates the control flow by an AND block abstraction. The input data item of the abstracted task is *"CRM data"* while the output data is *"Cause variable index"* which is an aggregation of *"Customer complaint index"* and *"Customer duration index"*. $O_S$ and $O_T$ are enriched by the abstract workflow tasks for all workflow streams that can be identified. The representation mapper uses the abstracted workflow tasks to align $O_S$ and $O_T$ at a higher level of abstraction. Thus, a workflow stream can be aligned via abstraction to the closest ancestor $\hat{x}_a \in O_S$ of its abstracted workflow task $x_a \in O_S$ that is part of the mapping, i.e. $f(\hat{x}_a) \in O_T$.

---

² SESE regions of a workflow are either a single workflow task or a larger fragment enclosed by corresponding split and join connectors [26].

In addition to the mapping of concepts that are common to both ontologies, we seek *structural analogies* in the ontologies to identify further mapping candidates. Gentner [8] defines analogy as an alignment process between two structured representations. As a starting point, we have chosen to analyse ancestor-descendant structures in the ontologies based on results of research on ontology alignment [21]. Ancestors of similar descendants, based on lexical similarity, become mapping candidates. In case a pair has a similar ancestor and a similar descendant with intermediate items in the target ontology the analogy detection method inserts intermediate items into the source ontology to alleviate the mapping of siblings of the descendant from the source ontology [5]. During the semi-automatic process of ontology engineering, it is decided which candidates become part of the actual mapping.

## 5    The Transfer Process

The transfer process aims to brigde the gap between the domains. It comprises two phases namely build time and transfer time.

The *build time* is the phase where transfer knowledge is created. The result of the build time is the representation mapper as described in Sect. 4. The phase includes two steps namely to enrich the ontologies and to create the analogical mapping. First, the existing case bases $CB_S$ and $CB_T$ are analysed to derive abstraction rules and to enrich the ontology with abstract workflow tasks as described above. At the moment, we identify workflow streams and the according abstract workflow tasks in a manual engineering process. More generally, abstraction tasks could be learned (compare recent work on learning adaptation operators [20]). Next, the analogical mapping is constructed following the ontology alignment methods decribed in Sect. 4.

The *transfer time* is the phase where the transfer knowledge is applied to the workflows from the source domain. We operationalize the transfer knowledge into a set of abstraction and generalization operators $OPS$, which transform workflows still within the source domain. The transfer process for a workflow $W_0$ is a search for operators $o_1$, $o_2$, ..., $o_n$ to form a transformation path $W_0 \Rightarrow^{o_1} W_1 \Rightarrow^{o_2} ... \Rightarrow^{o_n} W_f$ with the goal that the resulting workflow $W_f$ uses only vocabulary that is aligned to the target domain. Next, $W_f$ is translated directly into a workflow $W_f'$ in the target domain by replacing each activity and data object following the representation mapper. Please note that multiple transformation paths may exist for a workflow and that the translated workflows are likely to be on a high conceptual level. At the moment, we conduct a complete search for all transformation paths. This implies that there is a potential to create redundant cases which are structurally distinct. The phenomenon has been discussed in the literature on workflows, referred to as "workflow paraphrases" [28] or "variability" [10]. It occurs frequently in repositories of workflows that have been designed by human modeling experts. In our sample target domain, we consider it an advantage to achieve a variety of solutions. It could be useful to create additional opportunities by executing multiple workflows for the same problem.

## 6  Evaluation

We have conducted a preliminary experiment on initially five sample workflows with the aim to test the feasibility of our approach. We have chosen churn management as a source domain $D_S$ and the loosely related domain opportunity management as $D_T$. The experiment includes ratings from a CRM expert of the eleven workflows in the target case base $CB'_T$ that have been created by transferring the sample workflows from the source domain.

The experimental data includes two small case bases $CB_S$ with three workflow samples on churn management and $CB_T$ with two workflow samples on opportunity management. The workflow samples have been modeled in BPMN [4] following textual descriptions on typical churn and opportunity management processes. We retrieved the textual descriptions for $CB_S$ from SAP help[3]. The opportunity management samples originate from a book on Salesforce [29] and from a tutorial on lead management[4].

The experiment comprises the two phases build time and transfer time. During build time, two ontologies $O_S$ with finally 48 concepts and $O_T$ with finally 47 concepts have been engineered. $O_S$ includes four abstracted workflow tasks that have been derived directly from the workflow samples via their sub-workflows. $O_S$ has been enriched by three additional abstracted workflow tasks for workflow streams that have been identified in the workflow samples by the ontology engineers. The names for the latter tasks are taken from a reference process model on churn management from the literature [31]. Analog, $O_T$ has been enriched by one additional abstract workflow task following the nomenclature of a reference process model on opportunity management from the literature [9]. Table 1 lists the results of constructing the representation mapper for the workflow tasks. Only *"Customer profiling"* (line 4) is an abstracted task. The other tasks are generalized concepts in both domains. The mapping participants of data items are depicted in Table 2.

During transfer time, the three churn management workflows have been transformed using the representation mapper. We fully expanded the search

**Table 1.** Workflow tasks that are part of the mapping.

|   | Churn task in $O_S$ | Lead task in $O_T$ | Type of structural analogy |
|---|---|---|---|
| 1 | Analysis | Analysis | Direct overlap |
| 2 | Marketing action | Marketing action | Direct overlap |
| 3 | Data mining task | Data mining task | Direct overlap |
| 4 | Customer profiling | Customer profiling | Direct overlap |
| 5 | Customer scoring | Customer scoring | Inserted as intermediate concept |
| 6 | Preparatory analysis | Transform data | Via similar descendant |

---

[3] http://help.sap.com, last visit May 14, 2016.
[4] https://rdatascientist.wordpress.com/2015/08/15/, last visit May 14, 2016.

**Table 2.** Data items that are part of the mapping.

|   | Churn data in $O_S$ | Lead data in $O_T$ | Type of structural analogy |
|---|---|---|---|
| 1 | Analysis result | Analysis result | Direct overlap |
| 2 | CRM data | CRM data | Direct overlap |
| 3 | Customer list | Customer list | Direct overlap |
| 4 | Customer groups | Customer groups | Direct overlap |
| 5 | Classification result | Classification result | Inserted as intermediate concept |

space with the result that each source workflow achieved two target workflows. The first workflow resulted from preferring abstraction over generalization operators, the second vice versa.

We simulated the use of the transferred workflows for modeling support as follows. We have chosen manually a workflow stream from the target domain to refine every abstracted task. Since the size of our experimental data is quite limited this has led to five further target workflows. Figure 2 illustrates a sample workflow that results from transferring a churn management workflow with preference on abstraction operators. The workflow describes a three-step analysis of customer data in order to create new sales opportunities. It starts with the task *"Customer scoring"* that analyses the CRM data to filter out promising customers. *"Customer profiling"* is the task to acquire additional data on the customers. Finally, *"Customer segmentation"* is performed to identify the most promising customers. The abstracted task *"Customer profiling"* has been replaced by a workflow stream from $CB_T$, including the specialization from *"Customer list"* in the main workflow to *"Lead list"* in the sub-workflow. The modeler would probably propagate the same specialization to the main workflow, change some further data items and fill the black box for the abstracted task *"Customer segmentation"*, which has not be refined so far because the input data item of the candidate workflow stream from $CB_T$ does not match the input of the abstracted workflow task.

The eleven newly created target workflows from $CB'_T$ have been rated by an expert with a Likert scale for the estimated usefulness for the purpose of modeling support. The range is from a score of 1 for "unusable" to 5 for "extremely helpful". The results are shown in Table 3. Workflow S3 from the source domain results only in 3 target workflows since workflow 13 from the target domain does not contain any abstracted workflow task. Workflow 4 has a relatively low score because the order of tasks is not appropriate. The expert felt irritated with workflow 6 which contains two parallel tasks that apply a neural network to the same input data. This duplicate is a result of the sparse target ontology, which contains only one classification task namely neural networks. The results do not show a clear preference for the level of abstraction or for the preferred operators. However, the illustrating samples have been rated quite high, which provides a first hint for the general feasibility of the approach.

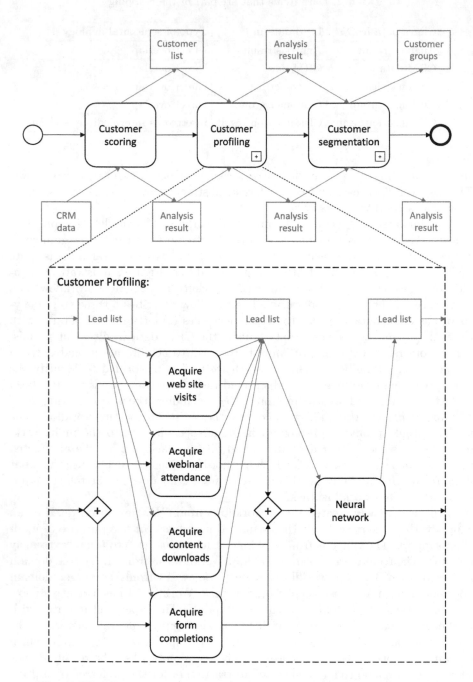

**Fig. 2.** Workflow on opportunity management as a sample transfer result.

**Table 3.** Score for the target workflows resulting from the expert rating.

| No in $CB_T'$ | No in $CB_S$ | Preferred operators | Level of abstraction | Score |
|---|---|---|---|---|
| 3 | S1 | Abstraction | Unaltered | 4.5 |
| 4 | S1 | Abstraction | Refined | 3 |
| 5 | S1 | Generalization | Unaltered | 3.5 |
| 6 | S1 | Generalization | Refined | 3 |
| 7 | S2 | Abstraction | Unaltered | 4 |
| 8 | S2 | Abstraction | Refined | 4 |
| 9 | S2 | Generalization | Unaltered | 5 |
| 10 | S2 | Generalization | Refined | 5 |
| 11 | S3 | Abstraction | Unaltered | 4 |
| 12 | S3 | Abstraction | Refined | 4.5 |
| 13 | S3 | Generalization | Unaltered | 4 |

# 7   Discussion and Conclusion

We have introduced a novel approach to transfer learning for process-oriented case-based reasoning and demonstrated its feasibility with a first lab experiment. Ontology alignment has been adopted to bridge the transfer distance between loosely related domains by a representation mapper. In particular, generalization and abstraction have been proposed to align workflow fragments in cases where a direct alignment is not feasible. Structural analogies in the vocabulary have been investigated in order to provide further transfer knowledge to be used by the representation mapper. The implementation is ongoing. The work is a first step towards an extension of the POCBR methods investigated in the research community so far.

Obviously, there are many open issues that might stipulate further research. The representation mapper requires improvement and a formative evaluation with a larger experimental base. The role of standard ontologies could be investigated as well as more sophisticated structure mapping approaches than our straight-forward analogical mapping. It is an intriguing open research issue which further methods of ontology alignment and beyond are promising to enrich the ontologies by useful transfer knowledge, such as mappings using further lexical and structural features [21] or machine learning approaches [6]. More sophisticated mapping methods will be investigated in our future work. A mapper could hypothesize correspondences between source and target concepts, for example, by using the ontologies $O_S$ and $O_T$ as previously described to match names, input and output data items for abstracted workflows tasks, or structural properties such as the same number of input and output data items. For each hypothesis, a mapping strength value could be determined. Specific matching rules [7,8] could be defined, for example, the rule that one source ontology concept must always be mapped to the same target ontology concept. Finally, the global mapping could be constructed such that it is consistent and maximizes an evaluation score that considers frequency numbers and the mapping strengths.

The subsequent step of our future work is to consider the *run time*, i.e. the phase where the transferred workflows in $CB'_T$ are used. For modeling support, a workflow $W'_f$ can be directly suggested to a user. Alternatively, a sequence of generalization and abstraction operators $o'_1, o'_2, \ldots, o'_m$ in the target domain can be searched to be applied "inversely" as specialization and refinement operators to $W'_f$. Currently, the latter is not yet implemented. A first idea is to employ methods of compositional adaptation using workflow streams [18].

A further interesting direction of research is to address the transfer of adaptation knowledge, such as process-oriented adaptation cases or workflow streams.

In addition, the impact of the transfer distance can be studied, for instance by varying the distance of workflow objectives or vocabulary. As a first step, such investigations require to develop a formal measure for the transfer distance between two domains. Encouraged by our preliminary results, we believe that TL for POCBR is a challenging new field with a reasonable chance of success and with a high impact for practical issues in business process management.

# References

1. Workflow Management Coalition Terminology and Glossary (Document No. WFMC-TC-1011) (1999)
2. Aha, D.W., Molineaux, M., Sukthankar, G.: Case-based reasoning in transfer learning. In: McGinty, L., Wilson, D.C. (eds.) ICCBR 2009. LNCS (LNAI), vol. 5650, pp. 29–44. Springer, Heidelberg (2009). doi:10.1007/978-3-642-02998-1_4
3. Bergmann, R., Wilke, W.: On the role of abstraction in case-based reasoning. In: Smith, I., Faltings, B. (eds.) EWCBR 1996. LNCS, vol. 1168, pp. 28–43. Springer, Heidelberg (1996). doi:10.1007/BFb0020600
4. Decker, G., Dijkman, R., Dumas, M., Garca-Bauelos, L.: The business process modeling notation. In: Hofstede, A.H.M., Aalst, W.M.P., Adams, M., Russell, N. (eds.) Modern Business Process Automation, pp. 347–368. Springer, Berlin (2009)
5. Dou, D., McDermott, D., Qi, P.: Ontology translation on the semantic web. In: Meersman, R., Tari, Z., Schmidt, D.C. (eds.) OTM 2003. LNCS, vol. 2888, pp. 952–969. Springer, Heidelberg (2003). doi:10.1007/978-3-540-39964-3_60
6. Ehrig, M., Staab, S., Sure, Y.: Bootstrapping ontology alignment methods with APFEL. In: Gil, Y., Motta, E., Benjamins, V.R., Musen, M.A. (eds.) ISWC 2005. LNCS, vol. 3729, pp. 186–200. Springer, Heidelberg (2005). doi:10.1007/11574620_16
7. Falkenhainer, B., Forbus, K.D., Gentner, D.: The structure-mapping engine: algorithm and examples. Artif. Intell. **41**(1), 1–63 (1989)
8. Gentner, D.: Structure-mapping: a theoretical framework for analogy. Cogn. Sci. **7**(2), 155–170 (1983)
9. Hickfang, O., Jacobshagen, M., Ritzerfeld, H.: Phasen im leadmanagementprozess. Technical report, Bitkom, Bundesverband Informationswitschaft, Telekommunikationund neue Medien e.V. (BITKOM), Albrechtstrae 10, 10117 Berlin (2006). https://www.bitkom.org/Bitkom/Publikationen/Leitfaden-Phasen-im-Leadmangement-Prozess.html

10. Hidders, J., Dumas, M., van der Aalst, W.M.P., ter Hofstede, A.H.M., Verelst, J.: When are two workflows the same? In: Proceedings of the 2005 Australian Symposium on Theory of Computing, CATS 2005, vol. 41, pp. 3–11. Australian Computer Society Inc., Darlinghurst, Australia (2005)
11. Klenk, M., Aha, D.W., Molineaux, M.: The case for case-based transfer learning. AI Mag. **32**(1), 54–69 (2011)
12. Klenk, M., Forbus, K.D.: Exploiting persistent mappings in cross-domain analogical learning of physical domains. Artif. Intell. **195**, 398–417 (2013)
13. Könik, T., O'Rorke, P., Shapiro, D.G., Choi, D., Nejati, N., Langley, P.: Skill transfer through goal-driven representation mapping. Cogn. Syst. Res. **10**(3), 270–285 (2009)
14. Kudenko, D.: Special issue on transfer learning. KI **28**(1), 5–6 (2014)
15. Kuhlmann, G., Stone, P.: Graph-based domain mapping for transfer learning in general games. In: Kok, J.N., Koronacki, J., Mantaras, R.L., Matwin, S., Mladenič, D., Skowron, A. (eds.) ECML 2007. LNCS (LNAI), vol. 4701, pp. 188–200. Springer, Heidelberg (2007). doi:10.1007/978-3-540-74958-5_20
16. Minor, M., Montani, S., Recio-García, J.A.: Editorial: process-oriented case-based reasoning. Inf. Syst. **40**, 103–105 (2014)
17. Montani, S., Leonardi, G., Quaglini, S., Cavallini, A., Micieli, G.: Mining and retrieving medical processes to assess the quality of care. In: Delany, S.J., Ontañón, S. (eds.) ICCBR 2013. LNCS (LNAI), vol. 7969, pp. 233–240. Springer, Heidelberg (2013). doi:10.1007/978-3-642-39056-2_17
18. Müller, G., Bergmann, R.: Workflow streams: a means for compositional adaptation in process-oriented case-based reasoning. In: Lamontagne, L., Plaza, E. (eds.) Case-Based Reasoning. Research and Development. LNCS, vol. 8766, pp. 315–329. Springer, Switzerland (2014)
19. Müller, G., Bergmann, R.: Generalization of workflows in process-oriented case-based reasoning. In: Proceedings of the Twenty-Eighth International Florida Artificial Intelligence Research Society Conference, FLAIRS 2015, pp. 391–396. AAAI Press. Hollywood (Florida), USA (2015)
20. Müller, G., Bergmann, R.: Learning and applying adaptation operators in process-oriented case-based reasoning. In: Hüllermeier, E., Minor, M. (eds.) ICCBR 2015. LNCS (LNAI), vol. 9343, pp. 259–274. Springer, Heidelberg (2015). doi:10.1007/978-3-319-24586-7_18
21. Noy, N.F.: Ontology Mapping. In: Staab, S., Studer, R. (eds.) Handbook on Ontologies. International Handbooks on Information Systems, pp. 573–590. Springer, Heidelberg (2009)
22. Pan, S.J., Yang, Q.: A survey on transfer learning. IEEE Trans. Knowl. Data Eng. **22**(10), 1345–1359 (2010)
23. Perkins, D.N., Salomon, G., Press, P.: Transfer of learning. In: International Encyclopedia of Education, 2 edn. Pergamon Press (1992)
24. Polyvyanyy, A., Smirnov, S., Weske, M.: Business process model abstraction. In: Brocke, J., Rosemann, M. (eds.) Handbook on Business Process Management 1, Introduction, Methods, and Information Systems. International Handbooks on Information Systems, 2nd edn, pp. 147–165. Springer, Heidelberg (2015)
25. Ragni, M., Strube, G.: Cognitive complexity and analogies in transfer learning. KI **28**(1), 39–43 (2014)
26. Reichert, M., Weber, B.: Enabling Flexibility in Process-Aware Information Systems: Challenges, Methods, Technologies. Springer, Heidelberg (2012)
27. Richter, M.M., Weber, R.O.: Case-Based Reasoning - A Textbook. Springer, Heidelberg (2013)

28. Schumacher, P.: Workflow extraction from textual process descriptions. Ph.D. thesis, Goethe University Frankfurt am Main (2015)
29. Taber, D.: Salesforce. Com Secrets of Success: Best Practices for Growth and Profitability. 2nd revised edn. Prentice Hall (2013)
30. Taylor, M.E., Stone, P.: Transfer learning for reinforcement learning domains: A survey. J. Mach. Learn. Res. **10**, 1633–1685 (2009)
31. Tecklenburg, T.: Churn-Management im B2B-Kontext. Gabler, Wiesbaden (2008)
32. Thorndike, E.L., Woodworth, R.S.: The influence of improvement in one mental function upon the efficiency of other functions. (I). Psychol. Rev. **8**(3), 247–261 (1901)

# Case Completion of Workflows
# for Process-Oriented Case-Based Reasoning

Gilbert Müller$^{(\boxtimes)}$ and Ralph Bergmann

Business Information Systems II, University of Trier, 54286 Trier, Germany
{muellerg,bergmann}@uni-trier.de
http://www.wi2.uni-trier.de

**Abstract.** Cases available in real world domains are often incomplete
and sometimes lack important information. Using incomplete cases in a
CBR system can be harmful, as the lack of information can result in
inappropriate similarity computations or incompletely generated adaptation knowledge. Case completion aims to overcome this issue by inferring missing information. This paper presents a novel approach to case
completion for process-oriented case-based reasoning (POCBR). In particular, we address the completion of workflow cases by adding missing
or incomplete dataflow information. Therefore, we combine automatically learned domain specific completion operators with generic domain-
independent default rules. The empirical evaluation demonstrates that
the presented completion approach is capable of deriving complete workflows with high quality and a high degree of completeness.

**Keywords:** Process-oriented case-based reasoning · Workflows · Workflow completion · Case completion · Completion operators · Completion
rules

## 1 Introduction

In Case-based Reasoning (CBR), knowledge is distributed over four knowledge
containers [19], i.e., the case base, the vocabulary, the similarity measure, and
the adaptation knowledge. If any of these knowledge containers is not sufficiently
specified, this incomplete container can be a burden for the entire CBR application unless another container is able to cover this knowledge gap. Consequently,
either the CBR system will not work or propose solutions with major mistakes.
Thus, lessening the containers' knowledge gap has been extensively investigated,
e.g., by learning of similarity measures [24] or adaptation knowledge [2,7,11,27].

It has been shown that incomplete cases, i.e., cases with missing information, also lead to such a critical knowledge gap [5]. A case base of incomplete
cases naturally results in a retrieval of incomplete cases that do not provide all
required information. Furthermore, similarity assessment is hampered because
necessary attributes are missing. Moreover, incomplete cases may also affect the
adaptation knowledge container, if adaptation knowledge is learned automatically from the case base. In this case the adaptation knowledge is also incomplete.

© Springer International Publishing AG 2016
A. Goel et al. (Eds.): ICCBR 2016, LNAI 9969, pp. 295–310, 2016.
DOI: 10.1007/978-3-319-47096-2_20

Consequently, adaptation capabilities are affected because adaptation quality is reduced or adaptation is prevented. Hence, incomplete cases may have a critical impact on the entire utilisation of the CBR application. Thus, various approaches have been presented that address case completion [1,6,10,20].

In this paper, we investigate this issue within process-oriented CBR (POCBR) [14], which deals with CBR applications for process-oriented information systems. POCBR aims at assisting domain experts in their work with workflows, in particular by supporting workflow reuse. Two important problems of workflow reuse are the retrieval of similar workflows from potentially large repositories [4] as well as the adaptation of workflows [15]. Recently, work has been presented for POCBR, where adaptation knowledge is also learned automatically from the case base [13,15]. The issue of incomplete cases also exists in the field of POCBR. Our investigations showed that workflows in existing workflow repositories are mostly incomplete. Whilst the control-flow of workflows is mostly fully specified, the dataflow usually is not completely defined or not existent at all. The IWi process model corpus [25], for example, contains more than 4000 process models resulting from various domains and sources. However, in most of these process models the dataflow is not complete. Consequently, employing such incomplete workflow cases may significantly affect retrieval and adaptation capabilities. This problem also arises in textual process descriptions. In the cooking domain, for example, textual descriptions of recipes contain not a full description of the dataflow, i.e., the output of tasks is usually missing. Thus, this issue has also recently been addressed by workflow extraction from textual process descriptions [21].

This paper presents a new approach for completing the dataflow of workflows automatically. Figure 1 summarizes the idea of completing workflow cases to lessen the knowledge gap in the case base as well as in the adaptation knowledge. Prior to inserting workflows into the case base, automatic case completion is applied. Case completion makes use of domain independent default completion rules as well as domain specific completion operators, automatically learned from the case base. Subsequently, adaptation knowledge is learned based on complete cases and thus retrieval as well as adaptation can be based on complete cases resulting in more complete adaptation knowledge.

In order to automatically learn completion operators, an initial case base has to be established that consists of workflows containing substantially defined dataflows. In application scenarios in which available workflows significantly lack the dataflow information, a small set of fully specified workflows may need to be completed manually to be used as training data. The learned completion operators can then be used to automatically complete the remaining workflow cases.

In the next section, our previous work is summarized providing the fundamentals for the presented approach. Then, workflow completion operators are introduced (see Sect. 3) and Sect. 4 describes how they can be learned from the case base automatically. Our novel approach to complete workflow cases based on the learned completion operators is presented in Sect. 5. Finally, we illustrate the feasibility of our approach by an evaluation (see Sect. 6) and sum up the paper with conclusions and future work.

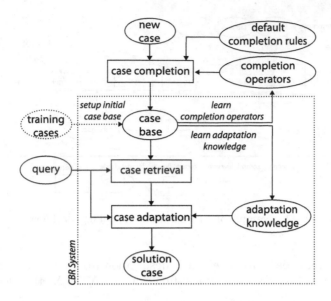

**Fig. 1.** Framework of case completion adopted from [5]

## 2    Foundations

We now briefly introduce relevant previous work in the field of POCBR.

### 2.1    Workflows

Broadly speaking, a *workflow* consists of a set of *activities* (also called *tasks*) combined with *control-flow structures* like sequences, parallel (AND) or alternative (XOR) branches, as well as repeated execution (LOOP). In addition, tasks exchange certain *data items*, which can also be of physical matter, depending on the workflow domain (e.g., ingredients in the cooking domain). Tasks, data items, and relationships between the two of them form the *dataflow*.

We illustrate our approach in the domain of cooking recipes (see example workflow in Fig. 2). A cooking recipe is represented as a workflow describing the instructions for cooking a particular dish [23]. Here, the tasks represent the cooking steps and the data items refer to the ingredients being processed by the cooking steps. An example cooking workflow for a sandwich recipe is illustrated in Fig. 2. Based on our previous work [15,17,18] we now introduce the relevant formal workflow terminology.

**Definition 1.** *A workflow is a directed graph $W = (N, E)$ where $N$ is a set of nodes and $E \subseteq N \times N$ is a set of edges. The nodes can either be data nodes $N^D$, task nodes $N^T$, or control-flow nodes $N^C$, i.e., $N = N^D \cup N^T \cup N^C$. In addition, we call $N^S = N^T \cup N^C$ the set of sequence nodes. Edges can be either control-flow edges $E^C \subseteq N^S \times N^S$, which define the order of the sequence nodes or dataflow edges $E^D \subseteq (N^D \times N^S) \cup (N^S \times N^D)$, which define how the data is shared between the tasks, i.e., $E = E^C \cup E^D$.*

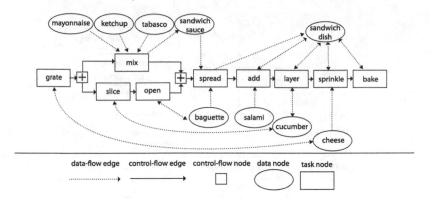

**Fig. 2.** Example of a block-oriented cooking workflow

Furthermore, nodes have a semantic description from a semantic meta data language $\Sigma$, which is assigned by the function $S : N \rightarrow \Sigma$, thus leading to a semantic workflow [4].

The control-flow edges $E^C$ of a workflow induce a strict partial order on the sequence nodes $N^S$. Thus, we define $s_1 < s_2$ for two sequence nodes $s_1, s_2 \in N^S$ as a transitive relation that expresses that $s_1$ is executed prior to $s_2$ in $W$.

We further denote $t^D$ as the set of data nodes produced by a task $t$, i.e., $t^D = \{t \in N^T | \exists (t, d) \in E^D\}$, and $t_D$ as the set of data nodes consumed by the task $t$, i.e., $t_D = \{t \in N^T | \exists (d, t) \in E^D\}$. Data nodes that are produced by a task, but not consumed by this task are referred to as creator data nodes $N^{D*} = \{d \in N^D | \exists t \in N^T : d \in t^D \wedge d \notin t_D\}$. Moreover, data nodes $d_1, d_2 \in N^D$ are dataflow connected $d_1 \ltimes d_2$ if there exists a task that consumes data node $d_1$ and produces data node $d_2$. Moreover, $d_1 \ltimes^+ d_2$ denotes that $d_1, d_2 \in N^D$ are transitively dataflow connected:

$$d_1 \ltimes d_2, \text{iff } \exists t \in N^T : (d_1 \in t_D \wedge d_2 \in t^D) \tag{1}$$

$$d_1 \ltimes^+ d_2, \text{iff } d_1 \ltimes d_2 \vee \exists d \in N^D : (d_1 \ltimes d \wedge d \ltimes^+ d_2) \tag{2}$$

## 2.2   Block-Oriented Workflows

We now restrict the workflow representation to block-oriented workflows [15], i.e., workflows in which the control-flow structures form blocks of nested workflows with an opening and closing control-flow element. These blocks must not be interleaved.

**Definition 2.** *A block-oriented workflow is a workflow in which the control-flow nodes $N^C = N^{C*} \cup N^{C^*}$ define the control-flow blocks. Each control-flow block has an opening node from $N^{C*}$ and a related closing node from $N^{C^*}$ specifying either an AND, XOR, or LOOP block. These control-flow blocks may be nested but must not be interleaved and must not be empty.*

Figure 2 shows an example block-oriented workflow, containing a control-flow block with an opening AND control-flow node and a related closing AND control-flow node (denoted by [+]).

As we aim at improving the completeness of incomplete workflows, we introduce a terminology of *consistent block-oriented workflows* to define our perception of complete workflows.[1] According to Davenport, "[...] a process is simply a structured, measured set of activities designed to produce a specific output [...]" [8]. In the following, these specific workflow outputs are denoted as $W^O \subseteq N^D$. In the cooking domain, the specific output is the particular dish produced, i.e., "sandwich dish" in Fig. 2. Hence, for a *consistent workflow*, we require that each ingredient must be contained in the specific output, as otherwise the ingredient as well as the related tasks would be superfluous. Further, identical AND/OR branches as well as several creator tasks producing the same data output usually are not very plausible and must consequently not exist in a consistent workflow. Thus, we define consistent workflows for this paper as follows.

**Definition 3.** *A block-oriented workflow is* consistent, *if and only if the following conditions hold. Each produced data node is contained in the specific output of the workflow. Thus, each data node must be transitively dataflow connected to the specific output $W^O$, i.e., $\forall d \in N^D \exists o \in W^O : d \bowtie^+ o$ and further each task has at least one input and one output data object, i.e., $t_D \neq \emptyset$ and $t^D \neq \emptyset$. Additionally, the workflow must not have identical XOR/AND branches[2] as well as several creator tasks $t_1, t_2$ producing the same data output, i.e., $t_1^D = t_2^D$, unless they are contained in different AND branches (see footnote 2).*

### 2.3 Taxonomies

The vocabulary of our POCBR system consists of taxonomical representation of terms of the tasks and data objects as previously defined by [17].

**Definition 4.** *A taxonomy $\psi$ is a tree of a partially ordered set of semantic terms $\Gamma = \{\gamma_0, \ldots, \gamma_n\}$, whereas $\gamma_i \sqsubset \gamma_j$ denotes that $\gamma_j$ is more general than $\gamma_i$ ($\gamma_i$ is more specific than $\gamma_j$). Further, $\gamma_i \sqsubseteq \gamma_j$ holds, iff $\gamma_i \sqsubset \gamma_j \lor \gamma_i = \gamma_j$.*

We use two distinct taxonomies, one for the task nodes $\psi_{\text{tasks}}$ (preparation steps) and one for the data nodes $\psi_{\text{data}}$ (ingredients), i.e., $\Sigma = \Gamma_{\psi_{\text{tasks}}} \cup \Gamma_{\psi_{\text{data}}}$. Hence, the function $S$ assigns to each node $n \in N$ an appropriate term from $\Gamma_{\psi_{\text{tasks}}}$ or $\Gamma_{\psi_{\text{data}}}$.[3] As an example, Fig. 3 shows the ingredients taxonomy, in which $beef \sqsubset meat$ holds. Please note that in taxonomies, the leaf nodes represent concrete entities that may occur in executable workflows. For example, a recipe may include potatoes and beef as potential ingredients, but usually not terms from

---

[1] The terms *completeness* and *consistency* used here with respect to workflows must not be confused with the use of those terms within logics — here, we mean something different, as defined below.

[2] Within the same control-flow block.

[3] We omit the index if it is obvious which ontology is referenced.

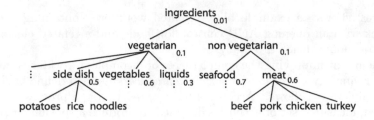

**Fig. 3.** Example for a data taxonomy

the inner nodes, such as vegetarian or meat. An inner node $\gamma$ represents a generalized term that stands for the set of more specific terms below it. For example, the generalized term *vegetarian* stands for the set $\{potatoes, rice, noodles\}$. Further on in the paper we use inner nodes in generalized completion operators to represent that an arbitrary ingredient from the set of its specializations can be chosen. Each inner node $\gamma$ is additionally annotated with a value $sim_\psi$ defining the similarity between all child terms of $\gamma$, (e.g., $sim_\psi(\text{meat}) = 0.6$ in Fig. 3).

## 3   Workflow Completion Operators

For completing the dataflow of workflows we now introduce completion operators that define valid flows of the data for a particular task. More precisely, the operator defines which input data nodes can be processed by a task and which output data nodes result after the task has been executed. A completion operator is represented as a workflow as defined in Definition 5.

**Definition 5.** *A completion operator $o$ is a workflow $o = (N_o, E_o)$ consisting of a single task $t$ (also denoted as $t_o$), i.e., $N_o^T = \{t\}$ and $N_o^C = \emptyset$, together with the set of input data nodes $t_D$ and output data nodes $t^D$, i.e., $N_o^D = t_D \cup t^D$.*

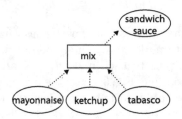

**Fig. 4.** Example completion operator

An example completion operator is illustrated in Fig. 4. It describes that mayonnaise, ketchup, and tabasco can be mixed to produce a sandwich sauce. Operators can also contain generalized terms of tasks and data nodes. Figure 5 shows such a generalized operator. It describes that mayonnaise and any sauce can be combined together with an arbitrary flavoring such that a sandwich sauce is produced.

We now describe under which circumstances an operator is applicable for a task $t$ of workflow $w$. Therefore, we introduce two mapping functions $m_{in}$ : $t_{oD} \rightarrow N^D$ and $m_{out}$ : $t_o^D \rightarrow N^D$, whose existence is a precondition for the application of an operator on a task $t$ (see Definition 6). The mappings enforce that the operator data nodes map the current corresponding input and output data nodes of task $t$. Furthermore, the labels of the operator data nodes must be identical or more general than those labels of the mapped data nodes in the workflow. Furthermore, as $m_{in}$ is a total mapping function, it requires that all input data nodes of the operator $o$ are contained in the workflow $w$ (it is not necessary that they are connected to $t$). In contrast to $m_{in}$, $m_{out}$ does not enforce that all output data nodes of the operator are contained in the workflow (in this case the corresponding output is going to be created).

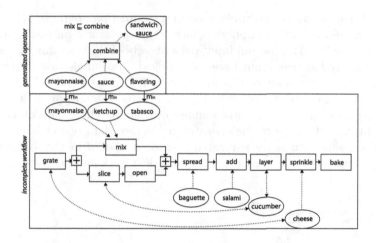

**Fig. 5.** Example mapping between generalized operator and incomplete workflow

**Definition 6.** *Let $m_{in} : t_{oD} \rightarrow N^D$ be a total injective mapping for a task $t$ in workflow $w = (N, E)$ s.t. $\forall d \in t_D : \exists d' \in t_{oD} : d = m_{in}(d')$ and $\forall d \in t_{oD} : S(m_{in}(d)) \sqsubseteq S(d)$. Further, $m_{out} : t_o^D \rightarrow N^D$ is a partial injective mapping s.t. $\forall d \in t^D : \exists d' \in t_o^D : d = m_{out}(d')$ and $\forall d \in t_o^D : S(m_{out}(d)) \sqsubseteq S(d)$.*

Based on these mappings we will now explain the applicability of an operator $o$ on task $t$ in Definition 7.

**Definition 7.** *An operator $o = (N_o, E_o)$ with operator task $o_t \in N_o^T$ is applicable for a task $t$ under the mapping pair $(m_{in}, m_{out})$ iff $S(t) \sqsubseteq S(o_t)$ and the mapping functions $m_{in} : t_{oD} \rightarrow N^D$ and $m_{out} : t_o^D \rightarrow N^D$ exist as previously defined.*

Applying a completion operator can result in multiple different workflows, because different data objects might be considered as input or output of a task, i.e., several possible mappings exist. We define a possible workflow resulting from the application of a completion operator with operator task $o_t \in N_o^T$ under $(m_{in}, m_{out})$ as follows. The mapping describes the input and output data links that have to be constructed in the dataflow of $W'$ if not already present. Additionally, newly added output data nodes also have to be linked to the output of the task $t$.

**Definition 8.** $W' = (N', E')$ *is a possible workflow after applying completion operator* $o = (N_o, E_o)$ *with operator task* $o_t \in N_o^T$ *under* $(m_{in}, m_{out})$ *on the task* $t$ *of workflow* $W = (N, E)$ *iff* $N' = N \cup N^{D+}$ *and* $E' = E \cup E^{D+}$. *Here,* $N^+$ *is the set of output data nodes produced by the operator application, i.e.,* $N^{D+} = \{d \in o_t^D \mid \nexists d' \in N^D : S(d') \sqsubseteq S(d)\}$. $E^{D+}$ *defines the new input and output data links generated by the operator such that* $\forall d \in t_{oD} : (m_{in}(d), t) \in E^{D+}$ *and* $\forall d \in t_o^D : (t, m_{out}(d)) \in E^{D+}$ *and* $\forall d \in N^{D+} : (t, d) \in E^{D+}$.

We will now illustrate the application of a completion operator by an example. Suppose we are going to apply a generalized operator $o$ on task $t$ "mix" of workflow $w$ (see Fig. 5). First, all input data objects of the operator that are not already connected as task input have to be linked. Mayonnaise and ketchup, for example, are already connected as input to task $t$, but tabasco still has to be linked (see grey marked data link in Fig. 6). Finally, the output data object(s) have to be either linked in the same manner or if no matching data object exists for this output data node of the operator $N_o^T$ a new data object is created. In our example, sandwich sauce is created as a data node $d$ in the workflow and linked as an output to the task $t$ (see grey marked workflow elements in Fig. 6).

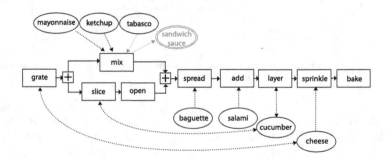

**Fig. 6.** Example result for an operator application

**Algorithm** COMPLETE_WF($W$,$O$);
**Input:** Workflow $W = (N, E)$ and completion operators $O$
**Output:** Set of completed workflow solutions $SOLUTIONS$
SOL=$\{W\}$;
**for each** $task\ t \in N^T$ *(tasks ordered by control-flow, i.e., first task is processed first)* **do**
    Init newSOL = $\emptyset$;
    **for each** *state* $s \in SOL$ **do**
        **for each** *operator* $o$ *with mapping pair* $(m_{in}, m_{out}) \in CO(t)$ **do**
            $s'$ := successor state of $s$ by applying $o$ under $(m_{in}, m_{out})$;
            **if** $\neg$ *(s' can never become consistent)* **then**
                $newSOL = newSOL \cup \{s'\}$;
    SOL=newSOL;
$SOL' = applyDefaultRules(SOL)$;
**return** $selectCompletedWorkflow(SOL')$;

**Algorithm 1.** Dataflow completion

## 4   Learning Workflow Completion Operators

To avoid modelling completion operators by hand, we learn them automatically from the workflows from the case base. The idea is that for each complete task $t$ (i.e., $t_d \neq \emptyset$ and $t^d \neq \emptyset$) of workflow $W = (N, E)$ in the case base a corresponding operator $o = (N_o, E_o)$ is constructed, such that $t$ defines the operator task $t_o$, i.e., $N_o^T = \{t\}$. The data nodes connected to this task $N_o^D = \{d \in N^D | t : d \in t_D \vee d \in t^D)\}$ represent the corresponding input and output data nodes of the operator task $t$, i.e., $E_o^D = \{(d,t) \cup (t,d) \in E^D | d \in N_o^D\}$.

Next, the learned completion operators are generalized in order to increase their applicability. This means that each task description is generalized to a level of the taxonomy which has a similarity value of at most $\theta$ assigned, i.e., $sim_\psi(S(n)) < \theta$. Here, $\theta$ is a threshold value that is used to control the degree of generalization performed.

When completion operators are constructed from several workflows, we can have several different operators for the same task label, i.e., several different operators for the task "mix". However, a single operator can also be derived from several different workflows. Those operators are stored only once. In addition, a score value $Score(o)$ is recorded that counts how often an operator has been constructed from the case base. This score value reflects a kind of significance, which is used later to prefer the application of highly scored operators during completion.

## 5   Workflow Completion

We now introduce the approach to complete the dataflow of incomplete workflows. First the learned domain specific operators are applied (Sect. 5.1). Subsequently, we apply some generic default rules (Sect. 5.2) for those tasks that could not be completed appropriately by a domain specific operator.

## 5.1 Applying Learned Completion Operators

Applying the learned completion operators aims to complete the dataflow of the workflow in such a manner that a best possible completed workflow w.r.t. consistency is produced. The algorithm is based on a breadth first search (see Algorithm 1). Starting from the first task node of the workflow the dataflow is expanded by each *applicable operator* resulting in multiple successor states. For all these successor states the dataflow of the next task w.r.t. the control-flow is expanded by all *applicable operators* and so forth. The set of applicable completion operators for a task $t$ together with the mapping pair under which they are applicable is denoted as $CO(t)$.

In order to reduce run time as well as memory consumption of the algorithm, we withdraw current solutions as soon as it is ensured that they can never become consistent according to Definition 3, i.e., if they contain several creator tasks with identical output data nodes or identical XOR or AND branches. It turned out that this is sufficient to restrict the search space to a feasible level.

After the application of domain-specific operators a set $SOL$ is created in which for some tasks the completion of the dataflow could be achieved. However, for some tasks no applicable completion operator may exist. Thus, we apply a set of default completion rules on those tasks as described in the next section.

## 5.2 Applying Default Completion Rules

For all workflow solutions $S = (N, E) \in SOL$ we now apply default completion rules on each task that has no output data nodes, i.e., $t^D = \emptyset$ (see *applyDefaultRules* $(SOL)$ in Algorithm 1). Thereby, tasks that are incomplete are completed by applying the generic default rules. These default rules represent common, domain independent heuristics for producing a plausible output for a task.

First, from the available set of operators $O$, we determine which data nodes are used as creator data nodes $D^* = \bigcup_{o \in O} N_o^{D*}$ in the set of operators, i.e., those data nodes which are produced by a task but not consumed by this task, e.g., sandwich sauce. Further, $t^{D-} = \{d \in N^D | \nexists t' \in N^T \wedge d \in t'^D \wedge t' < t\}$ defines the set of unproduced data objects w.r.t. task $t$, i.e., those data objects that have not been produced by a preceding task $t'$ of $t$. We refer to such data nodes as unproduced data nodes.

Next, we apply a default rule base for each task $t$ that has no output, i.e., $t^D = \emptyset$, ordered w.r.t. the control-flow (see rule base in Fig. 7). The rule base is applied in a top-down manner, such that only the first applicable rule is fired to determine the output of task $t$. If $t$ has only one input data object, it is set as the output. Otherwise if $t$ has only one creator data node that exists in the current workflow (i.e. contained in $N^{D*}$) as an input, it is set as the output. The third rule describes that if exactly one input $d$ of $t$ is used by a successor task as an input again, $d$ is the output. The next rule sets the output of the task $t$ as exactly this data node $N^D$ that has not been produced by a prior task and which is the

1: **IF** $t^D = \emptyset \wedge \exists! d \in t_D$ **THEN** $t^D := \{d\}$
2: **IF** $t^D = \emptyset \wedge \exists! d \in (t_D \cap N^{D*})$ **THEN** $t^D := \{d\}$
3: **IF** $t^D = \emptyset \wedge \exists! d \in t_D : \exists t' \in N^T \wedge t < t' \wedge d \in t'_d$ **THEN** $t^D := \{d\}$
4: **IF** $t^D = \emptyset \wedge \exists! d \in t^{D-} : \exists t' \in N^T \wedge t < t' \wedge d \in t'_d$ **THEN** $t^D := \{d\}$
5: **IF** $t^D = \emptyset \wedge \exists! d \in t^{D-} : (\exists d^* \in D^* : S(d) = S(d^*)) \wedge (\exists t' \in N^T \wedge t < t' \wedge d \in t'_d)$
**THEN** $t^D := \{d\}$

**Fig. 7.** Default rule base ($\exists! x$ denotes that exactly one $x$ exists)

only one used by the closest possible successor task as an input again. Finally, rule 5 describes the same condition as in rule 4 but restricts the possible output rules to those with labels of the creator data nodes in $D^*$ determined from the available set of operators.

### 5.3   Heuristical Solution Selection

After completing the workflow according to the previous sections, we now introduce a heuristical selection method which aims at selecting a single solution from $SOL'$ (see function $selectCompletedWorkflow(SOL)$ in Algorithm 1). As we aim to identify consistent workflows or at least workflows with most completed dataflow we apply the following steps.

1. Initialize $SOL'$ with the set of completed solutions $SOL$.
2. If $SOL'$ contains at least one consistent workflow, then all other workflows which are not consistent are removed from $SOL'$.
3. Next, only retain those workflow solutions in $SOL'$ with the largest number of complete tasks, i.e., tasks which have at least one input and one output data node.
4. Finally, rank the remaining workflows in $SOL'$ according to the sum of the score values of the completion operators that have been applied to produce them. Select the workflow with the highest score.

## 6   Evaluation

The described approach to workflow completion has been implemented as component of the CAKE framework [3]. We investigated, whether the dataflow could be completed such that the result is consistent (Hypothesis H1). Moreover, we validated whether the completed dataflow of the workflows are of acceptable quality (Hypothesis H2).

**H1.** The presented workflow completion approach constructs consistent workflows in most cases.
**H2.** The presented workflow completion approach completes the dataflow such a manner that the resulting workflows are of high quality.

## 6.1   Evaluation Data

We manually constructed 61 sandwich recipe workflows from the textual recipe descriptions from Internet sources with an average size of 20 nodes and 36 edges [16]. All these workflows are consistent, i.e., their dataflow is completely defined and of good quality as the dataflow is constructed according to proper interpretation of the textual recipe. Altogether, they contain 178 different ingredients and 79 tasks. For ingredients and tasks, a taxonomy was manually constructed. The extracted workflows contained AND, XOR, as well as LOOP structures. These 61 workflows serve as our training case base $CB_{train}$. They further represent golden standard workflows, i.e., as comparison workflows which should be reconstructed by the automatic dataflow completion of the workflows. Previous work (e.g., [18]) showed that such defined consistent workflows are convenient for retrieval and adaptation in POCBR. We performed an ablation study by automatically removing dataflow from the training workflows to construct incomplete test workflows. An investigation of textual recipes showed that for each step the input data objects are usually always explicitly contained in the instructions, while the output data objects are commonly missing. Thus, we removed all output data links from the workflows, resulting in a set of 61 incomplete workflows as our test case base $CB_{test}^{I}$. We also generated a second test case base $CB_{test}^{II}$ with a larger deconstruction of the workflow, i.e., we constructed workflows such that they only consist of input data nodes and only input data links to this task by which the corresponding input data is consumed for the first time w.r.t. the control-flow. Consequently, only the ingredients contained in the ingredient list are contained in the workflows' dataflow. As quality measures of an automatically completed workflow and a golden standard workflow we employ the precision and recall measures as defined by Schumacher et al. [22].

$$precision_T(t, t^*) = \frac{|t^D \cap t^{*D}| + |t_D \cap t_D^*|}{|t^D| + |t_D|} \tag{3}$$

$$recall_T(t, t^*) = \frac{|t^D \cap t^{*D}| + |t_D \cap t_D^*|}{|t^{*D}| + |t_D^*|} \tag{4}$$

$Precision_T$ defines the percentage of the data links of task $t$ that are contained in the golden standard task $t^*$, while $recall_T$ defines the percentage of data links of the golden standard task $t^*$ that have been reconstructed.

$$precision(W, W^*) = \frac{1}{|N^T|} + \sum_{i=0}^{|N^T|} precision_t(t_i, t_i^*) \tag{5}$$

$$recall(W, W^*) = \frac{1}{|N^T|} + \sum_{i=0}^{|N^T|} recall_t(t_i, t_i^*) \tag{6}$$

$Precision$ (see Eq. 5) defines how many of the tasks' data links are contained in the golden standard workflow $W^*$, while recall (see Eq. 6) defines how many of the tasks' data links of the golden standard workflow $W^*$ have been reconstructed. In both equations $t_i$ and $t_i^*$ refer to the same task but w.r.t. reconstructed workflow and golden standard workflow respectively.

$$F1(W, W^*) = 2 \cdot \frac{precision(W, W^*) \cdot recall(W, W^*)}{precision(W, W^*) + recall(W, W^*)} \tag{7}$$

The F1 measure aggregates both, the precision and recall values in a single score, which is basically the harmonic mean of precision and recall.

## 6.2   Experimental Evaluation and Results

We employed the leave-one-out evaluation principle to evaluate our presented approach. More precisely, we selected each workflow $w$ as a test workflow from the case base $CB_{test}^I$ or $CB_{test}^{II}$ respectively. Next, we learned all operators from the training case base $CB_{train}$ (without the corresponding training workflow of test workflow $w$) resulting in a set of operators $O_w$. In average for each of the test workflows 283 operators have been generated.

Next we applied our presented approach by completing the dataflow of workflow $w$ using the set of operators $O_W$ learned from the other workflows. In average 6.41 completion operators for $CB_{test}^I$ and 5.26 completion operators for $CB_{test}^{II}$ have been applied for each workflow. In average 2.0 solutions in $SOL'$ have been generated by $CB_{test}^I$ and 3.0 solutions in $SOL'$ by $CB_{test}^{II}$ of which the presented heuristic (see Sect. 5.3) identified a single solution. After completing the dataflow of the workflow, 44 test workflows (72.13 %) have been transformed to consistent workflows for $CB_{test}^I$. However, for the test case base $CB_{test}^{II}$ only 4 test workflows (6.56 %) have been transformed to consistent workflows. Thus, Hypothesis H1 has been mostly confirmed for $CB_{test}^I$ but not for $CB_{test}^{II}$. This shows that consistent solutions can only be constructed if enough information either in the incomplete workflow description or in the learned completion operators is contained.

**Table 1.** Quality compared to golden standard workflow

|  | $CB_{test}^I$ | | $CB_{test}^{II}$ | |
|---|---|---|---|---|
|  | Uncompleted case | Completed case | Uncompleted case | Completed case |
| Precision | 1.00 | 0.98 (−0.02) | 1.00 | 0.85 (−0.15) |
| Recall | 0.59 | 0.98 (+0.39) | 0.46 | 0.74 (+0.28) |
| F1 | 0.74 | 0.98 (+0.24) | 0.63 | 0.79 (+0.16) |

Moreover, we investigated the quality of the completed workflow by comparing it to the corresponding training workflow $w_{train}$. The results are illustrated in Table 1 showing the average precision, recall, and F1 score values for $CB_{test}^I$ and $CB_{test}^{II}$. We can see that while the precision has not been significantly reduced (low rate of wrong data links) the recall has significantly been increased from 0.6 to 0.98 resulting also in a weighty increase of the F1 score. Further, Table 1 shows for $CB_{test}^{II}$ that the recall (+0.27) and F1 score (+0.16) have significantly been increased. In contrast to $CB_{test}^{I}$ the precision score (−0.15) indicates that data links are more likely not in alignment with the golden standard workflow. This implies that the quality of the completion significantly relies on the available knowledge in the incomplete case. However, in total hypothesis H2 is confirmed as the F1 score has been significantly increased in both scenarios.

# 7   Conclusions and Future Work

We presented a novel approach for the completion of workflow cases which lessens the knowledge gap in the case base and increases the capabilities of POCBR applications. The evaluation showed that if the knowledge gap in the cases is small the completion algorithm is able to identify consistent solutions. Previous work has showed that such consistent workflows are convenient for retrieval and adaptation in POCBR [18]. Further, the quality of the completed dataflow is high even if the knowledge gap in the workflow case description is larger.

Various approaches have been presented for case completion in CBR [1,6,10, 20]. However, in POCBR case completion has only been marginally investigated so far. Schumacher [21] as well as Dufour-Lussier et al. [9], for example, investigated the automatic extraction of workflows from textual process descriptions. As such texual descriptions usually lack of information, both employ linguistic methods to complete workflow cases. Workflow completion has been also considered for Workflow Management Systems (WFMS) with the goal to complete the workflows such that they become executable [12,26].

Future work will comprise an extended evaluation towards other domains and further varying characteristics of the case base. Moreover, we plan to investigate methods in case of inefficient run time or memory consumption, e.g., for larger workflows or less completed workflows. Finally, we aim to increase the case completion of workflows with regard to consistency using additional default completion rules and by abstracting the learned completion operators.

**Acknowledgements.** This work was funded by the German Research Foundation (DFG), project number BE 1373/3-1.

# References

1. Bach, K., Althoff, K.D., Satzky, J., Kroehl, J.: CookIIS mobile: a case-based reasoning recipe customizer for android phones. In: Petridis, M., Roth-Berghofer, T., Wiratunga, N. (eds.) Proceeding of UKCBR 2012, 11 December, Cambridge, United Kingdom, pp. 15–26. Engineering and Mathematics, University of Brighton, UK, School of Computing (2012)
2. Badra, F., Cordier, A., Lieber, J.: Opportunistic adaptation knowledge discovery. In: McGinty, L., Wilson, D.C. (eds.) ICCBR 2009. LNCS (LNAI), vol. 5650, pp. 60–74. Springer, Heidelberg (2009). doi:10.1007/978-3-642-02998-1_6

3. Bergmann, R., Gessinger, S., Görg, S., Müller, G.: The collaborative agile knowledge engine CAKE. In: Proceedings of the 18th International Conference on Supporting Group Work, pp. 281–284. ACM (2014)
4. Bergmann, R., Gil, Y.: Similarity assessment and efficient retrieval of semantic workflows. Inf. Syst. **40**, 115–127 (2014)
5. Bergmann, R., Wilke, W., Vollrath, I., Wess, S.: Integrating general knowledge with object-oriented case representation and reasoning. In: Burkhard, H.D., Wess, S. (eds.) 4th German Workshop: Case-Based Reasoning - System Development and Evaluation, pp. 120–126. Humboldt-Universität Berlin, Informatik-Berichte Nr. 55 (1996)
6. Burkhard, H.: Case completion and similarity in case-based reasoning. Comput. Sci. Inf. Syst. **1**(2), 27–55 (2004)
7. Craw, S., Wiratunga, N., Rowe, R.: Learning adaptation knowledge to improve case-based reasoning. Artif. Intell. **170**(16–17), 1175–1192 (2006)
8. Davenport, T.: Process Innovation: Reengineering Work Through Information Technology. Harvard Business Review Press, Boston (2013)
9. Dufour-Lussier, V., Ber, F.L., Lieber, J., Nauer, E.: Automatic case acquisition from texts for process-oriented case-based reasoning. Inf. Syst. **40**, 153–167 (2014)
10. Hanft, A., Schfer, O., Althoff, K.D.: Integration of drools into an OSGI-based BPM-platform for CBR. In: Minor, M., Montani, S., Recio-Garcia, J.A. (eds.) Workshop on Process-Oriented Case-Based Reasoning, ICCBR 2011, 19th, September 12–15, Greenwich, London, United Kingdom. University of Greenwich (2011)
11. Hanney, K., Keane, M.T.: The adaptation knowledge bottleneck: how to ease it by learning from cases. In: Leake, D.B., Plaza, E. (eds.) ICCBR 1997. LNCS, vol. 1266, pp. 359–370. Springer, Heidelberg (1997). doi:10.1007/3-540-63233-6_506
12. Koehler, J., Tirenni, G., Kumaran, S.: From business process model to consistent implementation: a case for formal verification methods. In: Proceedings of EDOC 2002, 17–20 September 2002, Lausanne, Switzerland, p. 96. IEEE Computer Society (2002)
13. Minor, M., Görg, S.: Acquiring adaptation cases for scientific workflows. In: Ram, A., Wiratunga, N. (eds.) ICCBR 2011. LNCS (LNAI), vol. 6880, pp. 166–180. Springer, Heidelberg (2011). doi:10.1007/978-3-642-23291-6_14
14. Minor, M., Montani, S., Recio-Garca, J.A.: Process-oriented case-based reasoning. Inf. Syst. **40**, 103–105 (2014)
15. Müller, G., Bergmann, R.: Workflow streams: a means for compositional adaptation in process-oriented CBR. In: Lamontagne, L., Plaza, E. (eds.) ICCBR 2014. LNCS (LNAI), vol. 8765, pp. 315–329. Springer, Heidelberg (2014). doi:10.1007/978-3-319-11209-1_23
16. Müller, G., Bergmann, R.: CookingCAKE: a framework for the adaptation of cooking recipes represented as workflows. In: Kendall-Morwick, J. (ed.) Workshop Proceedings, ICCBR 2015, Frankfurt, Germany, 28–30 September 2015. CEUR Workshop Proceedings, vol. 1520, pp. 221–232 (2015). CEUR-WS.org
17. Müller, G., Bergmann, R.: Generalization of workflows in process-oriented case-based reasoning. In: Russell, I., Eberle, W. (eds.) Proceedings, FLAIRS 2015, Hollywood, Florida, 18–20 May 2015, pp. 391–396. AAAI Press (2015)
18. Müller, G., Bergmann, R.: Learning and applying adaptation operators in process-oriented case-based reasoning. In: Hüllermeier, E., Minor, M. (eds.) ICCBR 2015. LNCS (LNAI), vol. 9343, pp. 259–274. Springer, Heidelberg (2015). doi:10.1007/978-3-319-24586-7_18
19. Richter, M.M., Weber, R.O.: Case-Based Reasoning - A Textbook. Springer, Heidelberg (2013)

20. Roth-Berghofer, T.R.: Explanations and case-based reasoning: foundational issues. In: Funk, P., González Calero, P.A. (eds.) ECCBR 2004. LNCS (LNAI), vol. 3155, pp. 389–403. Springer, Heidelberg (2004). doi:10.1007/978-3-540-28631-8_29

21. Schumacher, P.: Workflow extraction from textual process descriptions. Dissertation, Johann Wolfgang Goethe Universität Frankfurt/Main (2016)

22. Schumacher, P., Minor, M., Schulte-Zurhausen, E.: Extracting and enriching workflows from text. In: IEEE 14th International Conference on Information Reuse & Integration, IRI 2013, San Francisco, CA, USA, 14–16 August 2013, pp. 285–292. IEEE (2013)

23. Schumacher, P., Minor, M., Walter, K., Bergmann, R.: Extraction of procedural knowledge from the web: a comparison of two workflow extraction approaches. In: Mille, A., Gandon, F.L., Misselis, J., Rabinovich, M., Staab, S. (eds.) Proceedings of WWW 2012, Lyon, France, 16–20 April 2012 (Companion Volume), pp. 739–747. ACM (2012)

24. Stahl, A.: Learning of Knowledge-Intensive Similarity Measures in Case-Based Reasoning. Dissertation, Technische Universität Kaiserslautern (2004)

25. Thaler, T., Dadashnia, S., Sonntag, A., Fettke, P., Loos, P.: The IWi process model corpus. Technical report, Institute for Information Systems (IWi) at the German Research Center for Artificial Intelligence (DFKI), October 2015

26. Trcka, N., van der Aalst, W.M.P., Sidorova, N.: Workflow completion patterns. In: IEEE Conference on Automation Science and Engineering, CASE 2009, Bangalore, India, 22–25 August 2011, pp. 7–12. IEEE (2009)

27. Wilke, W., Vollrath, I., Bergmann, R.: Using knowledge containers to model a framework for learning adaptation knowledge. In: Wettschereck, D., Aha, D.W. (eds.) ECML 1997. MLNet Workshop Notes - Case-Based Learning: Beyond Classification of Feature Vectors, pp. 68–75. NCARAI, Washington, D. C., USA (1997)

# Refinement-Based Similarity Measures for Directed Labeled Graphs

Santiago Ontañón$^{(\boxtimes)}$ and Ali Shokoufandeh

Computer Science Department, Drexel University, Philadelphia, PA, USA
santi@cs.drexel.edu, ashokouf@drexel.edu

**Abstract.** This paper presents a collection of similarity measures based on refinement operators for *directed labeled graphs* (DLGs). We build upon previous work on refinement operators for other representation formalisms such as feature terms and description logics. Specifically, we present refinement operators for DLGs, which enable the adaptation of three similarity measures to DLGs: the anti-unification-based, $S_\lambda$, the property-based, $S_\pi$, and the weighted property-based, $S_{w\pi}$, similarities. We evaluate the resulting measures empirically comparing them to existing similarity measures for structured data.

**Keywords:** Similarity assessment · Refinement operators · Labeled graphs

## 1 Introduction

Similarity assessment plays a key role in CBR [1], where new problem instances are typically solved by selecting, adapting or interpolating the solutions to the most similar training instances to the problem at hand. This paper presents a collection of similarity measures based on refinement operators [16] for *directed labeled graphs* (DLGs). DLGs are a representation formalism to which many other data representations can be translated. Thus, defining similarity measures for the general case of DLGs is an important step toward defining general case-based reasoning systems capable of dealing with complex structured data.

In this work, we build upon previous work on similarity measures based on refinement operators, which we introduced first in the context of feature terms [20], and was later extended to description logics [24,26], and partial-order plans [25]. A refinement operator is a function that, for a given term in a specific representation language, can generate variations of such term that are either more specific or more general (refinements). This idea can be exploited to define similarity measures, for example by using a refinement operator to generalize two terms until reaching a shared term, and then measuring the number of times a refinement operator had to be applied for two terms to meet. Moreover, similarity measures defined in this way are agnostic to the underlying representation formalism, and only depend on the existence of appropriate refinement

© Springer International Publishing AG 2016
A. Goel et al. (Eds.): ICCBR 2016, LNAI 9969, pp. 311–326, 2016.
DOI: 10.1007/978-3-319-47096-2_21

operators. In this paper, we present refinement operators for DLGs, over which we define three similarity measures: the anti-unification-based similarity, $S_\lambda$, the property-based similarity, $S_\pi$, and the weighted property-based similarity $S_{w\pi}$.

We evaluate the resulting similarity measures by comparing them to existing similarity measures for structured data. Moreover, the software artifacts associated with work in this paper has been implemented in an open-source library called $\rho G$ (RHOG)[1].

The remainder of this paper is organized as follows. Section 2 presents some background on existing similarity measures for structured data. After that, Sect. 3 introduces some necessary notation and background on directed labeled graphs. Section 4 presents a collection of refinement operators for DLGs, and Sect. 5 employs those operators to define three similarity measures. Finally, Sect. 6 empirically evaluates the resulting similarity measures.

## 2 Background: Similarity Measures for Structured Data

A significant amount of the work on similarity for structured data is based on adapting classic ideas of similarity (or distance) between basic data types, such as real numbers or sets. For example, the Jaccard index ($J(A, B) = \frac{|A \cup B|}{|A \cap B|}$) has been adapted many times to define similarity measures for structured data. Similarity measures between sequences, such as the Edit distance [18], have also been adapted to structured representations like trees [4].

Similarity measures for Horn Clauses have been investigated in Inductive Logic Programming (ILP). Hutchinson [14] presented a distance based on the least general generalization (*lgg*) of two clauses, which is analogous to the Jaccard index (with the *lgg* playing the role of the intersection, and the size of the variable substitutions as a measure of the difference in size between the intersection and the union). Another influential similarity measure for Horn Clauses is that in RIBL (Relational Instance-Based Learning) [7]. RIBL's measure follows a "hierarchical aggregation" approach (known as the "local-global" principle [11]): the similarity of two objects is a function of the similarity of their attributes (repeating this recursively). In addition to being only applicable to Horn Clauses, RIBL implicitly assumes that values "further" away from the root of an object will play a lesser role in similarity. This procedure makes RIBL not appropriate for objects with circularities. Other Horn Clauses similarity measures include the work of Bisson [5], Nienhuys-Cheng [19], and of Ramon [23].

In their work on description logics (DLs), González-Calero et al. [13] present one of the earliest similarity measures for DLs (by first preprocessing instances to remove circularities). More recently, Fanizzi et al. [8] presented a similarity measure based on the idea of a "committee of concepts". They consider each concept in a given ontology to be a feature for each individual (belonging or not to that concept). The ratio of concepts that two individuals share corresponds to

---

[1] A Refinement-Operator Library for Directed Labeled Graphs: https://github.com/santiontanon/RHOG.

their similarity. This idea has been further developed by d'Amato [6]. SHAUD, presented by Armengol and Plaza [2], is a similarity measure also following the "hierarchical aggregation" approach but designed for the feature logics (a.k.a., Feature Terms or Typed Feature Structures). SHAUD also assumes that the terms do not have circularities. Borgmann and Stahl [3] present a similarity measure for object-oriented representations based on the concepts of intra-class and inter-class similarity, defined in a recursive way, also following "hierarchical aggregation", making it more appropriate for tree representations. In the CBR community, similarity measures based on graph structures have been studied for objects such as software UML designs [27].

Another related area is that of kernels for structured representations, which allow the application of techniques such as Support Vector Machines to structured data. Typically, kernels for graphs are based on the idea of finding common substructures between two graphs. For example, Kashima et al. [15] present a kernel for graphs based on random walks. Fanizzi et al. [9] also studied how to encapsulate their similarity measure for description logics into a kernel. For a survey on kernels for structured data the reader is referred to [12].

The work presented in this paper builds upon recent work on similarity measures using *refinement operators*. The key idea of these measures is to define similarity solely relying on the existence of a refinement operator for the target representation formalism. In this way, by just defining refinement operators for different representation formalisms, the same similarity measure can be used for all of these formalisms. Similarity measures for *feature terms* [20], description logics [24, 26], and partial-order plans [25] have been defined in this framework.

In summary, there has been a significant body of work on similarity assessment for structured representations, but the work has been carried out independently for different representations. The main contribution of this work is to extend the general framework of similarity measures based on refinement operators to *directed labeled graphs* by introducing appropriate refinement operators, showing that the ideas generalize to a wide variety of representation formalisms.

## 3   Directed Labeled Graphs

Let us introduce some basic notation and definitions. We use capital letters to represent sets and lower case letters to represent the elements of those sets, curly brackets for sets, and square brackets for ordered sequences. We use the power notation $2^V$ to represent the set of all possible subsets of a given set $V$.

**Definition 1 (Directed Labeled Graph).** *Given a finite set of labels $L$, a directed labeled graph $g$ is defined as a tuple $g = \langle V, E, l \rangle$, where:*

- *$V = \{v_1, ..., v_n\}$ is a finite set of vertices,*
- *$E = \{(v_{i_1}, v_{j_1}), ..., (v_{i_m}, v_{j_m})\}$ is a finite set of edges,*
- *the function $l : V \cup E \to L$ assigns a label from $L$ to each vertex and edge.*

**Definition 2 (Connected DLG).** *A DLG $g = \langle V, E, l \rangle$ is connected when given any two vertices $v_1, v_2 \in V$ there is a sequence of vertices $[w_1, ..., w_k]$, such that $w_1 = v_1$, $w_k = v_2$, and $\forall 1 \leq j < k : (w_j, w_{j+1}) \in E \vee (w_{j+1}, w_j) \in E$.*

**Definition 3 (Bridge).** *An edge $e \in E$ is a bridge of a connected DLG $g = \langle V, E, l \rangle$ if the graph resulting from removing e from g, $g' = \langle V, E \setminus \{e\}, l \rangle$ is not connected. We will use bridges(g) to denote the set of all the bridges in g.*

In the remainder of this document we will only consider connected DLGs. Moreover, we will consider two types of DLGs:

– *Flat-labeled DLGs (FDLG)*: where the set of labels $L$ is a plain set without any relation between the different labels.
– *Order-labeled DLGs (ODLG)*: where the set of labels $L$ is a partially ordered set via a relation $\preceq$ such that for any $a, b, c \in L$, we have that:
  - $a \preceq a$,
  - $a \preceq b \wedge b \preceq a \implies a = b$,
  - $a \preceq b \wedge b \preceq c \implies a \preceq c$, and
  - there is a special element $\top \in L$ such that $\forall a \in L : \top \preceq a$.

Intuitively, $\langle L, \preceq \rangle$ can be seen as a multiple-inheritance concept hierarchy with a single top label $\top$. Also, when $a \preceq b$ but $b \not\preceq a$, we write $a \prec b$.

## 3.1   Directed Labeled Graphs Subsumption

**Definition 4 (Flat Subsumption).** *Given two DLGs, $g_1 = \langle V_1, E_1, l_1 \rangle$ and $g_2 = \langle V_2, E_2, l_2 \rangle$, $g_1$ is said to* subsume *$g_2$ (we write $g_1 \sqsubseteq g_2$) if there is a mapping $m : V_1 \to V_2$ such that:*

– $\forall (v, w) \in E_1 : (m(v), m(w)) \in E_2$,
– $\forall v \in V_1 : l_1(v) = l_2(m(v))$, and
– $\forall (v, w) \in E_1 : l_1((v, w)) = l_2((m(v), m(w)))$.

**Definition 5 (Subsumption Relative to $\preceq$).** *Given two order-labeled DLGs, $g_1 = \langle V_1, E_1, l_1 \rangle$ and $g_2 = \langle V_2, E_2, l_2 \rangle$, and $\preceq$, the partial order among the labels in $L$, $g_1$ is said to* subsume *$g_2$ relative to $\preceq$ (we write $g_1 \sqsubseteq_{\preceq} g_2$) if there is a mapping $m : V_1 \to V_2$ such that:*

– $\forall (v, w) \in E_1 : (m(v), m(w)) \in E_2$,
– $\forall v \in V_1 : l_1(v) \preceq l_2(m(v))$, and
– $\forall (v, w) \in E_1 : l_1((v, w)) \preceq l_2((m(v), m(w)))$.

The mapping $m$ between vertices, induces a mapping $m_e$ between edges. Thus we will write $m(e)$ to denote $(m(v), m(w))$. Two graphs are equivalent, $g_1 \equiv g_2$, if $g_1 \sqsubseteq g_2$ and $g_2 \sqsubseteq g_1$. Moreover, if $g_1 \sqsubseteq g_2$ but $g_2 \not\sqsubseteq g_1$, then we write $g_1 \sqsubset g_2$.

Intuitively, *subsumption* embodies the idea of *more general than*, i.e., $g_1$ subsumes $g_2$ if $g_1$ is more general than $g_2$ (in flat subsumption, this corresponds to checking whether $g_1$ is isomorph to any subgraph of $g_2$). The *subsumption* relative to $\preceq$ is a more general relation when the subgraph is required to just have labels that are smaller according to $\preceq$ than in the supergraph. The intuition behind this is that if labels represent concepts such as *vehicle* and *car*, and the $\preceq$ relation captures concept generality (*vehicle* $\preceq$ *car*), then the subgraph can have concepts that are more general or equal to those of the supergraph (i.e., a vertex labeled as *vehicle* in the subgraph can be mapped to a vertex labeled as *car* in the supergraph, since *vehicle* is a more general concept than *car*).

## 3.2   Object Identity

*Object identity* (OI) [10] is an additional constraint on the mapping $m$ often employed for subsumption. The intuition behind object identity is that "objects denoted with different symbols must be distinct". When applied to subsumption ($\sqsubseteq$ or $\sqsubseteq_\preceq$) over graphs, this translates to an additional constraint over the mapping $m : V_1 \rightarrow V_2$, namely that $v_1 \neq v_2 \implies m(v_1) \neq m(v_2)$.

## 3.3   Unification and Antiunification

**Definition 6 (Unification).** *Given two graphs, $g_1$ and $g_2$, and a subsumption relation $\sqsubseteq$ (which can be any of the ones defined above), $g$ is a* unifier *of $g_1$ and $g_2$ (we write $g = g_1 \sqcup g_2$) if $g_1 \sqsubseteq g$, $g_2 \sqsubseteq g$, and $\nexists g' \sqsubset g : g_1 \sqsubseteq g' \wedge g_2 \sqsubseteq g'$.*

In other words, a unifier of two graphs is a most general graph that is subsumed by two other graphs. The analogous operation is that of *anti-unification*.

**Definition 7 (Anti-unification).** *Given two graphs, $g_1$ and $g_2$, and a subsumption relation $\sqsubseteq$ (which can be any of the ones defined above), $g$ is an* anti-unifier *of $g_1$ and $g_2$ (we write $g = g_1 \sqcap g_2$) if $g \sqsubseteq g_1$, $g \sqsubseteq g_2$, and $\nexists g' \sqsupset g : g' \sqsubseteq g_1 \wedge g' \sqsubseteq g_2$.*

In other words, an anti-unifier of two graphs is the most specific graph that subsumes both of them. Notice that in general neither unification or anti-unification are unique.

# 4   Refinement Operators

A refinement operator for DLGs is a function that can generate variations of a DLG $g$ that are either more specific or more general than $g$ (refinements).

**Definition 8 (Downward Refinement Operator).** *A* **downward refinement operator** *over a quasi-ordered set $(G, \sqsubseteq)$ is a function $\rho : G \rightarrow 2^G$ such that $\forall g' \in \rho(g) : g \sqsubseteq g'$.*

**Definition 9 (Upward Refinement Operator).** *An* **upward refinement operator** *over a quasi-ordered set $(G, \sqsubseteq)$ is a function $\rho : G \rightarrow 2^G$ such that $\forall g' \in \rho(g) : g' \sqsubseteq g$.*

In the context of this paper, a downward refinement operator generates elements of $G$ that are "more specific" (upward refinement operators generate elements of $G$ that are "more general"). Moreover, refinement operators might satisfy certain properties of interest for the purposes of similarity assessment:

- A refinement operator $\rho$ is **locally finite** if $\forall g \in G$ $\rho(g)$ is finite.
- A downward operator $\rho$ is **complete** if $\forall g_1, g_2 \in G$ $g_1 \sqsubset g_2 : g_1 \in \rho^*(g_2)$.
- An upward operator $\gamma$ is **complete** if $\forall g_1, g_2 \in G$ $g_1 \sqsubset g_2 : g_2 \in \gamma^*(g_1)$.
- A refinement operator $\rho$ is **proper** if $\forall g_1, g_2 \in G$ $g_2 \in \rho(g_1) \Rightarrow g_1 \not\equiv g_2$.

where $\rho^*$ means the *transitive closure* of a refinement operator. Intuitively, *locally finiteness* means that the refinement operator is computable, *completeness* means we can generate, by refinement of $g_1$, any element of $G$ related to a given element $g_1$ by the order relation $\sqsubseteq$, and *properness* means that a refinement operator does not generate elements which are equivalent to the element being refined. When a refinement operator is locally finite, complete and proper, it is *ideal*.

Notice that all the subsumption relations presented above satisfy the *reflexive*[2] and *transitive*[3] properties. Therefore, the pair $(G, \sqsubseteq)$, where $G$ is the set of all DLGs given a set of labels $L$, and $\sqsubseteq$ is any of the subsumption relations defined above is a *quasi-ordered* set. Thus, this opens the door to defining refinement operators for DLGs. Intuitively, a downward refinement operator for DLGs will generate *refinements* of a given DLG by either adding vertices, edges, or by making some of the labels more specific, thus making the graph more specific.

## 4.1   Refinement of Flat-Labeled DLGs

We define refinement operators as a set of *rewriting rules*. A typical rewriting rule is composed of three parts: the original graph (above the line), the refined graph (below the line), and the applicability conditions. The following four rewriting rules define a downward refinement operator $\rho_f$ for flat-labeled DLGs:

**(R0)** Top operator (adds one vertex to an empty graph):

$$\begin{bmatrix} V = \emptyset, E = \emptyset, \\ v_* \notin V, a \in L \end{bmatrix} \frac{\langle V, E, l \rangle}{\left\langle V \cup \{v_*\}, E, l'(x) = \begin{cases} a & \text{if } x = v_* \\ l(x) & \text{otherwise} \end{cases} \right\rangle}$$

**(R1)** Add vertex operator with outgoing edge:

$$\begin{bmatrix} v_* \notin V, v_1 \in V, \\ a \in L, b \in L \end{bmatrix} \frac{\langle V, E, l \rangle}{\left\langle V \cup \{v_*\}, E \cup \{(v_*, v_1)\}, l'(x) = \begin{cases} a & \text{if } x = v_* \\ b & \text{if } x = (v_*, v_1) \\ l(x) & \text{otherwise} \end{cases} \right\rangle}$$

---

[2] A graph trivially subsumes itself with the mapping $m(v) = v$.

[3] If a graph $g_1$ subsumes $g_2$ through a mapping $m_1$, and $g_2$ subsumes $g_3$ through a mapping $m_2$, then we know $g_1$ subsumes $g_3$ via the mapping $m(v) = m_2(m_1(v))$.

**(R2)** Add vertex operator with incoming edge:

$$
\begin{bmatrix} v_* \notin V, v_1 \in V, \\ a \in L, b \in L \end{bmatrix} \quad \cfrac{\langle V, E, l \rangle}{\left\langle V \cup \{v_*\}, E \cup \{(v_1, v_*)\}, l'(x) = \begin{cases} a & \text{if } x = v_* \\ b & \text{if } x = (v_1, v_*) \\ l(x) & \text{otherwise} \end{cases} \right\rangle}
$$

**(R3)** Add edge operator:

$$
\begin{bmatrix} v_1 \in V, v_2 \in V, \\ (v_1, v_2) \notin E, a \in L \end{bmatrix} \quad \cfrac{\langle V, E, l \rangle}{\left\langle V, E \cup \{(v_1, v_2)\}, l'(x) = \begin{cases} a & \text{if } x = (v_1, v_2) \\ l(x) & \text{otherwise} \end{cases} \right\rangle}
$$

Although we do not provide proofs in this paper for the sake of space (proofs in [21]), it can be shown that the downward refinement operator $\rho_f$ defined by the rewrite rules R0, R1, R2, and R3 above is locally finite, and complete for the quasi-ordered set $\langle G, \sqsubseteq \rangle$, where $\sqsubseteq$ represents flat subsumption. Moreover, $\rho_f$ is ideal (locally finite, complete, and proper) when we impose Object Identity.

Given a DLG $g = \langle V, E, l \rangle$, the following two rewriting rules define an upward refinement operator $\gamma_f$ for flat-labeled DLGs:

**(UR0)** Remove Non-Bridge (removes a non-bridge edge of the graph):

$$
\begin{bmatrix} e \in E, \\ e \notin bridges(\langle V, E, l \rangle) \end{bmatrix} \quad \cfrac{\langle V, E, l \rangle}{\langle V, E \setminus \{e\}, l \rangle}
$$

**(UR1)** Remove Leaf (vertex connected to the graph by a single edge):

$$
\begin{bmatrix} v \in V, \\ E_v = \{e = (v_1, v_2) \in E \mid v = v_1 \lor v = v_2\}, \\ |E_v| \leq 1 \end{bmatrix} \quad \cfrac{\langle V, E, l \rangle}{\langle V \setminus \{v\}, E \setminus E_v, l \rangle}
$$

Notice that $E_v$ is the set of edges that involve $v$, and by enforcing $|E_v| \leq 1$, we are basically selecting only those vertices $v \in V$ that are either: (a) connected to the rest of the graph by one single edge (when $|E_v| = 1$), or (b) when the graph is just composed of a single vertex and no edges (when $E_v = \emptyset$).

The upward refinement operator $\gamma_f$ defined by the above rewrite rules UR0 and UR1 is locally finite and complete for the quasi-ordered set $\langle G, \sqsubseteq \rangle$, with $\sqsubseteq$ representing flat subsumption. Moreover, $\gamma_f$ is ideal (locally finite, complete and proper) when we impose the Object Identity constraint.

## 4.2   Refinement of Order-Labeled DLGs

Assuming the set $\langle L, \preceq \rangle$ is a partial order with a top element $\top \in L$, and given a DLG $g = \langle V, E, l \rangle$, the following rewriting rules define an downward refinement operator, $\rho_\preceq$ for order-labeled DLGs:

**(R0O)** Top operator (adds one vertex to an empty graph):

$$\begin{bmatrix} v_* \notin V, \\ V = \emptyset, E = \emptyset \end{bmatrix} \frac{\langle V, E, l \rangle}{\left\langle V \cup \{v_*\}, E, l'(x) = \begin{cases} \top & \text{if } x = v_* \\ l(x) & \text{otherwise} \end{cases} \right\rangle}$$

**(R1O)** Add vertex operator with outgoing edge:

$$\begin{bmatrix} v_* \notin V, \\ v_1 \in V \end{bmatrix} \frac{\langle V, E, l \rangle}{\left\langle V \cup \{v_*\}, E \cup \{(v_1, v_*)\}, l'(x) = \begin{cases} \top & \text{if } x = v_* \\ \top & \text{if } x = (v_1, v_*) \\ l(x) & \text{otherwise} \end{cases} \right\rangle}$$

**(R2O)** Add vertex operator with incoming edge:

$$\begin{bmatrix} v_* \notin V, \\ v_1 \in V \end{bmatrix} \frac{\langle V, E, l \rangle}{\left\langle V \cup \{v_*\}, E \cup \{(v_*, v_1)\}, l'(x) = \begin{cases} \top & \text{if } x = v_* \\ \top & \text{if } x = (v_*, v_1) \\ l(x) & \text{otherwise} \end{cases} \right\rangle}$$

**(R3PO)** Add edge operator:

$$\begin{bmatrix} v_1 \in V, v_2 \in V, \\ (v_1, v_2) \notin E \end{bmatrix} \frac{\langle V, E, l \rangle}{\left\langle V, E \cup \{(v_1, v_2)\}, l'(x) = \begin{cases} \top & \text{if } x = (v_1, v_2) \\ l(x) & \text{otherwise} \end{cases} \right\rangle}$$

**(R4O)** Refine vertex label (relative to $\preceq$):

$$\begin{bmatrix} v_1 \in V, \\ a = l(v_1), b \in L, a \prec b, \\ \nexists c \in L : a \prec c \prec b \end{bmatrix} \frac{\langle V, E, l \rangle}{\left\langle V, E, l'(x) = \begin{cases} b & \text{if } x = v_1 \\ l(x) & \text{otherwise} \end{cases} \right\rangle}$$

**(R5O)** Refine edge label (relative to $\preceq$):

$$\begin{bmatrix} e \in E, \\ a = l(e), b \in L, a \prec b, \\ \nexists c \in L : a \prec c \prec b \end{bmatrix} \frac{\langle V, E, l \rangle}{\left\langle V, E, l'(x) = \begin{cases} b & \text{if } x = e \\ l(x) & \text{otherwise} \end{cases} \right\rangle}$$

The downward refinement operator $\rho_{\preceq}$ defined by the rewrite rules R0O, R1O, R2O, R3O, R4O, and R5O above is locally finite and complete for the quasi-ordered set $\langle G, \sqsubseteq_{\prec} \rangle$, where $\sqsubseteq_{\prec}$ represents subsumption relative to the partial order $\prec$. Moreover, $\rho_{\preceq}$ is ideal (locally finite, complete, and proper) when we impose the Object Identity constraint.

The following rules define an upward operator, $\gamma_{\preceq}$ for order-labeled DLGs:

**(UR0O)** Generalize vertex label:

$$\begin{bmatrix} v_1 \in V, \\ a = l(v_1), b \in L, b \prec a, \\ \nexists c \in L : a \prec c \prec b \end{bmatrix} \frac{\langle V, E, l \rangle}{\left\langle V, E, l'(x) = \begin{cases} b & \text{if } x = v_1 \\ l(x) & \text{otherwise} \end{cases} \right\rangle}$$

**(UR1O)** Generalize edge label:

$$
\begin{bmatrix}
e \in E, \\
a = l(e), b \in L, b \prec a, \\
\nexists c \in L : a \prec c \prec b
\end{bmatrix}
\quad
\frac{\langle V, E, l \rangle}{\left\langle V, E, l'(x) = \begin{cases} b & \text{if } x = e \\ l(x) & \text{otherwise} \end{cases} \right\rangle}
$$

**(UR2O)** Remove Non-Bridge (removes a top non-bridge edge of the graph):

$$
\begin{bmatrix}
e \in E, \ l(e) = \top, \\
e \notin bridges(\langle V, E, l \rangle)
\end{bmatrix}
\quad
\frac{\langle V, E, l \rangle}{\langle V, E \setminus \{e\}, l \rangle}
$$

**(UR3O)** Remove Leaf (top vertex connected to the graph by a single edge):

$$
\begin{bmatrix}
v \in V, l(v) = \top, \\
E_v = \{e = (v_1, v_2) \in E \mid v = v_1 \ \lor \ v = v_2\}, \\
|E_v| \leq 1, \ \forall e \in E_v : l(e) = \top
\end{bmatrix}
\quad
\frac{\langle V, E, l \rangle}{\langle V \setminus \{v\}, E \setminus E_v, l \rangle}
$$

The upward refinement operator $\gamma_{\preceq}$ defined by the rewrite rules UR0O, UR1O, UR2O, and UR3O above is locally finite and complete for the quasi-ordered set $\langle G, \sqsubseteq_{\prec} \rangle$ $\gamma_{\preceq}$ is ideal when we impose the Object Identity constraint. Formal proofs can be found in the documentation of the $\rho G$ library [21].

## 5    Refinement-Based Similarity Measures

Graph subsumption introduces a concept of *information order* between graphs: if a graph $g_1$ subsumes another graph $g_2$, then all the information in $g_1$ is also in $g_2$. Thus, if we find the most specific graph $g$ that subsumes two other graphs $g_1$, and $g_2$, then $g$ captures the information that $g_1$ and $g_2$ have in common. The intuition of the first similarity function we define is to first compute such $g$, and then use it to numerically quantify the amount of information in $g$: the more information they share, the more similar they are. We will use the intuition that each time we apply a downward refinement operation, we introduce one new piece of information, so the length of the refinement path between $g_\top$ (the most general graph with respect to subsumption) and a given graph $g$ gives us a measure of the amount of information contained in it.

### 5.1    Anti-unification-based Similarity

**Definition 10 (Anti-unification-based Similarity).** *Given two graphs $g_1$, and $g_2$, a refinement operator $\rho$ and a subsumption relation $\sqsubseteq$, the anti-unification-based similarity $S_\lambda$ is defined as:*

$$
S_\lambda(g_1, g_2) = \frac{|g_\top \xrightarrow{\rho} (g_1 \sqcap g_2)|}{|g_\top \xrightarrow{\rho} (g_1 \sqcap g_2)| + |(g_1 \sqcap g_2) \xrightarrow{\rho} g_1| + |(g_1 \sqcap g_2) \xrightarrow{\rho} g_1|}
$$

*where $|g_1 \xrightarrow{\rho} g_2|$ represents the length of a path that starts in $g_1$ and goes to $g_2$ by repeated application of the refinement operator $\rho$.*

Intuitively, this measures the amount of information shared between $g_1$ and $g_2$ (size of their anti-unifier), and normalizes it by the total amount of information: shared information ($|g_\top \xrightarrow{\rho} (g_1 \sqcap g_2)|$), information in $g_1$ but not in $g_2$ ($|(g_1 \sqcap g_2) \xrightarrow{\rho} g_1|$), and information in $g_2$ but not in $g_1$ ($|(g_1 \sqcap g_2) \xrightarrow{\rho} g_2|$). The reader is referred to our previous work [20] for a more in-depth description and analysis of the anti-unification-based similarity.

One interesting thing about $S_\lambda$ is that there is nothing specific to directed labeled graphs in this formulation. In other words, the refinement operator allows us to abstract away from the underlying representation formalism.

## 5.2   Property-Based Similarity

The key idea of the *property-based similarity* measure is to decompose each graph into a collection of smaller graphs (which we will call *properties*), and then count how many of these properties are shared between two given graphs. The key advantage of this similarity measure is that each of these properties can be seen as a *feature*, and thus, we can apply feature weighting methods in order to improve accuracy in the context of machine learning methods. Let us first briefly explain how to decompose a graph into a collection of properties (operation, which we call *disintegration*). The reader is referred to our previous work [20], for a full description.

**Graph Disintegration.** Consider a refinement path $g_0 = [g_\top, ..., g_n]$ between the most general graph $g_\top$ and a given graph $g_n$, generated by repeated application of an upward refinement operator (going from $g_n$ to $g_\top$). The intuition of graph disintegration is the following: each time an upward refinement operator is applied to a graph $g_{i+1}$ to generate a more general graph $g_i$, a *piece of information* is removed, which $g_i$ does not have, and $g_{i+1}$ had. We would like the *disintegration* operation to decompose graph $g_n$ into exactly $n$ properties, each of them representing each of the pieces of information that were removed along the refinement path. In order to do this, the *disintegration* operation uses the *remainder* operation (introduced by Ontañón and Plaza [20] for feature terms):

**Definition 11 (Remainder).** *Given two graphs $g_u$ and $g_d$ such that $g_u \sqsubseteq g_d$, the remainder $r(g_d, g_u)$ is a graph $g_r$ such that $g_r \sqcup g_u \equiv g_d$, and $\nexists g \in G$ such that $g \sqsubset g_r$ and $g \sqcup g_u \equiv g_d$.*

In other words, the remainder is the most general graph $g_r$ such that when unifying $g_r$ with the most general of the two graphs ($g_u$), recovers the most specific of the two graphs ($g_d$). Given two graphs $g_i$, and $g_{i+1}$, such that $g_i \in \gamma(g_{i+1})$, then $r(g_{i+1}, g_i)$ is precisely the graph that captures the piece of information that $\gamma$ "removed" from $g_{i+1}$.

**Definition 12 (Disintegration).** *Given a refinement path $p = [g_0 = g_\perp, ..., g_n]$, a disintegration of the graph $g_n$ is the set $D(p) = \{r(g_{i+1}, g_i) | 0 \le i < n\}$.*

In practice, given a graph $g$ and complete upward refinement operator $\gamma$, we can generalize $g$ by successive application of $\gamma$ (selecting one of the generalizations $\gamma$ produced stochastically), and use the *remainder* operation to generate a property at each step[4]. We will write $D_\gamma(g)$ to denote a disintegration generated in this way. A more detailed discussion on disintegration and remainder, in the context of refinement graphs can be found in our previous work [22].

## Property-Based Similarity Definition

**Definition 13 (Property-Based Similarity).** *Given two graphs $g_1$ and $g_2$, a complete upward refinement operator $\gamma$, and a subsumption relation $\sqsubseteq$, the property-based similarity measure, $S_\pi$ is defined as follows:*

$$S_\pi(g_1, g_2) = \frac{|\{\pi \in P \mid \pi \sqsubseteq g_1 \ \wedge \ \pi \sqsubseteq g_2\}|}{|P|}$$

*where $P = D_\gamma(g_1) \cup D_\gamma(g_2)$.*

In other words, $S_\pi(g_1, g_2)$ is defined as the number of properties that are shared between both graphs divided by the number of properties that at least one of them have. In certain situations, it can be shown that $S_\pi(g_1, g_2)$ and $S_\lambda(g_1, g_2)$ are equivalent [20]. The intuition behind this is that the number of properties they share should be equivalent to the length of the refinement path from $g_\top$ to their anti-unification. Moreover, if a training set is available, we can use it to define $P$ as the union of the disintegration of all the graphs in the training set (which is the procedure used in our experiments).

The main advantage of $S_\pi$ with respect to $S_\lambda$ is that it allows for weighting the contribution of each property in the similarity computation, and thus, the similarity measure can be fitted to a given supervised learning task. As in our previous work [20], we see each property as a binary feature, and use Quinlan's *information gain* to define a weight for each property. Given such a set of weights, we can now define the *weighted property-based similarity* as follows:

**Definition 14 (Weighted Property-Based Similarity).** *Given two graphs $g_1$ and $g_2$, a training set $T = \{(g'_1, y_1), ..., (g'_n, y_n)\}$, a complete upward refinement operator $\gamma$, and a subsumption relation $\sqsubseteq$, the property-based similarity measure, $S_\pi$ is defined as follows:*

$$S_{w\pi}(g_1, g_2) = \frac{\sum_{\pi \in P \ such \ that \ \pi \sqsubseteq g_1 \ \wedge \ \pi \sqsubseteq g_2} w(\pi)}{\sum_{\pi \in P} w(\pi)}$$

*where $P = D_\gamma(g_1) \cup D_\gamma(g_2)$, and weights $w$ are generated using $T$.*

Intuitively, $S_{w\pi}$ is equivalent to $S_\pi$, except that $S_{w\pi}$ counts the sum of the weights of the properties, whereas $S_\pi$ counts the number of properties. Also,

---

[4] Notice that it is possible to compute the *remainder* without the need to actually perform any type of unification operation, which can be computationally expensive.

notice that in practice, we can precompute the weights for all the properties resulting from disintegrating all the graphs in the training set, and we would only need to compute weights during similarity assessment if the disintegration of either $g_1$ or $g_2$ yields a property that no other graph in the training set had.

# 6   Experimental Evaluation in Standard Datasets

In order to evaluate the similarity measures defined in this paper, we used two standard datasets from the literature of structured machine learning[5]:

- **Trains**: We used Muggleton's train generator[6], which generates trains based on the original dataset presented by Michalski [17], to generate datasets consisting of 10, 100 and 900 trains, represented using Horn clauses. Trains are to be classified in one of two possible classes (*east* or *west*). We translated it into labeled graphs by using the feature term version translated by Ontañón and Plaza [20]. Each instance represented was to a DLG by defining one vertex per variables in the feature term, and one edge per feature. The resulting graphs have between 14 and 29 vertices, and the set of labels contains 55 labels. An example graph from this dataset can be seen in Fig. 1.
- **Demospongiae**: The Demospongiae data set is a relational data set from the UCI machine learning repository where each instance is a tree. It contains 503 instances belonging to 8 different solution classes. The original dataset is represented using feature terms, and was converted to a DLG using the same procedure as the Trains dataset. The resulting graphs have between 21 to 59 vertices, and the set of labels contains 641 labels (the set of labels is large, since some correspond to numbers, which our framework considers as different labels, children of the "number" label). We also report results in a subset of 280 sponges, commonly used in the literature.

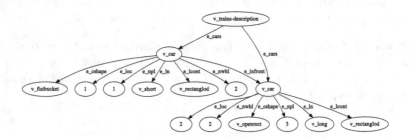

**Fig. 1.** Example graph from the Trains data set with 15 vertices.

We tested the similarity measures using a *nearest neighbor* classifier and a leave-one-out evaluation method. We also tested the following baselines:

---

[5] Datasets can be downloaded in *dot*, *GML* and *GraphML* format from https://github.com/santiontanon/RHOG.

[6] http://www.doc.ic.ac.uk/~shm/progol.html.

**Table 1.** Classification accuracy (in % of correct predictions), and time (in seconds) to predict the label of one instance.

| | Maj. | BoL | RIBL | Kashima | $S_\lambda$ | | $S_\pi$ | | $S_{w\pi}$ | |
|---|---|---|---|---|---|---|---|---|---|---|
| | | | | | $\rho_f$ | $\rho_\preceq$ | $\rho_f$ | $\rho_\preceq$ | $\rho_f$ | $\rho_\preceq$ |
| Classification Accuracy | | | | | | | | | | |
| Trains(10) | 50.00 | 50.00 | 60.00 | 20.00 | 50.00 | 60.00 | 40.00 | 30.00 | 60.00 | **70.00** |
| Trains(100) | 52.00 | 64.00 | 60.00 | **72.00** | 55.00 | 70.00 | 61.00 | 71.00 | 66.00 | 71.00 |
| Trains(900) | 50.00 | 75.78 | 64.22 | 68.44 | 62.89 | 80.88 | 67.20 | 74.00 | 86.11 | **87.55** |
| Demosp(280) | 41.79 | 93.93 | 88.21 | 90.71 | 96.07 | 92.86 | 95.36 | 92.50 | 95.71 | **97.50** |
| Demosp(503) | 23.36 | 86.68 | 82.31 | 83.10 | **93.18** | 84.69 | 89.07 | 87.08 | 88.87 | 90.66 |
| Time | | | | | | | | | | |
| Trains(10) | - | 0.00 | 0.05 | 0.06 | 1.79 | 2.62 | 0.01 | 0.08 | 0.02 | 0.10 |
| Trains(100) | - | 0.00 | 0.05 | 0.08 | 5.64 | 11.47 | 0.02 | 0.42 | 0.09 | 0.59 |
| Trains(900) | - | 0.01 | 0.31 | 0.51 | 50.75 | 115.97 | 0.30 | 7.43 | 0.44 | 11.81 |
| Demosp(280) | - | 0.02 | 0.18 | 0.21 | 365.54 | 121.82 | 0.11 | 2.77 | 0.18 | 3.71 |
| Demosp(503) | - | 0.03 | 0.32 | 0.51 | 629.04 | 213.96 | 0.57 | 4.68 | 0.75 | 6.78 |

- **Majority Class** (Maj.): predicts the most common class in the training set.
- **Bag-of-labels Similarity** (BoL): this similarity measure transforms each graph into a vector of size $|L|$, where each element of the vector corresponds to each of the labels in $L$. A given element of the vector is set to 1 if the graph contains the corresponding label, and 0 otherwise. Then, similarity between two graphs is assessed using the *cosine similarity* between their two corresponding vectors.
- **RIBL** [7]: RIBL is a well known similarity measure for first order logic (FOL). RIBL requires examples to be represented in FOL and not as DLGs. We converted the datasets to FOL predicates directly from their original feature term representation without loss of information.
- **Kashima** [15]: a random-walk graph kernel. Kashima's kernel has two parameters: $\gamma$, the probability of a random walk to end, and a kernel to assess similarity among the labels in the graph. We set $\gamma = 0.1$ in our experiments, since it gave the best results overall. We used the label partial-order to define a kernel for the labels of the graph.

Table 1 reports classification accuracy and execution time for all the datasets and similarity measures used in our experiments. The first thing we see is that $S_{w\pi}$ achieves the highest classification accuracy overall than any other similarity measure in our experiments. The second things we see is that the refinement-based measures tend to be more accurate when used in conjunction with $\rho_\preceq$ in the trains datasets, but not in the Demospongiae datasets. Moreover, in general, computation time also increases when using $\rho_\preceq$, since the refinement graph is more fine-grained, and thus refinement paths tend to be longer. This is not the case in Demospongiae because of the large number of labels caused by the numeric constants, which makes $\rho_f$ generate too many refinements. We also see that in the small trains dataset (with 10 instances), most measures achieve very low performance (although given the small size of the dataset, it is difficult to know if low results are due to statistical coincidences). Another interesting fact is that one of our baselines (BoL) worked surprisingly well, and outperformed many other complex similarity measures (such as RIBL, SHAUD or $S_\pi$). Finally, concerning computation time, we can see that $S_\lambda$ is computationally infeasible for large datasets (it took several days to execute for the Demospongiae datasets). $S_\pi$ is a computationally fast approximation, but results show that its performance is lower than $S_\lambda$. $S_{w\pi}$ achieves a good balance of performance and accuracy. The increased time reported for the trains (900) and Demospongiae (503) datasets is due to a necessary step where the union $P$ of all the properties has to be computed, which requires a quadratic number of subsumption tests as a function of the size of the training set to remove duplicates. In practice this step only has to be done once, but we did it for each instance, since we used a leave-one-out procedure.

# 7    Conclusions and Future Work

This paper has presented a collection of similarity measures for directed labeled graphs (DLGs) based on refinement operators. Specifically, we have defined four refinement operators for DLGs. Using those operators, we adapted three similarity measures from our previous work [20] for DLGs. We evaluated these similarity measures in several datasets with promising results. The refinement operators and similarity measures presented in this paper have been incorporated into an open-source library called $\rho$G (RHOG).

Similarity measures based on refinement operators are interesting since the same similarity formulation can be used for different representation formalisms, given an appropriate refinement operator. In this paper we have shown evidence of this by adapting three measures previously defined for feature logics to DLGs by just defining two new refinement operators.

As part of our future work, we would first like to devise refinement operators for other representation formalisms to allow the wide-spread application of these similarity measures. Second, we would like to study ways in which we could improve the computational efficiency of the proposed methods. Two lines of work we are currently pursuing are (1) accelerating subsumption testing (the major bottleneck), and (2) numerical approximations of these similarity measures.

**Acknowledgements.** This research was supported by grant IIS-1551338 from the National Science Foundation (NSF).

# References

1. Aamodt, A., Plaza, E.: Case-based reasoning: foundational issues, methodological variations, and system approaches. Artif. Intell. Commun. **7**(1), 39–59 (1994)

2. Armengol, E., Plaza, E.: Relational case-based reasoning for carcinogenic activity prediction. Artif. Intell. Rev. **20**(1–2), 121–141 (2003)
3. Bergmann, R., Stahl, A.: Similarity measures for object-oriented case representations. In: Smyth, B., Cunningham, P. (eds.) EWCBR 1998. LNCS, vol. 1488, pp. 25–36. Springer, Heidelberg (1998). doi:10.1007/BFb0056319
4. Bille, P.: A survey on tree edit distance and related problems. Theoret. Comput. Sci. **337**(1), 217–239 (2005)
5. Bisson, G.: Learing in FOL with a similarity measure. In: Proceedings of AAAI 1992, pp. 82–87 (1992)
6. d'Amato, C.: Similarity-based learning methods for the semantic web. Ph.D. thesis, Università degli Studi di Bari (2007)
7. Emde, W., Wettschereck, D.: Relational instance based learning. In: Saitta, L. (ed.) Machine Learning - Proceedings 13th International Conference on Machine Learning, pp. 122–130. Morgan Kaufmann Publishers (1996)
8. Fanizzi, N., d'Amato, C., Esposito, F.: Induction of optimal semi-distances for individuals based on feature sets. In: Proceedings of 2007 International Workshop on Description Logics, CEUR-WS (2007)
9. Fanizzi, N., d'Amato, C., Esposito, F.: Learning with Kernels in description logics. In: Železný, F., Lavrač, N. (eds.) ILP 2008. LNCS (LNAI), vol. 5194, pp. 210–225. Springer, Heidelberg (2008). doi:10.1007/978-3-540-85928-4_18
10. Ferilli, S., Fanizzi, N., Di Mauro, N., Basile, T.M.: Efficient theta-subsumption under object identity. In: Workshop AI*IA 2002, pp. 59–68 (2002)
11. Gabel, T.: Learning similarity measures: strategies to enhance the optimisation process. Diploma thesis, kaiserslautern university of technology (2003)
12. Gärtner, T., Lloyd, J.W., Flach, P.A.: Kernels for structured data. In: Matwin, S., Sammut, C. (eds.) ILP 2002. LNCS (LNAI), vol. 2583, pp. 66–83. Springer, Heidelberg (2003). doi:10.1007/3-540-36468-4_5
13. González-Calero, P.A., Díaz-Agudo, B., Gómez-Albarrán, M.: Applying DLs for retrieval in case-based reasoning. In: Proceedings of the 1999 Description Logics Workshop (DL 1999) (1999)
14. Hutchinson, A.: Metrics on terms and clauses. In: Someren, M., Widmer, G. (eds.) ECML 1997. LNCS, vol. 1224, pp. 138–145. Springer, Heidelberg (1997). doi:10.1007/3-540-62858-4_78
15. Kashima, H., Tsuda, K., Inokuchi, A.: Marginalized Kernels between labeled graphs. In: Proceedings of the Twentieth International Conference (ICML 2003), pp. 321–328. AAAI Press (2003)
16. van der Laag, P.R.J., Nienhuys-Cheng, S.H.: Completeness and properness of refinement operators in inductive logic programming. J. Logic Program. **34**(3), 201–225 (1998)
17. Larson, J., Michalski, R.S.: Inductive inference of VL decision rules. SIGART Bull. **63**(63), 38–44 (1977)
18. Levenshtein, V.I.: Binary codes capable of correcting deletions, insertions, and reversals. Sov. Phys. Dokl. **10**, 707–710 (1966)
19. Nienhuys-Cheng, S.-H.: Distance between herbrand interpretations: a measure for approximations to a target concept. In: Lavrač, N., Džeroski, S. (eds.) ILP 1997. LNCS, vol. 1297, pp. 213–226. Springer, Heidelberg (1997). doi:10.1007/3540635149_50
20. Ontanón, S., Plaza, E.: Similarity measures over refinement graphs. Mach. Learn. **87**, 57–92 (2012)
21. Ontañón, S.: RHOG: a refinement-operator library for directed labeled graphs. arXiv preprint cs.AI (2016). http://arxiv.org/abs/1604.06954

22. Ontañón, S., Plaza, E.: Refinement-based disintegration: an approach to re-representation in relational learning. AI Commun. **28**(1), 35–46 (2015)
23. Ramon, J.: Clustering and instance based learning in first order logic. AI Commun. **15**(4), 217–218 (2002)
24. Sánchez, D., Batet, M., Isern, D., Valls, A.: Ontology-based semantic similarity: a new feature-based approach. Expert Syst. Appl. **39**(9), 7718–7728 (2012)
25. Sánchez-Ruiz, A.A., Ontañón, S.: Least common subsumer trees for plan retrieval. In: Lamontagne, L., Plaza, E. (eds.) ICCBR 2014. LNCS (LNAI), vol. 8765, pp. 405–419. Springer, Heidelberg (2014). doi:10.1007/978-3-319-11209-1_29
26. Sánchez-Ruiz, A.A., Ontañón, S., González-Calero, P.A., Plaza, E.: Refinement-based similarity measure over DL conjunctive queries. In: Delany, S.J., Ontañón, S. (eds.) ICCBR 2013. LNCS (LNAI), vol. 7969, pp. 270–284. Springer, Heidelberg (2013). doi:10.1007/978-3-642-39056-2_20
27. Wolf, M., Petridis, M.: Measuring similarity of software designs using graph matching for CBR. In: Workshop Proceedings of AISEW 2008 at ECAI 2008. IOS Press (2008)

# FEATURE-TAK - Framework for Extraction, Analysis, and Transformation of Unstructured Textual Aircraft Knowledge

Pascal Reuss[1,2(✉)], Rotem Stram[1], Cedric Juckenack[1], Klaus-Dieter Althoff[1,2],
Wolfram Henkel[3], Daniel Fischer[3], and Frieder Henning[4]

[1] German Research Center for Artificial Intelligence, Kaiserslautern, Germany
reusspa@uni-hildesheim.de
[2] Institute of Computer Science, Intelligent Information Systems Lab,
University of Hildesheim, Hildesheim, Germany
[3] Airbus Operations GmbH, Kreetslag 10, 21129 Hamburg, Germany
[4] Lufthansa Industry Solutions, Hamburg, Germany
http://www.dfki.de
http://www.uni-hildesheim.de

**Abstract.** This paper describes a framework for semi-automatic knowledge extraction for case-based diagnosis in the aircraft domain. The available data on historical problems and their solutions contain structured and unstructured data. To transform these data into knowledge for CBR systems, methods and algorithms from natural language processing and case-based reasoning are required. Our framework integrates different algorithms and methods to transform the available data into knowledge for vocabulary, similarity measures, and cases. We describe the idea of the framework as well as the different tasks for knowledge analysis, extraction, and transformation. In addition, we give an overview of the current implementation, our evaluation in the application context, and future work.

## 1 Introduction

The amount of experience knowledge is huge in many companies. They store historical data about projects, incidents, occurred problems and their solutions, and much other information. All this can be used to gather experience to solve future problems. Because the aircraft domain is a technical domain, much information is clearly structured like attribute-value pairs, taxonomies, and ontologies. But there also exists information in form of free text written by cabin crew members, pilots, or maintenance technicians. Examples of free text are the cabin and pilot logbook, customer service reports, and maintenance reports. To use this information in the context of a case-based reasoning (CBR) system, they have to be analyzed and transformed into useful knowledge for vocabulary, similarity measures, and cases. While the structured information can be transformed with little to moderate effort, sometimes it can even be used without transformation, the unstructured information in free texts can only be analyzed and transformed

© Springer International Publishing AG 2016
A. Goel et al. (Eds.): ICCBR 2016, LNAI 9969, pp. 327–341, 2016.
DOI: 10.1007/978-3-319-47096-2_22

with high effort from a knowledge engineer. To support a knowledge engineer at this task, we developed a framework that combines several methods from natural language processing (NLP) and CBR to automate the transformation process. The framework is called FEATURE-TAK, a **F**ramework for **E**xtraction, **A**nalysis, and **T**ransformation of **U**nstructu**RE**d **T**extual **A**ircraft **K**nowledge. While parts of the framework are not new to the research community, other parts were developed or improved in-house to bridge the gap to an automated knowledge transformation directly usable in CBR systems. In addition, the combination of the NLP and CBR tasks as well as the underlying methodology and the benefit for knowledge transformation for CBR systems is a new approach and will help to reduce the creation and maintenance effort of CBR systems. In the following section we describe the project context in which the use case of the framework occurred and several basics. Section 3 contains related work about the topic, while Sect. 4 gives an overview of the framework and detailed information about the individual NLP and CBR tasks. We also describe our current implementation status and evaluation in Sect. 4.4. At the end we summarize our paper and give an outlook on future work.

## 2    OMAHA Project

The OMAHA project is supported by the Federal Ministry of Economy and Technology in the context of the fifth civilian aeronautics research program [9]. The high-level goal of the OMAHA project is to develop an integrated overall architecture for health management of civilian aircraft. The project covers several topics like diagnosis and prognosis of flight control systems, innovative maintenance concepts and effective methods of data processing and transmission. A special challenge of the OMAHA project is to outreach the aircraft and its subsystems and integrating systems and processes in the ground segment like manufacturers, maintenance facilities, and service partners. Several enterprises and academic and industrial research institutes take part in the OMAHA project: the aircraft manufacturer Airbus (Airbus Operations, Airbus Defense & Space, Airbus Group Innovations), the system and equipment manufacturers Diehl Aerospace and Nord-Micro, the aviation software solutions provider Linova and IT service provider Lufthansa Systems as well as the German Research Center for Artificial Intelligence and the German Center for Aviation and Space. In addition, several universities are included as subcontractors. The OMAHA project has several different sub-projects. Our work focuses on a sub-project to develop a cross-system integrated system health monitoring (ISHM). The main goal is to improve the existing diagnostic approach with a multi-agent system (MAS) with several case-based agents to integrate experience into the diagnostic process and provide more precise diagnoses and maintenance suggestions. In this context we have to acquire cases from historical data, which contains a high number of free texts. Therefore, the development of an approach to analyze and transform this free text is required.

## 2.1 SEASALT

The SEASALT (Shared Experience using an Agent-based System Architecture Layout) architecture is a domain-independent architecture for extracting, analyzing, sharing, and providing experiences [6]. The architecture is based on the Collaborative Multi-Expert-System approach [3, 4] and combines several software engineering and artificial intelligence technologies to identify relevant information, process the experience and provide them via an user interface. The knowledge modularization allows the compilation of comprehensive solutions and offers the ability of reusing partial case information in form of snippets. Figure 1 gives an overview over the SEASALT architecture.

The SEASALT architecture consists of five components: the *knowledge sources*, the *knowledge formalization*, the *knowledge provision*, the *knowledge representation*, and the *individualized knowledge*. The *knowledge sources* component is responsible for extracting knowledge from external knowledge sources like databases or web pages and especially Web 2.0 platforms, like forums and social media platforms. These knowledge sources are analyzed by so-called Collector Agents, which are assigned to specific Topic Agents. The Collector Agents collect all contributions that are relevant for the respective Topic Agent's topic. The *knowledge formalization* component is responsible for formalizing the extracted knowledge from the Collector Agents into a modular, structural representation.

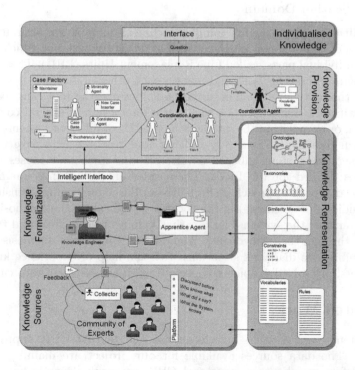

**Fig. 1.** Overview of the SEASALT architecture

This formalization is done by a knowledge engineer with the help of a so-called Apprentice Agent. This agent is trained by the knowledge engineer and can reduce the workload for the knowledge engineer. The *knowledge provision* component contains the so called Knowledge Line. The basic idea is a modularization of knowledge analogous to the modularization of software in product lines. The modularization is done among the individual topics that are represented within the knowledge domain. In this component a Coordination Agent is responsible for dividing a given query into several sub queries and pass them to the according Topic Agent. The agent combines the individual solutions to an overall solution, which is presented to the user. The Topic Agents can be any kind of information system or service. If a Topic Agent has a CBR system as a knowledge source, the SEASALT architecture provides a so-called Case Factory for the individual case maintenance. Several Case Factories are supervised by a so-called Case Factory Organization to coordinate the maintenance of the overall multi-agent system. The *knowledge representation* component contains the underlying knowledge models of the different agents and knowledge sources. The synchronization and matching of the individualized knowledge models improves the knowledge maintenance and the interoperability between the components. The *individualized knowledge* component contains the web-based user interfaces to enter a query and present the solution to the user [5,6,19].

## 2.2 Application Domain

The aircraft domain is a highly complex technical domain. An aircraft consists of hundreds of components, which consists of dozens of systems, which contains dozens of individual parts, called Line Replacement Units (LRU). These systems and LRUs are interacting with and rely on each other. Therefore, it is not easy to identify the root cause of an occurred fault, because the root cause can either be found within a single LRU, or within the interaction of several components of a system, or even within the interaction of LRUs of different systems. Finding cross-system root causes is a very difficult and resource expensive task. The existing diagnosis system onboard an aircraft can track root causes based on causal rules defined for the LRUs. These rules are not always unambiguous, because the diagnosis approach is effect-driven. Based on a comprehensible effect (visible, audible, or smellable) in the cockpit or the cabin, the diagnosis system tries to determine the system behavior that belongs to the effect and traces the root cause through the defined rules. The use of CBR for the diagnosis can help to clear ambiguous diagnosis situations with the help of experience knowledge from successfully solved problems, especially with cross-system root causes.

## 3  Related Work

Many systems with textual knowledge use the textual CBR approach, like [11,21,24]. The data sources available for our project are mainly structured data, therefore we choose a structural CBR approach. But the most important information about an occurred fault can be found in fault descriptions and logbook entries, which are free text. We decided to use a hybrid approach with the combination of structural CBR and NLP techniques to integrate all available information. There also exist several frameworks and toolkits for natural language processing like Stanford CoreNLP [17], Apache Lucene [18], GATE [12], and SProUT [13]. All these frameworks provide several algorithms and methods for NLP tasks, but do not link them directly to be used for CBR systems. Several methods from these frameworks are used by our framework, too. But we also combine them with techniques from association rule mining, case-based reasoning, and techniques developed in-house to have a direct use for knowledge modeling in CBR systems. There is extensive research pertaining to adjustment of feature weights in the past years and it is still an important topic. Wettschereck and Aha compared different feature weighting methods and developed five dimensions to describe these methods: Model, weight space, representation, generality and knowledge. [26] According to their work, our approach uses a wrapper model to optimize the feature weights iteratively during the training phases. The weight space is continuous, because the features of our problem vary in their relevance for different diagnoses. Our knowledge representation is a structural case structure with attribute-value pairs and this given structure is used for feature weighting. We are using case specific weights to set the weights for each diagnosis individually. This way we are able gain more precise results during the retrieval. Our approach for feature weighting is knowledge intensive, because we are using domain-specific knowledge to differentiate between individual diagnoses and setting case specific weights. An approach that addresses the same problem as our approach is presented by Sizov, Ozturk and Styrak [22]. They analyze free text documents from aircraft incidents to identify reasoning knowledge that can be used to generate cases from these text documents. While we are using a structural approach with attribute-value pairs and try to classify and map the identified relevant knowledge to attributes, the approach from Sizov and his colleagues uses a so-called text reasoning graph to represent their cases. The approach uses the same NLP techniques like the Standford CoreNLP Pipeline as our approach to analyze and preprocess the text documents. While we are using Association Rule Mining algorithms like Apriori and FP-Grwoth to identify associations between collocations and keywords, their approach uses pattern recognition to identify causal relations.

## 4   FEATURE-TAK

This section describes FEATURE-TAK, an agent-based **F**ramework for **E**xtraction, **A**nalysis, and **T**ransformation of Unstructe**RE**d Textual **A**ircraft **K**nowledge. We will describe the idea, the agent-based architecture and the individual tasks of the framework. We will support the description with a running example.

## 4.1    Problem Description and Framework Idea

Airbus databases contain a lot of data about historical maintenance problems and their solutions. These data sets are currently used for maintenance support in various situations. To use this information within our case-based diagnosis system, they have to be analyzed to find relevant pieces of information to be transformed into knowledge for CBR systems. The data sets from Airbus contain different data structures. Technical information can mostly be found in attribute-value pairs, while logbook entries, maintenance, and feedback are stored in form of free text articles. Based on a first data analysis, we choose a structural approach for our CBR systems. Information like fault codes, aircraft type and model, ATA (Air Transport Association) chapter and fault emitter are important information and can easily be used within a structural approach. Over time the main use case changed within the OMAHA project and the new data to be transformed has free text components. Therefore, we have to use the structured information as well as the free text. To transform the relevant information in the free texts into useful knowledge for our structural CBR system, we had to adapt and combine techniques from NLP and CBR. The idea is to develop a framework to combine several techniques and automatize the knowledge transformation. This framework could be used for knowledge acquisition and maintenance for CBR system in the development phase or for existing CBR systems.

## 4.2    Framework Architecture

The framework consists of five components: data layer, agent layer, CBR layer, NLP layer and interface layer. The data layer is responsible for storing the raw data and the processed data for each task. In addition, domain specific information like abbreviations and technical phrases are stored in this layer to be accessible for the other components. The agent layer contains several software agents. For every task an individual agent is responsible. All task agents communicate with a central supervising agent. This supervising agent coordinates the workflow. For visualization and communication purposes for the user, this layer also contains an interface agent. For each task an agent is spawned when starting the framework, but during the workflow additional agents can be spawned to support the initial agents with huge data sets. The NLP layer contains algorithms and methods like part of speech tagging, lemmatization, abbreviation replacement and association rule mining. These algorithms are used by the agents to execute their assigned tasks. The algorithms could either be third party libraries or own implementations. The fourth layer is the CBR layer and is responsible for the communication with a CBR tool like myCBR or jColibri. It contains methods to add keywords to the vocabulary, extend similarity measures and generate cases from the input data sets. The last layer contains the graphical user interface of the framework. This user interface can be used to configure the workflow, select input data, and start the workflow. In addition, the user interface presents the results of each task to the user and shows the status of the software agents.

## 4.3   Framework Tasks

In this section we will describe the eight tasks of the framework in more detail. These tasks and their interaction are defined based on the existing input data and the required data structure for our CBR systems. Based on our initial idea and the experience from the input data analysis, we had to regroup the tasks and their substeps. As input for the workflow a data set with free text components, for example a CSV file, or a pure free text document is possible. In addition to the data sets, a file with mapping information, an abbreviations file, and files with domain specific white and black lists are used. The data sets are first transformed into an internal representation of our case structure based on the information in the mapping file. It is not required to have information for every attribute in the data set or to use all information in the data set. The complete case structure for our use case consists of 72 attributes with different data types and value ranges and the mapping process adapts dynamically to the input information. The complete workflow with all tasks and possible parallelization is show in Fig. 2. In the following, the individual tasks will be described. As an example to illustrate the tasks, a free text problem description will be used:

– 'One hour before departure, cabin crew informed maint that the FAP was frozen.'

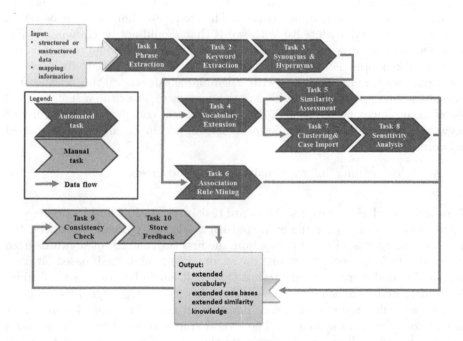

**Fig. 2.** Workflow task and possible parallelization

**Collocation Extraction.** The first task is the identification and extraction of phrases from the free text components of the input data. The idea is to find recurring combinations of words based on standard english grammar and domain-specific terms. For the phrases an acceptable length between 2 and 6 words is chosen. This length is based on manual analysis of free texts and domain-specific phrases. There are no domain-specific phrases with more than 6 words and the correct identification of phrases with more than 6 words only reaches 30 percent and generates not much additional benefit. This task has three substeps: part of speech tagging, multi-word abbreviation identification, and phrase extraction. First, the free text is tagged to identify nouns, verbs, and adjectives. The next step is to identify multi-word abbreviations, because the longform of these abbreviations counts as phrases, too. The last step is to identify phrases based on the tagging information and the word position in a sentence. This task was set as the initial task, because for a successful phrase identification the whole free text is required. The result of this tasks is a modified free text, reduced by multi-word abbreviations and found phrases. For our example, the identified phrases are

– 'One hour', 'cabin crew', and 'flight attendant panel'.

**Keyword Extraction.** The second task is the extraction of keywords from the remaining text and consists also of three substeps: stopword elimination, lemmatization, and single-word abbreviation replacement. As input for this task, the modified text from task one is used. The stopword elimination is based on common english and a white list with words that should not be eliminated. The second substep identifies abbreviations in the remaining words and replaces them with their longform. For all words the lemmata are determined. We replaced the former stemming algorithm with a lemmatization algorithm, because the lemmatization algorithm is considering the context of a word in a sentence and therefore produces better results. The result of the complete task is a list of keywords, reduced to their base form. According to our example, the extracted keywords are

– 'before', 'departure', 'inform', 'maintenance', and 'freeze'.

**Synonyms and Hypernyms.** The third task is responsible for identifying synonyms and hypernyms for the extracted keywords and phrases. Therefore, the input for this task is a list of phrases from the first task and list of keywords from the second task. For every keyword the synonyms are identified, based on common english and domain-specific terms. One challenge in this task, is to consider the context and word sense of a keyword to identify the right synonyms. Therefore, we are using the part of speech information and a blacklist of words, that should not be used as synonyms. The second step is to identify the hypernyms for all keywords. There are two goals for this task. The first goal is to enrich the vocabulary of our CBR systems and the second goal is to use the synonyms and hypernyms to enhance our similarity measures by extending or generating taxonomies. The result of this task is a list of keywords and their synonyms and hypernyms. In our example, the result could be as follows:

- before: earlier, once
- departure: exit, movement, withdrawal
- inform: describe, make known
- maintenance: support, administration
- freeze: stop, paralyze, stuck, immobilize
- flight attendant panel: monitor, display.

**Vocabulary Extension.** This task consists of adding the extracted keywords, phrases, synonyms, and hypernyms to the vocabulary of the CBR systems. The first step is to remove duplicate words and phrases to avoid redundant knowledge. The second step is to check the list of keywords against the list of phrases to identify keywords which occur as phrases. We want to slow down the growth of the vocabulary and therefore we identify keywords that are only occur as part of a collocation. These keywords are not added to the vocabulary. If a keyword occurs without the context of a collocation, it will be added.

**Similarity Measures.** When describing faults there are terms that are easily predictable and their similarity can be modeled by experts. However, when confronted with a large amount of manually inserted text from many sources it is virtually impossible to predict every concept that may appear, and how it stands in relation to other concepts. Therefore, this task is responsible for setting initial similarity values for newly discovered concepts and extends existing similarity measures. The first substep is to set similarity values between the newly added keywords and phrases and their synonyms. Therefore, the existing similarity matrices are extended and a symmetric similarity is proposed. The value itself could be configured, but we assume an initial similarity for synonyms of 0.8, based on the assumption that the similarity measures can take values from the [0;1] interval. The second step is to use the keywords, phrases, and hypernyms to extend or generate taxonomy similarity measures. The hypernyms serve as inner nodes, while the keywords and the synonyms are the leaf nodes. Keywords and their synonyms are sibling nodes if they have the same hypernym. This second step provides the possibility to model or extend similarity measures based on the layers of a taxonomy and therefore less similarity values have to be set. For values of keywords and phrases that could not be assigned to a taxonomy, no initial similarity value could be set, than 0. To overcome this hurdle, we employ social network analysis (SNA) methods to supplement the similarity between each two values of a given attribute. SNA is based on graph theory and utilizes the structure of the data and the relationships between the different items to reach conclusions about it, and has been used previously to measure the similarity of objects [2,15]. It is useful for our purposes since besides the structure of the data, which is readily available, no additional information is required. Our data consist of attribute-value pairs, representing different concepts of a fault. We want to compute the similarity degree between the different values, for instance between two systems. In order to do so, we see our data as a weighted bipartite graph. On the left side are all the values of a given attribute, and on the right side the case diagnoses. Nodes A and B are then connected if a value represented by A appeared in a case that received the diagnosis represented by B. To eliminate multi-edges, edge weights represent the number of connections between each node pair. Since we are only interested in the similarity of nodes of type value, we perform weighted one-mode projection (WOMP) [16] on the left side of the graph. The resulting edge weights of the WOMP are the similarity degree between the nodes, and between the values they represent.

**Association Rule Mining.** This task is used to analyze the keywords and phrases and find associations between the occurrence of these words within a data set as well as across data sets. Using association rule mining algorithms like the Apriori [1] or the FP-Growth [10] algorithm, we try to identify reoccurring associations to determine completion rules for our CBR systems to enrich the query. An association between keywords or phrases exists, when the combined occurrence exceeds a given threshold. For example, a combination between two keywords that occurs in more than 80 % of all analyzed documents, may be used as a completion rule with an appropriate certainty factor. To generate only completion rules with a high significance, a larger number of data sets have to be mined. Therefore, a minimum number of data sets has to be defined. Based on manual analysis of data sets in collaboration with aircraft experts, we assume a minimum of 10000 datasets and a confidence of 90 % will generate rules with the desired significance in the aircraft domain. In the aircraft domain many causal dependencies between systems, status, context and functions exist. Association rule mining can help identify this dependencies in an automated way to avoid the high effort from manually analyzing the data sets.

**Clustering and Case Generation.** This task is responsible for generating a case from each input data set and storing it in a case base. To avoid a large case base with hundreds of thousands of cases, we cluster the incoming cases and distribute them to several smaller case bases. Generating an abstract case for each case base, a given query can be compared to the abstract cases and this way a preselection of the required case bases is possible. The first substep uses the mapping document to map the content of the document to a given case structure. The data from the documents are transformed into values for given attributes. In collaboration with experts from Airbus and Lufthansa we identified the aircraft type and the ATA chapter as the two most discriminating features of the cases. Therefore, the clustering algorithm uses these features to distribute the cases on the different case bases. For each aircraft type (A320, A330, A340, etc.) a set of case bases will be created and each set will be separated by the ATA chapter. The ATA chapter is a number with four or six digits and is used to identify a component of an aircraft. The cases are discriminated by the first two digits of the ATA chapter, which identify the component, while the other digits are used to specify a system of the component.

**Sensitivity Analysis.** In this task the feature weights for the problem description of the given case structure are determined. Not all attributes are equal. In retrieval tasks some attributes are more important to determine which objects are relevant, but how do we identify these attributes, and what is their degree of importance? Can some attributes be detrimental to retrieval? To answer these questions we used sensitivity analysis, and developed a method to calculate a relevance matrix of attributes. In our data each case has a diagnosis, and a diagnosis set consists of all the cases with the given diagnosis. While some attributes may be important to determine whether or not a case belongs to set A, other attributes might be more important for set B. This is why we have a relevance matrix, and not a vector. Our method is based on work done by [20, 25], and includes three phases: 1. the static phase, where all attributes have the same weight for all diagnosis sets, and is used as a baseline to measure the contribution of the next two phases, 2. the initial phase, which includes a statistical analysis of the data set, and functions as the starting point of the next phase, 3. the training phase, where the values are optimized. The idea behind the training phase is that in a retrieval task there are two reasons for a false positive: first, the weights of attributes with a similar value are too high, and second, the weights of attributes with dissimilar values are too low. Much like the training phase of artificial neural networks, the contribution of each attribute to the error is calculated and propagated back through the weights, updating them accordingly. Within the OMAHA project, the analysis will be performed offline and the resulting relevance matrix will be embedded within the retrieval task. A more detailed description of the sensitivity analysis can be found in [23].

### 4.4   Current Implementation

This section describes the current implementation status of our framework. FEATURE-TAK is an agent-based framework that uses the scalability of multi-agent systems and parallelization possibilities. In addition, an agent-based framework could easily be integrated into the multi-agent system for case-based diagnosis developed within the OMAHA project. For the implementation of the agents the JADE framework [8] was used. Currently, seven agents are implemented: supervising agent, gui agent, collocation agent, keyword agent, synonym agent, vocabulary agent and cluster agent. The supervising agent is the central coordinator of the framework and routes the communication between the other agents. The gui agent controls the user interface of the framework. He receives the input data and sends the information to the supervising agent. He also presents the interim results to the user and shows the status of the workflow. The collocation agent uses the Stanford CoreNLP library [17], a suite of NLP tools for part-of-speech tagging and collocation extraction. The keyword agent uses Apache Lucene [18] and the Stanford CoreNLP library for stopword elimination and lemmatization. The abbreviation replacement of both agents is an own implementation based on domain-specific abbreviations from Airbus and Lufthansa. For the synonym identification the synonym agent is using WordNet [14] and its databases of common english synonyms and hypernyms. The results of the first three tasks are passed to the vocabulary agent. This agent uses the myCBR [7] API to access the knowledge model of the CBR systems. The last implemented agent is the cluster agent, which generates the cases from the data sets and distributes them to the different case bases. The clustering algorithm is an own implementation and the myCBR API is used to pass the generated cases to the correct case base. The functionality for similarity assessments, association rule mining and sensitivity analysis are implemented, but not integrated into the framework yet. Different import mechanisms are implemented to process data from CSV files and text files like word documents or PDF files. Because of the different content and data structures of the documents, the data is processed differently for each document type. CSV files and result sets are processed row-wise, while text documents are processed in the whole. The mapping file is written in XML format and contains information about which column in a CSV file or result set should be mapped to which attribute in the case structure. For text documents the mapping is far more difficult and not completely implemented yet.

### 4.5  Evaluation

The current implementation of the framework was tested with a CSV file containing 300 real world data sets. On these data sets five tasks were computed: collocation extraction, keyword extraction, synonym search, vocabulary extension, and case generation and clustering. The following results were generated:

- Collocation extraction: 2465 phrases extracted, 2028 distinct phrases
- Keyword extraction: 8687 keywords extracted, 1464 distinct keywords
- Synonym search: 21285 synonyms identified, 3483 distinct synonyms
- Vocabulary extension: 4621 concepts added to the vocabulary
- Case generation and clustering: 300 cases distributed over 8 case bases.

The results of the workflow were evaluated by experts from Airbus Operations GmbH and Lufthansa Industry Solutions. The extracted collocations and keywords were compared against the original fault description, while the synonyms were checked for adequate word sense. The added concepts were checked to identify duplicates or false entries. The following graphic illustrates the evaluation results (Fig. 3).

While we have good results for the collocation and keyword extraction, we have poor results for the synonyms identification. The reason is that our word sense disambiguation is just based on black and white lists and therefore our synonym task identifies a great number of synonyms with inappropriate word sense. Therefore, the word sense disambiguation has to be improved with state of the art approaches. In addition, we conducted a performance evaluation of the

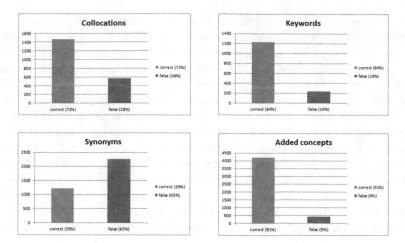

**Fig. 3.** Evaluation results

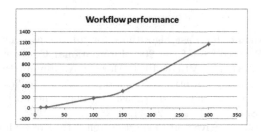

**Fig. 4.** Workflow performance

implemented tasks. Therefore, we run the workflow with different sized CSV files: 10, 20, 100, 150, and 300 data sets. Figure 4 shows the results. The y axis contains the time in seconds and the x axis the number of data sets. With an increasing number of data sets, the computation time appeared to grow exponentially. We identified the myCBR tool as the main cause for this performance problem during the task of the vocabulary extension.

## 5  Summary and Outlook

In this paper we describe the concept and implementation of our framework FEATURE-TAK. The framework was developed to transform textual information in the aircraft domain into knowledge to be used by structural CBR systems. We give an overview of the framework architecture and describe the individual tasks in more detail. In addition, we describe the status of our current implementation. The newly improved version is still in an evaluation process and will be tested with a larger data set based on historical problem data from Airbus. We will test the framework with input data of more than 65.000 single data sets.

Based on the evaluation results we will improve the framework methods. A specific challenge is the word sense disambiguation. We will address this challenge using pattern recognition and neural networks. In addition, we will integrate the remaining functionality into the framework and connect it with the corresponding agents. After the complete implementation of the framework, it will be integrated into the diagnosis system to provide the frameworks functionality for knowledge modeling and maintenance purposes. In addition to improvement on the semantic level, we also will improve the performance and scalability of the framework to support the computation of large data sets. For further development we plan to modularize and generalize the tasks and substeps to get a framework with domain-independent and domain-specific components, that could be configured for the use in different domains. We also want to support an interface for different additional NLP or CBR methods and tools, to provide the user with a greater variety on analysis, extraction and transformation possibilities.

# References

1. Agrawal, R., Srikant, R.: Fast algorithms for mining association rules in large databases. In: Proceedings of the 20th International Conference on Very Large Data Bases, pp. 487–499. VLDB 1994. Morgan Kaufmann Publishers Inc., San Francisco, CA, USA (1994). http://dl.acm.org/citation.cfm?id=645920.672836
2. Ahn, Y., Ahnert, S., Bagrow, J., Barabasi, A.: Flavor network and the principles of food pairing. Sci. Rep. **1**, 1–7 (2011)
3. Althoff, K.D.: Collaborative multi-expert-systems. In: Proceedings of the 16th UK Workshop on Case-Based Reasoning (UKCBR-2012), Located at SGAI International Conference on Artificial Intelligence, Cambride, p. 1, 13 December 2012
4. Althoff, K.D., Bach, K., Deutsch, J.O., Hanft, A., Mänz, J., Müller, T., Newo, R., Reichle, M., Schaaf, M., Weis, K.H.: Collaborative multi-expert-systems - realizing knowledge-product-lines with case factories and distributed learning systems. In: Baumeister, J., Seipel, D. (eds.) KESE @ KI 2007. Osnabrück, September 2007
5. Althoff, K.D., Reichle, M., Bach, K., Hanft, A., Newo, R.: Agent based maintenance for modularised case bases in collaborative mulit-expert systems. In: Proceedings of the AI2007, 12th UK Workshop on Case-Based Reasoning (2007)
6. Bach, K.: Knowledge acquisition for case-based reasoning systems. Ph.D. thesis, University of Hildesheim, Dr. Hut Verlag Mnchen (2013)
7. Bach, K., Sauer, C.S., Althoff, K.D., Roth-Berghofer, T.: Knowledge modeling with the open source tool MYCBR. In: Proceedings of the 10th Workshop on Knowledge Engineering and Software Engineering (2014)
8. Bellifemine, F., Caire, G., Greenwood, D.: Developing Multi-agent Systems with JADE. Wiley, New York (2007)
9. BMWI: Luftfahrtforschungsprogramms v (2013). www.bmwi.de/BMWi/Reda ktion/PDF/B/bekanntmachung-luftfahrtforschungsprogramm-5,property=pdf, bereich=bmwi2012,sprache=de,rwb=true.pdf
10. Borgelt, C.: An implementation of the fp-growth algorithm. In: Proceedings of the 1st International Workshop on Open Source Data Mining: Frequent Pattern Mining Implementations, OSDM 2005, NY, USA, pp. 1–5 (2005). http://doi.acm. org/10.1145/1133905.1133907

11. Ceausu, V., Desprès, S.: A semantic case-based reasoning framework for text categorization. In: Aberer, K., Choi, K.-S., Noy, N., Allemang, D., Lee, K.-I., Nixon, L., Golbeck, J., Mika, P., Maynard, D., Mizoguchi, R., Schreiber, G., Cudré-Mauroux, P. (eds.) ASWC/ISWC -2007. LNCS, vol. 4825, pp. 736–749. Springer, Heidelberg (2007). doi:10.1007/978-3-540-76298-0_53

12. Cunningham, H., Maynard, D., Bontcheva, K., Tablan, V., Aswani, N., Roberts, I., Gorrell, G., Funk, A., Roberts, A., Damljanovic, D., Heitz, T., Greenwood, M.A., Saggion, H., Petrak, J., Li, Y., Peters, W.: Text Processing with GATE (Version 6) (2011). http://tinyurl.com/gatebook

13. Drożdżyński, W., Krieger, H.-U., Piskorski, J., Schäfer, U.: SProUT – a general-purpose NLP framework integrating finite-state and unification-based grammar formalisms. In: Yli-Jyrä, A., Karttunen, L., Karhumäki, J. (eds.) FSMNLP 2005. LNCS (LNAI), vol. 4002, pp. 302–303. Springer, Heidelberg (2006). doi:10.1007/11780885_35

14. Feinerer, I., Hornik, K.: Wordnet: WordNet Interface, r package version 0.1-11 (2016). https://CRAN.R-project.org/package=wordnet

15. Jeh, G., Widom, J.: Simrank: a measure of structural-context similarity. In: Proceedings of the Eighth ACM SIGKDD International Conference on Knowledge Discovery and Data Mining, pp. 538–543 (2002)

16. Lapaty, M., Magnien, C., Vecchio, N.D.: Basic notions for the analysis of large two-mode networks. Soc. Netw. **30**, 31–48 (2008)

17. Manning, C.D., Mihai, S., Bauer, J., Finkel, J., Bethard, S., McClosky, D.: The stanford corenlp natural language processing toolkit. In: Proceedings of the 52nd Annual Meeting of the Association for Computational Linguistics: System Demonstrations, pp. 55–60 (2014)

18. McCandless, M., Hatcher, E., Gospodnetic, O.: Lucene in Action, 2nd edn. Manning Publications Co., Greenwich (2010)

19. Reuss, P., Althoff, K.D.: Explanation-aware maintenance of distributed case-based reasoning systems. In: Learning, Knowledge, Adaptation. Workshop Proceedings, LWA 2013, pp. 231–325 (2013)

20. Richter, M.: Classification and learning of similarity measures. In: Concepts, Methods and Applications Proceedings of the 16th Annual Conference of the Gesellschaft fr Klassifikation e.V, pp. 323–334 (1993)

21. Rodrigues, L., Antunes, B., Gomes, P., Santos, A., Carvalho, R.: Using textual CBR for e-learning content categorization and retrieval. In: Proceedings of International Conference on Case-Based Reasoning (2007)

22. Sizov, G., Öztürk, P., Štyrák, J.: Acquisition and reuse of reasoning knowledge from textual cases for automated analysis. In: Lamontagne, L., Plaza, E. (eds.) ICCBR 2014. LNCS (LNAI), vol. 8765, pp. 465–479. Springer, Heidelberg (2014). doi:10.1007/978-3-319-11209-1_33

23. Stram, R., Reuss, P., Althoff, K.D., Henkel, W., Fischer, D.: Relevance matrix generation using sensitivity analysis in a case-based reasoning environment. In: Proceedings of the 25th International Conference on Case-based Reasoning, ICCBR 2016. Springer Verlag (2016)

24. Weber, R., Aha, D., Sandhu, N., Munoz-Avila, H.: A textual case-based reasoning framework for knowledge management applications. In: Proceedings of the ninth german Workshop on Case-Based Reasoning, pp. 244–253 (2001)

25. Wess, S.: Fallbasiertes Problemlsen in wissensbasierten Systemen zur EntscheidungsunLersttzung und Diagnostik. Ph.D. thesis, TU Kaiserslautern (1995)

26. Wettschereck, D., Aha, D.: Feature weights. In: Proceedings of the First International Conference on Case-Based Reasoning, pp. 347–358 (1995)

# Knowledge Extraction and Annotation for Cross-Domain Textual Case-Based Reasoning in Biologically Inspired Design

Spencer Rugaber, Shruti Bhati[✉], Vedanuj Goswami, Evangelia Spiliopoulou,
Sasha Azad, Sridevi Koushik, Rishikesh Kulkarni, Mithun Kumble,
Sriya Sarathy, and Ashok Goel

School of Interactive Computing,
College of Computing Georgia Institute of Technology,
Atlanta, GA 30308, USA
Shruti.bhati@gatech.edu

**Abstract.** Biologically inspired design (BID) is a methodology for designing technological systems by analogy to designs of biological systems. Given that knowledge of many biological systems is available mostly in the form of textual documents, the question becomes how can we extract design knowledge about biological systems from textual documents for potential use in designing engineering systems? In earlier work, we described how annotating biology articles with partial Structure-Behavior-Function models helps users access documents relevant to a given design problem and understand the biological systems for potential transfer of their causal mechanisms to engineering problems. In this paper, we present an automated technique instantiated in the IBID system for extracting partial SBF models of biological systems from their natural language documents for potential use in biologically inspired design.

## 1 Background, Motivations and Goals

Biologically inspired design is a well-known design paradigm that uses nature as a source of practical, efficient and sustainable solutions to stimulate design of technological systems (Benyus 1997; Vincent and Mann 2002). Recently biologically inspired design has grown into a movement with an increasing number of engineering and system designers looking towards nature as a source of ideas (Lepora et al. 2013). Biologically inspired design entails cross-domain analogies: It views nature as a library of design cases and biologically inspired design as a process of abstracting, transferring and adapting designs of biological systems into designs of technological systems.

Biologically inspired design is also related to Textual Case Based Reasoning (TCBR) because knowledge of many biological systems is available mainly in the form of textual documents. Textual information typically is hard to process by

A. Goel et al. (Eds.): ICCBR 2016, LNAI 9969, pp. 342–355, 2016.
DOI: 10.1007/978-3-319-47096-2_23

computers due to its relatively unstructured format and the numerous possible variations in its interpretation. Hence, it is often necessary to introduce additional processing to extract structured knowledge from textual documents. As a result, TCBR is commonly combined with techniques from information retrieval, natural language processing, text mining, and knowledge discovery (Weber et al. 2005). The question then becomes how can we extract design knowledge about biological systems from textual documents for potential use in designing engineering systems?

In earlier work, we observed that when engaged in biologically inspired design, design teams typically searched the Web for biology articles describing systems that might inspire solutions to their problems (Vattam and Goel 2013a). We also found that the design teams typically struggled to locate biology articles relevant to their problems because search engines are not designed specifically for cross-domain retrieval, and, in particular, keywords that describe biology articles do not capture the design semantics of the biological systems described in the articles. Thus, in earlier work (Vattam and Goel 2013b), we presented an interactive system called Biologue that annotated biology articles with partial Structure-Behavior-Function (SBF) models (Goel et al. 2009) of the biological systems described in the articles. We found that the SBF annotations on the biology articles enhanced the precision and relevance of retrieved articles.

However, the semantic annotations in Biologue were handcrafted, which raised the issues of scalability and repeatability. If we are to make the interactive retrieval not only relevant and precise but also scalable and repeatable, then we must develop computational techniques for automatically extracting the partial SBF models of biological systems from their natural language descriptions. The objective of this paper is to describe a preliminary, high-level computational process for extracting structures, behaviors and functions of biological systems from textual documents. This process is embodied in the Intelligent Biologically Inspired Design (IBID) system presently under development.

## 2 The Problem: An Illustrative Example

Consider an engineer interested in improving water harvesting for a village in an arid region. Suppose that the engineer seeks inspiration from nature. Darkling beetles that live in the Namib Desert, one of the hottest places on Earth, survive by using their shells to draw water from periodic fog-laden winds. Thus, two beetle species from the genus *Onymacris* have been observed to fog-bask on the ridges of sand dunes (Norgaard and Danke 2010). How might our hypothetical engineer find biology articles describing the fog-harvesting processes of the beetles? How might the engineer confirm that the retrieved descriptions are relevant to her design problem? How might the engineer build a deep enough understanding of the beetles' fog-harvesting mechanisms to support application to her problem?

The engineer might conduct a literature survey using a web search engine. However, the current search technology for conducting this kind of literature survey is plagued by several problems (Vattam and Goel 2013a). First, biologically inspired design by definition entails cross-domain knowledge transfer from biology systems to technological designs. However, current search engines are not designed to support cross-domain search. Second, engineers and biologists speak different languages, and most engineers are novices in biology. This makes it difficult for engineers to interpret a biology article or even to form an effective query. Third, current search engines use keywords to filter their results. However, the keywords do not capture a deep understanding of the user's query of the design problem. Thus, the keyword-based search typically results in imprecise results, including voluminous hits on unrelated documents.

Even when a search engine notes an appropriate article, at best it highlights contents words that match keywords in the query. The engineer must still expend effort to understand the article sufficiently to determine relevance. Hence, there are two opportunities to improve engineer productivity: increase the precision of the set of retrieved articles and facilitate relevance checking by improved annotation.

## 3    Our Approach to Developing a Solution

IBID uses a representation for complex systems called Structure-Behavior-Function models (Goel et al. 2009). SBF models consist of three main parts. The Function submodel of a system is an abstraction over the system's actions on its external environment. The Structure sub-model expresses its physical components and the connections among them. The Behavior sub-model describes the causal mechanisms that arise from the interactions among the structural components and that accomplish the system's functions. We have previously used SBF models extensively in building theories of analogical design. In particular, Goel and Bhatta (2004) showed that domain-specific SBF models can be abstracted into Behavior-Function design patterns for cross-domain analogical transfer. We have also developed several tools to support biologically inspired design including DANE, a library of SBF models of biological systems (Goel et al. 2011) and Biologue (Vattam and Goel 2013b) mentioned above.

In a preprocessing phase in IBID, SBF models of biological systems are extracted from articles describing them, and the articles are annotated by the extracted models. While the functions in the extracted SBF models are expressed in a domain-independent controlled vocabulary, the structural components are expressed in a biology-specific vocabulary. Thus, in the current version of IBID, design queries are made by specifying the desired functions (in the domain-independent controlled vocabulary of functions), possibly augmented with a specification of biology-specific structural components. Biology articles are retrieved based on the match with the functional and structural annotations on the articles, thereby increasing precision. The retrieved articles are annotated with SBF model elements, thereby making it easier to evaluate the relevance of articles.

The extracted SBF model of the biological case contains pointers back into the document from which it was extracted for each of its model elements. When the document is retrieved, the pointers can be used to annotate how a segment of text contributes to the model. For example, a biological process described in an article can have a function, such as *transport*, highlighted in yellow, and the object of transport, *water*, highlighted in green. Thus the engineer reading the article can more quickly determine the relevance of the article.

### 3.1   Example Continued

In order to elaborate on the analysis and extraction of the aforementioned SBF models, we consider the article (Norgaard and Danke 2010). Following is a text snippet from the article:

"The mechanism by which fog water forms into large droplets on a beaded surface has been described from the study of the elytra of beetles from the genus *Stenocara*. The structures behind this process are believed to be hydrophilic peaks surrounded by hydrophobic areas; water carried by the fog settles on the hydrophilic peaks of the smooth bumps on the elytra of the beetle and form fast-growing droplets that - once large enough to move against the wind - roll down towards the head."

IBID identifies the following structural component from the snippet:
**Structure:**

- Name: elytra
- Properties: hydrophobic, hydrophilic, smooth
- Parts: grooves

Here is an example of one behavior extracted from this snippet.

**Behavior:**

- Predicate - move
- Cause - "water carried by the fog settles on the hydrophilic peaks of the smooth bumps on the elytra of the beetle and form fast-growing droplets that - once large enough"
- Effect - "roll down towards the head"

## 4   IBID

IBID is a web application that retrieves and annotates biology articles in support of BID. IBID uses SBF models and controlled vocabularies to facilitate its retrieval and annotation. Several other aspects of IBID are worth noting.

- **Natural language processing (NLP):** When it analyzes an article, IBID makes use of common NLP technology including parsing, part-of-speech tagging, and word-sense-disambiguation, to detect salient sections of the document. Technical vocabulary is detected by use of one or more domain-dependent taxonomies. For example, to analyze a document about the water-harvesting behavior of beetles, biological taxonomies from the fields of entomology and morphology might be used.

- **Taxonomies of Structures, Behaviors and Functions:** For the IBID system, we developed a domain-independent taxonomy of functions (Spiliopoulou et al. 2015). Our function ontology combines elements of the function ontology in the Functional Basis (Hirtz et al. 2002) with elements in AskNature's function ontology for biomimicry (Deldin et al. 2002). This is important because the domain-independence of the function taxonomy enables a cross-domain matching between functions delivered by biological systems and the functions desired in engineering problems. IBID also uses Vincent's (Vincent 2014) biology-specific taxonomy of structural components and connections. Finally, IBID uses a subset of Khoo et al.'s taxonomy of behavioral patterns (Khoo et al. 1998, 2000). These taxonomies of structures, behaviors and functions play a role in IBID similar to that of Schank's (1972) conceptual dependency in semantic processing: They help generate top-down expectations for completing the schemas corresponding to the elements in the taxonomies.
- **Semantic annotation:** Using the vocabulary and relations present in the SBF taxonomy, the textual content of a document can be semantically annotated. *Semantic annotation* is a technique that helps to add semantics to unstructured documents (Davies et al. 2006). In particular, IBID makes use of VerbNet (Kipper et al. 2008), a knowledge base of common verbs and their expected role-fillers. VerbNet further improves IBID's word-sense disambiguation, and its frames serve as the first level of IBID's semantic processing. For example, VerbNet was used to determine the roles (Names, Properties, and Parts) used in the Structure frame for elytra presented in the last section.
- **Faceted search:** When engineers query IBID's repository of biology articles, they do so using faceted search (Prieto-Diaz 1991). A *faceted search* interface provides an orthogonal set of controlled vocabularies, one for each dimension of the search space. These include the expected *title*, *author*, and *publication date* dimensions. More important, however, for achieving precision, is its use of dimensions for *structure*, *behavior* and *function*. For example, by selecting *water* as a structural element, the engineer can focus her search on specific kinds of biological processes.

The following subsections describe IBID's system architecture, computational process, data model, use cases and current status. The section concludes by relating the IBID approach to CBR.

## 4.1 IBID System Architecture

IBID uses a classical client–server architecture. The web client uses dynamic HTML, CSS and Javascript to support user query construction and perusal of results. The server is written in PHP, with analysis performed by a Java servlet. Extracted knowledge is stored in a MySQL database. In addition to the search-and-perusal scenario, IBID also supports two other uses cases: file upload and analysis and taxonomy management.

In Fig. 1, the user interacts with the interface of the tool via a web browser. The PHP server acts as an intermediary between the client side and the Java

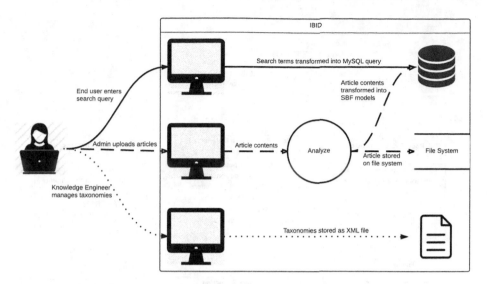

**Fig. 1.** IBID's conceptual architecture

semantic analyzer. The bold line in the figure depicts the user searching for models. The dashed line represents a scenario where an administrator can upload and analyze articles. The dotted line depicts a knowledge engineer managing taxonomies.

Both text and PDF files can be uploaded. Each file uploaded is sent to the Java analyzer. The file is parsed, and semantically processed to produce Structure-Behavior-Function models that are then stored into the MySQL database.

The core Java analyzer uses a number of NLP techniques to extract an SBF model of each biological process described in the article being analyzed. First, the article is broken down into sentences, each of which is parsed. The parse graph thus obtained is further processed by different modules to extract the Structure, Behavior and Function sub-models.

### 4.2   IBID's Computational Process for Extracting SBF Models

Figure 2 illustrates IBID's computational processes for extracting partial SBF models of biological systems from natural language documents. Initially, IBID breaks down the input text file into individual sentences and uses the Stanford parser (De Marneffe et al. 2011) to generate a parse tree for each.

**Function Extraction:** Function extraction focuses on the predicates present in each sentence. Using VerbNet's application programming interface (API), one or more frames are constructed for each sentence. The most relevant frame is selected and used to populate the SBF Function sub-model. The predicates

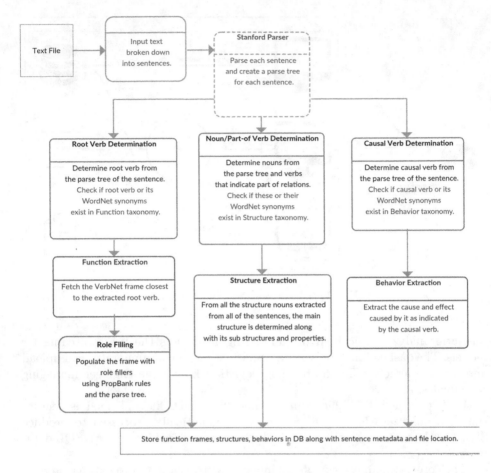

**Fig. 2.** IBID's computational processes for extracting SBF models of biological systems

selected have to be part of IBID's Function taxonomy. If this is not the case, a bespoke algorithm is applied to find the nearest match.

**Behavior Extraction:** The behavior of the biological system is captured in the form of causal chains: actions, effects and their causes. The action is the predicate in the sentence. The sentence is then matched to compiled patterns for causal chains to determine whether or not one is present.

**Structure Extraction:** Using the functional root verb in each sentence, the related subject and object are determined. Using WordNet (Miller 1995; Fellbaum 1998), synonyms and hypernyms are mapped into an ontology of biological components and connections due to Vincent (2014); only matches above a preset but tunable threshold are considered for further processing. The nouns thus found are designated as structural components of the SBF Structure sub-model.

While the extracted structures, behaviors and functions are composed from multiple sentences in the input text, the extracted SBF models are at least partially domain specific. In particular, while the extracted functions are expressed in the domain-independent control vocabulary of functions, the extracted structures are expressed in a biology-specific vocabulary of structural components and connections. Thus, the user must do additional processing to extract domain-independent behavior-function patterns (Goel and Bhatta 2004) for transfer to engineering design problems.

## 4.3   IBID's Data Model

Figure 3 is a detailed enhanced entity–relationship diagram of IBID's data model. There are three groups into which all the tables have been arranged: articles, taxonomies and models. The group on the right corresponds to the tables related to articles. While the actual document contents are stored in the file system, IBID retains key information in its database (metadata and unique IDs) that are used during retrieval. The group at the bottom consists of three tables that store the taxonomies for the SBF sub-models. The group of tables to the left contains the stored models. As part of document analysis, IBID extracts SBF function, structure and behavior sub-models. The function information is stored in a format similar to VerbNet's frames. Structure is stored in a custom format in the structure entity table, and the causal behavior information is stored in the causality table.

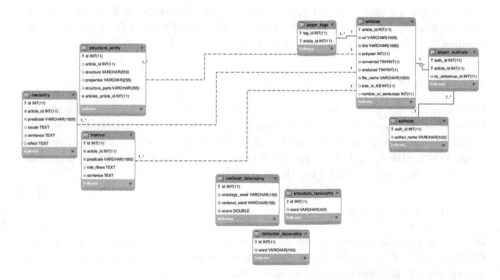

**Fig. 3.** Enhanced Entity-Relationship diagram for IBID

## 4.4   IBID Use Cases

The IBID tool has three main use cases: faceted search performed for an end-user engineer, document upload and analysis performed by a system administrator, and taxonomy management performed by a knowledge engineer.

Fig. 4 shows the first use case, the retrieval of several documents in which a biological process is accomplished via movement.

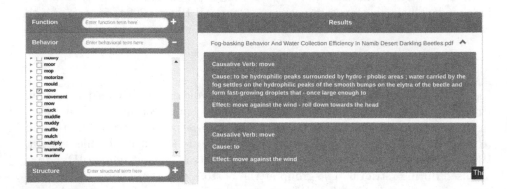

**Fig. 4.** Screenshot showing the Behavior cause and effects related to *move*.

When a user clicks on any term in the menu on the left side, the system expands its search and adds more synonyms and hypernyms for the term. All of these terms are then searched for in the database, and links to the documents that contain a related model are returned. The user can click on one of the links to peruse the document. The system highlights the relevant portion of the document to make it easier for the user to understand the results.

## 4.5   Current Status

IBID is a working prototype with all the above mentioned features implemented. Knowledge engineers can configure domain-specific vocabularies. System administrators can upload documents and analyze them. The analyzed articles are then tagged with SBF model elements. Once these models are stored, an engineer can search for and retrieve matching documents.

## 4.6   IBID and CBR

IBID serves as case based system in two ways. First, during the analysis phase, IBID extracts SBF models from documents. It uses the models as indices into its repository. It also searches for similar SBF models already in the repository. If similar cases exist, IBID stores the biological processes using the same SBF index.

Second, IBID acts as a CBR system during user search. By extracting SBF models for various biological processes, IBID indexes these processes. Using these SBF model indexes, IBID treats the various biological processes as cases. The cases stored by IBID can be retrieved using the index by searching either for Structures, Behaviors or Functions. The search results show the biological process that consist of the searched structure/behavior/function. This is the case-retrieval phase. From there, based on the biological processes retrieved, the user can adapt the process to her engineering problem taking advantage of her increased understanding of how the biological process works.

# 5   Validation

Our strategy for validating IBID has three main parts: (i) reliance on past work, (ii) execution of IBID on a large corpus of biology articles and inspection of results, and (iii) comparison with human performance.

## 5.1   Strategy 1: Reliance on Past Work

IBID assumes the following:

- Biologically Inspired Design is an effective design technique (Vincent and Mann 2002)
- SBF is robust in representing mechanisms in engineering and biology
- SBF models can improve search effectiveness for use in Biologically Inspired Design

## 5.2   Strategy 2: Execution of IBID on a Corpus of Articles

IBID's taxonomy of functions contain 8 functions at the top level of the hierarchy, with about 50 functions in all, with more than 45,000 hypernyms/synonyms. Its taxonomy of structural components and connections contains more than 200 elements. Thus, IBID is not a small system.

The IBID corpus of biology articles contains 255 journal papers and is a superset of Biolgue's corpus. We were able to successfully execute IBID on all biology articles in its corpus. Manual inspection of the SBF models extracted by IBID indicates that the models are incomplete but not incorrect. In particular, one way in which the SBF models are incomplete is that at present IBID does not fully relate the extracted structures, behaviors and functions with one another. For example, some of the structural elements it extracts do not appear to play any role in the accomplishment of the system functions. On the other hand, IBID presently does not always extract all the behaviors it should. Thus, IBID provides an automated computational technique for abstracting only partial SBF models of biological systems described in textual documents, storing the partial SBF models as biological design cases, and indexing the case by both their functions and structural elements in support of interactive biologically inspired design.

### 5.3   Strategy 3: Comparison with Human Performance

Our third strategy for ongoing evaluation of IBID focuses on comparing the quality of the SBF models extracted by IBID with those extracted by human experts. Ideally, the model extracted by IBID should be equivalent to the SBF model extracted by a human expert (a criteria that only a few practical CBR systems meet for tasks as complex as automated construction of a case library). Thus, we measured the Cohen's Kappa (Cohen 1960; Landis and Koch 1977) coefficient pairwise for models produced by a group of human evaluators. Preliminary Kappa results indicated an agreement of 0.559. A closer inspection uncovered three explanations:

First, there is no unique "best" SBF model for a complex biological system. There are always differences among SBF models generated by human experts as well, even when the human evaluators identify the same mechanisms. Thus, it is not easy to define and measure the degree of similarity between two SBF models, given that the same mechanism may be described in a different way in the models.

Second, for the above reason the Kappa coefficient is not the best measure for measuring the quality of SBF models extracted by IBID. In order to resolve these problems, we now use the Weighted Kappa Coefficient (Cohen 1968) that weighs each part of the model according to the reviewers' agreement on that part. Thus, when a word is described as a function by all the reviewers it is weighted more than when only half of the reviewers agree. Those weights are used later in order to calculate the similarity between IBID's extracted model and the models that humans' extracted.

Finally, as mentioned above, IBID extracts only partial SBF models: It does not presently fully integrate the structures, the behaviors and functions into a complete SBF model. We expect that the quality of the extracted SBF models to improve once IBID starts exploiting the constraints that full integration of SBF models will impose on decisions about individual structures, behaviors and functions.

## 6   Related Work

The IBID project relates to efforts in case-based reasoning, natural language processing, biologically inspired design, and computational creativity. In research on textual case-based reasoning, Weber et al. (2001) propose a knowledge management framework for acquiring cases from human experts as well as natural language documents. Bruninghaus and Ashley (1998, 2006) describe a technique for predicting the outcome of a legal case given a brief textual summary of the case facts. Schumacher et al. (2012) present a technique for extracting procedural knowledge from natural language documents available on the web. Sizov et al. in (2014, 2015) describe a technique for extracting causal relational graphs from natural language documents. Our work is related to the above research. The behaviors in SBF models can be viewed as graphs representing causal processes;

it differs from earlier work in that (i) IBID extracts multiple kinds of knowledge (structures, behaviors, functions) and (ii) it extracts SBF models for cross-domain analogical transfer from biological systems to engineering design.

In research on natural language processing, Berant et al. (2014) developed a system that answers multiple choice questions based on natural language paragraphs describing biological processes. Although their representation of causal processes is similar to that of IBID, their system uses manually preprocessed questions and answers. In research on biologically inspired design, Cheong and Shu (2012) have used natural language processing techniques to extract and categorize causally related biological functions. Finally, in research on computational creativity, Jursic et al. (2012) describe a process to identify and explore terms that relate different domains. In IBID, the taxonomy of functions provides the cross-domain words that lead to knowledge transfer from biology to engineering.

## 7  Conclusion

IBID is an interactive system for finding and semantically annotating biology articles relevant to a design problem. In a preprocessing phase, IBID extracts partial SBF models of biological systems from biology articles and uses the SBF models as annotations on the biology articles. Then, when the user specifies particular design functions of interest, IBID retrieves both the matching SBF models and the relevant biology articles. Thus, the ontology of functions acts as a cross-domain bridge between biological systems and engineering problems.

Work on IBID faces several types of challenges including disambiguating different senses of a word describing a function, a behavior or a structure; distinguishing between biological systems and other processes described in an article; improving the quality of extracted SBF models to match that of manually extracted models; and using the SBF models for supporting case-based reasoning in biologically inspired design. As mentioned earlier, IBID at present extracts only partial SBF models in that it does not fully integrate the structures, the behaviors and functions into a complete SBF model. The quality of the extracted SBF models should improve when IBID starts exploiting the constraints that full integration of SBF models will impose on decisions about individual structures, behaviors and functions.

**Acknowledgements.** We are grateful to Julian Vincent for sharing his ontology for describing biological systems (Vincent 2014); IBID uses Vincent's ontology of biological components and connections. We also wish to acknowledge the contributions of Arvind Jaganathan, Nilesh More, and Sanjana Oulkar to the development of IBID.

## References

Benyus, J.: Innovation Inspired by Nature. William Morrow, New York (1997)
Vincent, J.F., Mann, D.L.: Systematic technology transfer from biology to engineering. Philos. Trans. Roy. Soc. Lond. A Math. Phys. Eng. Sci. **360**(1791), 159–173 (2002)

Lepora, N.F., Verschure, P., Prescott, T.J.: The state of the art in biomimetics. Bioinspir. Biomim. **8**(1), 013001 (2013)

Weber, R.O., Ashley, K., Bruninghaus, S.: Textual case-based reasoning. Knowl. Eng. Rev. **20**(03), 255–260 (2005)

Vattam, S.S., Goel, A.K.: An information foraging model of interactive analogical retrieval. In: Proceedings of 35th Annual Meeting of the Cognitive Science Society (2013a)

Vattam, S.S., Goel, A.K.: Biological solutions for engineering problems: a study in cross-domain textual case-based reasoning. In: Proceedings of 21st International Conference on Case-Based Reasoning, pp. 343–357 (2013b)

Goel, A.K., Rugaber, S., Vattam, S.: Structure, behavior, and function of complex systems: the structure, behavior, and function modeling language. Artif. Intell. Eng. Des. Anal. Manufact. **23**(01), 23–35 (2009)

Norgaard, T., Dacke, M.: Fog-basking behaviour and water collection efficiency in Namib Desert Darkling beetles. Front. Zool. **7**(1), 1 (2010)

Goel, A., Bhatta, S.: Use of design patterns in analogy-based design. Adv. Eng. Inf. **18**(2), 85–94 (2004)

Goel, A.K., Vattam, S., Wiltgen, B., Helms, M.: Cognitive, collaborative, conceptual and creative - four characteristics of the next generation of knowledge-based cad systems: a study in biologically inspired design. Comput. Aided Des. **44**(10), 879–900 (2011)

Spiliopoulou, E., Rugaber, S., Goel, A., Chen, L., Wiltgen, B., Jagannathan, A.K.: Intelligent search for biologically inspired design. In: Proceedings of the 20th International Conference on Intelligent User Interfaces Companion, pp. 77–80. ACM, March 2015

Hirtz, J., Stone, R., McAdams, D., Szykman, S., Wood, K.: A functional basis for engineering design: reconciling and evolving previous efforts. Res. Eng. Des. **13**(2), 65–82 (2002)

Deldin, J.M., Schuknecht, M.: The AskNature database: enabling solutions in biomimetic design. In: Goel, A., McAdams, D., Stone, R. (eds.) Biologically Inspired Design, pp. 17–27. Springer, London (2014)

Vincent, J.F.: An ontology of biomimetics. In: Goel, A., McAdams, D., Stone, R. (eds.) Biologically Inspired Design: Computational Methods and Tools, pp. 269–285. Springer, London (2014)

Khoo, C., Kornfit, J., Oddy, R.N., Myaeng, S.H.: Automatic extraction of cause-effect information from newspaper text without knowledge-based inferencing. Lit. Linguist. Comput. **13**, 177–186 (1998)

Khoo, C.S.G., Chan, S., Niu, Y.: Extracting causal knowledge from a medical database using graphical patterns. In: Proceedings of 38th Annual Meeting of the Association for Computational Linguistics (ACL2000), pp. 336–343 (2000)

Schank, R.: Conceptual dependence: a theory of natural language understanding. Cogn. Psychol. **3**(4), 532–631 (1972)

Davies, J., Studer, R., Warren, P.: Semantic Web Technologies: Trends and Research in Ontology-Based Systems. Wiley, New York (2006)

Kipper, K., Korhonen, A., Ryant, N., Palmer, M.: A large-scale classification of English verbs. Lang. Resour. Eval. **42**(1), 21–40 (2008)

Prieto-Diaz, R.: Implementing faceted classification for software reuse. Commun. ACM **34**(5), 88–97 (1991)

De Marneffe, M.C., MacCartney, B., Manning, C.: Generating typed dependency parses from phrase structure parses. In: Proceedings of European Language Resources and Evaluation Conference (LREC 2006) (2006)

Miller, G.A.: WordNet: a lexical database for English. Commun. ACM **38**(11), 39–41 (1995)

Fellbaum, C. (ed.): WordNet: An Electronic Lexical Database. MIT Press, Cambridge (1998)

Cohen, J.: A coefficient for agreement of nominal scales. Educ. Psychol. Meas. **20**(1), 37–46 (1960)

Landis, J., Koch, G.: The measurement of observer agreement for categorical data. Biometrics **33**(1), 159–174 (1977)

Cohen, J.: Weighted kappa: nominal scale agreement provision for scaled disagreement or partial credit. Psychol. Bull. **70**(4), 213 (1968)

Weber, R., Aha, D.W., Sandhu, N., Munoz-Avila, H.: A textual case-based reasoning framework for knowledge management applications. In: Proceedings of the 9th German Workshop on Case-Based Reasoning. Shaker Verlag, pp. 244–253 (2001)

Bruninghaus, S., Ashley, K.: Developing mapping and evaluation techniques for textual case-based reasoning. In: Workshop Notes of the AAAI 1998 Workshop on Textual CBR (1998)

Bruninghaus, S., Ashley, K.D.: Progress in textual case-based reasoning: predicting the outcome of legal cases from text. In: Proceedings of the National Conference on Artificial Intelligence (Vol. 21, No. 2, p. 1577). Menlo Park, CA; Cambridge, MA; London; AAAI Press; MIT Press, 1999, July 2006

Schumacher, P., Minor, M., Walter, K.: Extraction of procedural knowledge from the web: a comparison of two workflow extraction approaches. In: Proceedings of the 21st International Conference on World Wide Web, pp. 739–747 (2012)

Sizov, G., Ozturk, P., Styrac, J.: Acquisition and reuse of reasoning knowledge from textual cases for automated analysis. In: Lamontagne, L., Plaza, E. (eds.) ICCBR 2014. LNCS, vol. 8765, pp. 465–479. Springer, Heidelberg (2014)

Sizov, G., Öztürk, P., Aamodt, A.: Evidence-driven retrieval in textual CBR: bridging the gap between retrieval and reuse. In: Hüllermeier, E., Minor, M. (eds.) ICCBR 2015. LNCS (LNAI), vol. 9343, pp. 351–365. Springer, Heidelberg (2015). doi:10. 1007/978-3-319-24586-7_24

Berant, J., Srikumar, V., Chen, P.C., Vander Linden, A., Harding, B., Huang, B., Manning, C.D.: Modeling biological processes for reading comprehension. In: EMNLP, October 2014

Cheong, H., Shu, L.H.: Automatic extraction of causally related functions from natural-language text for biomimetic design. In: ASME 2012 International Design Engineering Technical Conferences and Computers and Information in Engineering Conference, pp. 373–382. American Society of Mechanical Engineers, August 2012

Jursic, M., Cestnik, B., Urbancic, T., Lavrac, N.: Cross-domain literature mining: Finding bridging concepts with CrossBee. In: Proceedings of the 3rd International Conference on Computational Creativity, pp. 33–40, May 2012

# Predicting the Electricity Consumption of Buildings: An Improved CBR Approach

Aulon Shabani[1], Adil Paul[2], Radu Platon[3], and Eyke Hüllermeier[2(✉)]

[1] Polytechnic University of Tirana, Tirana, Albania
aulon.shabani@fie.upt.al
[2] Department of Computer Science, Paderborn University, Paderborn, Germany
{adil.paul,eyke}@upb.de
[3] Natural Resources Canada, CanmetENERGY, Varennes, Canada
radu.platon@nrcan.gc.ca

**Abstract.** Case-based reasoning has recently been used to predict the hourly electricity consumption of institutional buildings. Past measurements of the building's operation are modeled as cases and, combined with forecast weather information, used to predict the electricity demand for the next six hours. Elaborating on this idea, we present an improved CBR approach that yields more accurate predictions of energy consumption. In particular, we develop improved (local) similarity measures specifically tailored for this kind of application, and combine these measures with a regression-based method for similarity learning. Moreover, we incorporate a simple procedure for case adaptation. Experimental results for a real case study confirm a significant improvement in predictive accuracy compared to previous approaches.

## 1 Introduction

Buildings are major energy users, being responsible for more than one-third of the world's total energy consumption [1]. In North America (U.S. and Canada) alone, institutional and commercial buildings account for 40 % of total energy use [3]. A significant proportion of a building's energy consumption is used to operate increasingly complex systems and technologies, such as advanced mechanical heating, ventilation and air conditioning (HVAC) systems and thermal storage systems, designed to store energy for proper subsequent use.

Building operation and control need to be improved in order to reduce energy use, which becomes more and more a priority due to increasing energy prices and operation costs. The use of intelligent technologies enabling buildings to become proactive, by adapting their operation according to changing operational and environmental conditions can have a major impact on energy consumption. According to the Energy Star Program, energy consumption of commercial and institutional buildings can be reduced by up to 35 % by using intelligent technologies and by modifying control practices [5].

Forecasting building energy use is critical for optimizing the management of thermal energy storage systems and for improving control and operation

© Springer International Publishing AG 2016
A. Goel et al. (Eds.): ICCBR 2016, LNAI 9969, pp. 356–369, 2016.
DOI: 10.1007/978-3-319-47096-2_24

sequences in order to reduce energy consumption. It also enables energy use monitoring in order to identify periods of excessive consumption. Estimating the electricity consumption ahead of time enables improved planning of the operation of thermal energy storage devices linked to electrically-driven HVAC systems, optimizing their use and reducing peak loads and costs.

Different predictive models have been proposed for building energy use, mostly based on data-driven (machine learning) methods that require a significant amount of a building's historical operational data. However, data of that kind is not available for all buildings, such as in the case of new and retrofit buildings that underwent major changes to the point that previous data is no longer representative of current operation. As argued by Platon et al. [19], case-based reasoning offers a quite appealing alternative, not only due to being more transparent than black-box models like neural networks, but also due to its ability to operate with even little experience, and to learn and improve predictive accuracy as more data becomes available. Adding to this, we like to mention the potential of CBR to properly adapt predictions from previous to similar problems (such as retrofit buildings).

Recently, first promising results could indeed be achieved with a CBR model for predicting electricity use in an institutional facility over a time horizon of 6 h [19,20]. However, the predictive error of that model was still almost twice as high as that of a neural network, which severely hampers the willingness of building owners and operators to adopt this type of model: as decisions regarding building operation and control are made using the predicted energy consumption, the accuracy of the model is crucial for optimal operation and planning. Therefore, this paper presents various improvements made to the CBR model that led to a significant increase in predictive accuracy.

The rest of the paper is organized as follows. We start with a short overview of related work on energy prediction, prior to recalling the CBR model of [19,20] in Sect. 3. Our improved approach in then presented in Sect. 4. In Sect. 5, this approach is empirically evaluated using data from an institutional Canadian facility located in Calgary, prior to concluding the paper in Sect. 6.

## 2    Predicting Energy Demand in Buildings

Different types of methods for predicting energy demand in buildings have been proposed in the literature, including model-based approaches, statistical time series analysis, and machine learning methods.

Model-based approaches make use of a building's characteristics, such as total heating and cooling demand, thermal characteristics of walls, windows, other material proprieties, solar radiation, etc., in order to develop mathematical models for the simulation of the building's energy performance. Typical examples of such approaches are DOE-2, BLAST, EnergyPlus (a combination of DOE-2 and BLAST), SPARK, and TRNSYS; for a detailed description of the most commonly used simulation tools, we refer to [8].

The design of simulation models is a costly and time-consuming process, which requires a significant amount of expert knowledge. As an alternative, machine learning methods such as Artificial Neural Networks (ANN) and Support Vector Machines (SVM) can be used to induce models for energy demand prediction in a data-driven way, i.e., on the basis of energy demand observed in the past. For example, Azadeh et al. [4] train multi-layer perceptrons for predicting annual energy consumption of high energy consumers in the industrial sector. Likewise, Gonzalez and Zamarreno [12] predict energy consumption using a recurrent neural network. Using real data and taking forecast temperature values as attributes, highly precise results are achieved. Hybrid approaches combining simulation models with neural networks can be found in the literature, too, for example to predict energy consumption of a passive solar building [14]. Examples of prediction methods based on SVMs include [17,18]. A detailed review of machine learning methods for the prediction of a building's energy consumption is provided by [23].

As already mentioned in the introduction, CBR has been put forward as yet another alternative for the purpose of predicting a building's energy consumption more recently [15,16,19,20]. Compared to standard (model-based) machine learning methods like ANN and SVM, case-based reasoning arguably comes with a number of advantages. In particular, since CBR is an inherently incremental process, it is able to adequately deal with an initial absence of historical consumption data, while continuously improving when more data becomes available over time. Moreover, CBR appears to be especially appealing for realizing knowledge transfer from one building to another, i.e., for exploiting data about one building to improve predictions for different yet similar buildings. First results on the use of CBR for energy prediction are promising and adhere to the limits recommended by the ASHRAE (American Society of Heating, Refrigerating and Air-Conditioning Engineers) [2]. More details about CBR for energy prediction are provided in the following section.

## 3   CBR for Predicting Electricity Consumption

Since our work mainly builds on [19,20], we devote this section to a short overview of these approaches, prior to presenting our improved method in Sect. 4. Platon et al. are interested in predicting hourly energy consumption based on historical measurements. To this end, they proceed from a case representation as shown in Table 2. Each case provides information about the development of 10 variables $V_1, \ldots, V_{10}$ (see Table 1) measured over 9 h. The query case contains values of these variables for the current hour ($t_0$) as well as the previous two hours ($t_{-1}$ and $t_{-2}$). Moreover, for the two variables air temperature and humidity, it contains predicted values over a period of 6 h. The goal is to predict the electricity consumption over these 6 h. The source case (memorized in the past) comprises the same information, though with real (instead of forecast) values for temperature and humidity; besides, the values for the target variable, electricity consumption, are given, too.

In the following, we denote by $x_{i,j}$ the value of the variable $V_i$ at time point $t_j$ in the source case ($1 \leq i \leq 10$, $-2 \leq j \leq 6$), and by $p_j$ the value of the

**Table 1.** Variable description and measurement unit

| | Variable | Unit |
|---|---|---|
| $V_1$ | Forecast outside air temperature | (°C) |
| $V_2$ | Forecast outside air relative humidity | (%) |
| $V_3$ | Air handling unit 2 supply hot air temperature | (°C) |
| $V_4$ | Air handling unit 3 supply hot air temperature | (°C) |
| $V_5$ | West wing air handling unit supply cold air temperature | (°C) |
| $V_6$ | Air handling unit 4 supply cold air temperature | (°C) |
| $V_7$ | Chiller outlet water temperature | (°C) |
| $V_8$ | Chiller outlet water flow rate | (l/s) |
| $V_9$ | Boiler outlet water temperature | (°C) |
| $V_{10}$ | Boiler outlet water flow rate | (l/s) |

**Table 2.** Example of a query and a source case. Numbers in blue in the query case are forecast. Numbers in gray in the source case are known but not used for comparison with the source case (for which they are not given).

**query case**

| | date and time | $V_1$ | $V_2$ | $V_3$ | $V_4$ | $V_5$ | $V_6$ | $V_7$ | $V_8$ | $V_9$ | $V_{10}$ | $P$ |
|---|---|---|---|---|---|---|---|---|---|---|---|---|
| $t_6$ | 2014-04-07 15:00 | 10 | 32.2 | | | | | | | | | ? |
| $t_5$ | 2014-04-07 14:00 | 9 | 39.9 | | | | | | | | | ? |
| $t_4$ | 2014-04-07 13:00 | 8 | 32.2 | | | | | | | | | ? |
| $t_3$ | 2014-04-07 12:00 | 9 | 33.8 | | | | | | | | | ? |
| $t_2$ | 2014-04-07 11:00 | 10 | 34.6 | | | | | | | | | ? |
| $t_1$ | 2014-04-07 10:00 | 11 | 29.4 | | | | | | | | | ? |
| $t_0$ | 2014-04-07 09:00 | 12 | 29.9 | 29.4 | 28.4 | 15.8 | 24.9 | 30.5 | -.05 | 67.3 | 76.2 | 203.1 |
| $t_{-1}$ | 2014-04-07 08:00 | 12 | 31.2 | 29.6 | 17.4 | 10.1 | 21.4 | 32.7 | -.05 | 65.2 | 76.2 | 203.8 |
| $t_{-2}$ | 2014-04-07 07:00 | 11 | 31.0 | 28.3 | 9.7 | 10.6 | 22.4 | 30.8 | -.04 | 66.3 | 73.2 | 197.6 |

**source case**

| | date and time | $V_1$ | $V_2$ | $V_3$ | $V_4$ | $V_5$ | $V_6$ | $V_7$ | $V_8$ | $V_9$ | $V_{10}$ | $P$ |
|---|---|---|---|---|---|---|---|---|---|---|---|---|
| $t_6$ | 2014-03-06 14:00 | 12 | 29.9 | 25.4 | 24.7 | 17.9 | 22.2 | 31.2 | -.04 | 63.6 | 77.1 | 202.9 |
| $t_5$ | 2014-03-06 13:00 | 12 | 29.9 | 28.4 | 27.4 | 16.7 | 25.2 | 31.2 | -.04 | 65.8 | 78.1 | 204.6 |
| $t_4$ | 2014-03-06 12:00 | 12 | 29.9 | 31.1 | 29.1 | 17.4 | 26.1 | 31.4 | -.05 | 69.1 | 77.1 | 205.1 |
| $t_3$ | 2014-03-06 11:00 | 11 | 20.5 | 27.4 | 27.4 | 19.8 | 22.6 | 29.4 | -.04 | 71.1 | 76.8 | 202.1 |
| $t_2$ | 2014-03-06 10:00 | 12 | 28.4 | 29.2 | 28.8 | 15.9 | 24.2 | 30.6 | -.05 | 67.8 | 76.0 | 203.8 |
| $t_1$ | 2014-03-06 09:00 | 11 | 29.3 | 27.4 | 25.4 | 17.8 | 26.9 | 31.5 | -.05 | 66.3 | 77.2 | 204.6 |
| $t_0$ | 2014-03-06 08:00 | 12 | 29.4 | 29.2 | 28.4 | 16.8 | 22.9 | 31.5 | -.05 | 69.3 | 77.2 | 204.2 |
| $t_{-1}$ | 2014-03-06 07:00 | 11 | 31.2 | 29.8 | 17.6 | 10.1 | 22.4 | 32.7 | -.05 | 65.2 | 76.2 | 204.8 |
| $t_{-2}$ | 2014-03-06 06:00 | 10 | 31.0 | 28.1 | 9.8 | 11.6 | 21.4 | 31.8 | -.04 | 65.3 | 73.2 | 199.8 |

consumption $P$ at time $t_j$. The corresponding values for the query case are denoted $y_{i,j}$ and $q_j$. The measurements of each variable $V_i$ over time are collected in the time series $x_i$ and $y_i$, respectively (corresponding to individual columns in the case representation). The combination of all values are referred to as $X$ (source case) and $Y$ (query case), respectively.

Similarity between cases is derived in two steps. First, given a new query case, only those previous cases are considered that fulfill the following properties: The time $t_0$ differs by at most one hour, and the absolute temperature at $t_0$ differs by at most $2°C$. Since the temperature and the time of the day are two very important properties, this can be seen as a prefiltering of presumably irrelevant cases (the similarity of which is formally set to 0).

For all other cases, the similarity is defined as a weighted average of the similarities of the different (input) variables:

$$\text{CS}(X,Y) = \sum_{i=1}^{M} v_i \cdot \text{VS}'_i(x_i, y_i), \tag{1}$$

where $M = 10$ is the number of variables, $v_i \geq 0$ is the weight of the variable $V_i$, and $\text{VS}'_i(x_i, y_i)$ the (local) similarity of the cases on that variable. As illustrated in Fig. 1, variable similarity is defined as

$$\text{VS}'_i(x_i, y_i) = \begin{cases} 0 & \text{if } D_w(x_i, y_i) > d^i_{max} \\ \frac{D_w(x_i,y_i)-d^i_{min}}{d^i_{max}-d^i_{min}} & \text{if } d^i_{min} \leq D_w(x_i, y_i) \leq d^i_{max} \\ 1 & \text{if } D_w(x_i, y_i) < d^i_{max} \end{cases} . \tag{2}$$

Here, $d^i_{min}$ and $d^i_{max}$ are variable-specific thresholds specifying what can be seen as completely similar and completely dissimilar (cf. Table 3), and $D_w$ is the weighted Euclidean distance:

$$D_w(x_i, y_i) = \frac{\sqrt{\sum_{j=-2}^{n} w_j (x_{i,j} - y_{i,j})^2}}{\sum_{j=-2}^{n} w_j}, \tag{3}$$

where $n = 0$ or $n = 6$ (depending on the variable), and the weights $w_j = 1 + j/3$ for $j \in \{-2, -1, 0\}$ and $w_j = 1 - j/7$ for $j \in \{1, \ldots, 6\}$ are such that observations closer to the current time $t_0$ have a higher influence.

At prediction time, given a query case $Y$, those previous cases $X_1, \ldots, X_K$ with similarity $\text{CS}(X_k, Y) > 0.8$ are retrieved from the case base, and predictions of energy consumption are obtained as weighted averages of the consumptions observed for these cases:

$$\hat{q}_j = \frac{\sum_{k=1}^{K} \text{CS}(X_k, Y) \cdot p_{k,j}}{\sum_{k=1}^{K} \text{CS}(X_k, Y)}, \tag{4}$$

where $p_{k,j}$ is the consumption for case $X_k$ at time $j \in \{1, \ldots, 6\}$.

As commonly done in the electricity and energy domain[1], predictive performance is measured in terms of the CV-RMSE (Coefficient of Variation Root

---

[1] ANSI/BPI-2400-S-2012 Standard Practice for Standardized Qualification of Whole-House Energy Savings Predictions by Calibration to Energy Use History.

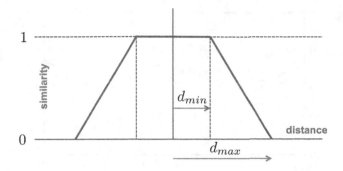

**Fig. 1.** Transformation of Euclidean distance into similarity.

Mean Square Error): With $\{\hat{q}_t \mid t \in T\}$ a set of predicted consumptions (for a single but possibly also for several query cases) and $\{q_t \mid t \in T\}$ the corresponding observed values, this measure is defined as

$$\text{CV-RMSE} = \frac{\sqrt{\frac{1}{|T|-1}\sum_{t=1}^{|T|}(q_t - \hat{q}_t)^2}}{\bar{q}} \times 100, \tag{5}$$

where $\bar{q}$ is the mean of true values.

## 4   Improved CBR Model

Building on the CBR model as outlined in the previous section, we devised a number of improvements that will be described in the following.

### 4.1   Variable Similarity

According to (2), the similarity between two measurement sequences on a variable is a non-linear transformation of the Euclidean distance between these two sequences. While Euclidean distance is an established and reasonable measure, it arguably fails to properly account for the *trend* in the corresponding time series. Needless to say, looking at the trend is important when it comes to extrapolating into the future. For example, Fig. 2 shows the time series for a specific variable (amplitude) and three cases. According to Euclidean distance, the first one (green, solid line) is as similar to the second (blue, short dashes) as to the third one (orange, long dashes). Looking at the trend, however, the third one appears to be much more relevant. In particular, the third case seems to be much more amenable to adaptation (cf. Sect. 4.3 below).

To capture the trend of time series, we define a second (variable) similarity measure based on the well-known cosine similarity [9,10]: A sequence of values $x_i = (x_{i,-2}, x_{i,-1}, \ldots, x_{i,n})$ is considered as a bundle of two-dimensional vectors[2]

---

[2] In our case, the difference between time steps, $\Delta t_j$, is always 1, because measurements are made on an hourly basis.

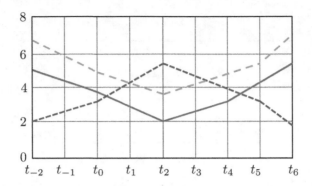

**Fig. 2.** Example of time series with different shape. (Color figure online)

$$\left\{ (\Delta x_j, \Delta t_j)^\top = (x_{i,j+1} - x_{i,j}, 1)^\top \mid j = -2, \ldots, n-1 \right\},$$

and similarity is defined as the averaged (normalized) angle between the corresponding vectors (cf. Fig. 3):

$$\mathrm{VS}_i''(x_i, y_i) = \frac{1}{\pi(n+2)} \sum_{j=-2}^{n-1} \cos^{-1} \left( \frac{\Delta x_j \Delta y_j + 1}{\sqrt{(\Delta x_j)^2 + 1}\sqrt{(\Delta y_j)^2 + 1}} \right) \qquad (6)$$

Finally, we define a new variable similarity measure in terms of a weighted average of the original measure (2) and the new (trend-based) similarity (6), where the weights have been determined empirically:

$$\mathrm{VS}_i(x_i, y_i) = 0.3 \, \mathrm{VS}_i'(x_i, y_i) + 0.7 \, \mathrm{VS}_i''(x_i, y_i) \qquad (7)$$

## 4.2   Case Similarity

According to (1), the similarity between two cases is defined as a weighted average of the variable similarities. In previous work, the flexibility of weighting has actually not been exploited, i.e., all weights were simply set to the same value $v_i = 1/M$. However, since different variables are obviously of different importance, a generalization of this approach is desirable.

The determination of optimal variable weights $v_i$ is closely connected to the problem of learning similarity measures, which has been studied intensively in CBR [11,21,22]. More specifically, the problem is to optimally combine given local (variable) similarities into a global (case) similarity [6]. To solve this problem, we take advantage of the fact that, according to (1), the combination is a *linear* one, i.e., global similarity is a linear (convex) combination of local similarities.

Concretely, we formalize the problem of learning weights $v_i$ for variables $V_i$ as a problem of *linear regression*: For every pair of cases $X$ and $Z$ from our case base, we can compute the (local) variable similarities

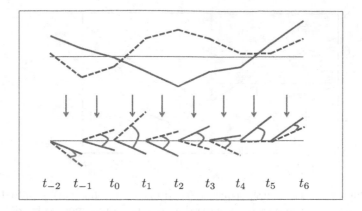

**Fig. 3.** Representation of time series as a bundle of vectors. The similarity for each pair of vectors depends on the angle between them (0 for an angle of $\pi$, 1 for an angle of 0). These similarities are averaged to obtain the overall similarity.

$$(s_1, \ldots, s_M) = \Big( \text{VS}_1(x_1, z_1), \ldots, \text{VS}_M(x_M, y_M) \Big) \in [0,1]^M.$$

Moreover, we can compute a similarity $s_{out}$ on the consumptions measured for $X$ and $Z$, again using the transformation (2) of their Euclidean distance, with proper choices of $d_{min}$ and $d_{max}$.[3] Ideally, the (global) case similarity is close to this value, i.e., $s_{out} \approx \text{CS}(X, Z)$. Therefore, the weights $v_j$ in (1) should be such that

$$\sum_{j=1}^{M} v_j \cdot s_j \approx s_{out}. \tag{8}$$

As already said, an (approximate) Eq. (8) can be derived for each pair of cases from the case base, and each such equation can be seen as a training example for a (multivariate) linear regression problem, with the values of the input variables given by $(s_1, \ldots, s_M)$ and the value of the output variable by $s_{out}$. Thus, optimal weights can simply be found by solving this regression problem; more specifically, since the weights, which correspond to the regression coefficients, must be non-negative and sum up to 1, a constrained regression problem needs to be solved.

### 4.3 Adaptation

According to (4), similar cases retrieved from the case base are used in the prediction step without any adaptation. As a potential improvement, we propose a method for adaptation that is inspired by the idea of amplitude transformation [7]. More specifically, assuming that the future relation of energy consumption for two cases will approximately equal the relation in the past, the energy consumption of a source case retrieved from the case base is shifted by a proportional factor prior to using it for prediction.

---

[3] We used $d_{min} = 15$ and $d_{max} = 35$.

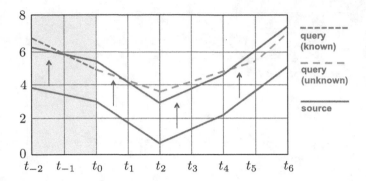

**Fig. 4.** Adaptation: The source sequence (solid line on the bottom) is shifted upward, so that the mean in the past (first three time points, gray region) coincides with the mean on the query (dashed line); the future values of the query need to be predicted and are therefore shown in gray.

**Table 3.** Variable thresholds and weights

| Variable | $d_{min}$ | $d_{max}$ | weight |
|---|---|---|---|
| Forecast outside air temperature | 2 | 6 | 0.1961 |
| Forecast outside air relative humidity | 10 | 25 | 0.1540 |
| Air handling unit 2 supply hot air temperature | 2 | 6 | 0.0001 |
| Air handling unit 3 supply hot air temperature | 2 | 6 | 0.0001 |
| West wing air handling unit supply cold air temperature | 2 | 6 | 0.009 |
| Air handling unit 4 supply cold air temperature | 2 | 6 | 0.1065 |
| Chiller outlet water temperature | 2 | 15 | 0.1075 |
| Chiller outlet water flow rate | 5 | 30 | 0.0001 |
| Boiler outlet water temperature | 2 | 15 | 0.3064 |
| Boiler outlet water flow rate | 5 | 30 | 0.1284 |

Recall that the values $q_{-2}, q_{-1}, q_0$ for electricity consumption are assumed to be known for the query case (while consumption needs to be predicted for the six hours ahead), and let $p_{k,-2}, p_{k,-1}, p_{k,0}$ denote the consumption of the $k^{th}$ neighbor in the past three hours. We then replace each of the future values $p_{k,j}$ ($j = 1, \ldots, 6$) of that case by

$$p_{k,j} \cdot \left( \frac{q_{-2} + q_{-1} + q_0}{p_{k,-2} + p_{k,-1} + p_{k,0}} \right)$$

before using it for prediction in (4); see Fig. 4 for an illustration.

### 4.4 Other Modifications

Instead of retrieving all past cases with a similarity $CS(X, Y)$ exceeding a fixed similarity threshold (of 0.8), we fix the number of neighbors to be used for prediction to $K = 50$ and retrieve the $K$ most similar ones (if there are less than $K$ cases with a similarity $> 0$, these cases are all retrieved).

# 5  Experiments

## 5.1  Data

Data was collected from an institutional building facility located in Calgary (Alberta, Canada) for working days between $1^{st}$ of January 2013 and $9^{th}$ of May 2014 (with some missing data from $29^{th}$ of March to $1^{st}$ of May). The building has a total floor space of 16,800 $m^2$ and houses mainly office and storage spaces. The HVAC equipment consists of 5 air handling units served by a one chiller and 3 natural gas boilers. The data consists of hourly averages of measurements related to the operation of the chiller, boilers and air handling units, the building electricity consumption, and weather information—current and forecast values of outside air temperature and relative humidity (see list of variables in Table 1). Building operating modes corresponding to office working and non-working hours were identified. The building consumes approximately 80 % more electricity during working hours—7 AM to 5 PM—than during non-working hours; only the model developed using working-hours measurements is presented in this paper.

## 5.2  Methods

Our CBR approach was implemented as described in the previous section. Under certain circumstances, it may happen that the case base does not contain a single case that is similar (i.e., has a similarity $> 0$) to the query case. In such a situation, our method yields the current consumption (i.e., the consumption $q_0$ at time $t_0$) in the query case as a default prediction for $\hat{q}_j$ for the next time points ($j = 1, \ldots, 6$); this predictor will also be used as one of our baselines (see below). For learning the weights of variables in the case similarity measure (cf. Sect. 4.2), we constructed a set of training data by randomly sampling 10,000 pairs of cases from the case base. The weights obtained by linear regression, which are shown in Table 3, are plausible and indeed give the highest importance to those variables that are intuitively deemed most relevant.

We compare our CBR approach with a number of other methods that are used as baselines to compete with. The first three baselines are extremely simple, and they all forecast a constant value for the six hours prediction horizon. They predict, respectively,

- the average consumption of all past cases stored in the case base;
- the average consumption $(q_{-2} + q_{-1} + q_0)/3$ of the past three hours in the query case;
- the current consumption $q_0$ in the query case.

Moreover, following [19], we also included an artificial neural network (ANN), namely a multilayer perceptron with one hidden layer consisting of 10 neurons, trained using the back propagation algorithm with Levenberg-Marquardt optimization. As input, the network takes the measurement values of the current and past two hours of a case, including the energy consumption (hence 33 values in total), and as output, it produces predictions of the energy consumption for the next six hours.

## 5.3   First Experiment

In the first experiment, the data is separated into two parts: a set of past cases with measurements from the first $m$ months of 2013 (where $m \in \{4, 6, \ldots, 12\}$) that corresponds to our case base and serves as training data for the ANN, and the remaining set of future cases till September 2014 that serves as test data. Performance is reported in Table 4 in terms of the CV-RMSE (5) on the test data. As can be seen, our CBR approach compares quite favorably and is much better than the baselines. The performance of the ANN is even slightly better if enough training data is available, but CBR seems to have advantages if training data is sparse.

**Table 4.** Results of the first experimental study in terms of CV-RMSE (%), with the best performance highlighted in bold font.

| training data | baseline 1 | baseline 2 | baseline 3 | ANN | CBR |
|---|---|---|---|---|---|
| 01/2013 – 04/2013 | 9.97 | 9.55 | 8.80 | 8.99 | **7.94** |
| 01/2013 – 06/2013 | 10.15 | 9.50 | 8.65 | 8.15 | **7.39** |
| 01/2013 – 08/2013 | 10.23 | 9.41 | 8.58 | 7.63 | **7.45** |
| 01/2013 – 10/2013 | 10.12 | 9.35 | 8.66 | **7.31** | 7.69 |
| 01/2013 – 12/2013 | 10.35 | 9.54 | 8.73 | **6.17** | 6.55 |

## 5.4   Second Experiment

In the second experiment, we applied our CBR approach in an online setting, in which prediction and learning (case memorization) are interleaved: Cases are considered in a sequence one by one, and at each time step $t$,

- a prediction of the consumption for the $t^{th}$ case is obtained based on the previous $t - 1$ cases already stored in the case base,
- the true consumption is revealed, and the cumulative error (CV-RMSE on the first $t$ cases) is updated,
- the new case is added to the case base.

As can be seen in Fig. 5, the performance is relatively poor in the beginning, when only few cases are available, but quickly improves and then reaches a level similar to the error (around 6.3 %) in the previous experiment. This is a significant improvement compared to the previous CBR approach, for which the error is twice as high [19, 20].

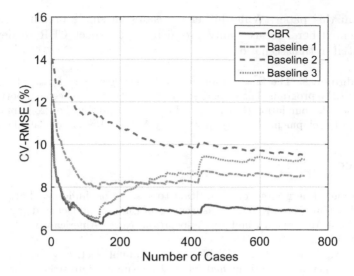

**Fig. 5.** Performance of the CBR model in the online setting.

## 6 Conclusion

This paper presents the application of a CBR model for predicting the hourly electricity consumption of an institutional building located in Calgary, Canada. The model uses measurements related to the building operation, as well as measured and forecast weather information to predict the building electricity consumption for the next 6 h. It is based on a previous CBR approach applied to the same problem, however, modifications and extensions related to variable and case similarities, case selection and adaptation resulted in significant predictive accuracy improvements: The model has a test error approximately twice as low compared to the previous approach. This is important, as predictive accuracy is critical in enabling operators to take the appropriate operation and control decisions that ultimately result in reduced building energy consumption.

There are several directions to be pursued in future work. First, there is probably still some scope to further improve predictive accuracy. Perhaps more interestingly, however, we also plan to apply the approach of *credible case-based inference* [13], which allows for predicting confidence intervals instead of only point values. Thus, in our case, predictions will be intervals of the form $[q_t^{low}, q_t^{up}]$, coming with the guarantee that the true consumption will lie in that range with high probability. Predictions of that kind, reflecting uncertainty in a proper way, can usefully support safety-critical decisions, for example regarding peak loads.

Second, going beyond a single building, we plan to extend our approach toward knowledge (case) transfer between different building. As already mentioned, CBR appears to be especially suitable for realizing this kind of transfer learning, which, as a critical step, requires a reasonable approach to *inter-building* case adaptation in addition to the simpler *intra-building* case adaptation as presented in this paper. While hitherto results on single buildings, including those presented here, are certainly promising, we expect CBR to develop its true potential in that scenario.

**Acknowledgments.** The first author gratefully acknowledges financial support by the EUROWEB+ program. The second and fourth author have been supported by the German Research Foundation (DFG), and the third author by the Program of Energy Research and Development (PERD) operated by Natural Resources Canada (NRCan).

# References

1. International Energy Agency: Transition to sustainable buildings. Paris (2013)
2. ASHRAE: Guideline 14–2002, Measurement of Energy and Demand Savings (2002)
3. Continental Automated Buildings Association: North American Intelligent Buildings Roadmap. Ottawa (2011)
4. Azadeh, A., Ghaderi, S.F., Sohrabkhani, S.: Annual electricity consumption forecasting by neural network in high energy consuming industrial sectors. Energy Convers. Manage. **49**(8), 2272–2278 (2008)
5. CABA: North American Intelligent Buildings Roadmap (2011)
6. Cheng, W., Hüllermeier, E.: Learning similarity functions from qualitative feedback. In: Althoff, K.-D., Bergmann, R., Minor, M., Hanft, A. (eds.) ECCBR 2008. LNCS (LNAI), vol. 5239, pp. 120–134. Springer, Heidelberg (2008). doi:10.1007/978-3-540-85502-6_8
7. Chu, K.K.W., Wong, M.H.: Fast time-series searching with scaling and shifting. In: Proceedings of the 18th ACM SIGMOD-SIGACT-SIGART Symposium on Principles of Database Systems, PODS 1999, pp. 237–248. ACM, New York (1999)
8. Crawley, D.B., Hand, J.W., Kummert, M., Griffith, B.T.: Contrasting the capabilities of building energy performance simulation programs. Build. Environ. **43**(4), 661–673 (2008)
9. Dohare, D., Devi, V.S.: Combination of similarity measures for time series classification using genetic algorithms. In: IEEE Congress of Evolutionary Computation (CEC), pp. 401–408, June 2011
10. Dong, Y., Sun, Z., Jia, H.: A cosine similarity-based negative selection algorithm for time series novelty detection. Mech. Syst. Sign. Process. **20**(6), 1461–1472 (2006)
11. Gabel, T., Godehardt, E.: Top-down induction of similarity measures using similarity clouds. In: Hüllermeier, E., Minor, M. (eds.) ICCBR 2015. LNCS (LNAI), vol. 9343, pp. 149–164. Springer, Heidelberg (2015). doi:10.1007/978-3-319-24586-7_11
12. Gonzlez, P.A., Zamarreo, J.M.: Prediction of hourly energy consumption in buildings based on a feedback artificial neural network. Energy Buildings **37**(6), 595–601 (2005)
13. Hüllermeier, E.: Credible case-based inference using similarity profiles. IEEE Trans. Knowl. Data Eng. **19**(5), 847–858 (2007)
14. Kalogirou, S.A., Bojic, M.: Artificial neural networks for the prediction of the energy consumption of a passive solar building. Energy **25**(5), 479–491 (2000)
15. Monfet, D., Arkhipova, E., Choiniere, D.: Evaluation of a case-based reasoning energy prediction tool for commercial buildings (2013)
16. Monfet, D., Corsi, M., Choinire, D., Arkhipova, E.: Development of an energy prediction tool for commercial buildings using case-based reasoning. Energy Buildings **81**, 152–160 (2014)

17. Niu, D., Wang, Y., Wu, D.D.: Power load forecasting using support vector machine and ant colony optimization. Expert Syst. Appl. **37**(3), 2531–2539 (2010)
18. Pai, P.-F., Hong, W.-C.: Forecasting regional electricity load based on recurrent support vector machines with genetic algorithms. Electr. Power Syst. Res. **74**(3), 417–425 (2005)
19. Platon, R., Dehkordi, V.R., Martel, J.: Hourly prediction of a building's electricity consumption using case-based reasoning, artificial neural networks and principal component analysis. Energy Buildings **92**, 10–18 (2015)
20. Platon, R., Martel, J., Zoghlami, K.: CBR model for predicting a building's electricity use: on-line implementation in the absence of historical data. In: Hüllermeier, E., Minor, M. (eds.) ICCBR 2015. LNCS (LNAI), vol. 9343, pp. 306–319. Springer, Heidelberg (2015). doi:10.1007/978-3-319-24586-7_21
21. Stahl, A.: Learning similarity measures: a formal view based on a generalized CBR model. In: Muñoz-Ávila, H., Ricci, F. (eds.) ICCBR 2005. LNCS (LNAI), vol. 3620, pp. 507–521. Springer, Heidelberg (2005). doi:10.1007/11536406_39
22. Stahl, A., Gabel, T.: Optimizing similarity assessment in case-based reasoning. In: Proceedings of the 21st National Conference on Artificial Intelligence AAAI 2006 (2006)
23. Zhao, H., Magouls, F.: A review on the prediction of building energy consumption. Renew. Sustain. Energy Rev. **16**(6), 3586–3592 (2012)

# Case Representation and Retrieval Techniques for Neuroanatomical Connectivity Extraction from PubMed

Ashika Sharma[1,2(✉)], Ankit Sharma[1], Dipti Deodhare[2], Sutanu Chakraborti[1],
P. Sreenivasa Kumar[1], and P. Partha Mitra[3]

[1] Department of Computer Science,
Indian Institute of Technology Madras, Chennai, India
{ashika,sutanuc,ankit,psk}@cse.iitm.ac.in
[2] Center for Artificial Intelligence and Robotics, DRDO, Bangalore, India
{ashika,dipti}@cair.drdo.in
[3] Cold Spring Harbor Labs, Cold Spring Harbor, New York, USA
mitra@cshl.edu

**Abstract.** PubMed is a comprehensive database of abstracts and references of a large number of publications in the biomedical domain. Curation of structured connectivity databases creates an easy access point to the wealth of neuroanatomical connectivity information reported in the literature over years. Manual curation of such databases is time consuming and labor intensive. We present a Case Based Reasoning (CBR) approach to automatically compile connectivity status between brain region mentions in text. We focus on the Case Retrieval part of the CBR cycle and present three Instance based learning techniques to retrieve similar cases from the case base. These techniques use varied case representations ranging from surface level features to richer syntax based features. We have experimented with diverse similarity measures and feature weighting schemes for each technique. The three techniques have been evaluated and compared using a benchmark dataset from PubMed and it was found that the one using deep syntactic features gives the best trade off between Precision and Recall. In this study, we have explored issues pertaining to representation of, and retrieval over textual cases. It is envisaged that the ideas presented in the paper can be adapted to needs of other textual CBR domains as well.

**Keywords:** Case representation · Case retrieval · Connectivity extraction · Instance based learning

## 1 Introduction

PubMed is a comprehensive database of abstracts and references of a large number of publications in the biomedical domain. Curation of structured connectivity databases creates an easy access point to the wealth of neuroanatomical connectivity information reported in the literature over years. Manual curation

© Springer International Publishing AG 2016
A. Goel et al. (Eds.): ICCBR 2016, LNAI 9969, pp. 370–386, 2016.
DOI: 10.1007/978-3-319-47096-2_25

of such databases is time consuming and labor intensive. In this study we propose a Case Based Reasoning (CBR) approach for extracting connectivity between brain region mentions. The corpus of natural language sentences from PubMed literature has been used to build the case base. The problem component of the case is a text sentence with a pair of brain region mentions and the solution component is a class label/value (positive or negative) denoting the connectivity status between the brain regions as depicted by the sentence.

Case retrieval involves reminding of the most similar experiences of the past from the case memory. This is a crucial part of the CBR cycle as it significantly impacts the performance of the CBR system. Starting with the set of problem-solution pairs in the case base, given a new case (sentence), we retrieve the most similar cases in the case base and use their solution components to derive the solution for the new case.

The task of finding solution to a new case is posed as a Classification task. The new case is assigned a solution based on the class labels of the retrieved cases. For the case retrieval process to be efficient, it is required to optimize the representation of the problem component of the case. Textual case bases need to deal with good amount of unstructured text. To handle the natural language sentences in our case base, we have explored surface level features and deeper features for representing the cases. We propose three different case representations for Instance Based Learning techniques using diverse similarity measures and feature weighing schemes, and present a comparative empirical evaluation across them. Our solution serves as a decision support tool for neuroscientists to discover new connections from text.

Section 2 introduces the problem domain, the motivation and challenges involved in solving the problem that the paper addresses. Section 3 gives a detailed description of the CBR approach for the problem, covering the three proposed techniques. Section 4 talks about the empirical evaluation describing the dataset, evaluation measures and results. Section 5 explains the relevance of the proposed techniques to CBR approach.

## 2    Domain Description

In the area of neuroscience, researchers are interested in the map of connections between various regions of the brain of an organism. Such a map of neural connections within an organism's nervous system, typically the brain is called a Connectome. Figure 1 shows the Connectome of the Rat brain organized as a matrix. Rows and columns represent brain regions and each cell of the matrix shows the strength of the connection between corresponding brain regions. Neuro-scientists perform tract tracing experiments, where a tracer is injected into a brain region and the axonal pathways followed by the tracer are used to infer connections between different regions.

Experimental results reported by researchers are published in journals and conferences. Large amount of information about brain region connections is present in the literature, but is not centralized for access. The challenge is,

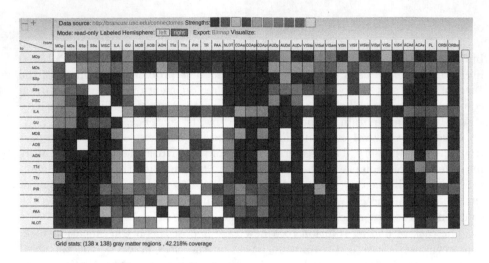

**Fig. 1.** Connectome [2] of rat cerebrum showing the matrix of connections between different regions of the rat cerebrum

these research findings are reported in natural language and there is a need for sophisticated Natural Language Processing (NLP) techniques to convert them into a structured database of connections.

Let us consider a sentence from a PubMed article which represents a case in our case base. The text of the sentence which forms the problem part of the case is, *"The projections from the **subiculum** to the **hypothalamus** were comprehensively examined in the rat."* The words in bold represent brain region mentions. Solution to this case is '*Positive*' since the two brain regions in the sentence are connected. These sentences report research findings of people from all over the world. Since they are authored by human beings in natural language, there exists different patterns in which connectivity information is conveyed. Some of the patterns in the sentences (The terms case and sentence are used interchangeably in this paper) present in the case base are stated below:

- BrainRegion1 *is connected to* BrainRegion2
- BrainRegion1 *efferent neurons in the* BrainRegion2
- *... projections from* BrainRegion1 *to* BrainRegion2 *...*
- BrainRegion1 *inputs to* BrainRegion2
- BrainRegion1 *was found to receive afferents from* BrainRegion2 *...*

The challenge is to automatically recognize the different patterns of connectivity and also the variations of a specific pattern in order to retrieve similar cases from the case base. It is assumed that Named Entity Recognition (NER) of brain region mentions is performed on the case base as part of preprocessing. Additionally, connected brain region pairs are assumed to be local to a sentence.

# 3   CBR Approach

Instance Based Learning algorithms work on the principle that similar examples have similar labels. This idea fits into the "similar problems have similar solutions" philosophy of CBR. The techniques presented in this paper compacts the set of cases in the case base by transforming them into a handful number of patterns as described in Sect. 2. These patterns essentially capture the existence of connectivity between the brain region mentions in the sentences of the case base and will guide the case retrieval process to efficiently find matching cases. Relevance of attribute/features is considered during case retrieval by weighing the features appropriately. The three techniques proposed differ in the case representation schemes, associated similarity measures and feature weighing policies used for case retrieval from the textual case base. The case representation should aid automatic generation of good connectivity patterns from the case base. The proposed case representations can be broadly categorized based on the type of features used, which are explained below.

## 3.1   Surface Level Case Representation

Surface level representation schemes consider words appearing in sentences as features and ignore any higher level links or deeper syntactic relations between the words. Two different representations based on the set of words used as features are described below:

1. *Bag-of-Word (BoW) representation*
   Every sentence in the case base is represented as a vector of words containing the left, middle and right contexts as inspired by Agichtein *et al.* [1]. The words in the sentence are stemmed and a vector space model is built for each of the contexts. Every word is represented by its normalized term frequency in the corresponding context. An $n$-word window has been considered for the left and right contexts. For the sentence in Sect. 2, considering $n=4$, the left, middle and right vectors are represented as *"[The:0.97, project:0.17, from:0.16, the:0.97]"*, *"[to:0.43, the:0.91]"* and *"[were:0.26, comprehens:0.01, examin:0.06, in:0.97]"* respectively.
2. *Connectivity-Word (CW) representation*
   Important words that emphasize the existence of connectivity between brain regions in the sentences have been identified from the case base with the help of domain experts. Some of these words are: *Afferent, Connect, Innervate, Originate, Pathway, Project, Receive, Input etc.* Let us call them as Connectivity Words. In this representation, each of the Connectivity Words has been used as a feature and a binary valued feature vector has been built based on the presence/absence of these connectivity words in the input sentence. *"[connect:0, project:1, input:0, send:0]"* The idea behind this representation is to consider only useful (in terms of connectivity relation) words in the sentence and filter the remaining words as noise.

**Pattern Generation Mechanism and Similarity Measures.** The classifier that assigns the solution to a new case is built as a Pattern matching system that labels the case as positive if the pattern formed with the corresponding brain region pair is connected, else as negative. Patterns used by the classifier are induced from the cases in the case base and are used to classify a new case by finding similar cases. Patterns of connectivity in sentences need to be automatically identified. These patterns will aid the classifier to learn the label for a new case. Pattern identification is achieved by clustering the vector representation of the sentences as described by Agichtein *et al.* [1].

The following similarity measure is used for forming clusters among the sentences in the BoW representation,

$$Match(v_i, v_j) = w_l(l(v_i).l(v_j)) + w_m(m(v_i).m(v_j)) + w_r(r(v_i).r(v_j)) \qquad (1)$$

where $v_i$, $v_j$ represent the vector representations of cases $i$ and $j$ respectively and $l,m,r$ represent the left, middle and right context vectors of the corresponding sentence. $w_l, w_m, w_r$ represent the weights assigned to left, middle and right vectors respectively. Equation 1 calculates the similarity between input vectors as a weighted dot product. It was found that the middle context had more words indicative of connectivity. Based on this, weights for the contexts have been fixed at ($w_l = 0.3, w_m = 0.6, w_r = 0.1$). Further fine tuning may be achieved by cross validation. The CW representation uses the following similarity measure,

$$Match(v_i, v_j) = v_i.v_j \qquad (2)$$

where $v_i$, $v_j$ represent the connectivity word vector representations of cases $i$ and $j$ respectively. Equation 2 calculates the similarity between input vectors as a dot product.

Single pass clustering [3] with a similarity threshold $\tau_{sim}$ has been used for clustering the vectors. Once clusters are formed, the centroid of each cluster is found by taking the mean of all the vectors in the cluster. Each centroid is a potential Connectivity pattern.

## 3.2   Link Parse Based Case Representation

The Surface Level Case Representation schemes described in Sec. 3.1 use words as features to represent a textual case. These features are shallow and induce noise into the representation by including words that are not relevant for identifying connectivity. In the Link Parse based Case Representation, the linkages/dependencies between words are used as features. The sentences in the case base are subjected to syntactic analysis by parsing them using link parser, which is a syntactic parser based on Link Grammar [10]. Given a sentence, the link parser assigns to it a syntactical structure consisting of a set of labeled links connecting pairs of words. The link parser output of a sentence is a planar graph, with links as edges labeled by connectors and words as nodes.

Let us consider the sentence *"The BR1 also projects to the BR2."*. Here BR1 and BR2 represent the two Brain Regions in the sentence. The link parse output

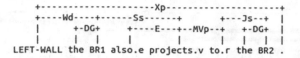

```
    +-------------------------Xp------------------------+
    +-----Wd----+------Ss------+            +---Js--+ |
    |       +-DG+      +----E--+--MVp--+      +-DG+  |
    |       |   |      |       |       |    | |   | |
    LEFT-WALL the BR1 also.e projects.v to.r the BR2 .
```

**Fig. 2.** Link parse for *"The BR1 also projects to the BR2."* showing linkages

depicting the linkages in the sentence is shown in Fig. 2. The connectors on the
links between the words depict the relationship between words. In this example,
link 'Ss' connects subject noun 'BR1' to verb 'projects' and 'Dg' connects deter-
miner 'the' to noun 'BR1' etc. A detailed description of the links can be found
in the Link Parser website [7].

Suchanek *et al.* [11] define a *Pattern* as a linkage in which two entities have
been replaced by placeholders. In Fig. 2 the placeholders are the two brain regions
BR1 and BR2, and the linkages in this sentence represent a *Pattern*. Suchanek
*et al.* [11] also defined the concept of *Bridge*, to be the shortest path between the
two placeholders in a *Pattern*. In Fig. 2, the Bridge is, "BR1 → projects → to →
BR2". Words in a Bridge are generalized by substituting nouns and adjectives
by the corresponding part-of-speech labels. Two *Patterns* match, if their underlying bridges are same, although nouns and adjectives are allowed to change.
Figure 3 depicts two patterns with equivalent bridges. In the first pattern, word
*'connections'* is present as a noun and in the second, word *'projections'* is present
as a noun.

Bridge captures the relationship between placeholders by considering only
the words and links on the shortest path and drops noisy words (words not relevant to describe the relation). The constituents of a Bridge originally proposed
by Suchanek *et al.* [11] is a sequence of nodes and edges on the shortest path
between placeholders. This idea has been extended to represent shortest path as
a set of *'Quadruples'* between two brain regions. Each quadruple is of the form
*<Link, Left word, Right word, Context>* and describes an edge on the shortest

```
"BR1 connections of the BR2 in the squirrel monkey."

  +---------------------------------Xp---------------------------------+
  |               +-------------------Sp--------------------+          | |
  +-------Wd------+         +---Js--+     +-----Ju----+     |          | |
  |    +---AN---+---Mp--+  +-DG+-Mp+  +---Dmu--+      |          | |
  |    |       |       |   |   |  |   |        |      |          | |
  LEFT-WALL BR1 connections.n of the BR2 in the squirrel.n-u monkey.v .


"BR1 projections of six BR2 in the owl monkey, Aotus trivirgatus."

  +--------------------------------------Xp----------------------------------------+
  |                +------------------Sp-----------------+                          |
  +-------Wd------+         +--Js--+   +---Js--+         |                          |
  |    +---AN---+---Mp---+  +DmC+-Mp+  +-Ds-+  |                          |
  |    |       |        |   |   |  |   |    |  |                          |
  LEFT-WALL BR1 projections.n of six BR2 in the owl.n monkey.v [,] [Aotus] [trivirgatus] .
```

**Fig. 3.** Two sentences having equivalent bridge, *"BR1 → <noun> → of → BR2"*

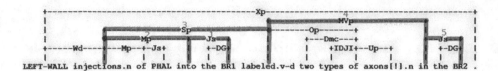

**Fig. 4.** Link parse for sentence *"Injections of PHAL into the BR1 labeled two types of axons in the BR2"* with Bridge highlighted in bold

path. *Link* is the connector on the edge, *Left* and *Right* words represent left and right nodes that the edge connects, *Context* takes as value one of $\{1, 2, 3\}$ which stands for *Left, Middle* and *Right* contexts respectively and refers to the context in the sentence where the edge occurs. For extracting connectivity relation between brain region mentions, context in which the edge occurs was found to be important. An edge is considered to occur in Middle context if the associated Left word comes before BR2 and the Right word comes after BR1.

Consider the sentence *"Injections of PHAL into the BR1 labeled two types of axons in the BR2"*. Figure 4 shows the link parse output for this sentence with the bridge marked in bold. The quadruple based bridge representation for this sentence is,

**Bridge:** [['Js', 'into', 'BR1', 1], ['Mp', 'n', 'into', 1], ['Sp', 'n', 'label', 2], ['MVp', 'label', 'in', 2], ['Js', 'in', 'BR2', 2]]

**Pattern Generation Mechanism and Similarity Measures.** In this case representation scheme, Bridge represents the pattern depicting the connectivity between brain regions. To generate the set of patterns in the case base, a unique Bridge Id is assigned for every sentence based on the underlying bridge. As the words and links in a bridge are generalized, multiple sentences have equivalent bridges and hence the same bridge id. This will cluster multiple sentences having the same bridge together. Each such cluster represents one unique pattern.

The feature vector in this representation is a bridge which is a list of quadruples. Similarity between two bridges has been defined using a modified Edit distance measure. Edit distance as described by Jurafsky *et al.* [5] originally measures the extent of dissimilarity between strings by calculating the minimum number of basic operations like insertion, deletion and substitution required to transform one string to another. The dynamic programming algorithm for Edit distance is modified to calculate the similarity between the quadruple representations of two bridges. The Bridge Edit Distance is illustrated in Algorithm 1. Length of a bridge is defined as the number of quadruples in it. Let us assume the length of $bridge_1$ to be $m$ and that of $bridge_2$ to be $n$.

Insertion and deletion penalties are fixed at 1 for matching two quadruples. Substitution penalties are given below.

- 0.3 for one word
- 0 if links and corresponding left and right words match.

---

**Algorithm 1.** Bridge Edit Distance

---
1: **procedure** BRIDGE EDIT DISTANCE($bridge_1$, $bridge_2$)
2:    **for** i = 0 to m **do**
3:       **for** j = 0 to n **do**
4:          **if** ($i == 0$ ) **then**
5:             $matrix(i, j) = j$
6:          **else if** ($j == 0$ ) **then**
7:             $matrix(i, j) = i$
8:          **else**
9:             $Cost = Substitute(quadruple_{i-1}, quadruple_{j-1})$
10:            $matrix(i, j) = min(matrix(i, j\text{-}1)\text{+}1, matrix(i\text{-}1, j)\text{+}1, matrix(i\text{-}1, j\text{-}1)\text{+}Cost)$
11:    **return** $matrix(i, j)$

---

- 0.3 if links match, but only one of the corresponding words match.
- 0.6 if links match, but none of the corresponding words match.
- 1 if links do not match, but only one of the corresponding words match.
- 2 if neither links, nor, both the corresponding words match.

Equality check for match does not distinguish between singular and plural forms of the links. For example, $Jp$ and $Js$ are considered as same links, $Mp$ and $Ms$ as same links.

Edit distance between two bridges $B_1$ and $B_2$ computed using the above algorithm is normalized as follows:

$$Normalized\ Edit\ Distance(B_1, B_2) = \frac{Bridge\ Edit\ Distance\ (B_1, B_2)}{2 * min(l_1, l_2) + |l_1 - l_2|} \quad (3)$$

where the denominator represents the maximum possible edit distance between two bridges with length $l_1$ and $l_2$.
Similarity between two bridges $B_1$ and $B_2$ is calculated as,

$$Similarity(B_1, B_2) = 1 - Normalized\ Edit\ Distance(B_1, B_2) \quad (4)$$

The following example illustrates the similarity computation between two bridges. Let us consider a sentence S1, "*All nerves, including those innervating the BR1, project to the BR2 and are somatopically organized*".

Bridge B1 for S1:

*[['Os', 'innerv', 'BR1', 1], ['Mg', 'those', 'innerv'], ['Op', 'includ', 'those', 1], ['MX\*p', 'n', 'includ', 1], ['Sp', 'n', 'and', 2], ['VJlpi', 'project', 'and', 2], ['MVp', 'project', 'to', 2], ['Js', 'to', 'BR2', 2]]*

Now, consider sentence S2, "*The BR1 also projects to the BR2*".
Bridge B2 for S2:

*[['Ss', 'BR1', 'project', 2], ['MVp', 'project', 'to', 2], ['Js', 'to', 'BR2', 2]*

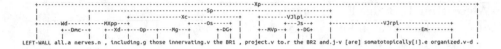

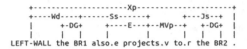

**Fig. 5.** Linkage for Sentence S1

**Fig. 6.** Linkage for Sentence S2

The linkages of Sentences S1 and S2 are depicted in Figs. 5 and 6 respectively. On applying Algorithm 1 to Bridges B1 and B2, the normalized edit distance is calculated as 0.509. Therefore, similarity is 0.490. Manual inspection of the sentences reveals that both sentences S1 and S2 are positive and both are following the pattern "BR → projects → to → BR2". But, Bridge B1 contains many noisy quadruples (that are not indicative of connectivity) in the left context, thereby reducing the similarity. To counter this, feature weighing is implemented where we are trying to assign weights to each quadruple based on its importance.

**Feature Weighing.** In the Link Parse based Case representation, features are the set of quadruples that make up the bridge of a sentence. A quadruple $Q$ is of the form $<Link, Left\ word, Right\ word, Context>$. The weight to be assigned to a quadruple is based on the importance of links and words present in it. Importance of a link or word $e$ in a particular context $c$ is calculated using Eq. 5. Weight of $e$ should represent its importance w.r.t depicting connectivity in the sentence. Given the dataset, we need to find how discriminating the link/word is, for connectivity extraction. A measure similar to entropy has been leveraged for this. Higher the discrimination ability, lower is the entropy of the word. Frequency of occurrence is used to scale the entropy to arrive upon the weight of the link/word. Similarly, the weight of all links and words in the case base are calculated for each of the Left, Middle and Right contexts.

$$weight(e) = log\left[Freq(e) * \frac{1}{Ent(e)}\right]$$

$$Freq(e) = log(frequency(e))$$

$$Ent(e) = -PlogP - NlogN$$

$$P = \frac{|Positive\ sentnces\ with\ e\ |}{|Positive\ sentences\ with\ e\ | + |Negative\ sentences\ with\ e|}$$

$$N = \frac{|Negative\ sentences\ with\ e\ |}{|Positive\ sentences\ with\ e\ | + |Negative\ sentences\ with\ e|}$$

$$(5)$$

The weight of the quadruple $Q = < link, lw, rw, cntxt >$ where $lw$ and $rw$ represent the left and right words respectively is calculated as in Eq. 6.

$$weight(Q) = \frac{weight(link) + weight(lw) + weight(rw) + weight(cntxt)}{4} \quad (6)$$

*Weighted quadruple Bridge* is derived by adding the quadruple weight calculated using Eq. 6 to the bridge.

The weighted quadruple Bridge WB1 for Sentence S1: *[['Os', 'innerv', 'BR1', 1], 0.37], [['Mg', 'those', 'innerv', 1], 0.44], [['Op', 'includ', 'those', 1], 0.59], [['MX\*p', 'n', 'includ', 1], 0.25], [['Sp', 'n', 'and', 2], 0.43], [['VJlpi', 'project', 'and', 2], 0.66], [['MVp', 'project', 'to', 2], 0.68], [['Js', 'to', 'BR2', 2], 0.44]]*

It is expected that weight of a noisy quadruple will be lower than the more informative ones, thereby enhancing the similarity value. As can be seen in the Weighted quadruple bridge for Sentence S1, quarduples like *['VJlpi', 'project', 'and', 2]* and *['MVp', 'project', 'to', 2]* which are important connectivity indicators have higher weights compared to quadruples *['MX\*p', 'n', 'includ', 1]* and *['Os', 'innerv', 'BR1', 1]*, which are not very informative.

Similarly, the weighted quadruple Bridge WB2 for Sentence S2: *[[['Ss', 'BR1', 'project', 2], 0.55], [['MVp', 'project', 'to', 2], 0.68], [['Js', 'to', 'BR2', 2], 0.54]]*

**Weighted Similarity Measure.** The weight of the quadruple can now be factored into the similarity computation of two bridges. This is implemented by modifying the *BridgeEditDistance* into *WeightedBridgeEditDistance* as described in Algorithm 2. It was observed that if quadruple occurs in left or right context then it is usually noisy. In the Weighted Bridge Edit distance algorithm, if the Quadruple belongs to the left or right context, instead of using a static cost for insertion/deletion, the corresponding weight of the quadruple is used as the insertion/deletion cost. This way, the penalty due to the presence of a noisy quadruple is controlled by giving it less weightage.

The procedure for *Substitute* is same as in Algorithm 1. Similarity is calculated as in Eq. 4. On applying Algorithm 2 to Weighted quadruple Bridges WB1 and WB2, the normalized edit distance and hence the similarity is 0.3281 and 0.671 respectively. For the same pair of sentences, we are able to achieve a higher similarity value which is more realistic compared to the unweighted similarity computation.

### 3.3 Pattern Confidence and Case Retrieval

Each of the case representation schemes generate patterns using various similarity measures as described above. Each pattern is further scored by a confidence value. Confidence of a pattern $P$ is defined as,

$$Conf(P) = \frac{P.positive}{(P.positive + P.negative)} \quad (7)$$

where $P.positive$ is the number of positive cases matching $P$ and $P.negative$ is the number of negative cases matching $P$ in the case base. Equation 7 checks that,

---

**Algorithm 2.** Weighted Bridge Edit Distance

---

1: **procedure** WEIGHTED BRIDGE EDIT DISTANCE($bridge_1$, $bridge_2$)
2:     **for** i = 0 to m **do**
3:         **for** j = 0 to n **do**
4:             **if** ($i == 0$ and $j == 0$ ) **then**
5:                 $matrix(i, j) = 0$
6:             **else if** ($i == 0$ ) **then**
7:                 $matrix(i, j) = weight(quadruple_{j-1})$
8:             **else if** ($j == 0$ ) **then**
9:                 $matrix(i, j) = weight(quadruple_{i-1})$
10:            **else**
11:                $Cost = Substitute(quadruple_{i-1}, quadruple_{j-1})$
12:                $matrix(i, j) = min(matrix(i, j\text{-}1) + weight(quadruple_{j-1}), matrix(i\text{-}1, j)$
13:                              $+ weight(quadruple_{i-1}), matrix(i\text{-}1, j\text{-}1) + Cost)$
14:     **return** $matrix(i, j)$

---

out of the total number of cases matching pattern $P$ in the case base, how many are positive. This is a measure of confidence of the pattern in generating positive cases. Thereby, it accounts for selectivity among patterns. Selectivity tunes the classifier to output only high confidence positive matches. Additionally, we also calculate the Coverage of a pattern w.r.t. other generated patterns.

$$Cov(P) = \frac{P.positive}{max(|p_i.positive|)} \qquad (8)$$

where $P.positive$ is the number of positive cases matching $P$ and the denominator indicates the maximum number of positive matches for any pattern $p_i$. $Cov(P)$ is used to scale Eq. 7 to find the total confidence of pattern $P$.

$$Confidence(P) = Conf(P) * Cov(P) \qquad (9)$$

Patterns are further filtered by applying a pattern confidence threshold $\tau_{pc}$ on $Confidence(P)$. Patterns falling below $\tau_{pc}$ can be ignored as they are either similar to the confident patterns or would be considerably noisy.

Now we have a final set of patterns that have captured the variations in which connectivity is reported in the case base. Given a new case, these patterns can be used to retrieve the most similar cases from the case base and classify the new case as positive or negative. The new case has to be represented in the form of the corresponding feature representation schemes. In case of BoW, in the form of left, middle and right context vectors by projecting them onto the corresponding vector spaces and for CW the boolean feature vector of connectivity words. Link parse representation generates the Weighted Quadruple Bridge for the new case. The most similar cases in the case base matching the new case are retrieved by finding similarity of the new case with the highly confident patterns selected in the previous step. Each of the three techniques use their corresponding similarity measures to retrieve similar cases. As a measure of quality, the confidence associated with the new case sentence and the patterns matching the new case are calculated. The following formula is used to calculate the confidence of the new case $C$ to have a connected pair of brain regions

$$Confidence(C) = \frac{\sum_{i=1}^{P} Confidence(P_i) * Similarity(P_i, C)}{\sum_{i=1}^{P} Confidence(P_i)} \qquad (10)$$

where $P = P_i$ is the set of patterns that matched $C$ with degree of match $Similarity(P_i, C)$ more than similarity threshold $\tau_{sim}$. $Confidence(P_i)$ is the confidence associated with pattern $P_i$ as calculated in Eq. 9.

The idea is, if the new case has been generated by several high confidence patterns, the confidence of the case itself will be high. As a last step, the confidence of the new case is thresholded by applying $\tau_{cc}$ which is the case confidence threshold. If a new case has confidence greater than $\tau_{cc}$, it is labeled as positive, else as negative.

Given the above case representation schemes with their corresponding similarity measures, the case base is setup as follows:

1. Represent sentences in the case base in the corresponding case representation scheme.
2. Generate underlying patterns from the cases in the case base.
3. Calculate confidence of patterns using Eq. 9 and select highly confident patterns.

Given a new case $C$, the following procedure is followed to find a solution for $C$:

1. Represent $C$ in the corresponding case representation scheme.
2. Calculate similarity of $C$ with each of the highly confident patterns generated from the case base in previous step.
3. Calculate the confidence of $C$ using Eq. 10 and assign positive label if $C$ is highly confident, else assign a negative label.

## 4 Experimental Evaluation

### 4.1 Dataset

French *et al.* [4] compiled the White Text Corpus from several abstracts of the Journal of Comparative Neurology on the PubMed. This dataset has been used to evaluate the proposed techniques. The corpus forms the case base and each sentence denotes the problem component of the case. The class label (Connected/Not connected), forming the solution part of the case is provided with every sentence in the corpus. Corpus sentences have variable number of brain region mentions. We analyzed the distribution of number of brain region mentions per sentence in the corpus. Figure 7 shows a histogram of this distribution. It was seen that around 52 % of the sentences have only two brain region mentions in them. For our experiments, we considered this subset of the WhiteText corpus with only two brain region mentions and we call this dataset the *2BRWhiteText*.

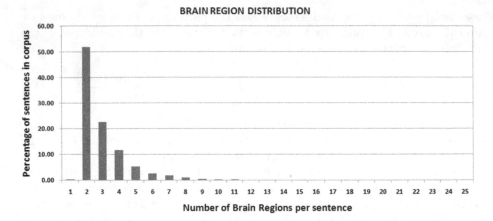

**Fig. 7.** Distribution of Brain Region mentions in WhiteText Corpus showing that more than 50 % of sentences have only two brain region mentions

Human annotators have categorized a given pair of brain regions in a sentence as being *Connected* if the same is evident in the language of the sentence. We are posing the problem in hand as a 2-class classification problem where the *Connected* category denotes the positive class and the *Not Connected* denotes the negative class.

### 4.2 Evaluation Measures

We have performed 10 runs of experiments with each run being a 90:10 train-test split of the corpus as described by Michalski *et al.* [8]. The *2BRWhiteText* corpus contains 3150 sentences, out of which, a different set of 315 sentences were included in test for each train-test split. 10-fold cross validation was performed on the train data of each train-test split, to choose the best values for the pattern confidence threshold $\tau_{pc}$, similarity threshold $\tau_{sim}$ and case confidence threshold $\tau_{cc}$. The final results reported are the average scores over the 10 train-test splits. Paired-t test is performed on the results of different techniques to check for statistical significance. We used Precision, Recall and F-Measure as the performance metrics for evaluation. Precision and Recall have been calculated using the following formulae:

$$Precision = \frac{|Relevant \cap Retrieved|}{|Retrieved|} \quad Recall = \frac{|Relevant \cap Retrieved|}{|Relevant|}$$

$$F_1 \ measure = 2.\frac{Precision.Recall}{Precision + Recall} \quad F_2 \ measure = \frac{5.Precision.Recall}{4.Precision + Recall}$$

*Retrieved* is the percentage of cases marked as connected by the algorithm in the test dataset. *Relevant* is the percentage of cases that have a connected brain region pair as identified by the human annotators. The ultimate goal of this system is to aid Neuro-scientists by collating experimental results pertaining to brain connectivity, published in various forums into a structured database. As the domain experts would be more interested in a system with high recall, the system has been tuned to be recall oriented. We have compared our techniques with the rule based method for discovering connectivity between brain regions proposed by Richardet *et al.* [9]. This is an alternative to the pattern based retrieval techniques proposed by the paper. RUle based Text Annotation (RUTA) method uses rules that were hand-crafted using the Apache UIMA RUTA workbench, Kluegl *et al.* [6]. Following is a sample RUTA rule.

   *(Projection "from" BrainRegionChunk "to") ->MARK(BRCooc, 1, 2) Brain-RegionChunk;* Here, *BrainRegionChunk* represents a brain region. If the rule matches the input, the matching text is annotated by the tag *BRCooc* denoting a connection between the brain region pair.

   To compare our approach with the Rule based approach, the RUTA rules were run on the test dataset considered in each of the experiments and the corresponding results were reported. The results of the various experiments are presented in Table 1.

### 4.3   Experimental Results

The BoW case representation, as shown in Table 1 has high recall as the patterns generated by this technique are quite generic in nature. The low precision is due to the influence of many noisy words in the feature representation. CW case representation improves on precision by identifying Connectivity word features in consultation with domain experts. CW achieves a 25 % boost in precision compared to BoW, but recall of CW is low indicating that, just the

**Table 1.** Precision (P), Recall (R), $F_1$ and $F_2$ measures computed on *2BRWhite-Text* dataset using Surface level and Link Parse based Case Representation techniques compared with Rule based UIMA RUTA. Results highlighted in bold indicate best performance w.r.t the considered evaluation measure.

| Approach | | $P$ in % | $R$ in % | $F_1$ score | $F_2$ score |
|---|---|---|---|---|---|
| Surface level Case Representation | *Bag-of-Word (BoW) features* | 29 | **91***  | 0.44 | 0.64 |
| | *Connectivity Word (CW) features* | 54 | 58 | **0.56** | 0.57 |
| Link Parse based Case Representation | *Weighted Quadruple Bridge (LPBridge) features* | 42 | 79 | 0.55 | **0.67*** |
| UIMA RUTA | *Rule based Similarity* | **89*** | 27 | 0.42 | 0.32 |

*indicates statistically significant results with $p < 0.05$ when compared with the other methods using pairwise t-test

connectivity word features are not sufficient to handle the variations in the natural language sentences. LP Bridge representation takes a more conservative syntax driven approach, enhancing precision over BoW by 13 % while still maintaining a recall of 79 %. This technique falls short of precision, but improves recall by 21 % as compared to the domain driven CW features. The precision of LP Bridge is affected by sentences on which Link parser fails to generate a parse tree and sentences that do not generate a Bridge. The precision of the Rule based method, RUTA is highest, but recall is significantly low. Reason for the reduced recall is that the rules are hand-crafted to cover only a few patterns. Sentences occurring according to any of the patterns covered by the rules are retrieved (high precision). But there are many more patterns or variations of the patterns which are not retrieved by the overly specific rules (low recall). Moreover, every pattern should be manually identified and a corresponding RUTA rule must be written. BoW has a high recall, meaning that, this technique retrieves many sentences, but not all of them are correct. A human using this technique will be overwhelmed by the number of false positives. On the other hand, CW retrieves many correct sentences as it has a high precision, but misses many sentences that are connected. The F-Measure is a more balanced and reliable statistic that measures both precision and recall simultaneously. $F_2$ score assigns more weight to recall as compared to precision and is more important in the context of this effort, as ours is a *recall oriented system*. The LP Bridge representation gives the maximum $F_2$ score, striking the best trade off between precision and recall. LP Bridge performs statistically significantly better than other techniques. $F_1$ score comparison shows that CW features is the winner; however, the LP Bridge is very close, falling short of CW by only 0.01.

## 5    Discussion and Outlook

The ideas presented in the paper are of general significance to Textual case bases. Knowledge in TCBR applications can be structured using statistical methods, background/domain information and linguistic information. The main thesis of this paper is to explore these three techniques. The Bag-Of-Word case representation uses statistical measures like frequency of occurrence of words as features to represent the cases. Connectivity Word representation leverages the domain inputs about the set of words indicative of connectivity in the case base. The Link parse based Bridge representation uses linguistic knowledge in the sentences by parsing them and using the syntactic features to represent the cases. TCBR applications can benefit from constructing features based on the representations proposed in this paper.

In all our techniques, the cases in the case base are abstracted into a set of patterns which are used to find similar cases during case retrieval. The advantage of our approach is that, it relies on these abstract patterns, which is a more interpretable model, unlike SVM or Kernel based methods which are also used for information extraction. Our models are not completely lazy, but have an eager component, that generalizes the cases into generic pattern representations that can be applied not only in exactly same contexts, but similar contexts. Once a new case arrives, our models obviate the need for making an assessment against every case in the case base, rather assessment is made against a pattern, which is a generalized expression of a cluster of cases. Once similar patterns are identified, the new case is classified by the algorithm. As part of future work, we would want to extend the solution to handle sentences with more than two brain regions. The Link parse based technique needs to be evolved to handle ungrammatical, ill-formed and complex sentences. We may explore the possibility of further specializing the query and comparing against specific cases stored under each pattern.

# 6   Conclusion

In this study, we have explored issues pertaining to representation of, and retrieval over textual cases. We have proposed three Instance based learning techniques using different Case representation schemes. Surface level statistical features and rich syntactic features using linguistic information have been explored along with different similarity measures to aid effective case retrieval. Performance of the three techniques have been compared and contrasted using a benchmark dataset. The link parse based syntactic approach is shown to achieve the best trade off between precision and recall. It is envisaged that the ideas presented in the paper can be adapted to needs of other textual CBR domains as well.

# References

1. Agichtein, E., Gravano, L.: Snowball: extracting relations from large plain-text collections. In: Proceedings of the Fifth ACM Conference on Digital Libraries, pp. 85–94. ACM (2000)
2. Connectome of Rat cerebrum from Brain Architecture Management System (BAMS). https://bams2.bams1.org/connections/grid/80/
3. Frakes, W.B., Baeza-Yates, R.: Information Retrieval: Data Structures and Algorithms. Prentice-Hall, Englewood Cliffs (1992)
4. French, L., Liu, P., Marais, O., Koreman, T., Tseng, L., Lai, A., Pavlidis, P.: Text mining for neuroanatomy using WhiteText with an updated corpus and a new web application. Front. Neuroinformatics **9**, 13 (2015). doi:10.3389/fninf.2015.00013. ISSN: 1662–5196
5. Jurafsky, D.: Speech and Language Processing. Pearson Education, India (2000)
6. Kluegl, P., Toepfer, M., Beck, P.D., Fette, G., Puppe, F.: UIMA Ruta: Rapid development of rule-based information extraction applications. Nat. Lang. Eng. **1**(1), 1–40 (2014)
7. Link Grammar Parser. http://www.link.cs.cmu.edu/link/
8. Michalski, R.S., Carbonell, J.G., Mitchell, T.M. (eds.): Machine Learning: An Artificial Intelligence Approach. Springer, Heidelberg (2013)
9. Richardet, R., Chappelier, J.C., Telefont, M., Hill, S.: Large-scale extraction of brain connectivity from the neuroscientific literature. Bioinformatics **31**(10), 1640–1647 (2015)

10. Sleator, D.D., Temperley, D.: Parsing English with a link grammar. arXiv preprint (1995)
11. Suchanek, F.M., Ifrim, G., Weikum, G.: LEILA: Learning to extract information by linguistic analysis. In: Proceedings of the 2nd Workshop on Ontology Learning and Population: Bridging the Gap Between Text and Knowledge, pp. 18–25 (2006)

# Compositional Adaptation of Explanations in Textual Case-Based Reasoning

Gleb Sizov[✉], Pinar Öztürk, and Erwin Marsi

Department of Computer Science, Norwegian University of Science and Technology,
Trondheim, Norway
{sizov,pinar,emarsi}@idi.ntnu.no

**Abstract.** When problem solving systems are deployed in real life, it is usually not enough to provide only a solution without any explanation. Users need an explanation in order to trust the system's decisions. At the same time, explanations may also function internally in the system's own reasoning process. One way to come up with an explanation for a new problem is to adapt an explanation from a similar problem encountered earlier, which is the idea behind the case-based explanation approach introduced by [29]. The original approach relies on manual construction of cases with explanations, which is difficult to scale up. In earlier work, therefore, we developed a system for automatic acquisition of cases with explanations from textual reports, including retrieval and adaptation of such cases [32,33]. In this paper, we improve the adaptation method by combining explanations from more than one case, which we call *compositional adaptation*. The method is evaluated on an incident analysis task where the goal is to identify the root causes of a transportation incident, explaining it in terms of the information contained in the incident description. The evaluation results show that the proposed approach increases both the recall and the precision of the system.

**Keywords:** Textual case-based reasoning · Compositional adaptation · Explanation · Incident analysis

## 1 Introduction

When problem solving systems are deployed in real life, it is usually not enough to provide only a solution. Users also expect an explanation of how the problem was solved, justifying the conclusions reached, in order to trust the system's solution. For example, in medical reasoning, the symptoms (problem) and the patient history are "indirectly" linked to the disease label (solution) via intermediate physiological states. In our own research we investigate an incident analysis task, which – similar to medical diagnosis – requires incidents to be explained by a root cause via proxy causes. Our goal is to automate a case-based problem-solving process for this task using free text incident reports as the source of cases. In addition, the system should be able to explain the reasoning behind the solution it proposes.

© Springer International Publishing AG 2016
A. Goel et al. (Eds.): ICCBR 2016, LNAI 9969, pp. 387–401, 2016.
DOI: 10.1007/978-3-319-47096-2_26

One way to construct an explanation for a new problem is to adapt explanations from similar problems encountered earlier. This case-based reasoning (CBR) approach to explanations has been introduced by [29] and relies on storing, indexing and retrieving "explanation patterns" (XPs), which are specific or generalised explanations of events. XPs can be tweaked to fit new situations as described in [31]. This approach was implemented in the SWALE system [16, 19], which can explain complex real-life problems such as the death of the racehorse Swale, which was successful in its career but died prematurely of unknown causes. SWALE (the system, not the horse) proposed several explanations for this event by retrieving and tweaking XPs from similar events, such as drug overdose in the "Janis Joplin" XP and a heart attack in the "Jim Fixx" XP. Explanations in SWALE are thus not only used by the system as a means of understanding the situation, but also to clarify the system's reasoning to the user.

The main limitation of the case-based explanation approach in SWALE and similar systems like Meta-AQUA [13], is that they rely on knowledge engineering to construct domain-specific explanation patterns, which is often both time-consuming and expensive. To avoid these costs, we proposed in previous work a method for reasoning with explanations contained originally in textual documents [33]. This approach is implemented in a textual CBR system that automatically extracts explanations from incident reports, using text mining methods, and reuses them to explain new transportation incidents. It includes methods for extraction, retrieval and adaptation of explanations contained in text. The advantage of this approach is that it uses existing incident reports as the source for constructing cases without any costly human involvement.

The adaptation method in our previous work adapted the explanation from the single most similar case. However, evaluation showed that use of only one case often yields an incomplete explanation. The reason is that often only a part of the previous case explanation is relevant for the current situation, which leaves much of the target problem unexplained. In the current paper, we therefore propose a compositional adaptation method that reuses explanations from multiple similar cases. Our main hypothesis is that combining and then adapting explanations from several previous cases will result in a larger, more complete explanation for the target problem, while at the same time increasing accuracy.

The rest of this paper is organized as follows. Section 2 provides an overview of related research on case-based explanations. In Sect. 3, we describe the incident analysis task. Section 4 provides an overview of the textual CBR approach for incident analysis proposed in our previous work. The compositional adaptation method, which constitutes the main contribution of this paper, is described in Sect. 5. The evaluation method and results are presented in Sect. 6. Finally, Sect. 7 outlines conclusions and future work.

## 2   Case-Based Explanations

In AI systems, explanations have two roles: *internal* and *external* [1]. Internal explanations are used in the reasoning process by the system itself, whereas external explanations are intended for users of the system to increase their understanding of how it solves a problem, help to understand the solution itself or how to use the solution. These two roles of explanations are not necessarily distinct: internal explanations can sometimes be used directly as, or translated into, external explanations.

Case-based explanation approaches can be divided into *knowledge-lean* and *knowledge-intensive* as suggested by Cunningham, Doyle and Loughrey [14]. Knowledge-intensive approaches generate explanations using rule-based or model-based inference. An example of knowledge-intensive explanations is the CREEK system [2] which connects observations in the case to the suggested solution though relations in the causal domain model, serving as internal and external explanation. Domain knowledge can also be applied to explain similarity between cases [8]. A combination of model-based and case-based reasoning is applied in the legal reasoning system by Brüninghaus and Ashley [10] where the solutions themselves are the explanations. The main limitation of the knowledge-intensive approach is the cost of the knowledge engineering required to acquire the rules and the general domain models underlying explanations.

Knowledge-lean approaches require far less knowledge engineering. They usually focus on retrieval, however, with limited adaptation capabilities. The most common type of knowledge-lean "explanation" is simply displaying the most similar case along with the solution, without any explicit explanation. This leaves it to the users to make their own judgement regarding the similarity between the target problem and the retrieved case. Since retrieval of similar cases is part of the CBR process, it serves as both internal and external explanation. Cunningham, Doyle and Loughrey [14] showed that this type of explanation increases the user's confidence in the solution. In addition, some CBR systems display, as the external explanation, cases that provide contrasting evidence for the suggested solution. For example, The Stamping Advisor [20] shows two cases similar to the target problem but with different solutions. Visualization of the case space and similarity can serve as an external explanation as well. Examples of this approach are the work by McArdle and Wilson [23] where similarity between cases is visualized in a two-dimensional space and the FormuCaseViz system [21] where visualization is used to spot the difference in attributes between cases and how they influence the solution.

In expert systems, explanations are often produced from a reasoning trace consisting of the rules used in the reasoning process, e.g. MYCIN [12]. In CBR, steps of the reasoning process can be also used for explanation. For example, the LID system [27] generates explanation as a hierarchy of general-to-specific categories containing the target case. Another example is the Top Case recommendation system [24], a conversational CBR system that asks users questions to learn about their preferences. Explanations show how the answers to the questions lead to the solution recommended by the system.

In general, the capability to generate explanations is an important advantage of CBR over sub-symbolic approaches such as statistical learning and neural networks. However, it often requires knowledge which is hard to acquire for real-life domains.

# 3   Incident Analysis Task

The research presented here is motivated by an *incident analysis* task where the goal is to identify and explain incident causes and contributing factors. The primary source of knowledge to accomplish this task is incident reports, in particular incident reports from the Transportation Safety Board of Canada (TSB)[1] that document aviation, marine and rail incidents. Each report provides a brief description of the incident in the *summary section* which is used as the input to the system, e.g.:

> The Bell 206L-3, was on a visual flight rules flight. Approximately 20 min after take-off, the needle on the engine oil pressure gauge started to fluctuate. As a precaution, the pilot landed the aircraft in a marsh and shut down the engine. After conducting a pre-flight inspection, the pilot started the engine and took off with the intention of landing on a road one kilometre away. Just before the helicopter reached the road, there was a fluctuation in the engine oil pressure and engine torque. Right after that, there was an explosion and the engine stopped.

Each report also contains an *analysis section* which explains why the incident happened, describing its causes and contributing factors, e.g.

> The area around bearings 6 and 7 had exceeded a temperature of 900°. The oil that burned away did not return to the tank and, after a short time, the oil level became very low, causing the engine oil pump to cavitate and the engine oil pressure to fluctuate. Furthermore, since the oil did not return to the tank, the oil temperature did not change, or at least not significantly, and the pilot falsely deduced that the engine oil pressure gauge was displaying an incorrect indication.

Finally, each report includes one or more sections outlining causes and contributing factors, which we consider as the *conclusions*, e.g.

- The bearings were destroyed for undetermined reasons, causing an engine failure.
- Moving the helicopter towards the road when the engine was showing signs of malfunction contributed to the failure of bearings 6 and 7.
- During the auto-rotation, the helicopter was not levelled at the time of the landing, which resulted in a hard landing.

Usually incident analysis is accomplished by human experts who posses substantial knowledge and experience in the domain. It is a challenging task for an AI system, especially when the only source of domain knowledge is in textual form. The task of analysing airplane crashes given stories and data about previous crashes is mentioned as an example of a *cognitive understanding* in [30]. The

---

[1] TSB reports are available at: http://www.tsb.gc.ca/eng/rapports-reports.

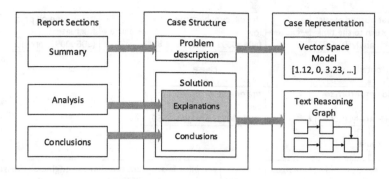

**Fig. 1.** Case representation

authors argue that explanations are at the heart of understanding and propose a revised Turing test based on the ability of an AI system to explain its decisions and actions. More recently, a workshop dedicated to this task was organised as part of the International Conference on Case-Based Reasoning [9]. There is also a challenge paper encouraging the automatic reuse of knowledge from incident reports through natural language processing techniques [35].

## 4    Textual CBR for Incident Analysis

In our previous work we proposed a textual CBR system that automatically explains a new incident by adapting an explanation from a previous incident as described in incident reports from TSB [33]. These reports have a fairly consistent structure with *summary, analysis* and *conclusion* sections, which are mapped to the problem description and solution parts in the case structure. As shown in Fig. 1, the summary section is mapped to the problem description part of the case, while analysis and conclusion sections are both mapped to the solution part. The conclusion section outlines the root causes and contributing factors, whereas the analysis section comprises an explanation chain connecting features in the problem description to the causes through intermediate reasoning steps. Explanations in our system play both the internal and external role: they are used internally in the system to generate new solutions and, at the same time, they provide the user with an explanation of the system's decisions.

Different representations are used for the problem and the solution parts of cases. A problem description is represented with a vector space model (VSM) to facilitate efficient retrieval of similar cases. VSMs are the most common text representation in information retrieval and are based on the frequency of occurrence of words in documents. The solution part is represented by a Text Reasoning Graph (TRG) [33], which is a graph-based representation of explanations as causal chains of states and events, connecting pieces of information in the problem description to conclusions. As can be seen in Fig. 2, nodes in the TRG are phrases and sentences extracted from the report, while edges correspond to

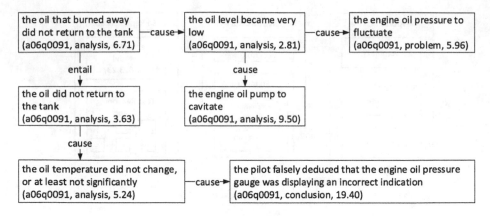

**Fig. 2.** Example of a TRG that represents an explanation, which is part of the case solution.

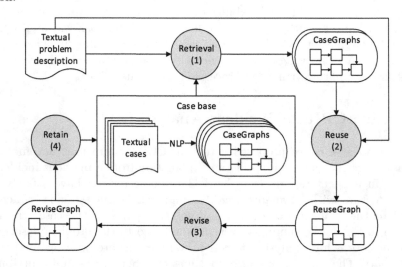

**Fig. 3.** Analysis cycle in textual case-based reasoning

causal and (textual) entailment relations. The TRG is extracted automatically using a variety of natural language processing techniques including syntactic parsing, causal relation extraction and textual entailment recognition. The complete extraction process is described in [33].

The reasoning process in our system follows the classical CBR cycle of retrieval, reuse, revise and retain steps as shown in Fig. 3. The outputs of the first three steps are referred to as CaseGraph, ReuseGraph and ReviseGraph respectively. All of these graphs capture case solutions at different stages of the process using the TRG representation. The process starts with a textual description of a new incident, one or two paragraphs in length. The retrieval step identifies cases from the case-base that are similar to the new problem description. A certain number of most similar cases are retrieved. In the reuse step, solutions of the retrieved cases, CaseGraphs, are adapted to the new problem description producing the ReuseGraph. In the revise step, the adapted solution in the ReuseGraph is validated and corrected by human experts resulting in the Revise-Graph. Finally, in the retain step, the ReviseGraph along with the new problem description is added to the case base for future reuse.

The focus of the current paper is on the reuse step. In particular the method for combining and adapting explanations from several previous cases to generate the explanation for a new problem, i.e. multiple retrieved CaseGraphs are combined and adapted. For the the retrieval step we used a traditional information retrieval technique based on VSM and cosine similarity that was also used in our previous work [32]. The revise and the retain steps are left for future work.

## 5 Compositional Adaptation of Textual Explanations

Our compositional adaptation approach contains three steps, shown as steps 4, 5 and 6 in Fig. 4. The first step combines the CaseGraphs of the retrieved cases into a *local explanation model* (LEM). We refer to this model as *local* to highlight the idea that it only applies in the vicinity of the target case rather than globally over the problem space. When combining CaseGraphs, nodes that are paraphrases of each other are merged into a single node, thus creating a connected graph. Nodes that are not paraphrases are checked for textual entailment, adding the entailment relations between them when necessary. The result is thus a large TRG comprised of nodes and relations from several CaseGraphs.

The second step extracts those explanation chains from the LEM that explain parts of the target problem. This process starts by identifying *evidence nodes* $(N_{ev})$, which are nodes in the LEM that contain text snippets carrying the same or similar information as the sentences in the target problem description $(S_p)$. Textual entailment recognition is used to compare text snippets in the nodes with the sentences in the problem description. The following equation defines evidence nodes:

$$N_{ev} = \{TextEntail(s_p, n) \mid n \in LEM, s_p \in S_p\} \qquad (1)$$

Evidence nodes serve as the starting points for extracting explanation chains. The end points are the conclusion nodes derived from the conclusion sections of the report. In the incident analysis domain, the conclusion nodes correspond to root causes and contributing factors for the incident. Explanation chains are extracted as the shortest paths connecting evidence nodes $N_{ev}$ to conclusion nodes $N_c$ in the *LEM*:

$$Chains = \{ShortPath(n_{ev}, n_c, LEM) \mid n_{ev} \in N_{ev}, n_c \in N_c\} \qquad (2)$$

Since the LEM is composed of multiple CaseGraphs, the explanation chains may include nodes and relations from different cases, as shown in Fig. 5. The extracted explanation chains are combined into the ReuseGraph in the same way

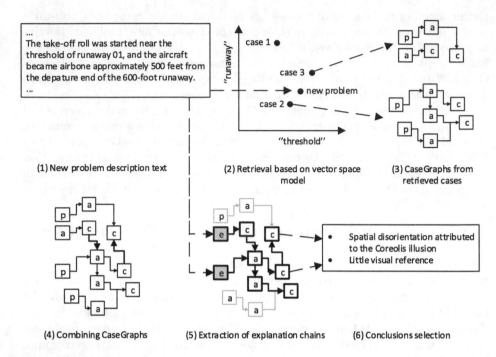

(1) New problem description text

(2) Retrieval based on vector space model

(3) CaseGraphs from retrieved cases

(4) Combining CaseGraphs

(5) Extraction of explanation chains

(6) Conclusions selection

**Fig. 4.** TCBR process with the results generated at each step. Node labels stand for the part of the source report where the information contained in the node appears: (p) problem description, (a) analysis, (c) conclusion. Evidence nodes are labelled with (e).

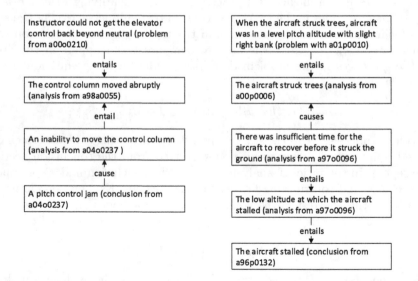

**Fig. 5.** Reasoning paths automatically extracted from the local explanation model.

CaseGraphs were combined earlier: nodes that are paraphrases of each other are merged into a single node and entailment relations are added for the rest of the nodes.

The conclusion selection step selects the best conclusions from the Reuse-Graph. Depending on the number of the CaseGraphs used in the adaptation process, the ReuseGraph may contain a large number of conclusion nodes and there is a need for a mechanism to rank and select the best ones. Previously proposed compositional adaptation methods select solution components that appear in more retrieved cases because they are conceived as more representative [4,7]. A similar heuristic is applied in our approach where conclusion nodes connected to more evidence nodes are preferred. This heuristic is formalized in the *conclusion support* measure:

$$Support(n_c) = \frac{|\{Connected(n_{ev}, n_c) \mid n_{ev} \in N_{ev}\}|}{|N_{ev}|} \tag{3}$$

where $n_c$ is the conclusion node and $N_{ev}$ is a set of evidence nodes. Conclusions with a support value above a set threshold are selected.

# 6   Evaluation

Our evaluation is aimed at verifying the hypothesis that combining and then adapting explanations from several previous cases will result in a larger, more complete explanation for the target problem, while at the same time increasing accuracy. In particular, we observe the performance of the system with different number of cases used in adaptation.

## 6.1   Evaluation Method

We evaluate explanations based on the conclusions they lead to. Our assumption is that the recall and the precision of the conclusions reflect the accuracy and the completeness of the explanations leading to them. The reason why we evaluate explanations indirectly through their conclusions is that it is a much more straightforward process than evaluation of the explanations directly. Each report contains a list of conclusions that can be used as the gold standard to compare the candidate conclusions to. However, we acknowledge the limitation of this evaluation approach and plan to investigate methods that evaluate explanations directly.

For evaluation, we are using 922 aviation, 375 marine and 298 rail incident reports from the TSB dataset as described in Sect. 3. Each report is 5–10 pages long and describes an incident including the analysis of its causes. Our system is evaluated separately on the aviation and marine report collections. The rail reports are used for optimizing system parameters: the support threshold in the conclusion support measure (see Sect. 5) and the similarity threshold in the textual entailment recognition component used for the extraction of CaseGraphs from text as described in our previous work [33]. The optimal values, 0.1 for the

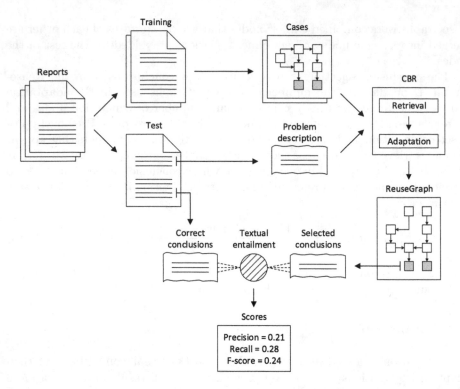

**Fig. 6.** Evaluation of the textual case-based explanation system

support threshold and 0.6 for the similarity threshold, were obtained through the grid search, maximizing the F-score. This experimental setup was chosen to demonstrate that our TCBR approach is general enough to be used in different domains (rail, aviation and marine), albeit for the same analysis task. Parameters obtained on the rail reports are used for aviation and marine reports without any changes.

Our evaluation procedure is visualized in Fig. 6. For evaluation, we split the collection of reports into test and training sets containing 20 % and 80 % of the reports, respectively, with no overlap. Our evaluation criteria are based on the comparison between the selected conclusions from the ReuseGraph and the actual, correct conclusions from the report. In particular, selected conclusions nodes from the ReuseGraph are matched with similar sentences from the conclusion section of the test report. One-to-one matching is performed using the Hungarian algorithm [17], which maximizes the number of matches while preventing multiple conclusion nodes to be matched with the same conclusion sentence and vice versa. Whether a node is similar to a sentence is determined by the textual entailment recognition component described in Sect. 6.2. The evaluation scores – precision, recall and F-score – are then computed based on the number of matched conclusion nodes ($N_m$), dividing it by the total number of the conclusion nodes ($N_c$) in the ReuseGraph and conclusion sentences ($S_c$) in the test report:

$$Precision = |N_m| \, / \, |N_c| \tag{4}$$

$$Recall = |N_m| \, / \, |S_c| \tag{5}$$

$$FScore = \frac{2 \cdot Precision \cdot Recall}{Precision + Recall} \tag{6}$$

### 6.2 Textual Entailment Recognition

The similarity between the conclusion nodes and conclusion sentences is determined by the textual entailment recognition component. In general, the task of textual entailment recognition (TER) is to determine whether the meaning of one text fragment is entailed by the meaning of another. There exist several methods of textual entailment in the natural language processing field [6]. The method used in our evaluation is based on the pairwise similarity between the words in two text fragments. The method includes the following steps:

1. LCH similarity is computed between each pair of words in the two text fragments [18]. LCH is a semantic word similarity measure based on the notion of shortest path between the corresponding two word senses in WordNet [25].
2. The Hungarian algorithm [17] is used to align similar words in two text fragments. The similarity between the fragments is then computed as the sum of similarities between the aligned words divided by the total number of words in these fragments.
3. The obtained similarity value is compared to a similarity threshold. The entailment holds for the fragments with a similarity value above the threshold, which was set to 0.6.

### 6.3 Results and Analysis

The evaluation results were obtained for four configurations of the system:

- **base**: Explanations chains are extracted directly from the CaseGraphs without combining them into a LEM first. All the conclusions are selected from the extracted chains without using the conclusion support measure. This configuration is analogous to our previous work on retrieval and adaptation of explanations from incident reports [33].
- **base + select**: Same as the *base* but the conclusions are selected using the conclusion support measure.
- **base + combine**: Same as the *base* but the CaseGraphs are combined into the LEM and the explanations are extracted from it.
- **base + combine + select**: The complete system as described in Sect. 5

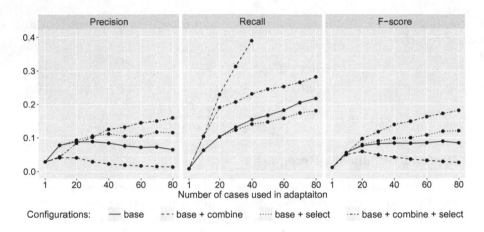

**Fig. 7.** Evaluation results on the aviation incident report collection

The number of cases retrieved and then adapted by the system ranges from 1 to 80 with a step size of 10. Although reusing more cases would likely lead to higher scores for some configurations, 80 cases seemed to be enough for observing the trends in the performance of different configurations.

The results presented in Figs. 7 and 8 show that increasing the number of reused cases from 1 to 10 leads to higher scores for all configurations. With more than 10 cases, the recall continues to increase while the precision goes down for the configurations without the *select* step. This decrease in precision can be explained by more distant cases being reused, thus higher chance for erroneous conclusions reached by the explanations. The *select* step is able to compensate for this by filtering out conclusions that are not supported by enough evidences. It is worth mentioning that the number of evidences increases with the number of cases used in adaptation, thus providing more information for the selection step.

The performance of the configurations with and without the *combine* step differs substantially as well. The recall of the configurations that include the *combine* step increases much faster than without it as more conclusions are reached through the LEM than from individual CaseGraphs. However, this gain in recall comes at a cost of the precision, which goes down for the *base + combine* configuration. This is due to the explanation chains going through many different cases in LEM, which are less consistent than the ones extracted from individual CaseGraphs, leading to wrong conclusions. The best performance is achieved by the *base + combine + select* configuration, where the *combine* step increases the recall while the *select* step increases the precision. The support threshold allows to balance the recall and the precision, achieving the highest F-score. Evaluation results obtained on the aviation and marine incident reports are consistent with each other, showing the same trends.

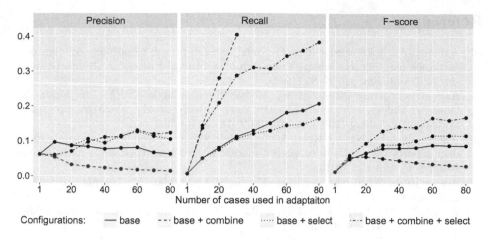

**Fig. 8.** Evaluation results on the marine incident report collection

The idea of automated incident analysis based on the incident report has been discussed in two challenge papers [9, 35]. The related research on the topic is dedicated to retrieval, classification and clustering of the reports and their parts [3, 11, 22, 26, 28, 34, 36]. These tasks are substantially different from the incident analysis task used in the current work, making it difficult to compare the results.

## 7    Conclusion and Future Work

This paper is a continuation of our work on automated incident analysis with textual CBR, focusing on adaptation of explanations extracted from text. Previously we introduced the Text Reasoning Graph representation to capture explanations contained in text, methods for automatic acquisition of explanations from text as well as retrieval and adaptation methods to reuse explanations from previous incidents to explain a new incident. Prior to the current work, our adaptation method was limited to adaptation of a single explanation, which was not sufficient to generate a complete explanation for a new incident, resulting in low recall. In the current work we proposed the compositional adaptation method which combines explanations from multiple cases to explain a new incident. Our evaluation results show that this approach increases both the recall and the precision of the system.

The proposed adaptation method suggests several directions for future work. In particular we would like to investigate the use of activation propagation in the TRG for adaptation, which might provide better control over the adaptation process. Another promising direction is a semi-automatic adaptation, where users can interact with the system to guide the adaptation process. Previous research in conversational CBR [5, 15] might provide clues for how to develop such a system.

# References

1. Aamodt, A.: A knowledge-intensive, integrated approach to problem solving and sustained learning. Knowledge Engineering and Image Processing Group. University of Trondheim, pp. 27–85 (1991)
2. Aamodt, A.: Knowledge-intensive case-based reasoning in CREEK. In: Funk, P., González Calero, P.A. (eds.) ECCBR 2004. LNCS (LNAI), vol. 3155, pp. 1–15. Springer, Heidelberg (2004). doi:10.1007/978-3-540-28631-8_1
3. Abedin, M.A.U., Ng, V., Khan, L.: Cause identification from aviation safety incident reports via weakly supervised semantic lexicon construction. J. Artif. Intell. Res. **38**(1), 569–631 (2010)
4. Adeyanju, I., Wiratunga, N., Recio-García, J.A., Lothian, R.: Learning to author text with textual CBR. In: ECAI, pp. 777–782 (2010)
5. Aha, D.W., Breslow, L.A., Muñoz-Avila, H.: Conversational case-based reasoning. Appl. Intell. **14**(1), 9–32 (2001)
6. Androutsopoulos, I., Malakasiotis, P.: A survey of paraphrasing and textual entailment methods. J. Artif. Intell. Res. **38**, 135–187 (2010)
7. Arshadi, N., Badie, K.: A compositional approach to solution adaptation in case-based reasoning and its application to tutoring library. In: Proceedings of 8th German Workshop on Case-Based Reasoning, Lammerbuckel (2000)
8. Bergmann, R., Pews, G., Wilke, W.: Explanation-based similarity: a unifying approach for integrating domain knowledge into case-based reasoning for diagnosis and planning tasks. In: Wess, S., Althoff, K.-D., Richter, M.M. (eds.) EWCBR 1993. LNCS, vol. 837, pp. 182–196. Springer, Heidelberg (1994). doi:10.1007/3-540-58330-0_86
9. Bridge, D., Gomes, P., Seco, N.: Analysing air incident reports: workshop challenge. In: Proceedings of the 4th Workshop on Textual Case-Based Reasoning (2007)
10. Brüninghaus, S., Ashley, K.D.: Combining case-based and model-based reasoning for predicting the outcome of legal cases. In: Ashley, K.D., Bridge, D.G. (eds.) ICCBR 2003. LNCS (LNAI), vol. 2689, pp. 65–79. Springer, Heidelberg (2003). doi:10.1007/3-540-45006-8_8
11. Carthy, J., Wilson, D.C., Wang, R., Dunnion, J., Drummond, A.: Using T-Ret system to improve incident report retrieval. In: Gelbukh, A. (ed.) CICLing 2004. LNCS, vol. 2945, pp. 468–471. Springer, Heidelberg (2004). doi:10.1007/978-3-540-24630-5_57
12. Clancey, W.J.: The epistemology of a rule-based expert system–a framework for explanation. Artif. Intell. **20**(3), 215–251 (1983)
13. Cox, M.T., Ram, A.: Introspective multistrategy learning: on the construction of learning strategies. Artif. Intell. **112**(1–2), 1–55 (1999)
14. Cunningham, P., Doyle, D., Loughrey, J.: An evaluation of the usefulness of case-based explanation. In: Ashley, K.D., Bridge, D.G. (eds.) ICCBR 2003. LNCS (LNAI), vol. 2689, pp. 122–130. Springer, Heidelberg (2003). doi:10.1007/3-540-45006-8_12
15. Gupta, K.M., Aha, D.W.: Conversation for textual case-based reasoning. In: Proceedings of the 4th Workshop on Textual Case-Based Reasoning (2007)
16. Kass, A.M., Leake, D.B.: Case-based reasoning applied to constructing explanations. In: Proceedings of a Workshop on Case-Based Reasoning, pp. 190–208. Holiday Inn., Clearwater Beach (1988)
17. Kuhn, H.W.: The Hungarian method for the assignment problem. Naval Res. Logistics Q. **2**(1–2), 83–97 (1955)

18. Leacock, C., Miller, G.A., Chodorow, M.: Using corpus statistics and WordNet relations for sense identification. Comput. Linguist. **24**(1), 147–165 (1998)
19. Leake, D.B.: Evaluating Explanations: A Content Theory. L. Erlbaum Associates, Hillsdale (1992)
20. Leake, D.B., Birnbaum, L., Hammond, K., Marlow, C., Yang, H.: An integrated interface for proactive, experience-based design support. In: Proceedings of the 6th International Conference on Intelligent User Interfaces, pp. 101–108. ACM (2001)
21. Massie, S., Craw, S., Wiratunga, N.: Visualisation of case-base reasoning for explanation. In: Proceedings of the ECCBR 2004 Workshops, Madrid, pp. 135–144. Citeseer (2004)
22. Massie, S., Wiratunga, N., Craw, S., Donati, A., Vicari, E.: From anomaly reports to cases. In: Weber, R.O., Richter, M.M. (eds.) ICCBR 2007. LNCS (LNAI), vol. 4626, pp. 359–373. Springer, Heidelberg (2007). doi:10.1007/978-3-540-74141-1_25
23. McArdle, G., Wilson, D.: Visualising case-base usage. In: Workshop Proceedings ICCBR, pp. 105–114 (2003)
24. McSherry, D.: Explanation in recommender systems. Artif. Intell. Rev. **24**(2), 179–197 (2005)
25. Miller, G.A.: WordNet: a lexical database for english. Commun. ACM **38**(11), 39–41 (1995)
26. Orecchioni, A., Wiratunga, N., Massie, S., Chakraborti, S., Mukras, R.: Learning incident causes. In: Proceedings of the 4th Workshop on Textual Case-Based Reasoning (2007)
27. Plaza, E., Armengol, E., Ontañón, S.: The explanatory power of symbolic similarity in case-based reasoning. Artif. Intell. Rev. **24**(2), 145–161 (2005)
28. Recio-Garcıa, J.A., Dıaz-Agudo, B., González-Calero, P.A.: Textual CBR in jCOLIBRI: from retrieval to reuse. In: Proceedings of the ICCBR 2007 Workshop on Textual Case-Based Reasoning: Beyond Retrieval, pp. 217–226. Citeseer (2007)
29. Schank, R.C.: Explanation: a first pass. In: Kolodner, J., Riesbeck, C. (eds.) Experience, Memory, and Reasoning, pp. 139–165. Lawrence Erlbaum associates, Hillsdale (1986)
30. Schank, R.C., Kass, A.: Inside Case-Based Explanation. Psychology Press, Hove (1994)
31. Schank, R.C., Leake, D.B.: Creativity and learning in a case-based explainer. Artif. Intell. **40**(1), 353–385 (1989)
32. Sizov, G., Öztürk, P., Aamodt, A.: Evidence-driven retrieval in textual CBR: bridging the gap between retrieval and reuse. In: Hüllermeier, E., Minor, M. (eds.) ICCBR 2015. LNCS (LNAI), vol. 9343, pp. 351–365. Springer, Heidelberg (2015). doi:10.1007/978-3-319-24586-7_24
33. Sizov, G., Öztürk, P., Štyrák, J.: Acquisition and reuse of reasoning knowledge from textual cases for automated analysis. In: Lamontagne, L., Plaza, E. (eds.) ICCBR 2014. LNCS (LNAI), vol. 8765, pp. 465–479. Springer, Heidelberg (2014). doi:10.1007/978-3-319-11209-1_33
34. Tirunagari, S., Hanninen, M., Stanhlberg, K., Kujala, P.: Mining causal relations and concepts in maritime accidents investigation reports. Int. J. Innovative Res. Dev. **1**(10), 548–566 (2012)
35. Toolan, F., Carthy, J., Drummond, A., Dunnion, J.: Automating incident analysis: a challenge paper. In: 2nd Workshop on the Investigation and Reporting of Incidents and Accidents, Williamsburg, Virginia, United States, pp. 99–109 (2003)
36. Tsatsoulis, C., Amthauer, H.A.: Finding clusters of similar events within clinical incident reports: a novel methodology combining case based reasoning and information retrieval. Qual. Saf. Health Care **12**(Suppl 2), ii24–ii32 (2003)

# Relevance Matrix Generation Using Sensitivity Analysis in a Case-Based Reasoning Environment

Rotem Stram[1]([✉]), Pascal Reuss[1,2], Klaus-Dieter Althoff[1,2], Wolfram Henkel[3], and Daniel Fischer[3]

[1] Knowledge Management Group,
German Research Center for Artificial Intelligence, Kaiserslautern, Germany
{rotem.stram,pascal.reuss,klaus-dieter.althoff}@dfki.de
[2] Institute of Computer Science, Intelligent Information Systems Lab,
University of Hildesheim, Hildesheim, Germany
[3] Airbus Operations GmbH, Kreetslag 10, 21129 Hamburg, Germany
{wolfram.henkel,daniel.fischer}@airbus.com

**Abstract.** Relevance matrices are a way to formalize the contribution of each attribute in a classification task. Within the CBR paradigm these matrices can be used to improve the global similarity function that outputs the similarity degree of two cases, which helps facilitate retrieval. In this work a sensitivity analysis method was developed to optimize the relevance values of each attribute of a case in a CBR environment, thus allowing an improved comparison of cases. The process begins with a statistical analysis of the values in a given dataset, and continues with an incremental update of the relevance of each attribute.

The method was tested on two datasets and it was shown that the statistical analysis performs better than evenly distributed relevance values, making it a suitable initial setting for the incremental update, and that updating the values over time gives better results than the statistical analysis.

**Keywords:** Relevance matrix · Sensitivity analysis · Case-based reasoning · Classification · Retrieval · Similarity

## 1 Introduction

Relevance matrices are an important tool to represent the contribution of each attribute in a classification task. In this paper we will explore their use in the retrieval task of a case-based reasoning (CBR) system, which is closely related to classification. This work is a contribution to the OMAHA project, with the goal of creating a CBR system for aircraft fault diagnosis for Airbus [14].

The idea behind CBR is that similar problems have similar solutions. The process is made of four steps: retrieve, reuse, revise, and retain [1], and lies heavily on methods humans use to solve problems. When a person is faced with

© Springer International Publishing AG 2016
A. Goel et al. (Eds.): ICCBR 2016, LNAI 9969, pp. 402–412, 2016.
DOI: 10.1007/978-3-319-47096-2_27

a problem she first compares the current situation to past experience. This action is parallel to the retrieval phase of CBR. Next, she adapts her previous experience to the current task, much like the reuse stage of CBR, and applies it. Her actions are either successful, or unsuccessful (revise step). If her actions are successful, she will remember this solution for future reference (retain step).

A dataset of past experiences, called cases, each comprising of a problem description and a solution, is the foundation of every CBR system. In our scenario of fault diagnosis each case is built out of a fault description comprising of a set of attributes and their values, a diagnosis, and the actions taken to fix the problem, namely the solution. The main focus of this work is the retrieval phase, where a list of cases from the dataset, which are similar to a new fault description entered by the user, are retrieved.

At the base of the case retrieval is the similarity measure, which comes in two forms: local and global. Local similarity measures how similar two attribute values are to each other, while global similarity measures the similarity between two cases (more specifically, two problem descriptions) and is calculated by amalgamating the local similarities. When calculating the global similarity, weights can be assigned to each attribute in the amalgamation function, and here lies the crux of our task.

Our goal is to find the relevance matrix for the attributes used by the cases in our case base. Each case has a diagnosis, and each attribute in the fault description should have a different weight for different diagnoses. We want to set the weights so that for a retrieval task only cases with a relevant diagnosis will be deemed similar enough to be retrieved. Precision is therefore more important in our case than recall.

Similar work has been done in the past by Wess and Richter [7,8,10], under the PATDEX/2 system. In this work, three phases were defined to assign values to the relevance matrix: Initial Phase, where starting values are set according to statistical analysis of the attribute values, Training Phase, where the weight values are optimized for classification, and the Application Phase, where weights are constantly updated according to the changing case base. Since both Wess and Richter only used binary attributes, their work is inapplicable to our system in its current form, but instead is used as the basis of this paper, and further developments, mainly in the second phase, will be discussed in the coming sections.

The remainder of this paper is organized as follows. In the next section an overview of the related work, namely finding and assigning weights to features in the CBR environment, is described. In Sect. 3 we define the keywords and basic formulas, which are important for this work. Section 4 describes the method that was developed and used. Then, Sect. 5 provides information on the experiments that were run and their results. Finally, a short discussion of this work is provided, along with ideas of how it can be further developed and used withing the OMAHA project.

## 2    Related Work

There were several researches done about adjustment of feature weights in the past years, and it is still an important topic. Wettschereck and Aha compared different feature weighting methods and developed five dimensions to describe these methods: Model, weight space, representation, generality and knowledge [11]. According to their work, our approach uses a wrapper model to optimize the feature weights iteratively during the training phases. The weight space is continuous, because the features of our problem vary in their relevance for different diagnoses. Our knowledge representation is a structural case structure with attribute-value pairs and this given structure is used for feature weighting. We are using case specific weights to set the weights for each diagnosis individually. This way we are able to gain more precise results during the retrieval. Our approach for feature weighting is knowledge intensive, because we are using domain-specific knowledge to differ between individual diagnoses and setting case specific weights.

The approach from Richter and Wess introduces a so-called relevance matrix to deal with irrelevant symptoms in the PATDEX/2 diagnosis system. These relevances are determined in context of special situations. This means that for every diagnosis individual symptom relevances are set. These initial relevances are computed from a given set of cases and improved during training phases [7]. It is important to note that Richter and Wess used only the binary attribute type, and that our work generalizes the analysis to any attribute type.

Another approach for learning feature weights is from Armin Stahl. He presents a framework for learning of similarity measures, which is able to learn local and global similarity measures as well as feature weights. A so-called similarity teacher rates the utility of case pairs with respect to a given query to define the correct order of retrieved cases to the query based on the utility of a case. Based on this correct case order, a so-called similarity learner is able to adjust the similarity measures or feature weights to minimize the error during the retrieval [9].

Zhang and Quang describe an approach for a maintenance system with weight adjustments. They propose a three-layered architecture for a case structure: attribute-value pair layer, problem layer and solution layer. Between the attribute-value pairs and the problem and between the problem and the solution a set of weights can be defined. The feature weights are adjusted based on the feedback of a user, who selects an appropriate solution for a given problem. For each selected solution and the corresponding problem the weights are adjusted [12,13].

David Aha developed the Case-based Learning Algorithm 4 (CBL 4) as an approach to learn the importance of features. The algorithm sets initial feature weights and then learns the new feature weights during a training phase. A shortcoming of the CBL 4 algorithm is the missing of context consideration. CBL 4 can learn feature weights only for all cases and not case or context specific [2].

## 3   Definitions

The retrieval task of ĊBR is based on comparing a new problem description with descriptions stored in the case base, and retrieving only those cases that are *sufficiently similar*, meaning their similarity score is above a predefined threshold. We define a case base $CB$ as $CB = \{c_1, c_2, ..., c_n\}$ a set of cases $c_i$, $n$ number of cases. A case is a tuple $c_i = (desc, sol, diag)$, where: $c_i.desc$ a problem description, $c_i.sol$ the description of how the problem was solved, and $c_i.diag$ the diagnosis of the case, i.e. the problem type, which also functions as its class. The difference between the solution and the diagnosis is that the diagnosis is a cluster or set of very similar problems, which may have been solved in different ways. The problem description is a function that maps a set of attributes $A = a_1, a_2, ..., a_m$, $m$ number of attributes, to their values, so that $c_i.desc(a_j)$ is the value of attribute $a_j$ under the case $c_i$. The solution $c_i.sol$ is a string attribute, while $c_i.diag$ takes symbolic values.

A local similarity is defined as the similarity function of two attribute values

$$sim_{local}(c_i.desc(a_j), c_k.desc(a_j)) = sim_{local}(a_{ij}, a_{kj}) \tag{1}$$

The global similarity function $sim_{global}(c_i, c_k)$ compares two cases, and is defined as the amalgamation of the local similarities:

$$sim_{global}(c_i, c_k) =$$
$$amal(sim_{local}(a_{i1}, a_{k1}), sim_{local}(a_{i2}, a_{k2}),$$
$$..., sim_{local}(a_{im}, a_{km})) \tag{2}$$

Since some attributes are more important than others to determine the similarity of two cases, a weight for each local similarity can be assigned:

$$sim_{global}(c_i, c_k) =$$
$$amal(w_1 \cdot sim_{local}(a_{i1}, a_{k1}), w_2 \cdot sim_{local}(a_{i2}, a_{k2}),$$
$$..., w_m \cdot sim_{local}(a_{im}, a_{km})) \tag{3}$$

Here $w_j$ is the weight of attribute $a_j$. Another improvement on the similarity function is to give the same attributes different weights under different diagnoses, so that if $c_k.diag = d$, $d \in D$ we have:

$$sim_{global}(c_i, c_k) =$$
$$amal(w_{d1} \cdot sim_{local}(a_{i1}, a_{k1}), w_{d2} \cdot sim_{local}(a_{i2}, a_{k2}),$$
$$..., w_{dm} \cdot sim_{local}(a_{im}, a_{km})) \tag{4}$$

where $w_{dj}$ is the weight of attribute $a_j$ under diagnosis $d$. This means that different weights are used for the amalgamation when comparing against cases from different diagnoses. The reasoning behind this is that the same attributes may be differently important to determine membership of different diagnosis sets. From here we come to the relevance matrix (Table 1), which is nothing more than the weights assigned to each attribute under the different diagnoses.

**Table 1.** The relevance matrix

| Diagnosis \ Attribute | $a_1$ | $\ldots$ | $a_m$ |
|---|---|---|---|
| $d_1$ | $w_{11}$ | $\ldots$ | $w_{1m}$ |
| $\vdots$ | $\vdots$ | $\ddots$ | $\vdots$ |
| $d_s$ | $w_{s1}$ | $\ldots$ | $w_{sm}$ |

Our goal is to find the optimal relevance matrix, so that for each new retrieval task only the relevant cases will be retrieved. In our case, relevant cases are those who share the same diagnosis.

## 4    Method

Following Wess' three phases as first defined for the PATDEX/2 system and discussed in the introduction, we will focus on the first two: the initial phase and the training phase. His methods were updated to accommodate different types of attributes, and the similarity functions in use.

### 4.1    The Initial Phase

The purpose of the initial phase is to determine the starting weight values for the optimization that is taking place in the training phase. The initial weights are calculated using a statistical analysis of the attribute values.

**Symbolic Attributes.** We begin by looking at symbolic attributes, where there is a finite set of possible values. We define a diagnosis set $C_j \subseteq CB$ as the set of all cases $c_i$ with $c_i.diag = j$. Let $|D| = s$ be the number of possible diagnoses, $a \in A$ an attribute, and $B = \{b_1, b_2, ..., b_t\}$ the set of *possible* values for attribute $a$. We also set $b_{ji}$ to be the number of appearances of value $b_i$ in $C_j$ under the attribute $a$.

The relative weight of value $b_i$ in the diagnosis set $C_j$ is calculated as:

$$w_{ji} = \frac{b_{ji}}{\sum_{x=1}^{s} b_{xi}} \tag{5}$$

The impact of the value on the diagnosis set is then $V_{ji} = \frac{b_{ji}}{|C_j|}$. To calculate the weight of attribute $a$ under the diagnosis set $C_j$, the following formula is used:

$$W_{ja} = \sum_{x=1}^{|B|} V_{jx} \cdot w_{jx} \tag{6}$$

**General Attributes.** Since possible attribute values in general is not limited to a finite set, the formula to calculate the attributes' weights needs to be adjusted. Let $a \in A$ be defined as before, $B = \{b_1, b_2, ..., b_t\}$ the set of *used* values under $a$ in all diagnosis sets, $B_j \subseteq B$ is the set of values that appear in $C_j$, $b_{ji} \in B_j$, $r \in \mathbb{R}$ a similarity threshold. The relative weight of $b_{ji}$ is then

$$w_{ji} = \frac{|\{b_{jx}|b_{jx} \in B_j \wedge sim_{local}(b_{ji}, b_{jx}) \geq r\}|}{|\{b_x|b_x \in B \wedge sim_{local}(b_{ji}, b_x) \geq r\}|} \tag{7}$$

Since we regard each value as unique, the impact of each value on $C_j$ is $V_{ji} = \frac{1}{|B_j|}$. The total weight of attribute $a$ under $C_j$ is

$$W_{ja} = \frac{1}{|B_j|} \cdot \sum_{i=1}^{|B_j|} w_{ji} \tag{8}$$

## 4.2   The Training Phase

After determining the initial weights our job is not done. In order to optimize the relevance matrix we need to train the system. For a given query a list of cases is retrieved, and sorted according to their similarity to the query, as given by the similarity function $sim_{global}$. We choose a threshold $\xi$, such that if $sim_{global}(query, case) \geq \xi$ the diagnosis of both *query* and *case* should be the same. In case $query.diag \neq case.diag$ and $sim_{global}(query, case) = \xi + \Delta$, we have a false positive results, which can be attributed to one of the following reasons [10]:

1. Both *query* and *case* contain the same problem description, such that $query.desc = case.desc$
2. The threshold $\xi$ is too low
3. The weight of the similar attributes of *query* and *case* is too high
4. The weight of the dissimilar attributes of *query* and *case* is too low

If reason 1 is the source of the false positive result and the problem descriptions are identical then we either have an inconsistency problem and *case* should be removed from the case base, or *case* is incomplete and should be updated. Neither of these scenarios are within the scope of this work. We will ignore reason 2 since we want to keep a fixed threshold, and will so focus on reasons 3 and 4. Our goal is then to strengthen the differences between *query* and *case*, and weaken their similarities.

Consider the following scenario: We have a query case $q$ and a retrieved case $c$ with $c.diag = d$, $q.desc \neq c.desc$ and

$$sim_{global}(q, c) = \xi + \Delta =$$
$$w_{d1} \cdot sim_{local}(a_{q1}, a_{c1}) + w_{d2} \cdot sim_{local}(a_{q2}, a_{c2}) +$$
$$... + w_{dm} \cdot sim_{local}(a_{qm}, a_{cm}) \tag{9}$$

We define a local similarity threshold $\xi'$ for the local similarity measure, and the following sets: $S = \{i|sim_{local}(a_{qi}, a_{ci}) \geq \xi'\}$ and $NS = \{j|sim_{local}(a_{qj}, a_{cj}) < \xi'\}$ the sets of those attributes which are similar and those that are not similar, respectively. We want to update the similarity measure so that $sim_{global}(q, c) \leq \xi$. In order to do so, the similarity formula is updated in the following way:

$$sim_{global}(q, c) =$$

$$\sum_{i \in S}(w_{di} - z_i) \cdot sim_{local}(a_{qi}, a_{ci}) + \sum_{j \in NS}(w_{dj} + y_j) \cdot sim_{local}(a_{qj}, a_{cj}) = \xi \quad (10)$$

This way the weight of similar attributes is reduced, while the weight of dissimilar attributes is increased for the diagnosis set of the retrieved case, thus increasing the dissimilarity between the two cases. In order to update the weights we need to find the values of $z_i$ and $y_i$. For simplicity reasons we set $\forall_{z_i, z_j} z_i = z_j, \forall_{y_i, y_j} y_i = y_j$ and $\sum_{i \in S} z_i = \sum_{j \in NS} y_j = Z$. This means that we need to find a value $Z$, such that

$$z_i = \frac{Z}{|S|}, y_i = \frac{Z}{|NS|} \quad (11)$$

This value is calculated with the help of the following formula:

$$Z = \frac{\Delta \cdot |S| \cdot |NS|}{|NS| \cdot \sum_{i \in S} sim_{local}(a_{qi}, a_{ci}) - |S| \cdot \sum_{j \in NS} sim_{local}(a_{qj}, a_{cj})} \quad (12)$$

The derivation of formula 12 is: let $t = sim_{global}(q, c) - \Delta$, $x_i = sim_{local}(a_{qi}, a_{ci})$, then:

$$\sum_{i \in S} x_i \cdot w_{di} + \sum_{j \in NS} x_j \cdot w_{dj} = t + \Delta \quad \Longrightarrow$$

$$\sum_{i \in S} x_i \cdot (w_{di} - z_i) + \sum_{j \in NS} x_j \cdot (w_{dj} + y_j) = t \quad \Longrightarrow$$

$$\sum_{i \in S} x_i \cdot (w_{di} - \frac{Z}{|S|}) + \sum_{j \in NS} x_j \cdot (w_{dj} + \frac{Z}{|NS|}) = t \quad \Longrightarrow$$

$$\sum_{i \in S}(x_i \cdot w_{di} - \frac{x_i \cdot Z}{|S|}) + \sum_{j \in NS}(x_j \cdot w_{dj} + \frac{x_j \cdot Z}{|NS|}) = t \quad \Longrightarrow$$

$$\sum_{i \in S} x_i \cdot w_{di} + \sum_{j \in NS} x_j \cdot w_{dj} - \frac{Z}{|S|} \cdot \sum_{i \in S} x_i + \frac{Z}{|NS|} \cdot \sum_{j \in NS} x_j = t \quad \Longrightarrow$$

$$t + \Delta - Z \cdot (\frac{\sum_{i \in S} x_i}{|S|} - \frac{\sum_{j \in NS} x_j}{|NS|}) = t \quad \Longrightarrow$$

$$\Delta = Z \cdot \frac{|NS| \cdot \sum_{i \in S} x_i - |S| \cdot \sum_{j \in NS} x_j}{|S| \cdot |NS|} \quad \Longrightarrow$$

$$Z = \frac{\Delta \cdot |S| \cdot |NS|}{|NS| \cdot \sum_{i \in S} x_i - |S| \cdot \sum_{j \in NS} x_j} \quad Q.E.D$$

For a falsely retrieved case, the diagnosis set's weights are updated as follows:

1. Find the sets $S$ and $NS$
2. Calculate $Z$ according to formula (12)
3. Calculate $z_i$, $y_j$ according to formula (11)
4. $\forall_{i \in S} w_i' = w_i - z_i$, $\forall_{j \in NS} w_j' = w_j + y_j$
5. Set all $w_i'$, $w_j'$ as the new weights of the respective attributes for the diagnosis set of the retrieved case

One epoch of the training procedure would then go as follows:

- For each case c in the case base
  - Build a query $q$ from the problem description of $c$
  - Use $q$ to retrieve similar cases
  - For each retrieved case $c_i$, $sim_{global}(q, c_i) > \xi$, $c_i.desc \neq c.desc$
    - Update the weights of the diagnosis set of $c_i.desc$

There are two main problems with this procedure; First, the weight update sequence depends both on the order of the queries and on the order of retrieved cases as the weights are updated. Another problem is that after the first weight update for a query $q$, the second update for a second falsely retrieved case may not be relevant any longer, as the weights of that diagnosis set may have already changed. A renewed retrieval for the same query after each weight change is computationally too expensive and might even produce an endless loop. In order to overcome these problems the training phase is changed so that weights are updated only once per query.

In the new version, an average $Z$ value, $aveZ$, is calculated for each diagnosis set as the mean $Z$ from all falsely retrieved items of this set within an entire epoch. The local update values are then: $z_i = \frac{aveZ}{|S|}$, $y_j = \frac{aveZ}{|NS|}$, where $|S|$ and $|NS|$ are the number of attributes that appeared more times in $S$ or in $NS$ respectively.

The attribute weights for each set are then updated only once per epoch: For each attribute $a_i$, let $|S_i|$, $|NS_i|$ be the number of retrieved cases of the diagnosis set where $a_i$ appeared under $S$ or $NS$ respectively. The weight of $a_i$ in the diagnosis set is then updated as follows:

$$
w_i' = \begin{cases}
w_i - z_i & \text{if } |NS_i| = 0 \\
w_i + y_i & \text{if } |S_i| = 0 \\
w_i - (1 - \frac{|S_i|}{|NS_i|}) \cdot z_i & \text{if } |S_i| < |NS_i| \\
w_i - (1 - \frac{|NS_i|}{|S_i|}) \cdot y_i & \text{if } |NS_i| < |S_i|
\end{cases}
\tag{13}
$$

This phase is closely related to the back-propagation training of an artificial neural network (ANN). In each retrieval or classification task the contribution of each attribute, much like each neuron, to the error is assessed, and its weight, similarly to activation weights, is updated accordingly.

## 5    Evaluation and Results

To test the newly developed method we used two datasets from the literature, since an aircraft-related one is not yet readily available. Both datasets were taken from the classification discipline, as each item is assigned to a class, something that is lacking in available CBR data.

The method was implemented in the Java programming language and relied heavily on the myCBR tool [3] both in the creation of the case bases, and for retrieval. The datasets which were used are the *Car Evaluation Data Set* [4] and the *Yeast Data Set* [5], both obtained from the UC Irvine Machine Learning Repository[1], and from here on referred to as Cars and Yeast respectively. These datasets were chosen thanks to the ease of converting them into a case base, and their size, which allowed testing in a timely manner.

The method's thresholds were set to the following values: $\xi = \xi' = 0.75$. The settings for the local similarities were set with the distance function "DIFFERENCE" and type "POLYNOMIAL_WITH".

Since both datasets have a relatively low number of data points, it was necessary to perform cross validation for the training phase. Each dataset was randomly divided into four subsets, and the initialization and training phases were performed four times on the training set, each time using a different subset as the test set.

Before the training phase two tests were performed: first, with a static and uniform weight for all attributes, and second using the weights from the initial phase. Following this, the weights were trained for 15 epochs on the training set. The recall and precision results of each epoch was documented, and the F-measure was calculated with $\beta = 0.3$, giving a heavier weight to the precision. The results can be seen in Fig. 1, and show the F-measure values as calculated from the test set.

As can be seen, the F-measure value of the Cars dataset was higher for the initial weights than the static ones. When the training phases begins the retrieval performance decreases drastically, and then slowly picks up and reaches values higher than those of the initial phase. The situation of the Yeast dataset is different, as the peak of the F-measure is reached after the first training epoch, and values then slowly decrease and approaches what seams to be a saturation point that is above both the static and the initial phase. What can be learned from these results is that the statistical analysis improves performance over the evenly distributed weights, while training improves performance over the statistical analysis.

## 6    Discussion and Future Work

Sensitivity analysis was used in this work to optimize the relevance matrix, which represents the importance of each attribute in a global similarity. The optimization method was developed as part of an aircraft fault diagnosis CBR

---

[1] http://archive.ics.uci.edu/ml.

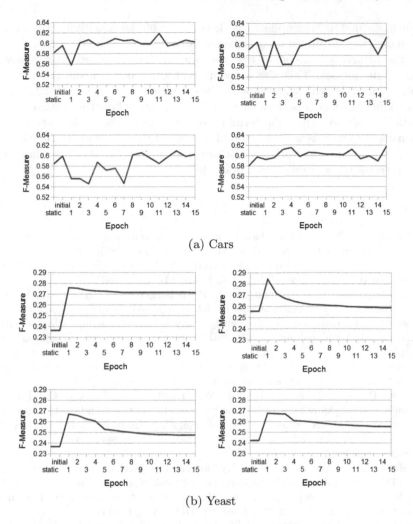

(a) Cars

(b) Yeast

**Fig. 1.** F-measure as a function of each setting and epoch. Each graph shows results of one cycle of cross-validation

tool, with the goal of improving the retrieval of past cases similar to a query case.

With the help of two datasets from the literature, the method was tested on three different settings: static and evenly distributed relevance values, initial statistically analyzed values, and 15 epochs of optimization. Performance of the different settings was measured with the help of the F-Measure, giving higher importance to precision than to recall.

It was shown that the initial values, obtained by statistical analysis, allowed the retrieval to perform better than the evenly distributed values. However, when the relevance matrix was updated with the help of the sensitivity analysis, the performance was improved even more for both datasets.

In the future, we intend to test the optimization method on a real-world aircraft fault dataset. This was not done yet since no such dataset is readily available, but this will change in the near future, as one is currently in the making. Once the test is performed we will better know how the system behaves, and will be able to adjust the termination criteria of the optimization cycles. The sensitivity analysis will be instated as part of the toolchain of the OMAHA project, which is a framework to transform a semi-structured dataset into cases, and to retrieve cases relevant to a query [6].

# References

1. Aamodt, A., Plaza, E.: Case-based reasoning: foundational issues, methodological variations, and system approaches. AI Commun. **7**(1), 39–59 (1994)
2. Aha, D. W. Case-based learning algorithms. In: Proceedings of the 1991 DARPA Case-Based Reasoning Workshop, vol. 1, pp. 147–158 (1991)
3. Bach, K., Sauer, C. S., Althoff, K. D., Roth-Berghofer, T.: Knowledge modeling with the open source tool myCBR. In: KESE@ ECAI (2014)
4. Bohanec, M., Rajkovic, V.: Knowledge acquisition and explanation for multi-attribute decision making. In: 8th Intl Workshop on Expert Systems and their Applications, pp. 59-78 (1988)
5. Horton, P., Nakai, K.: A probabilistic classification system for predicting the cellular localization sites of proteins. In: Ismb, vol. 4, pp. 109–115 (1996)
6. Reuss, P., Althoff, K.-D., Henkel, W., Pfeiffer, M., Hankel, O., Pick, R.: Semi-automatic knowledge extraction from semi-structured and unstructured data within the OMAHA project. In: Hüllermeier, E., Minor, M. (eds.) ICCBR 2015. LNCS (LNAI), vol. 9343, pp. 336–350. Springer, Heidelberg (2015). doi:10.1007/978-3-319-24586-7_23
7. Richter, M.M., Wess, S.: Similarity, uncertainty and case-based reasoning in PATDEX. In: Automated Reasoning, pp. 249-265. Springer, Heidelberg (1991)
8. Richter, M.M.: Classification and learning of similarity measures. In: Information and Classification, pp. 323–334. Springer, Heidelberg (1993)
9. Stahl, A.: Learning feature weights from case order feedback. In: Aha, D.W., Watson, I. (eds.) ICCBR 2001. LNCS (LNAI), vol. 2080, pp. 502–516. Springer, Heidelberg (2001). doi:10.1007/3-540-44593-5_35
10. Wess, S.: Fallbasiertes problemlösen in wissensbasierten systemen zur entscheidungsunterstützung und diagnostik. Ph.D thesis. TU Kaiserslautern (1995)
11. Wettschereck, D., Aha, D.W.: Weighting features. In: Veloso, M., Aamodt, A. (eds.) ICCBR 1995. LNCS, vol. 1010, pp. 347–358. Springer, Heidelberg (1995). doi:10.1007/3-540-60598-3_31
12. Zhang, Z., Yang, Q.: Towards lifetime maintenance of case base indexes for continual case based reasoning. In: Giunchiglia, F. (ed.) Artificial Intelligence: Methodology, Systems, and Applications. LNCS, vol. 1480, pp. 489–500. Springer, Heidelberg (1998)
13. Zhang, Z., Yang, Q. Dynamic refinement of feature weights using quantitative introspective learning. In: IJCAI, pp. 228–233 (1999)
14. German Aerospace Center - DLR, LuFo-Projekt OMAHA gestartet, http://www.dlr.de/lk/desktopdefault.aspx/tabid-4472/15942_read-45359/

# Combining Case-Based Reasoning
# and Reinforcement Learning for Tactical Unit
# Selection in Real-Time Strategy Game AI

Stefan Wender[✉] and Ian Watson

The University of Auckland, Auckland, New Zealand
{s.wender,ian}@cs.auckland.ac.nz

**Abstract.** This paper presents a hierarchical approach to the problems inherent in parts of real-time strategy games. The overall game is decomposed into a hierarchy of sub-problems and an architecture is created that addresses a significant number of these through interconnected machine-learning (ML) techniques. Specifically, individual modules that use a combination of case-based reasoning (CBR) and reinforcement learning (RL) are organised into three distinct yet interconnected layers of reasoning. An agent is created for the RTS game StarCraft and individual modules are devised for the separate tasks that are described by the architecture. The modules are individually trained and subsequently integrated in a micromanagement agent that is evaluated in a range of test scenarios. The experimental evaluation shows that the agent is able to learn how to manage groups of units to successfully solve a number of different micromanagement scenarios.

**Keywords:** CBR · Reinforcement learning · Game AI · Layered learning

## 1 Introduction

An area that has always been at the forefront of interesting AI utilization is games. Games provide a fertile breeding ground for new approaches and an interesting and palpable test area for existing ones. And as games such as checkers and chess are devised as high-level abstractions of mechanisms and processes in the real world, creating AI that works in these games can eventually lead to AI that solves real-world problems.

One of the most popular genres of computer video games is real-time strategy (RTS). RTS is a genre of computer video games in which players perform simultaneous actions while competing against each other using combat units. Often, RTS games include elements of base building, resource gathering and technological developments and players have to carefully balance expenses and high-level strategies with lower-level tactical reasoning. RTS games incorporate many different elements and are related to areas such as robotics and military simulations. RTS games can be very complex and, especially given the real-time

© Springer International Publishing AG 2016
A. Goel et al. (Eds.): ICCBR 2016, LNAI 9969, pp. 413–429, 2016.
DOI: 10.1007/978-3-319-47096-2_28

aspect, hard to master for human players. Since they bear such a close resemblance to many real-world problems, creating powerful AI in an RTS game can lead to significant benefits in addressing those related real-world tasks.

The creation of powerful AI agents that perform well in computer games is made considerably harder by the enormous complexity these games exhibit. The complexity of any board game or computer game is defined by the size of its state- and decision space. A state in chess is defined by the position of all pieces on the board while the possible actions at a certain point are all possible moves for these pieces. [14] estimated the number of possible states in chess as $10^{43}$. The number of possible states in RTS games is vastly bigger. [2] estimated the decision-complexity of the *Wargus* RTS game (i.e. the number of possible actions in a given state) to be in the 1,000s even for simple scenarios that involve only a small number of units. StarCraft, a pioneering commercial RTS game from 1998, is even more complex than Wargus, with a larger number of different unit types and larger combat scenarios on bigger maps, leading to more possible actions. [20] estimated the number of possible states in StarCraft, defined through hundreds of possible units for each player on maps that can have maximum dimension of $256 \times 256$ tiles, to be in excess of $10^{11500}$. In comparison, chess has a decision complexity of about 30.

The topic of this paper is the creation of an agent that focuses on the tactical and reactive tasks in RTS games, the so-called 'micromanagement'. Our agent architecture is split into several interconnected layers that represent different levels of the decision making process. The agent uses a set of individual CBR/RL modules on these different levels of reasoning in a fashion that is inspired by the layered learning model [16]. The combination of CBR and RL that is described in this paper is performed in order to enable the agent to address more complex problems by using CBR as an abstraction- and generalisation-technique.

## 2   Related Work

Creating the overall model as well as the individual sub-components of the architecture was influenced by previous research that evaluated the suitability of RL for the domain [21] and a combination of CBR and RL for small-to-medium-sized micromanagement problems [23].

**Reinforcement Learning.** The application of RL algorithms in computer game AI has seen a big increase in popularity within the past decade, as RL is very effective in computer games where perfect behavioural strategies are unknown to the agent, the environment is complex and knowledge about working solutions is usually hard to obtain. Recently, the *UCT* algorithm (*Upper Confidence Bounds applied to Trees*) [9], an algorithm based on Monte-Carlo Tree Search (MCTS), has lead to impressive results when applied to games. MCTS and UCT are closely related to RL which is partially based on Monte-Carlo methods.

[7] overcame this and described the use of heuristic search to simulate combat outcomes and control units accordingly. Because of the aforementioned lack in speed and precision of the StarCraft game environment, the authors first

created their own simulator, *SparCraft*, to evaluate their approach and later re-integrate the results into a game-playing agent. Apart from MCTS and UCT however, few of the new theoretical discoveries in RL have made it into game AI research. Most research in computer game AI, including this paper, works with the well-tested temporal difference (TD) RL algorithms such as Q-learning [19]. Q-learning integrates different branches of previous research such as dynamic programming and trial-and-error learning into RL. [3] extended an online Q-learning technique with CBR elements in order for the agent to adapt faster to a change in the strategy of its opponent. The resulting technique, *CBRetaliate*, tried to obtain a better matching case whenever the collected input reading showed that the opponent was outperforming it. As a result of the extension, the CBRetaliate agent was shown to significantly outperform the Q-learning agent when it came to sudden changes in the opponent's strategy.

**Case-Based Reasoning and Hybrid Approaches.** Using only RL for learning diverse actions in a complex environment quickly becomes infeasible and additional modifications such as ways of inserting domain knowledge or combining RL with other techniques to offset its shortcomings are necessary.

Combining CBR with RL has been identified as a rewarding hybrid approach [5] and has been done in different ways for various problems.

[8] extended the standard *GDA* algorithm presented in [12] into *Learning GDA*. LGDA was created by integrating CBR with RL, i.e. the agent tried to choose the best goal, based on the expected reward. While the integration of CBR and RL differs from the approach pursued in the CBR/RL modules in this paper, the online acquisition of knowledge using a CBR/RL approach is similar.

[11] described the integration of CBR and RL in a continuous environment to learn effective movement strategies for units in a RTS game. This approach was unique in that other approaches discretize these spaces to enable machine learning. As a trade-off for working with a non-discretized model, the authors only looked at the movement component of the game from a meta-level perspective where orders are given to groups of units instead of individuals and no orders concerning attacks are given.

An example of an approach which obtains knowledge directly from the environment is [4]. The authors used an iterative learning process that is similar to RL and employed that process and a set of pre-defined metrics to measure and grade the quality of newly-acquired knowledge while performing in the RTS game *DEFCON*. Similar to this approach, the aim in this paper and the CBR/RL modules created as part of it is to acquire knowledge directly through interaction with the game. The learning process is controlled by RL which works well in this type of unknown environment without previous examples of desired outcomes. CBR is then used for managing the acquired knowledge and generalising over the problem space.

**Hierarchical Approaches and Layered Learning.** Combining several ML techniques, such as CBR and RL, into hybrid approaches leads to more powerful techniques that can be used to address more complex problems. However, problems such as those simulated by commercial RTS games with many actors in diverse environments still need significant abstraction in order for agents to solve the problems they are confronted with. A common representation of the problems that are part of RTS games is in a hierarchical architecture [13].

An early application of hierarchical reasoning in RTS games was described in [6], where planning tasks in RTS games are divided into a hierarchy of three different layers of abstraction. This is similar to the structure identified in the next section, with separate layers for unit micromanagement, tactical planning in combat situations and high-level strategic planning. The authors used *MCPlan*, a search/simulation based Monte Carlo planning algorithm, to address the problem of high-level strategic planning.

Layered learning (LL) was devised for computer robot soccer, an area of research that pursues similar goals as RTS games and can be regarded as a simplified version of these combat simulations [16]. The main differences between the two are the less complex domain and less diverse types of actors in computer soccer. Additionally, computer soccer agents often compute their actions autonomously while RTS game agents orchestrate actions between large numbers of objects [13]. Because of the many similarities, LL makes an excellent, though as of now mostly unexplored, paradigm for a machine learning approach to RTS game AI. [10] combine both original and a concurrent LL approach [24] to create *overlapping layered learning* for tasks in the simulated robotic soccer domain. The original paradigm froze components once they had acquired learning for their tasks. The concurrent paradigm purposely kept them open during learning subsequent layers, thus finding a middle ground between freezing each layer once learning is complete and always leaving previously learned layers open.

## 3   A Hybrid Hierarchical CBR/RL Architecture

The hierarchical architecture and its constituent separate modules that address the micromanagement problem in RTS games are based on previous approaches described in [21,23]. Subdividing the problem enables a more efficient solution than when addressing the problem on a single level of abstraction, something which would either result in case representations which are too complex to be used for learning in reasonable time, or that require such a high level of abstraction that it prevents any meaningful learning process.

The structure of the core problems inherent in RTS games such as StarCraft, shown in Fig. 1, leads to most RTS agents being hierarchical [13]. The architecture we devised covers the micromanagement component of the game, enclosed in the solid red square shown in Fig. 1. Reconnaissance is currently not part of the framework, as the CBR/RL agent only works with units which are already visible.

Based on this task decomposition, three distinct organisational layers are identified. The *Tactical Level* is the highest organisational level and represents the entire world the agent has to address, i.e. the entire battlefield and the entire solid red square in the figure. The *Squad Level* is indicated by the dotted green square. Sub-tasks represented here concern groups of units, potentially spread over the entire battlefield. Finally, the *Unit Level* is the bottommost layer. This

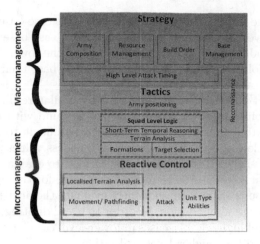

**Fig. 1.** RTS micromanagement tasks

layer covers pathfinding, works on a per-unit basis and is denoted by the dashed blue square in the diagram. Translating this layered problem representation into a CBR/RL architecture is done through a number of hierarchically interconnected case-bases. The approach to hierarchical CBR here is strongly inspired by that in [15], which describes a hierarchical CBR (HCBR) system for software design. One major difference between the approach described here and the one in [15] is that the use of RL for updating fitness values in the hierarchically interconnected case-bases means that each case-base has its own Adaptation-part of the CBR cycle [1]. Figure 2 shows the case-bases resulting from modeling the problem in this hierarchical fashion. Both the tactical level and the unit level are represented by a single case-base. The unit level is only responsible

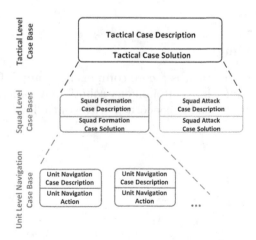

**Fig. 2.** Hierarchical structure of the case-bases

for *Navigation*. The intermediate squad level has one case-base for two possible actions on that level, *Attack* and *Formation*. Each case-base is part of a distinct CBR/RL module. Higher levels can then use the lower level components to interpret their solutions. As a result, higher levels base their learning process on the knowledge previously acquired on lower levels. RL relies in its learning process on the fact that similar actions lead to similar results. Otherwise the learning process continues until a stable policy is found with non-changing fitness values for state-action pairs. This would be difficult to achieve within a reasonable time if lower-level case-bases change fitness values at the same time as higher-level case-bases. Therefore, it was decided to evaluate and train lower level components first, retain the acquired knowledge for the respective tasks in the appropriate case-bases and subsequently evaluate the next-higher level using the lower-level cases as a foundation. In order to avoid diluting the learning- and evaluation process of higher levels, cases in lower-level case-bases are not changed once they are reused by a higher-level evaluation.

This evaluation and training procedure is not ideal since it partially negates the online learning characteristic of the CBR/RL agent. However, the alternative is a very noisy learning process that would seriously complicate the use of RL.

## 4    Lower-Level Modules

The individual modules that make up the overall architecture all follow a similar design and use a hybrid CBR/RL approach [23]. This section sums up the three lower-level modules (*Pathfinding*, *Attack* and *Formation*) and the MDP framework that is created for them [17]. All modules use a Q-learning algorithm to learn how to maximise the rewards for their respective tasks. Structure and implementation of the module for *Tactical Unit Selection* is described in detail in the next section. Underlying the decomposition into the modules described here is the analysis of tasks that are relevant to micromanagement in RTS games as displayed in Fig. 1.

### 4.1    Unit Pathfinding

Unit navigation and movement is a core component of any RTS game and also extends to other areas such as autonomous robotic navigation. This module is described in detail in [22] and is concerned with controlling a single agent unit (Table 1).

## States

**Table 1.** Navigation case-base summary

| Attribute | Description |
|---|---|
| Agent unit IM | Map with $7 \times 7$ fields containing the damage potential of adjacent allied units. |
| Enemy unit IM | Map with $7 \times 7$ fields containing the damage potential of adjacent enemy units. |
| Accessibility IM | Map with $7 \times 7$ fields containing true/false values about the accessibility. |
| Unit type | Type of a unit. |
| Last unit action | The last movement action taken. |
| Target position | Target position within the local $7 \times 7$ map |

**Actions.** The case solutions are concrete game actions. There currently are four *Move* actions for the four different cardinal directions, i.e. one for every $90°$.

**Reward Signal.** The compound reward $\mathcal{R}_{ss'}^{a}$, that is computed after finishing an action is based on damage taken during the action $\Delta h_{unit}$, the time the action took $t_a$ and the change in distance to the chosen target location $\Delta d_{target}$.

$$\mathcal{R}_{ss'}^{a} = \Delta h_{unit} - t_a + \Delta d_{target}.$$

### 4.2  Squad-Level Coordination

Squad-level modules define and learn how to perform actions that coordinate groups of units while re-using the pathfinding component on the lowest level of the architecture.

**Unit Formations.** Tactical formations are an important component in RTS games, which often resemble a form of military simulator and are heavily inspired by real-life combat strategy and tactics. The *Formation* module creates formations that are a variant of dynamic formations [18] and learn through CBR/RL the best unit-slot associations, i.e. which slot in the formation a certain unit is assigned to (Table 2).

**States**

Table 2. Formation state case description

| Category | Attribute | Type |
|---|---|---|
| Index | Unit # Agent | Integer |
| Unit | Type | Enum |
| | Health | Integer |
| | Position | Integer |
| Opponent | Attacking Damage towards the Formation Center from each of the 8 (Inter)Cardinal Directions | Integer |

**Actions.** Actions are an assignment of the controlled units to certain slots in the formation. This means that the available actions are basically a permutation of all available units over all available formation slots.

**Reward Signal.** The two main criteria for an effective formation-forming action were decided to be the speed with which the action is executed $t_{form}$ and, weighted slightly higher, the potential damage that units in the formation can deliver at any one point in time $d_{avg}$.

$$r_{form} = 1.5 * d_{avg} - t_{form}.$$

**Unit Attack.** The goal of using attacking units in the most efficient way is to focus on a specific opponent unit in order to eliminate it and, as a result, also eliminate the potential damage it can do to agent units. As part of this *Attack* component, it also was decided to simplify the module by giving all agent units assigned to a single *Attack* action the same target. More complex attacking behaviour can then be created by queuing several *Attack* action after another (Table 3).

**States**

Table 3. Attack state case description

| Category | Attribute | Type |
|---|---|---|
| Index | Units Opponent | Integer |
| Target Unit | Type | Enum |
| | Health | Integer |
| | Average Distance to Attackers | Integer |
| Agent | Combined Attacking Unit Damage | Integer |

**Actions.** The potential case solutions/actions for attack cases are the attack targets. This means that there is one solution for each attack target/enemy unit.

**Reward Signal.** The reward signal is composed of components for the time it takes to finish the attack action $t_{att}$, the damage done to the target $dam$ as well as the damage removed if the target is eliminated $dam_{elim}$.

$$r_{att} = dam + dam_{elim} - t_{att}.$$

**Unit Retreat.** While also a selectable action like Attack and Formation, *Retreat* does not use CBR/RL and thus doesn't have its own module in Fig. 2. The *Retreat* action is designed to avoid potential sources of damage. The *Retreat* action takes into account a larger area of the immediate surroundings of a unit when compared to these other actions, a $15 \times 15$ plot, compared to $7 \times 7$ used for pathfinding. In a two-step process that also takes into account the influence of neighbouring plots, the action selects the area with the lowest amount of enemy influence/damage potential.

## 5    Tactical Unit Selection

The *Tactical Unit Selection* component is structured in a way similar to that of lower-level components, based on a hybrid CBR/RL integration. Given the decomposition of the problem as described in Fig. 1, the task of the *Tactical Unit Selection* component is to find an ideal distribution of units among the three different modules on the level below, i.e. *Formation*, *Attack* and *Retreat*.

One major simplification that was introduced in order to avoid increasing the number of possible solutions exponentially and making learning infeasible with the current model is that all units assigned to *Attack* or *Formation* actions will perform the same action. This means that any unit assigned to an attack will attack the same target. Any unit assigned to a formation, will be part of the same formation.

### 5.1    Tactical Decision Making Model

The model used for the *Tactical Unit Selection* module, similar to those for *Formation* and *Attack* components, describes the problem in terms of an MDP. As this problem integrates the three lower-level modules, the model also combines elements of these modules.

**States.** *Tactical Unit Selection* states (or cases) are basically a combination of *Attack* and *Formation* states. However, some of the attributes that those state models use are part of both *Attack* and *Formation*, while others contain the same information but in less detail.

**Table 4.** Tactical state case description

| Category | Attribute | Type |
|---|---|---|
| Index | Units Agent | Integer |
| | Units Opponent | Integer |
| Unit Agent | Type | Enum |
| | Health | Integer |
| | Damage | Integer |
| | Quadrant | Integer |
| | Cooldown | Boolean |
| Unit Opponent | Type | Enum |
| | Health | Integer |
| | Damage | Integer |
| | Quadrant | Integer |
| | Average Distance | Integer |

The resulting composition of the case description of a *Tactical Unit Selection* state can be seen in Table 4.

Opponent units have two attributes containing different information (direction versus distance, relative to agent units) that indicate their position: *Quadrant* and *AverageDistance*. Agent units also have the *Quadrant* attribute to indicate their position relative to each other. The Boolean *Cooldown* value indicates if a unit's weapon is currently in cooldown or if it can be used. *Type* only distinguishes among *Melee*, *Ranged* and *Air* instead of specific unit types.

Given this composition, the dimensionality of the case description is considerably higher than for previous modules. For example, in a scenario with $n_a = 4$ agent units and $n_o = 5$ opponent units, case descriptions have $2 + 4*5 + 5*5 = 47$ attributes.

**Actions.** *Tactical Unit Selection* case solutions are distributions of the available agent units among the three available actions, i.e. triples $(n_a, n_f, n_r)$ that indicate how many units are assigned to each action type. The overall number of solutions for $n$ units distributed among the three categories is thus $\binom{3+n-1}{n}$. Given five agent units, the possible distributions for (*Attack*,*Formation*,*Retreat*) can be (5,0,0), (4,1,0) ... (0,0,5). For $n = 5$ units the number of solutions is therefore $\binom{3+n-1}{n} = 21$. This definition leads to a requirement for limiting the number of controlled units, if the number of learning episodes is to remain reasonable. The maximum number of agent and opponent units used in the evaluation scenarios was set to *ten*. By allowing a maximum of ten agent units in a game state, a single case can have at most $\binom{12}{10} = 66$ possible solutions.

**Reward.** The reward signal contains a negative component $t_{tac}$ for the time it takes for a *Tactical Unit Selection* action to complete, a negative component $dam_{opp}$ for the damage that agent units received while performing the last action and two positive components, $dam_{ag}$ for the damage done by agent units as well as $dam_{elim}$ for the summed-up damage potential of all opponent units eliminated during the last action. Additionally, a third negative component $dam_{loss}$ is added: this represents the damage potential lost when an agent unit is eliminated.

$$r_{tac} = dam_{ag} + dam_{elim} - dam_{opp} - dam_{loss} - t_{tac}.$$

Overall, the agent should attempt to choose solutions which eliminate opposing units quickly, while sustaining no (or only very little) damage to its own units.

## 5.2   CBR/RL Algorithm

Figure 3 shows a graphical representation of the steps and components involved in assigning actions to the available units. The algorithm chooses, from top to bottom, a *Tactical Unit Selection* unit distribution and, based on this distribution, an attack target, a formation unit-to-slot assignment as well as retreat destinations. Using the unit destinations computed through the lower-level components, the *Navigation* component then manages the unit movement. There

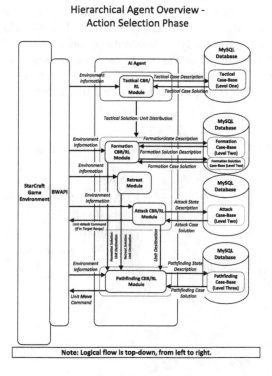

**Fig. 3.** Action selection using hierarchical CBR/RL for unit micromanagement

can be several *Navigation* actions until a unit reaches the destination assigned to it by one of the higher-level modules. There is always at most one action for *Attack*, *Formation* and *Retreat*, or zero, if no unit is assigned to a specific action category. The overall *Tactical Unit Selection* action is finished once all modules on lower levels indicate they are finished with their tasks.

## 6     Experimental Setup and Evaluation

Depending on the choice of parameters, large numbers of episodes can be required for finding optimal policies. Since this can easily become prohibitive if complex scenarios are used, a first step is an analysis of the case-base behaviour in a subset of the test scenarios, to find an appropriate threshold $\psi$ that determines how similar a retrieved case in the CBR component has to be. Using a low $\psi$ would mean that fewer cases are required to cover the entire case-space. However, this might lead to the retrieval of non-matching cases for a given situation and thus to sub-optimal performance due to a bad solution. Therefore, the selected $\psi$ should lead to an optimal trade-off between performance and learning time. A number of representative micromanagement combat situations were created for the evaluation, each one with the aim to win the overall scenario against the built-in AI while retaining as much of the agent's own force as possible. Unit numbers and types vary between scenarios, as does the layout of the environment. Unit types are limited to standard non-flying units. The chosen algorithmic parameters for the CBR and RL components are listed in Table 5. The parameters are similar to those used successfully for evaluation and training of the *Navigation*, *Attack* and *Formation* modules. Starting positions are always a random spread opposite each other and the map-size is $2048 \times 2048$ pixels, the smallest possible StarCraft map size. Every experiment was run five times and the results were averaged.

**Table 5.** Tactical decision making evaluation parameters

| Parameter | Values |
|---|---|
| Scenario | A(3vs5), B(6vs6), C(5vs5), D(4vs9), E(10vs10) |
| Number of games | 100–100,000 |
| Algorithm | One-Step Q-learning |
| Case-base similarity threshold $\psi$ *A, B* | 30 % − 95 % |
| Case-base similarity threshold $\psi$ *C, D, E* | 80 % |
| RL learning rate $\alpha$ | 0.1 |
| RL discount factor $\gamma$ | 0.8 |
| RL exploration rate $\epsilon$ | 0.8–0 |

## 6.1  Results

The first two scenarios were, as stated above, run with a number of different similarity thresholds $\psi$. Table 6 shows the results for *Scenario A*. Table 7 shows the results for *Scenario B*. The reward is normalized to a value between 0 % and 100 %. 0 % is achieved in a game in which agent units are eliminated without doing any damage. 100 % is a perfect game in which all opponents are eliminated without the agent units sustaining any damage. This allows to compare results of scenarios with different absolute values for maximum and minimum rewards.

As results in both the tables show, similarity thresholds between 80 % and 95 % lead to results that are roughly within a 10 % interval in terms of overall performance. However, the number of cases and, more importantly, the number of overall solutions increases significantly among the different thresholds. Therefore, it was decided to use a threshold of $\psi = 80 \%$ for the subsequent evaluation scenarios. Given the results from the case-base analysis, the number of training episodes was set based on the number of agent units. The number of training

**Table 6.** Tactical decision making evaluation scenario A

| $\psi$ | # Episodes | # Cases | # Solutions | # Actions | Max. % Reward |
|---|---|---|---|---|---|
| 95 % | 100, 000 | 2, 376.4 | 18, 853.0 | 47.32 | 92.48 % |
| 90 % | 60, 000 | 1, 265.2 | 9, 976.4 | 45.27 | 87.33 % |
| 85 % | 20, 000 | 366.8 | 2, 570.2 | 41.15 | 82.93 % |
| 80 % | 8, 000 | 192.0 | 1, 299.6 | 41.06 | 81.82 % |
| 70 % | 1, 500 | 52.6 | 293.4 | 35.43 | 70.16 % |
| 60 % | 800 | 39.2 | 224.6 | 30.96 | 60.73 % |
| 50 % | 500 | 29.2 | 156.2 | 23.7 | 52.79 % |
| 40 % | 300 | 17.6 | 95.8 | 11.49 | 33.35 % |
| 30 % | 100 | 13 | 69 | 10.36 | 25.47 % |

**Table 7.** Tactical decision making evaluation scenario B

| $\psi$ | # Episodes | # Cases | # Solutions | # Actions | Max. % Reward |
|---|---|---|---|---|---|
| 95 % | 160, 000 | 1570.4 | 31, 201.6 | 7.10 | 78.15 % |
| 90 % | 75, 000 | 699.8 | 12, 339.0 | 6.92 | 75.13 % |
| 85 % | 30, 000 | 324 | 5, 324.20 | 6.96 | 73.01 % |
| 80 % | 15, 000 | 259.8 | 3755.8 | 6.99 | 69.97 % |
| 70 % | 7, 500 | 159.4 | 2092.6 | 7.45 | 63.09 % |
| 60 % | 5, 000 | 95 | 1, 309.2 | 8.72 | 57.99 % |
| 50 % | 3, 000 | 63.2 | 801.4 | 9 | 55.19 % |
| 40 % | 2, 000 | 47.2 | 631.4 | 9.5 | 45.53 % |
| 30 % | 1, 500 | 38 | 517 | 10.18 | 43.44 % |

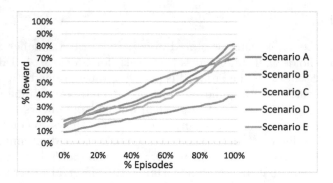

**Fig. 4.** Performance results for all scenarios

episodes is set to 15,000 for *Scenario C*, 10,000 for *Scenario D* and 50,000 for *Scenario E*. These comparably high amount of training episodes was chosen to ensure an optimal or near-optimal policy.

The results in Fig. 4 show that the hierarchical RL/CBR agent achieves a notable increase in average reward obtained for all five scenarios over the duration of their respective training runs. In terms of reward development, there is a difference between *Scenarios B* and *D* which use melee units only, and the other three scenarios. *Scenarios B* and *D* show an almost linear reward development over the time their respective experiments run. *Scenarios A, C* and *E*, which all use both melee and ranged units, show reward development curves that are more similar to those encountered in previous evaluations.

## 7   Discussion

*Scenarios A* and *B* have about ten *Tactical Unit Selection* actions (i.e. *Attack, Formation* or *Retreat*) in an average episode for the lowest, worst-performing setting of $\psi = 30\%$ where there is only a single case for each agent-opponent unit number combination. For higher thresholds, which allow for a more optimized performance, the number of actions diverges significantly. For *Scenario A*, the number of *Tactical Unit Selection* actions exceeds 40 for $\psi >= 80\%$. The reason for this is the learned hit-and-run strategy that performs best for the units in this particular scenario and which requires extensive use of *Retreat* actions. Lower similarity thresholds mean there is not enough distinction between inherently different cases, which in turn does not allow the agent to learn and effectively execute this hit-and-run strategy. The melee-unit-focused *Scenario B* teaches the agent a fundamentally different strategy, indicated by the average number of *Tactical Unit Selection* actions. For $\psi >= 70\%$, the average number of actions per game is below nine. This is due to the main strategy in this scenario, which is based on focusing attacks (covered by the *Attack* action) combined with minimal regrouping or retreating through *Formation* or *Retreat* actions. There is no use for extensive *Retreat* patterns since opponent- and agent unit types are identical, which means hit-and-run style attacks are useless.

The fact that agent and opponent use identical melee units in *Scenario B* also explains the difference in overall maximum rewards achieved. While the hit-and-run strategy allows the agent to achieve perfect or near-perfect rewards of more than 90 % for *Scenario A*, the average reward in *Scenario B* reaches a maximum value of just below 80 %. This is because attacking melee units with other melee units will always lead to suffering a certain amount of damage. The low number of actions required for optimal performance in *Scenario B* also means that it is easier to achieve good results in terms of average reward by using random untried solutions.

In all scenarios, the AI agent manages to obtain a significant improvement in the average reward. For all army compositions in the different scenarios, the agent finds optimal or near-optimal policies. Due to the unit types involved, *Scenario A* is the only scenario where the army composition theoretically allows a 'perfect game', i.e. eliminating all enemy units without sustaining damage. The agent manages to obtain more than 80 % average reward in this scenario. In *Scenarios C* and *E*, which both contain melee units that are harder to manage and are basically guaranteed to sustain damage when they attack, the agent manages to obtain above 75 % of the maximum possible reward. Even in *Scenario D*, which only uses melee units, the agent reaches nearly 70 % of the possible reward, pointing to effective use of focus-fire and manoeuvring.

When comparing the reward development of the different scenarios as depicted in Fig. 4, there is a difference between *Scenarios B* and *D* which use only melee units and the other three scenarios. This directly reflects the ideal behaviours in those scenarios and how these behaviours are reflected in action-selection policies. Optimal behaviour in a given scenario depends both on the layout of the scenario and on the agent and opponent army compositions.

# 8    Conclusion and Future Work

Overall, the results show that the hierarchical CBR/RL agent successfully learns the micromanagement tasks it was built to solve. The agent learns near-optimal policies in all evaluated scenarios which cover a range of in-game situations. The agent successfully re-uses the lower-level modules created for the squad-level tasks and the knowledge stored while training these modules.

One major restricting condition which was introduced to avoid a combinatorial explosion of possible solutions is limiting *Attack* and *Formation* to a single action for all units assigned to the appropriate category on the highest level. The evaluation of the hierarchical architecture showed that for the tested scenarios, the implementation achieved good to very good results on all occasions. However, it could already be observed that the performance suffered slightly for bigger scenarios when compared to the excellent results in scenarios with fewer units. One way to overcome this limitation would be to introduce another level above the currently highest level. The additional level would then simply perform a pre-allocation of all available units among several lower-level modules.

An important aspect which could be part of future work is the comparison of the approach presented here to other bot architectures. While this comparison will require additional logic to also address the strategic layer such a test could provide valuable insights into the power of adaptive online ML in relation to other ML, static and search-based approaches.

Currently there is a separate training phase for each of the lower-level modules. Creating modules which can be trained concurrently would be one way to accelerate the learning process. Other possible ways of improving performance would be through speeding up the individual CBR/RL components by employing better algorithmic techniques such as improved case-retrieval.

In summary, the key contribution of this paper is an integrated hierarchical CBR/RL agent which learns how to solve both reactive and tactical RTS game tasks. The creation of the individual hybrid CBR/RL modules for tasks in RTS game micromanagement is based on thorough analyses of TD RL algorithms, CBR behaviour and the relevant problem domain tasks. The resulting agent architecture acquires the required knowledge through online learning in the game environment and is able to re-use the knowledge to successfully solve tactical RTS game scenarios.

# References

1. Aamodt, A., Plaza, E.: Case-based reasoning: foundational issues, methodological variations, and system approaches. AI Commun. **7**(1), 39–59 (1994)
2. Aha, D.W., Molineaux, M., Ponsen, M.: Learning to win: case-based plan selection in a real-time strategy game. In: Muñoz-Ávila, H., Ricci, F. (eds.) ICCBR 2005. LNCS (LNAI), vol. 3620, pp. 5–20. Springer, Heidelberg (2005). doi:10.1007/11536406_4
3. Auslander, B., Lee-Urban, S., Hogg, C., Muñoz-Avila, H.: Recognizing the enemy: combining reinforcement learning with strategy selection using case-based reasoning. In: Althoff, K.-D., Bergmann, R., Minor, M., Hanft, A. (eds.) ECCBR 2008. LNCS (LNAI), vol. 5239, pp. 59–73. Springer, Heidelberg (2008). doi:10.1007/978-3-540-85502-6_4
4. Baumgarten, R., Colton, S., Morris, M.: Combining AI methods for learning bots in a real-time strategy game. Int. J. Comput. Games Technol. **2009**, 1–10 (2008)
5. Bridge, D.: The virtue of reward: performance, reinforcement and discovery in case-based reasoning. In: Muñoz-Ávila, H., Ricci, F. (eds.) ICCBR 2005. LNCS (LNAI), vol. 3620, pp. 1–1. Springer, Heidelberg (2005). doi:10.1007/11536406_1
6. Chung, M., Buro, M., Schaeffer, J.: Monte carlo planning in RTS games. In: Proceedings of the IEEE Symposium on Computational Intelligence and Games (2005)
7. Churchill, D., Saffidine, A., Buro, M.: Fast heuristic search for RTS game combat scenarios. In: Proceedings of the Eight Artificial Intelligence and Interactive Digital Entertainment International Conference (AIIDE 2012) (2012)
8. Jaidee, U., Muñoz-Avila, H., Aha, D.: Integrated learning for goal-driven autonomy. In: Proceedings of the Twenty-Second International Conference on Artificial Intelligence (IJCAI 2011) (2011)
9. Kocsis, L., Szepesvári, C.: Bandit based monte-carlo planning. In: Machine Learning ECML 2006, pp. 282–293 (2006)

10. MacAlpine, P., Depinet, M., Stone, P.: UT austin villa 2014: Robocup 3D simulation league champion via overlapping layered learning. In: Proceedings of the Twenty-Ninth AAAI Conference on Artificial Intelligence (AAAI) (2015)
11. Molineaux, M., Aha, D., Moore, P.: Learning continuous action models in a real-time strategy environment. In: Proceedings of the Twenty-First Annual Conference of the Florida Artificial Intelligence Research Society, pp. 257–262 (2008)
12. Muñoz-Avila, H., Aha, D., Jaidee, U., Klenk, M., Molineaux, M.: Applying goal driven autonomy to a team shooter game. In: Proceedings of the Florida Artificial Intelligence Research Society Conference, pp. 465–470 (2010)
13. Ontañón, S., Synnaeve, G., Uriarte, A., Richoux, F., Churchill, D., Preuss, M.: A survey of real-time strategy game AI research and competition in starcraft. IEEE Trans. Comput. Intell. AI Games **3**(4), 293–311 (2013)
14. Shannon, C.E.: Programming a computer for playing chess. In: Levy, D. (ed.) Computer Chess Compendium, pp. 2–13. Springer, New York (1950)
15. Smyth, B., Cunningham, P.: Déjà vu: A hierarchical case-based reasoning system for software design. In: ECAI, vol. 92, pp. 587–589 (1992)
16. Stone, P.: Layered Learning in Multiagent Systems: A Winning Approach to Robotic Soccer. MIT Press, Cambridge (1998)
17. Sutton, R.S., Barto, A.G.: Reinforcement Learning: An Introduction. MIT Press, Cambridge (1998)
18. Van Der Heijden, M., Bakkes, S., Spronck, P.: Dynamic formations in real-time strategy games. In: 2008 IEEE Symposium on Computational Intelligence and Games, pp. 47–54. IEEE (2008)
19. Watkins, C.: Learning from Delayed Rewards. Ph.d. thesis, University of Cambridge, England (1989)
20. Weber, B.: Integrating Learning in a Multi-Scale Agent. Ph.d. thesis, University of California, Santa Cruz (2012)
21. Wender, S., Watson, I.: Applying reinforcement learning to small scale combat in the real-time strategy game starcraft: broodwar. In: IEEE Symposium on Computational Intelligence and Games (CIG) (2012)
22. Wender, S., Watson, I.: Combining case-based reasoning and reinforcement learning for unit navigation in real-time strategy game AI. In: Lamontagne, L., Plaza, E. (eds.) ICCBR 2014. LNCS (LNAI), vol. 8765, pp. 511–525. Springer, Heidelberg (2014). doi:10.1007/978-3-319-11209-1_36
23. Wender, S., Watson, I.: Integrating case-based reasoning with reinforcement learning for real-time strategy game micromanagement. In: Pham, D.-N., Park, S.-B. (eds.) PRICAI 2014. LNCS (LNAI), vol. 8862, pp. 64–76. Springer, Heidelberg (2014). doi:10.1007/978-3-319-13560-1_6
24. Whiteson, S., Stone, P.: Concurrent layered learning. In: Proceedings of the Second International Joint Conference on Autonomous Agents and Multiagent Systems, pp. 193–200. ACM (2003)

# Defining the Initial Case-Base for a CBR Operator Support System in Digital Finishing
## A Methodological Knowledge Acquisition Approach

Leendert W.M. Wienhofen[1(✉)] and Bjørn Magnus Mathisen[1,2]

[1] SINTEF ICT, Sluppen, PO Box 4760, 7465 Trondheim, Norway
{leendert.wienhofen,bjornmagnus.mathisen}@sintef.no, bjornmm@idi.ntnu.no
[2] Department of Computer and Information Science,
Norwegian University of Science and Technology, 7491, Trondheim, Norway

**Abstract.** Case-based reasoning (CBR) literature defines the process of defining a case-base as a hard and time-demanding task though the same literature does not report in detail on how to build your initial case base. The main contribution of this paper is the description of the methods that we used in order to build the initial case-base including the steps taken in order to make sure that the quality of the initial case set is appropriate. We first present the domain and argue why CBR is an appropriate solution for our application. Then we detail how we created the case base and show how the cases are validated.

## 1 Introduction

Case-based reasoning (CBR) literature defines the process of defining a case-base as a hard and time-demanding task though do not report in detail on how to actually build your initial case base. Öztürk and Tideman say in their 2014 review paper [17]: "Initial population of a case base is a daunting task in classical CBR because it is manually crafted by knowledge engineers who make use of domain experts or written material to extract the case content. We believe case grounding problem is the reason why CBR has not seen wide-spread adoption in the industry - because manual extraction of cases from reports and records is costly and time consuming". In this context, we present a knowledge acquisition process that was applied to create an initial set of cases while constructing a CBR system in an industrial setting. We explain the domain in which we applied CBR and argue why it is an appropriate solution for our application. This is followed by a description of a methodological approach for building an initial case base. Revision and validation of the case base and the similarity features are presented in the discussion section.

The main contribution of this paper is the description of the methods that we used in order to build an initial case-base for our CBR system in an industrial domain, including the steps taken in order to make sure that the quality of the initial case set is appropriate.

© Springer International Publishing AG 2016
A. Goel et al. (Eds.): ICCBR 2016, LNAI 9969, pp. 430–444, 2016.
DOI: 10.1007/978-3-319-47096-2_29

## 1.1  Background

The case-based decision support system described in this paper is part of a project that is trying to increase the speed of digital conversion. Digital conversion is the process of cutting or milling various types of materials into shapes, based on a digital design. The speed is to be increased concerning the actual cutting speed as well as the time to shift between different jobs.

This paper will focus on the latter and the main objective, as set forward in the project proposal, is to decrease the time an operator uses between jobs by 80 %.

Digital conversion machines (such as shown in Fig. 1), also referred to as cutting tables, offer a plethora of different settings and the intervention is suggested to be an intelligent operator user interface to the conversion machine, based on case-based (CBR) and rule-based reasoning (RBR). By automating parts of the process relating to load shifts, the job for the operator will be easier and faster, with a lower margin for errors compared to the current situation.

The finished system should facilitate and automate learning from past experiences (meaning cutting/milling jobs with settings specific for a design and material) within a specific company. Future work will enable the system to share data between deployments of the system, so that even competing companies can share their experiences without sharing their competitive advantage.

A cutting table is a further development of a flatbed pen plotter, where the pen can be substituted by knives and millings bits, and the drawing paper by other types of material. Operations on the X and Y-axis (given a certain depth and pressure on the Z-axis) vary per material type. The optimal speed and acceleration for a given actuator depend not only on the material type, but also on the vendor (as quality can vary from vendor to vendor), the wear and tear of

**Fig. 1.** A digital conversion table from Esko Graphics (Copyright Esko Graphics)

the actuator as well as the complexity of the design to be cut/milled out from the material, just to name a few. Therefore, these cutting tables have a myriad of settings and require an experienced operator in order to get the best results. Most of the knowledge required for configuring the machine correctly is currently implicit, and knowledge transfer is typically done on a face-to-face basis between operators. Generally speaking, we can say that inexperienced operators do not dare to use the full potential of the table in fear of damaging the materials on which the cutting or milling operation is to be carried out on.

By making domain knowledge explicit in the form of a domain model with instances, an inexperienced user can find similar cases and re-use the settings. In our approach, we take this one step further by applying case-based reasoning, which automatically selects the most applicable case and related setting for the user so that the full potential of the machine can be used.

### 1.2 Case-Based Reasoning as an Enabler For experience Transfer

Based on interviews and observations at companies using digital conversion tables, we conclude that experience is typically not stored in a structured manner and knowledge transfer happens in an informal way between co-workers. Operators of these machines typically learn by doing, and because of this the full potential of a machine is not always reached, especially when operated by inexperienced users. Users report that they are afraid of breaking something when they apply parameters they are unsure of.

In some cases a note with settings is taped on the operator console, though these contain proven "safe settings" for a typical material and tool combination. Another company uses a whiteboard for settings, though it is rarely updated and personnel indicate that they actually do not use the settings that are noted there and rather trust their own feelings concerning the settings. There is no structured means of storing experiences among the companies that have been observed during the case study.

As the working situation is based on a known desired outcome, case based reasoning is an appropriate manner of addressing the problem at hand.

We intend to create a knowledge base where the digital finishing machine retains the settings, material type and other relevant parameters.

### 1.3 Distributed Case-Based Reasoning

In the digital finishing industry companies use many different material types, some on a more regular basis than others. As this is a very experience-based process, chances are that the proper expertise is not available in all companies. By providing access to case bases created in other (competing) companies, one can draw from the experience.

### 1.4 Related Work

It has been shown [12] that CBR is well suited as a means of decision support for operators in a manufacturing setting. CBR is a form of AI where the decision making support is based on a known outcome. It takes a case (which is the product to be made) as input and tries to find the most similar case in a case base. This means that cases with a similar profile are suggested.

Competing companies can help each other increase their efficacy by sharing case bases can be achieved via distributed case-based reasoning. Distributed CBR has been around for a while and is well described in among others [14,16,18]. However, it seems to be limited to non-competing companies, making knowledge sharing a clear-cut benefit. In order to avoid potential problem with patented designs, we decouple the geometry information from the design, reducing it to an indication of the complexity of the design based on a float where 0.0 is the least complex and 1.0 the most complex. From the technical point of view there is no real difference in the implementation.

Our CBR system will implement explanatory features enabling the operator to choose to either apply the suggested settings or retain the self-chosen setting based on the suggested settings and the corresponding explanation. Based on the interviews, we can state that it is important that the CBR system does not actually make a decision, rather suggests a decision based on the most similar case. This way the system supports the operator in his decision. The explanation helps the operator to understand why a certain proposal has been made by the system and therewith enables to operator to make an informed decision. The fundamental issues of explanations in CBR are described well in [21].

Aamodt and Plaza [2] have formalized Case-based reasoning for purposes of computer reasoning as a four-step process: Retrieve, reuse, revise and retain.

In order to retrieve a case, one needs to identify features, collect descriptors, interpret problem and infer descriptors. Prior to being able to do that, one needs to have a case base. It is of importance that the right features are extracted from a case as it will be the fundament for further reasoning. Case acquisition is often manually intensive. According to [10], a manually intensive approach for storing experiences of individuals has been widely used in many CBR applications. The general approach – as case bases are very domain specific- is to talk to a domain expert and extract which parameters are of most value and use that as a starting point. Getting the full picture, however, requires talking to more than one expert and an iterative approach in order to make sure that the right parameters are used for the case base. In the following sections we present such an approach.

The case quality needs to be safeguarded as the case base must contain a representative set of problem solution pairs from the domain at the initial stage of the CBR system. At the same time we need to ensure that the case-base yields high quality results. Little attention has been given to case-quality in the available literature, and therefore the CBR expands without inspecting itself [29]. We want to address the quality problem by making sure that both the initial case information as well as the cases to be learned will be initiated and checked by humans.

If the case template is wrong, the result will be wrong. There is a need to understand how the case template is defined, in practice. Next we will see who has addressed this central issue and what they can tell us about how to do address it.

Öztürk and Tideman's [17] statement "We believe case grounding problem is the reason why CBR has not seen wide-spread adoption in the industry - because manual extraction of cases from reports and records is costly and time consuming" is one of the main reasons why we report our approach related to knowledge acquisition. We do agree that it is a time consuming effort, though, when consulting existing literature for knowledge acquisition for CBR to learn how to extract and categorize the relevant information, we did not find any clear guidelines or methodological descriptions for case grounding. This might be an additional reason to why CBR has not seen a widespread adoption in the industry.

The recent trend is to (semi)-automate the case acquisition process [10, 15, 23, 24, 29]. The approach sketched by [10] is based on initiating the case base with random values, though still based on a formalized data-sheet template for case representation. However, there is no mentioning on how the template was established (the assumption is that domain experts have been asked). They state: "Case engineering is among the most complicated and costly tasks in implementing a case-based reasoning system".

The cases that are part of the case-base are supposed to yield solutions to the problems with minimal adaptation or human input. This is desirable as otherwise the major usefulness of a CBR system to reuse existing knowledge would be substantially harmed [8]. This implies that the case base must support this type of knowledge.

Richter [19] describes knowledge containers as keepers of case information. The first requirement is that the case base should only contain cases (p, s) where the utility of s is maximal or at least very good for the problem p. This is knowledge contained in the individual cases.

The case acquisition process itself, meaning the initiation of a case base, is not described though 4 different sources are mentioned in Richter's invited talk at ICCBR in 1995 [27]: domain knowledge, cases, similarity knowledge and adaptation knowledge.

The domain knowledge is what fills the template which can be used for matching cases. According to, template retrieval is similar to SQL queries in databases, where all cases fitting a template of parameters are retrieved. The main merit of using of template retrieval is that the faster retrieval and high currency by prevents irrelevant case from being considered in similarity matching.

Aamodt [1] described a framework for modeling the knowledge contents of CBR systems based on Richters knowledge containers. The model suggests decomposition in three perspectives. The power of using three perspectives (tasks, methods, and models) for knowledge level modeling lies in the interaction between the perspectives, and the constraints they impose on each other. However, there is no description on how to initiate the case base.

Cordier et al. [6] state that when there is a lack of domain knowledge, the system may infer a solution that is correct with respect to the knowledge base but not with the real world: making the results invalid in the real world. The FRAKAS system [7] is an approach for interactive domain knowledge acquisition.

Learning takes place during the use of the system and aims at acquiring domain or adaptation knowledge. The evaluation of the adapted solution may highlight that it does not meet the requirements of the target problem. In this situation, a reasoning failure occurs and is processed by a learning process. The expert is involved in the process of identifying inconsistent parts of the solution which helps to augment the knowledge base. The expert is involved in a simple manner to point out faulty knowledge and he/she may provide a textual explanation of the identified error to support complementary off-line knowledge acquisition. The approach defined here is interesting with respect to further population of a knowledge base, and a similar approach can be used both to fill the knowledge base once a basic case set has been established as well as a part of the regular learning curve (one of the 4 R's).

As in the CBR literature little is mentioned on how to populate the initial case-bases, we turn to the cognitive science domain where the fundamental concept is that "thinking can best be understood in terms of representational structures in the mind and computational procedures that operate on those structures."[1] Cognitive science in turn is related to the knowledge management and knowledge engineering field where extracting information from experts in order to create the foundation for among others expert systems has matured over the past decades. Watson [28] does describe how to apply knowledge management for CBR, however, it lacks detail on the establishing of the case base. Cognitive science is also mentioned in [20] and regarding representation of knowledge they state the following: "more generic issues of knowledge representation are seldom addressed". Followed by "The case base plays a special role because the cases can be entered without understanding them. The main point is that knowledge can be shifted between containers (their content is not invariant), which can be modeled using a learning process. In addition, the shifting can be done manually without the support of a learning method".

Our guiding motivating hypothesis is that an operator support system based on case-based reasoning can help speed up the cutting/milling process while maintaining satisfactory quality results.

As the intention is to create an operator support system using CBR, we need a formal representation of the cases. By creating a domain model, we separate domain knowledge from the operational knowledge, enable the reuse of domain knowledge and make domain assumptions explicit. Once the domain model is in place, we can also populate the case base with relevant cases. Finding out what a relevant case is and what needs to be represented in the domain model go hand in hand. Our second hypothesis is that a user-centered iterative approach is a good method to create a good formal representation as a basis for the operator support system.

---

[1] Thagard, Paul, Cognitive Science, The Stanford Encyclopedia of Philosophy (Fall 2008 Edition), Edward N. Zalta (ed.).

# 2   Method

While many publications (i.e. [3–5,9,26]) do describe the knowledge acquisition approach for their domain, most do it on a relatively technical level. We have applied several methods for knowledge acquisition and the focus has been on a user-centered iterative process. In the subsections below we give a brief explanation of these methods and highlight our experience with these forms of knowledge acquisition.

## 2.1   Research Method

To systematically guide our research in this project we used the design science research method is used according to [11], as depicted in Fig. 2. The research environment consists of machine supplier experts, as well as machine operators. The research is driven by the need to use the machines in an optimal manner, with the assumed outcome a more optimal operation and therewith cost reductions. The knowledge base is based on the existing literature on CBR and knowledge acquisition as well our own findings.

**RQ1:** What is the effect of introducing an expert system based on CBR on the effectiveness of operators?

**RQ2:** What is the effect of introducing a distributed expert system based on CBR?

User-centered design is conducted prior to the development of complex systems to ensure deep understanding of user and stakeholder roles. The aim is to ensure that system designed support the daily work of end users and the role of stakeholders [13,22]

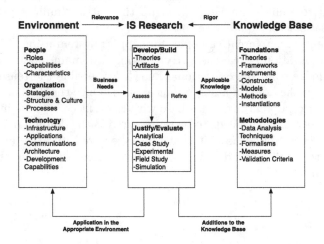

**Fig. 2.** Conceptual framework for IS research [11]

We have applied user-centered design in all activities in the iterative process of assessing and refining our artifacts, adding it to our knowledge base. The activities are carried out in close cooperation with real stakeholders by means of various methods for data collection, as described in the sections below.

## 2.2  Data Collection

In our study we focus on a single manufacturer of digital conversion tables. The study is based on design science research and evaluation research and has been implemented at 3 different locations that represent a typical customer of this manufacturer.

The intervention is the introduction of a distributed case-based decision support system to support operators to make the right decisions quicker and therewith both reduce the number of errors and speed up the full process of job shifting.

The artifact to be created for this intervention is a research challenge itself as populating a knowledge base is a non-trivial task. A step wise approach for populating a CBR knowledge base will be developed and the effect will be tested.

Some of the needed information can be retrieved from logs, though this is a non-validated information source.

The study is divided in two parts: data collection participants and intervention participants. The data collection methods are described in the sub-sections below.

In both cases the population is recruited the manufacturer of digital conversion tables- and the inclusion criterion is that the participant is currently a customer operating digital conversion machines. We focus on the data collection part in this paper.

For the data collection, we focus around the following questions

**DC1:** Which information to extract from the operators?
**DC2:** What is/are the bottleneck(s) in the load shift?
**DC3:** Which factors impact the time used?
**DC4:** What is the mean time?
**DC5:** Does the knowledge of an operator impact the operation? And in what way?
**DC6:** How much information are companies willing to share with competitors?
**DC7:** How and when to present suggestions from the expert system to operators?

In the sub-sections below we first present which methods we have used for the data collection and in Sect. 3 we provide the results of the activities.

**Observation.** The first data gathering activity was based on observations. The intention was to form a structure for later interviews and the first subject was asked to explain (while preparing and operating the cutting table) what he was doing and why he was doing it this way. The observer did not interfere with the process.

**Semi-structured Interviews.** We have conducted interviews at digital finishing companies in Norway, Belgium and The Netherlands. The interview subjects were mainly cutting table operators, though also managers/owners. As the companies were relatively small, the latter category also in all cases were table operators, yet not on a daily basis. The interview questions were based on the results from the observation session and have been expanded based on finding between the interviews. We used a set with main questions and expanded while commencing the interview.

**Questionnaire.** We have developed a questionnaire in order to map the time operators use when operating the machines. It was sent to 100 digital finishing companies throughout the world.

**Workshops.** The technology provider catered for a workshop with employees with a computer science background. During this workshop technical boundaries were explored and details regarding the integration of the operator support system discussed.

**Re-use of Available Data.** We have gained access to a product guide describing which tools can be used for which materials, and for some of these also a set of settings for certain material types. However, the settings are relatively conservative as they pertain to a material family. Specific materials use material specific settings which can be much faster than the material family setting. For the most used specific materials, specific settings are available. Also an operator manuals of the current Esko machines with i-cut software has been used as an information source.

## 3   Results

### 3.1   Case Study: As Is Situation

Input for the study uses the data gathering methods described above, in addition, one of the researchers took a table operator course to get a real hands-on feel of using the system.

In Fig. 3 you see the repetitive and cyclic process of enhancing the input, which can be mapped to the IS research part of Fig. 2; both Develop/build and justify/evaluate to ensure both relevance and rigor.

The methods have been applied to digital finishing companies in Norway, Belgium and The Netherlands.

Unfortunately, the response rate for the questionnaire was so low that we were unable to use the results as a pinpoint for the average type of operator and other information regarding machine use.

From the observation and interview activities, we learned that machine operation to a large degree is completely experience based and that the experience

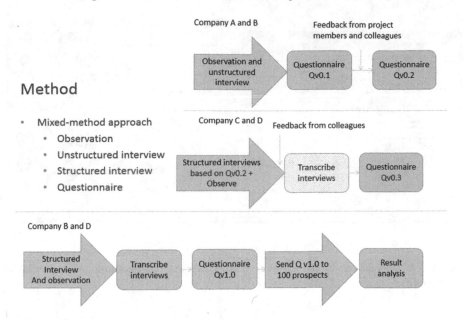

**Fig. 3.** Knowledge acquisition

transfer is sub-optimal. Some factors that influence the choices are the quality of the material that is used, the wear and tear of the used actuator, the desired output quality (not all customers demand a high quality finish) as well as the time available between jobs. An ideal situation according to one of the shop managers is that the machine is in use continuously. We did a test using optimal speed settings with new actuators and high quality material vs the regular settings with a new actuator and high quality material. We found that the cutting speed in this specific case was 13 min vs 22 min. This supported the assumption that the operators do not use the optimal settings and that an operator support system indeed can be useful. For this specific case, relating to RQ1, we can state that there is a good effect in using the operator support system recommendations.

Knowing the type of information the operators wish to use and how they wish to use it, we discussed the technical boundaries with the table and cutting table software provider. We gained access to subsets of the required information required to create an operator support system. All of the gathered information has been structured into a domain model. See the next subsection for more details.

## 3.2 Domain Model

In order to model the domain, we need to map domain knowledge (for an impression, see Fig. 4a). The main parameters that need to be contained can be summarized as such:

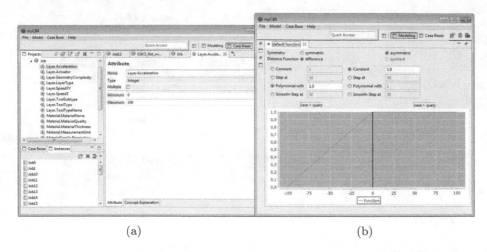

(a)                                                    (b)

**Fig. 4.** A screenshot of the domain model (a) and an example of a similarity measure (b).

*A cutting/milling* **job** *is performed by an* **operator** *on a* **cutting table** *which uses a* **tool set** *with different* **actuators** *on a* **material type** *following* **patterns** *stemming from a* **design**. *Considerations regarding the* **speed** *and* **quality** *of the job are done by the* **operator** *based on* previous experiences *and* **customer demands**.

The previous sentences describe what the domain model needs to include on an overall level. In short, it needs to include all relevant information for an operator to be able to do a job in the fastest possible manner or with the highest quality possible. These two are not always mutually exclusive, though high speeds can sometimes lead to a lower end-product quality. In some cases, the lower quality is still within the quality assurance threshold.

- Some questions that the operator support system needs to be able to help answer are: Which settings should I avoid to use?
- What is the most optimal setting for this particular job with regards to either quality or speed?
- What is the maximum speed I can use?
- Will these settings break stuff?
- Which settings should I change?
- Will this actuator (bit/blade) work with this material?
- What are the limitations of this tool applied on this material?

These questions imply that we need to know about the properties of the materials, design, tools and table. During the domain knowledge gathering process, we have identified the relevant terms to include in the domain model. Due to space restrictions, we do not include the domain model in this paper, though some of it can be seen in the screenshots from MyCBR.

One of the results from the interviews shows that operators are more likely to trust a recommendation if an explanation is given. If the settings are presented following a pattern such as "in a similar case we have successfully applied the following settings with a satisfactory quality" followed by a question if the operator wishes to use these settings instead, the operators responded positively. However, without such explanation, the operators would not simply accept new settings.

## 3.3   CBR

We have applied the domain model and created a CBR system prototype using MyCBR [25].

The initial case base has been made in close cooperation with experts from the company. Instances with proven cases in different levels of aptness have been entered. It is important to note that these cases are based on material family and not a specific instance of the material itself. As properties are supplier specific, different settings should be used. These settings will during the course of the use of the system be formed as cases.

The similarity features (Fig. 4b) are based on conversations with a tool and material experts. Each specific material combined with specific actuators have specific settings, also pertaining to the complexity of the output to be generated.

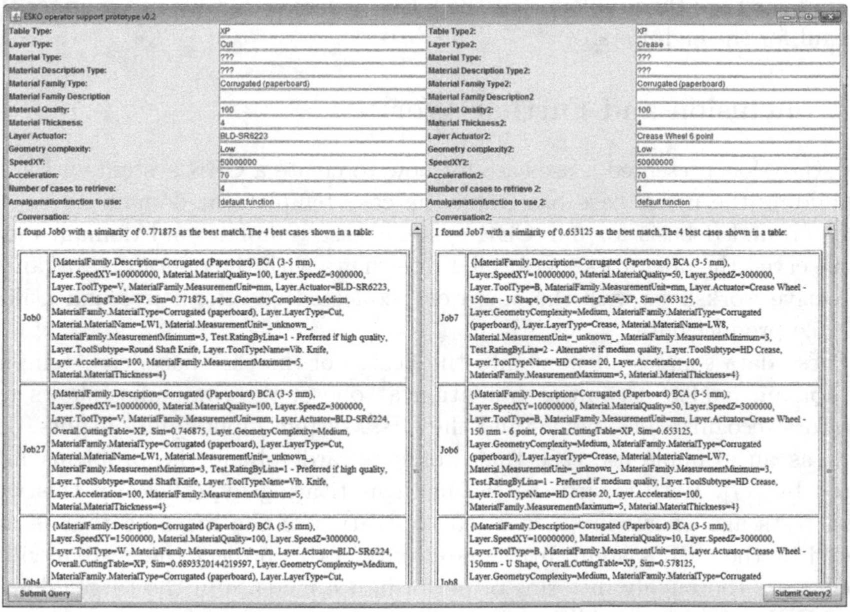

**Fig. 5.** Matching results screenshot

## 3.4   Validation

Testing has been done based on the different cases with each their rating. For material types two or more different cases have been entered in the initial case base, including an indication in the aptness. Similarity values and weights have been tuned in order to get the closest case to match. This was later tried with new cases and the results were satisfactory. A screenshot of matching results is shown in Fig. 5

## 4   Discussion and Lessons Learned

The variety of knowledge elicitation methods we have used and the variety of companies visited may seem like an too rigorous information gathering, though we feel that in our case this was the right thing to do. It is time and resource demanding, though by presenting the various approaches, we hope to contribute to the knowledge gap that seems to exist concerning creating an initial case-base. Different situations cater for different methods of knowledge elicitation, and in many cases, a less rigorous approach might be sufficient. Creating a sound and valid foundation for the case template and case base is resources demanding. However creating a CBR system that is neither valid or useful is even more resource demanding. In general we can recommend to talk to the system owner and a variety systems users multiple times in order to best understand the problem at stake and validate that the researchers (CBR system builders) really understand the problem that the CBR system is to solve in a manner that is useful for the end-users.

## 5   Conclusion and Further Work

This study has presented a use case for how to create a CBR system with focus on building the initial case base, and the case template or domain model. To create grounded basis for our CBR system, case template and domain model we; observed the operators, performed interviews with the operators, organized interactive workshops with the operators, collected questionnaires and utilized available product data.

These data sources all went into the design of the case base, case template and domain model. An initial validation at one of the companies shows that operators recognize and understand the CBR system inputs and outputs. This serves as an example use case that works toward solving the problems highlighted by [17]. With regards to the main motivating hypothesis of this work initial tests also shows increase in the operation of the machine that is augmented by the CBR system. In the next part of this project the system will be tested more thoroughly in terms of performance increase in the target domain of the CBR system. In addition, we will develop a method for abstracting and extracting high level knowledge from cases to be sent into a distributed case base to ensure both knowledge sharing across competing stakeholders while not disclosing competitive advantages.

**Acknowledgments.** The authors gratefully acknowledge the Norwegian Research Council and the BIA program for financial support of the project (partially through grant 235427) as well as the participating case companies, which together enabled this study.

# References

1. Aamodt, A.: Modeling the knowledge contents of CBR systems (2001)
2. Aamodt, A., Plaza, E.: Case-based reasoning: foundational issues, methodological variations, and system approaches. AI Commun. **7**(1), 39–59 (1994)
3. Bach, K.: Knowledge Acquisition for Case-Based Reasoning Systems. Dr. Verlag Hut, Munich (2012)
4. Bergmann, R., et al.: Developing Industrial Case-Based Reasoning Applications: The INRECA Methodology. LNAI, vol. 1612. Springer, Heidelberg (2003)
5. Bergmann, R.: Experience Management Foundations, Development Methodology, and Internet-Based Applications. Springer-Verlag, Heidelberg (2002)
6. Cordier, A., Fuchs, B., Lieber, J., Mille, A.: Failure analysis for domain knowledge acquisition in a knowledge-intensive CBR system. In: Weber, R.O., Richter, M.M. (eds.) ICCBR 2007. LNCS (LNAI), vol. 4626, pp. 463–477. Springer, Heidelberg (2007). doi:10.1007/978-3-540-74141-1_32
7. Cordier, A.: Interactive and opportunistic knowledge acquisition in case-based reasoning. Thesis (2008). https://tel.archives-ouvertes.fr/tel-00364368
8. Cunningham, P.: CBR: strengths and weaknesses. In: Pasqual del Pobil, A., Mira, J., Ali, M. (eds.) IEA/AIE 1998. LNCS, vol. 1416, pp. 517–524. Springer, Heidelberg (1998). doi:10.1007/3-540-64574-8_437. http://link.springer.com/chapter/10.1007/3-540-64574-8_437
9. Díaz-Agudo, B., González-Calero, P.A., Recio-García, J.A., Sánchez-Ruiz-Granados, A.A.: Building CBR systems with jcolibri. Sci. Comput. Prog. **69**(1), 68–75 (2007)
10. Dufour-Lussier, V., Le Ber, F., Lieber, J., Nauer, E.: Automatic case acquisition from texts for process-oriented case-based reasoning. Inf. Syst. **40**, 153–167 (2014). http://www.sciencedirect.com/science/article/pii/S0306437912001573
11. Hevner, A.R., March, S.T., Park, J., Ram, S.: Design science in information systems research. MIS Q. **28**(1), 75–105 (2004)
12. Hinkle, D., Toomey, C.: Applying case-based reasoning to manufacturing. AI Mag. **16**(1), 65 (1995). http://www.aaai.org/ojs/index.php/aimagazine/article/viewArticle/1124
13. Kubie, J., Melkus, L.A., Johnson, R.C., Flanagan, G.A.: User-Centred Design, 7th edn. CRC Press Inc., Boca Raton (1999)
14. Leake, D.B., Sooriamurthi, R.: When two case bases are better than one: exploiting multiple case bases. In: Aha, D.W., Watson, I. (eds.) ICCBR 2001. LNCS (LNAI), vol. 2080, pp. 321–335. Springer, Heidelberg (2001). doi:10.1007/3-540-44593-5_23
15. Manzoor, J., Asif, S., Masud, M., Khan, M.J.: Automatic case generation for case-based reasoning systems using genetic algorithms (2012)
16. Nagendra Prasad, M.V., Lander, S.E., Lesser, V.R.: On retrieval and reasoning in distributed case bases (1995)
17. Öztürk, P., Tidemann, A.: A review of case-based reasoning in cognition-action continuum: a step toward bridging symbolic and non-symbolic artificial intelligence. Knowl. Eng. Rev. **29**(01), 51–77 (2014). http://journals.cambridge.org/article_S0269888913000076

18. Plaza, E., McGinty, L.: Distributed case-based reasoning. Knowl. Eng. Rev. **20**(03), 261–265 (2005). http://journals.cambridge.org/article_S0269888906000683
19. Richter, M.M.: Knowledge containers. In: Readings in Case-Based Reasoning. Morgan Kaufmann Publishers (2003)
20. Richter, M.M., Aamodt, A.: Case-based reasoning foundations. Knowl. Eng. Rev. **20**(03), 203–207 (2005). http://journals.cambridge.org/action/displayAbstract? fromPage=online&aid=435273&fileId=S0269888906000695
21. Roth-Berghofer, T.R.: Explanations and case-based reasoning: foundational issues. In: Funk, P., González Calero, P.A. (eds.) ECCBR 2004. LNCS (LNAI), vol. 3155, pp. 389–403. Springer, Heidelberg (2004). doi:10.1007/978-3-540-28631-8_29
22. Shluzas, L.A., Steinert, M., Katila, R.: User-centered innovation for the design and development of complex products and systems. In: Leifer, L., Plattner, H., Meinel, C. (eds.) Design Thinking Research, pp. 135–149. Springer, Heidelberg (2014)
23. Shokouhi, S.V., Aamodt, A., Skalle, P.: A semi-automatic method for case acquisition in CBR a study in oil well drilling (2010)
24. Shokouhi, S.V., Skalle, P., Aamodt, A.: An overview of case-based reasoning applications in drilling engineering. Artif. Intell. Rev. **41**(3), 317–329 (2014)
25. Stahl, A., Roth-Berghofer, T.R.: Rapid prototyping of CBR applications with the open source tool myCBR. In: Althoff, K.-D., Bergmann, R., Minor, M., Hanft, A. (eds.) ECCBR 2008. LNCS (LNAI), vol. 5239, pp. 615–629. Springer, Heidelberg (2008). doi:10.1007/978-3-540-85502-6_42
26. Tautz, C.: Costumizing software engineering experience management systems to organizational needs. Fraunhofer-IRB-Verlag (2000)
27. Veloso, M., Aamodt, A.: ICCBR 1995. LNAI, vol. 1010. Springer, Heidelberg (1995). http://www.springer.com/computer/ai/book/978-3-540-60598-0
28. Watson, I.: Applying Knowledge Management: Techniques for Building Corporate Memories. Morgan Kaufmann, New York (2003)
29. Yang, C., Farley, B., Orchard, B.: Automated case creation and management for diagnostic CBR systems. Appl. Intell. **28**(1), 17–28 (2008)

# Author Index